Handbook of Textile Engineering

Handbook of Textile Engineering

Editor: Dick Baia

NY RESEARCH PRESS

New York

Published by NY Research Press
118-35 Queens Blvd., Suite 400,
Forest Hills, NY 11375, USA
www.nyresearchpress.com

Handbook of Textile Engineering
Edited by Dick Baia

© 2019 NY Research Press

International Standard Book Number: 978-1-63238-644-1 (Hardback)

Cataloging-in-Publication Data

Handbook of textile engineering / edited by Dick Baia.
p. cm.
Includes bibliographical references and index.
ISBN 978-1-63238-644-1
1. Textile fabrics. 2. Textile industry--Technological innovations. 3. Textile machinery.
I. Baia, Dick.
TS1765 .H36 2019
677--dc23

Contents

Preface

This book has been a concerted effort by a group of academicians, researchers and scientists, who have contributed their research works for the realization of the book. This book has materialized in the wake of emerging advancements and innovations in this field. Therefore, the need of the hour was to compile all the required researches and disseminate the knowledge to a broad spectrum of people comprising of students, researchers and specialists of the field.

Textile engineering has made significant progress in processing and manufacturing technology in the past few decades due to the adoption of computer-aided design and manufacturing. As a result of progress in polymer science and renewed interest in traditional textiles and handloom industries, a lot of research areas are being explored in this discipline. Research is also being conducted to incorporate elements of nanotechnology in the manufacturing process for better quality and ease. This book explores all the important aspects of textile engineering in the present day scenario. It strives to provide a fair idea about this discipline and to help develop a better understanding of the latest advances within this field. This book aims to equip students and experts with the advanced topics and upcoming concepts in this domain.

At the end of the preface, I would like to thank the authors for their brilliant chapters and the publisher for guiding us all-through the making of the book till its final stage. Also, I would like to thank my family for providing the support and encouragement throughout my academic career and research projects.

Editor

The Importance of *Aloe debrana* Plant as a Thickening Agent for Disperse Printing of Polyester and Cotton in Textile Industry

Sisay Awoke[1]*, **Yirga Adugna**[1], **Redwan Jihad**[2], **Habtam Getaneh**[3]

[1]*Department of Chemistry, College of Natural Science, Wollo University, Ethiopia*
[2]*Department of Textile Engineering, Wollo University, Ethiopia*
[3]*Department of Textile Biology, Wollo University, Ethiopia*

Abstract

Aloe debrana plant grows in several regions in South Wollo, Ethiopia, and has been used mostly as some of traditional medicine, and to stop breastfeeding. The technical feasibility of using *Aloe debrana* gel as a new thickener for printing polyester and cotton with disperse dyes was examined in 2013. The results indicated that the properties of the printed fabric samples (colour strength, K/S, overall fastness properties) were dependent on gel concentration, as the fixation conditions using direct heating technique. The optimum conditions for printing polyester fiber with disperse dyes using *Aloe debrana* gel as a thickener were as follows: 30 g/kg disperse dye, 75 g/kg urea, 10 g/kg citric acid, 500 g/kg *Aloe debrana* thickener and other water drying at 100°C for 3 min followed by direct fixation for 6 min at 180°C. A natural preservative lemon juice has high potential to preserve the gel more than two months. The results showed that no significant difference between fresh gel and preserved gel.

Keywords: Gel; Fastness; Dyes; Paste; Scattering; Absorption; Fabric

Introduction

Aloe plant belongs to *Aloaceae* family, and has more than 400 species. The species are easily recognized by their flower and rosettes, thick and succulent leaves, which are sometimes spotted. Most have leaves more or less V-shaped in cross section. The leaf margin is almost always armed with sharp teeth. The different aloe species have different flowering periods; some flower over an extended period, thereby offering a continuous food supply for nectar feeding birds and insects. Each flower is supported by a bract, the shape and size of which are important for the identification of the species. Flower colouration is most often red, orange or yellow, and rarely white [1,2].

Aloe species have been used for a long time in folk medicine for the treatment of constipation, burns and dermatitis. Recently, some species of *Aloe* have been used in a wide range of skin and hair care products, and also form the basis of health drinks and laxative drugs. The slimy gel inside the leaves consists of a complex mixture of polysaccharides, amino acids, minerals, trace elements and other biologically active substances, such as enzymes. Recent research has indicated that *Aloe* might kill the bacteria responsible for tuberculosis, Mycobacterium tuberculosis, and virus causing herpes genitals and inhibits growth of many common organisms such as yeasts, fungi, and bacteria infecting wounds [3].

Aloe debrana is known in Ethiopia and found in South Wollo, *Dessie zuria* and *Kalu woreda*, among other species in the genus Aloe. This plant need very less water for living and has resistance to diseases and insects. The *Aloe debrana* plant has long (up to 20 inches long and 5 inches wide) leaves. Each plant usually has 12-16 leaves and can be harvested every 6 to 8 weeks by removing 3 to 4 leaves per plant. The fresh parenchyma gel from the centre of the leaf is clear and shining [4].

Most areas in South Wollo, the land is a steep mountain slope and not well structured to plough for farmers and also highly exposed to erosion. *Aloe debrana* plant can grow in the place with eco-friendly and keep control of land conservation as well. The gel from this plant can be used for several purposes like dyeing fabrics.

Disperse dyes are mainly used in the dyeing and printing of polyester and its blended fabrics. Disperse print has excellent printing properties like fastness and colour strength and involves localized coloration. This is usually achieved by applying thickened pastes containing dyes or pigments onto a fabric surface according to a given colour design. In particular, the viscosity of a print paste is critical and determines the volume of paste transferred to the fabric and the degree to which it spreads on the surface fabric [2,5-7].

Textile printing is carried out from a paste preparation, which consists of the colouring materials (pigments/dyes), thickeners, binders (in pigment printing), and other ingredients. The thickeners control rheology or flow behaviour of the print formulation, which in turn controls presentation and migration resistance of the printed color thereby ensuring clear cut design and the right outline definition of prints [8,9].

The present work was undertaken with the aim of determining the ideal disperse printing conditions for attaining high performance polyester and cotton fabric prints using *Aloe debrana* gel as a new thickening agent. Moreover, as well known Ethiopia is currently one of fast growing African country in Industry, especially in Textile. All industries import these thickening agents from abroad. So, we believe that using this plant material as a thickening agent has benefits for the country by decrease foreign currency, because of high dominancy of *Aloe debrana* plants in all high lands of Ethiopia this work the community can secure food by selling the plant to the nearby industries, and it is non toxic to the workers in the factories and environmental friendly.

*Corresponding author: Sisay Awoke, Department of Chemistry, College of Natural Science, Wollo University, Ethiopia, E-mail: sisgenetre@gmail.com

Materials and Methods

Description of study area

The study areas were encompassing South Wollo Zone: *Dessie zuria* and *Kalu woreda*, and then two kebele from each *woreda*, Harawobelo and Harbu respectively were selected. The selection of kebele was depending on the dominant availability of *Aloe debrana* plants in that particular area (after conducting survey).

Data collection

The data collection was conducted in two phases and sampling method was purposive. In the first phase the survey was undertaken to identify the *Aloe* plant species based on their size and amount of the gel found inside. In the second phase, the plant species was collected from the selected areas forests of study areas. The amount of the gel found inside the *Aloe debrana* from *Kalu* was not enough for experimental use due to the hot season, as known the gel content is dependent on geographical location and weather condition. The sample collected from *Dessie zuriya* was good enough for experimental use due to high content of gel.

Materials

Bleached 100% polyester (250 g/m²), and 100% cotton (250 g/m²) woven fabrics were used.

Aloe debrana gel, Alginate and emulsion were used as thickening agent. The disperse dyes was Disperse Blue, and also chemicals; sodium bicarbonate, citric acid and urea were used.

Methods

Extraction of aloe gel: The gel of *Aloe debrana* was carefully extracted with the help of knife from the part of the plant as an eco-friendly and it was collected in the beaker. The beaker was immediately closed in order to protect it from contact of air, which is the cause of oxidation since the gel is sensitive. Using citric acid (lemon juice) as a natural preservative, the aloe extract was preserved for two months to evaluate the storing time.

Print paste and screen preparation: The printing pastes for disperse printing of polyester and Cotton was prepared by using fresh and preserved *Aloe debran* gel as a thickening agent. Each paste was prepared by using the following recipe: Disperse dye 30 g/Kg, Thickening agent 500 g/Kg, Urea 75 g/Kg, Citric acid 10 g//Kg, Water 385 mL/Kg as total weight of the paste was 1000 g. Here when *Aloe debrana* fresh gel was being used as thickening agent the water from the recipe was omitted since there is more water in the gel naturally, thus the addition of extra water had done the paste thinner as compared with other and brought less effective the paste while printing. Totally the numbers of pastes prepared were seven and by using these pastes the printing was performed on cotton and polyester fabrics. In addition to the above pastes recipe, pure *Aloe debrana* without acid and without urea pastes were prepared with the same concentration of other chemicals and also another paste was prepared with mixture of the synthetic and *Aloe debrana* gel thickener but without changing the concentration and composition of other recipe. The flat screen was already prepared by Bahir Dar University Institute of Textile Engineering with Ethiopian beautiful woman picture.

Printing procedure and testing: Printing was carried out using the flat screen technique by squishing a total of 20 g paste four times manually and then printed samples were dried at 100°C for 3 min and fixed by mini-dryer and supper heated steam at 150°C for 6 min. The samples were rinsed in cold water for 15 min and then in hot water at 60°C for 15 min, followed by soaping with an anionic detergent (2 g/L), then rinsed well and finally dried at 85°C for 5 min. The printed samples were evaluated for the depth of the prints (expressed as K/S, where K is the absorption coefficient and S is the scattering coefficient), as well as the fastness properties, e.g. washing and rubbing using standard methods.

Rubbing fastness and wash fastness: Color fastness to rubbing (dry and wet) was assessed manually by hand rubbing one sample ten times and grey scale as per ISO: 105-AO3-1995 extent of staining was used. Color fastness to washing of the dyed fabric samples was determined as per ISO: 105-AO2-1995 method using a launder-O-meter.

The k/s value of printed sample was determined by measuring surface reflectance of the samples using a computer-aided Macbeth 2020 plus reflectance spectrophotometer, using the following Kubelka Munk equation with the help of relevant software:

$$K/S = \frac{(1 - R\lambda_{max})^2}{2R\lambda_{max}}$$

where K the coefficient of absorption; S the coefficient of scattering and Rλmax the reflectance value of the sample at a particular wavelength, on which maximum absorption occurs for a particular dye/colour component.

Results and Discussion

From the selected areas based on different physical appearance (colour, leaf arrangement, height, edge, spot and stem) seven samples were taken to Addis Ababa herbarium. These samples were categorized into only two species, namely *Aloe debrana* and *Aloe camperi*. The two species have been planted in the garden of the herbarium and have been waiting for the flowering season of the plant to confirm their exact type with the help of flower. Since the ultimate goal of this study was to evaluate the performance of a thickening agent for disperse printing of polyester and cotton fabrics, as well as to search for the ideal printing paste components and fixation conditions for attaining darker prints with better overall fastness properties. *Aloe debrana* plant has been selected to precede this work since it has more gel compared from *Aloe camperi*.

From Table 1 it can be seen that using fresh *Aloe debrana* gel as a thickener of pastes to both cotton and polyester have good wash fastness, and rubbing fastness, the sample also have good colour staining fastness test because when they are compared with grey scale they show good result which means five. The decrement in k/s is also small in both samples which mean there is minimum release of dyes from printed samples, this indicates the dye is reacted with the textile substrate and it is the desired property. The change of k/s after washing indicates that there is an excess dyes on textile substrate before washing and the surface of printed sample have dull shade after washing which means there is an increase in brightness of printed sample so it indicates the decrement of k/s.

Alginate and Emulsion thickeners used as standard to compare the thickening potential of *Aloe debrana*. The change of k/s after washing shows increment using Alginate and some decrement with Emulsion of both polyester and cotton fabrics. The nature of Alginate has harsh feeling so the surface of printed sample shows dull shade than that of Emulsion before washing. The wash fastness and rubbing result on printed polyester shows very good results whereas on cotton fabric the

Print paste Thickener	Viscosity	Substrate	K/S Value (400 nm)		Wash Fastness		Rubbing		Handling
			Before Soaping	After Soaping	Colour Change	Colour Staining	Dry Rubbing	Wet Rubbing	
Aloe debrana	10.6	Polyester	1.2299	1.1911	5	5	5	5	S
		Cotton	0.4855	0.4028	5	5	5	5	S
Alginate	11	Polyester	1.0607	1.0962	5	5	5	5	S
		Cotton	0.1427	0.3411	2	2	2	2	-
Emulsion	11.4	Polyester	1.3161	1.1555	5	5	5	5	S
		Cotton	0.7116	0.3153	2	2	2	2	-
Aloe debrana-Alginate	10.8	Polyester	1.0962	1.0705	5	5	5	5	S
		Cotton	0.1715	0.2048	3	3	3	3	-
Aloe debrana-Emulsion	10.9	Polyester	2.6896	1.4238	5	5	5	5	S
		Cotton	1.8151	0.4479	3	3	3	3	-

Table 1: Pastes prepared with fresh *Aloe debrana* gel and other synthetic thickeners (Alginate and Emulsion).

Print paste Thickener	Viscosity	Substrate	K/S Value (400 nm)		Wash Fastness		Rubbing		Handling
			Before Soaping	After Soaping	Colour Change	Colour Staining	Dry Rubbing	Wet Rubbing	
Aloe debrana	10.7	Polyester	3.8369	1.2033	5	5	5	5	S
		Cotton	3.0815	1.5336	5	5	5	5	S

Table 2: Printing paste prepared with preserved *Aloe debrana* gel.

Print paste Thickener	Viscosity	Substrate	K/S Value (400 nm)		Wash Fastness		Rubbing		Handling
			Before Soaping	After Soaping	Colour Change	Colour Staining	Dry Rubbing	Wet Rubbing	
Aloe debrana without acid	10.6	Polyester	2.3033	1.6121	4/5	4/5	4/5	4/5	m
		Cotton	2.9316	1.1532	4	4	4	4	m
Aloe debrana without Urea	10.6	Polyester	5.9613	4.9388	4/5	4/5	4/5	4/5	m
		Cotton	2.4149	1.0306	4/5	4/5	4/5	4/5	m

Table 3: Printing paste prepared without acid and without urea using fresh *Aloe debrana* gel.

result is very poor. That is may be the fabric nature of polyester has hydrophobic end chemically and cotton has hydrophilic chemical end and the dye is hydrophobic which brought good chemical interaction with polyester and less with cotton. However the mixture of *Aloe debrana* with Alginate and Emulsion (*Aloe debrana*-Emulsion and *Aloe debrana*-Alginate) showed some change of k/s value after washing and there is some improvement of the wash fastness and rubbing on cotton when it is compared to that of pure alginate and emulsion, this indicates that *Aloe debrana* gel has a potential of changing the property of disperse dye to have good interaction with cotton.

The viscosity of *Aloe debrana* paste is good enough when compared with synthetic thickeners. Moreover, the viscosity of natural thickeners, *Aloe debrana*, improved when small amount of synthetic thickeners are added.

The gel was stored for two months within highly secured container by using lemon juice as preservative. The result in Table 2 indicated that the decrement in value of k/s of preserved *Aloe debrana* gel used for printing of cotton and polyester fabric is because of a removal of some un reacted dyes from the surface of the fabric which results to the increment of reflectance of the fabric. The sample has good fastness to washing and rubbing, it also shows good result in colour staining test. The comparison of the k/s value of preserved gel with fresh *Aloe debrana* increased that indicates the viscosity of the preserved gel is better than fresh gel. Even though the change in k/s value in Table 3 is the same trend with normal recipe of *Aloe debrana*, the result of washing and rubbing fastness was poor. This shows the presence of the acid and urea in the recipe improves the media and hygroscopic nature of the paste, respectively, which are very essential component of the print paste. The absences of both acid and urea do not affect the viscosity of the paste.

Heating temperature, fixation time and stirring time

Increasing heating temperature from 140°C up to 180°C was accompanied by a gradual sharp increase in the depth of the obtained disperse prints, facilitating the release of disperse dye from the thickener film, allowing sublimation and diffusion into the accessible area of the polyester and cotton structure, as well as enhancing the extent of dye retention and fixation, thereby giving rise to a more intense depth of shade and increase the swell ability of both the thickener film and the polyester structure.

The effect of heating fixation time on the extent of dyeing polyester and cotton fabrics is clear that prolonging the heating time up to 6 min at 180°C resulted in a remarkable increase in the depth of the obtained disperse prints. Decreasing in k/s value indicates that improvement of scattering of light after wash. That reflected the positive role of proper heating time on swelling the thickening film, enhancing the extent of release of dye molecules from the thickener film as well as dye sublimation, adsorption onto and diffusion within the substrate, thereby enabling the volatile components and dye vapour to be strongly adsorbed and retained by the hydrophobic polyester and hydrophilic cotton component. Expanded fixation time may results migration on cotton fabrics and fire hazard for polyester fabrics.

Stirring is the process of mixing the components of print paste; thickener, urea, acetic acid, disperse blue and water, to make solid paste. When the time of stirring increase the paste becomes uniform and good for printing, but when the stirring time is extended the paste prepared from *Aloe debrana* becomes enough solid for good performance or viscosity for printing whereas the synthetic thickeners, alginate and emulsion, becomes more solid which is not good for printing.

Effect of substrate and thickeners

Changes in the k/s values of the obtained prints along with their fastness properties as a function of type of substrate (fabric) are shown on tables, which demonstrates that the depth of the obtained prints was governed by the kind of substrate, its chemical structure, and its affinity for the used dye, the degree of penetration, interaction and subsequent fixation. Polyester and cotton have different nature chemically, the former is the polymer of non polar (hydrophobic) and the latter is polymer of polar (hydrophilic). *Aloe debrana* as a thickening agent in disperse printing results in good thickening efficiency as well as better depth and fastness properties of the obtained prints in comparison with the Alginate and Emulsion thickener.

Effect of gel concentration

Increasing the thickening agent (*Aloe debrana*) concentration from 300 g/kg up to 500 g/kg resulted in an improvement in the k/s values of the disperse prints, which could be discussed in terms of higher paste viscosity. Further increase in the concentration of *Aloe debrana* gel had no negative effect on the depth of the obtained prints and softness. The overall fastness properties of the printed polyester fabric samples showed that using *Aloe debrana* at 300 g/Kg as a thickening agent yielded the best improvement in the handling, sharpness and fastness properties of the printed samples. However, in the case of cotton fabrics numerically show less k/s than that of polyester it might be the chemical affinity of the gel with the cotton product, but still the gray scale and handling show the same result.

Conclusion

This research focused on the application of a new thickening agent based on *Aloe debrana* gel for disperse printing of polyester and cotton. On the basis of the experimental results, we have made the following conclusions: Changes in the k/s values of the obtained prints along with their fastness properties as a function of type of substrate (fabric) are shown, which demonstrates that the depth of the obtained prints was governed by the kind of substrate, its chemical structure, and its affinity for the used dye, the degree of penetration, interaction and subsequent fixation. Polyester and cotton have different nature chemically. Cotton has hydrophilic end chemical nature and it is not expected to have good interaction with hydrophobic dyes however using the *Aloe debrana* thickener with the hydrophobic dyes like disperse dyes showed very good interaction with cotton and printing was carried out with good performance. A natural preservative lemon juice has high potential to preserve the gel more than two months. The results showed that no significant difference between fresh gel and preserved gel. Apart from this, the chemical aspects of the gel with dye and cotton fabric need further investigation.

Acknowledgement

First of all, we would like to thank Wollo University, which provides conducive environment and budget to do this work without any problem. We are also pleased to thank South Wollo Zone Agriculture office, Bahir Dar University of Textile Institute, AddisAbaba University Herberium center. Lastly we also acknowledge all respected Departments for their cooperativeness during the work.

References

1. Nordal I, Sebsebe D, Stabbetorp O (1998) Endemism in groups of Ethiopian geophytes ('Liliiflorae'). Food and Agricultural Organization 247-258.

2. Ragheb AA (1994) Suitability of Nitfogen-Containing starch derivative in printing polyester fabrics with disperse dyes. American Dyestuff Reporter 83: 15-24.

3. Vazquez B, Avila G, Segura D, Escalante B (1996) Antiinflammatory activity of extracts from Aloe vera gel. J Ethnopharmacol 55: 69-75.

4. Dagne E, Bisrat D, Viljoen A, Van Wyk B (2000) Chemistry of Aloe species. Curr Org Chem 4: 1055-1078.

5. Enas MR, El Zairy (2011) New thickening agent based on aloe vera gel for disperse printing of polyester. AUTEX Res J 11.

6. Aspland JR (1997) Textile dyeing and coloration. AATCC, NC, US.

7. Ibrahim NA, El Zairy EMR (2009) Union disperse printing and UV-protecting of Wool/polyester blend using a reactive ß-cyclodextrin. Carb Polymers 76: 244-249.

8. Choudhury AKR (2006) Textile preparation and dyeing. Science publishers, Enfield, NH, USA.

9. Ibrahim NA, El Zairy MR, Abo-Shosha MH (1994) Preparation and rheology of new synthetic thickeners based on polyacrylic acid. Dyes and Pigments 24: 249-257.

2

The Effect of Warp and Weft Variables on Fabric's Shrinkage Ratio

Kadi N and Karnoub A*

Department of Textile and Spinning, Faculty of Mechanical Engineering, University of Aleppo, Syria

Abstract

This research deals with the fabric's shrinkage ratio in both directions (warp and weft), by studying the effect of each of the warp and weft variables on the shrinkage ratio. These variables are the type, density, Yarn tension and the count of both warp and weft yarns in addition to the weave structure.

After test different types of these variables and found that weft density relationship and weft count with shrinkage ratio toward warp is a positive relationship, while the relationship shows the amount of weave float and Yarn tension of both warp and weft thread with shrinkage ratio towards warp is an inverse relationship. The warp density and warp count with shrinkage ratio towards weft relationship is a positive relationship, while the relationship shows the amount of weave float and Yarn tension of both warp and weft thread with the shrinkage ratio towards weft is an inverse relationship. Finally, there is no effect to the type of weave as if it was a satin or twill on the proportion of fabric shrinkage in both directions.

Using the SPSS statistical software solutions have been modeling the previous results and get the formula to calculate the fabric shrinkage ratio that takes into account all previous variables.

Keywords: Fabric's shrinkage ratio; Fabric's actual length; Warp and weft variables

Introduction

Warp thread is exposed during the course of the formation of fabric for a number of tensile forces resulting from (the position of a backrest roll, open the shed, move the comb to insert a weft), as the weft yarn is exposed also to the tensile forces due to (pull of the weft regulator in order to store it, pull of the rapier for publication within the shed) [1].

Researcher Watson [2] found that fabric shrinkage ratio affected by the following variables:

1. The type, count and density of the warp yarns.

2. The type, count and density of the weft yarns.

3. Weave structure

But the warp and weft tension fixed at a certain value, but the researcher Nisbet [3] found that the effect of warp and weft tension is opposite effect on shrinkage ratio.

The researcher Moghe [4] calculates the percentage of fabric shrinkage towards the warp only when using wefts yarn of polyester and weft density of 28 to 44 (pick/inch) and so for plain weave 1/1, and found that when weft density increases the shrinkage ratio increases.

The researcher Edita [5] studied the relationship between weft density and shrinkage ratio for weft densities between 21 and 27 pick/cm, but passed the low densities, and concluded that the weft density relationship with the shrinkage ratio is a direct correlation.

The researchers Çeven and Özdemir [6] have calculated the shrinkage ratio towards the weft thread, used Chenille yarn following count (4, 4.5, 5, 5.5, 6) Nm on plain weave 1/1 and warp density of 10 to 30 thread/cm. The result of their research that a positive relationship between the warp density and the percentage of shrinkage towards the weft. The relationship between the weft count and the shrinkage ratio is opposite relationship, where the yarn count is metric (indirect count).

The study of the cross-section of the relationship of overlap between the warp and weft entrance to determine the shrinkage ratio, where researchers Vyšanská and Sirková [7] study conducted on one type of yarn and one density and plain weave 1/1, and reached the following equation:

$$warp\ distance = \frac{\left(\frac{100}{D0}.n1\right).\sqrt{4.(ds)^2 - (ds)^2}}{p2.\sqrt{4.(ds)^2 - (ds)^2} + d1.(n1 - p2)}$$

n1: distance between the center of warp thread and weaving axis

ds: the average diameter of warp and weft

p2: inverted weft density

The same of the previous principle, researchers Milašius and Vytautas [8] finding of what is known as a factor of weaving (F), which is a summary of the study of Interference warp and weft yarns, and which can predict the desired length to Interference between the yarns, also this study were the richest because of inclusion of a set of weaves are (1/1 plain, 1/2 twill, 3/1 twill, 2/2 twill, 2/2 ribes) and four types of weft yarn, however, that all these weaves used just for clothes do not include the curtains and furnishings weaves, where the amount of weave floating in this fabrics larger.

The researcher Černoša [9] and his research team have studied the effect of warp and weft tension on warp to weft variables and shrinkage ratio in both directions warp and weft, and They concluded that the increase in yarn tension lead to a decrease shrinkage ratio within the limits of elasticity of the thread, this search based on five types of weaves but on one type and one count of the warp and weft yarns.

Percentage of shrinkage have studied and calculate in two stages after the weaving and the second after immersion cloth in pure water

***Corresponding author:** Karnoub A, Department of Textile and Spinning, Faculty of Mechanical Engineering, University of Aleppo, Syria
E-mail: amerkarnoub@gmail.com

by researcher Rukuižienė [10] and his research team, studied has based on a Honeycomb weave and cotton weft and warp yarns, researcher concluded during the study that shrinkage relationship with the density and count is a positive relationship.

The researchers Rukuižienė and Milašius [11] calculates the percentage of shrinkage toward warp through three stages, the first stage is between warp width and comb width, and the second stage between comb width and woven point width, and the third stage between the fabric width during weaving and cloth after weaving stage and free it from the forces of tensile applied it in loom. But just for plain weave 1/1 and polyester yarn from count 70- 100- 150 den. They concluded that the fabric shrinks three phases in addition to that the increasing count direct thread increases the shrinkage ratio.

The percentage of shrinkage φ can calculate in the study of researcher Kumpikaitė [12] through a mathematical equation based on the density of warp and weft and linear distance between the warp and weft yarns and factor named Milašius's factor which expresses the inverted weave float, and the type of materials used.

$$\varphi = \sqrt{\frac{12}{\pi}} \frac{1}{p1} \sqrt{\frac{Tav}{\rho}} S2^{\frac{1}{1+2/3\sqrt{T1/T2}}} S1^{\frac{2/3\sqrt{T1/T2}}{1+2/3\sqrt{T1/T2}}}$$

T1: warp density T2: weft density

Tav: the arithmetic average of the fabric density

P1: Milašius factor for weaving

ρ: Density of raw materials

S1: warp settings in woven fabric S2: weft settings in woven fabric

The researcher Anderson [13] have presented a mathematical equation to calculate shrinkage percentage for three yarn types and three weaves are plain 1/1 and twill 1/3 and satin 1/7. This equation suggests that the shrinkage rate be greater when using plain 1/1 goes down when using twill 3/1 and be at the lowest value when using satin 1/7.

The aim of the research is that includes the effect of each warp and weft variables on the shrinkage ratio, the one hand, the weft density value range of 4 to 20 pick/cm, the warp density range is from 33 to 66 thread/cm, yarn tensile force ranges between 12 to 20 cN/Tex, weaves float from 2 to 20 and finally the types of thread for the weft yarn studied more than four types in terms of use, namely, (chenille -polypropylene with continuous filament-thread with continuous filament and amplifying by air ATY (jet)-cotton yarn turbine spinning) while the warp (mixed with a thread of cotton and viscose (trivera) -polyester DTY) with different magnitude of yarn count depends on type of yarn.

Materials and Methods of Search

The study and experiments were conducted on the loom model (GTM) from the production company (Picanol) Belgian, a loom capable of producing all type of fabrics, because it contains the Jacquard (a device to open the shed) (Figure 1).

Warp thread tension was adjusted by a device to measure yarn tension, its model (I1901) of production company (Schmidt) German.

Shrinkage rate define as the difference between length of thread before weaving and beyond, the experiments are divided into two types:

• determine the percentage of shrinkage towards the weft yarns: weft yarn identified before weaving by a device for determine

certain lengths of yarn has model (L232), this device has a wheel surroundings of one meter, but it is difficult to be carried out tests on the basis of that length of the thread before weaving 1m because the entire fabric width 1.4 m, and as the wheel of the machine is divided into 6 sections by 6 beams, which means that the distance between the crossbars are 16.67 cm we will put signs on the yarns at each edge of the symptoms of the cupboard to be determining the length of a string weft before weaving.

• determine the percentage of shrinkage towards the warp yarns: warp thread identified before weaving on warp beam before backrest roll order to determine the length before being subjected to any kind of friction with the loom parts, was chosen length lab 15 cm depending on the length laboratory at researcher Anderson [13], as evident in Figure 2 order to reduce error rate measurement.

To find out the effect of each of the variables on the fabric shrinkage rate has been changing the value of the variable according to the most widely used field with fixed the rest of the variables at a specific value, note that variables has been fixed as follows:

1   warp density: 66 thread/cm

2.   the type of warp yarn: thread Trivera

3.   warp count: 150 den

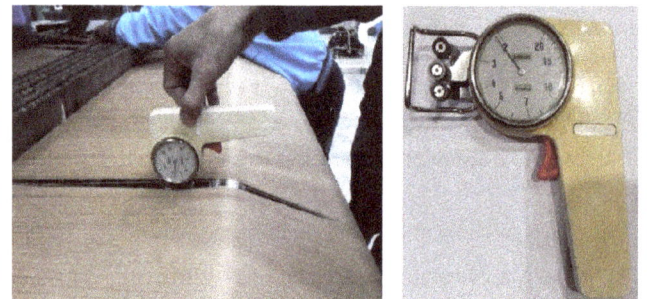

Figure 1: A device for measuring Yarn tension of thread.

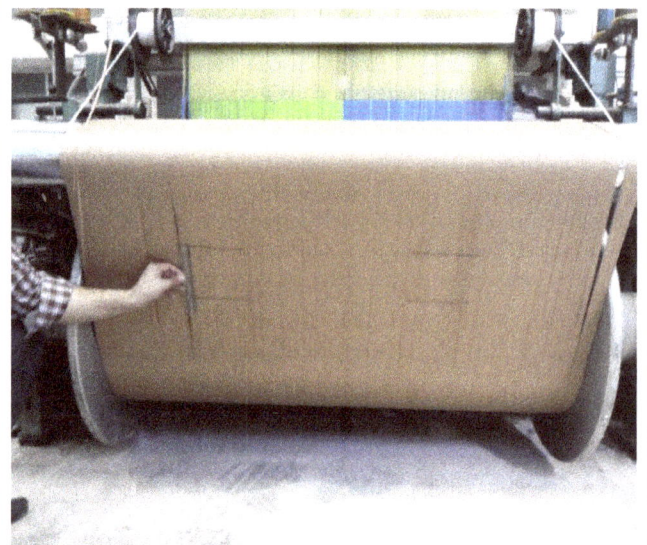

Figure 2: Reference Determination of the warp yarn length.

4. weft density: 12 thread/cm

5. the type of weft yarn: Polyester DTY

6. weft count: 150 den

7. tension applied to the warp yarns: 15 cN/Tex

8. tension applied to the weft yarns: 15 cN/Tex

9. weave structure: Satin 8

Results and Discussion

Effect of weft density on fabric shrinkage ratio toward warp

Densities tested were (4, 8, 12, 16, 20) thread/cm, because this range covers weft densities in all kinds of fabrics, even consisting of more than one type of weft yarn such as the curtains and furnishings (Table 1).

In definite of the Figure 3 that increase weft density will increases fabric shrinkage and reason for this is an increase in density of the weft increases length of the warp thread which interferes with the weft thread path. Comparing this result with results of the researcher Edita M note match outcomes, but the researcher Edita M studied the densities of high field only and therefore not able to infer the equation linking weft density and the percentage of shrinkage, while our study has larger weft density field.

Effect of weave type on fabric shrinkage ratio toward warp

Due to plain is one of the twill weaves, the comparison here will be between twills and satins, and will test two amount of floating for each of the two weaves (1/7, 1/15) (Figures 4 and 5).

Note match in the percentage of shrinkage towards warp yarns between each of the weave satin and twill for two tested float

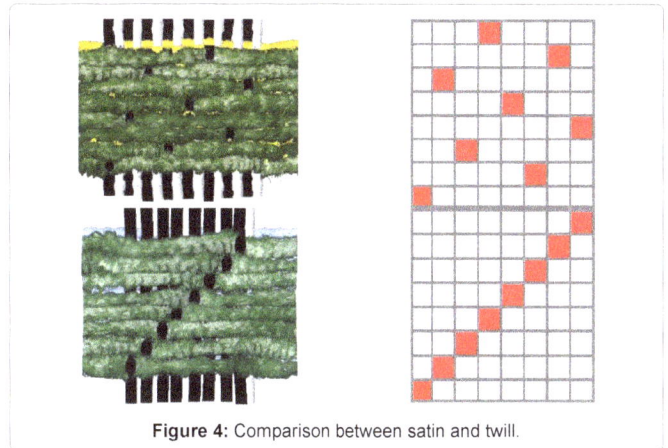

Figure 4: Comparison between satin and twill.

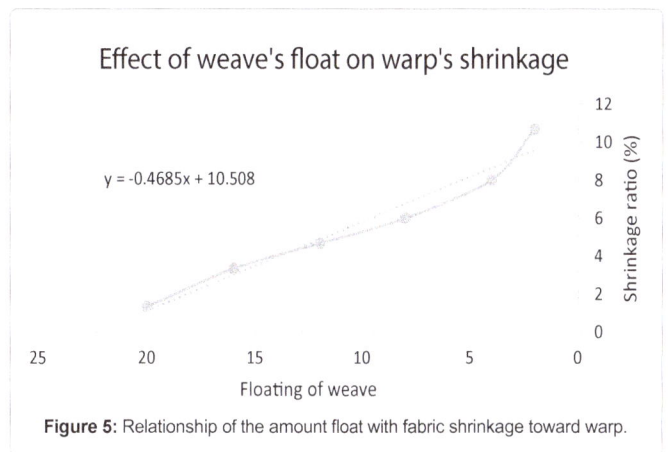

Figure 5: Relationship of the amount float with fabric shrinkage toward warp.

sample	Length lost (%)	Length after test (cm)	Length before test (cm)	Type of weave
B 001	6	14.1	51	Satin
B 002	6	14.1	51	Twill

Table 2: Shrinkage ratios toward the warp yarns for satin and twill 7/1.

sample	Length lost (%)	Length after test (mm)	Length before test (mm)	Type of weave
B 003	3.33	14.5	51	Satin
B 004	3.33	14.5	51	Twill

Table 3: Shrinkage ratios toward the warp yarns for satin and twill 15/1.

sample	Length lost (%)	Length after test (cm)	Length before test (cm)	Weft density (pick/cm)
A 001	3.33	14.1	51	4
A 002	4.67	14.3	51	8
A 003	6	14.1	51	12
A 004	7.33	13.9	51	16
A 005	8.67	13.7	51	20

Table 1: Shrinkage ratios toward warp for each of weft densities.

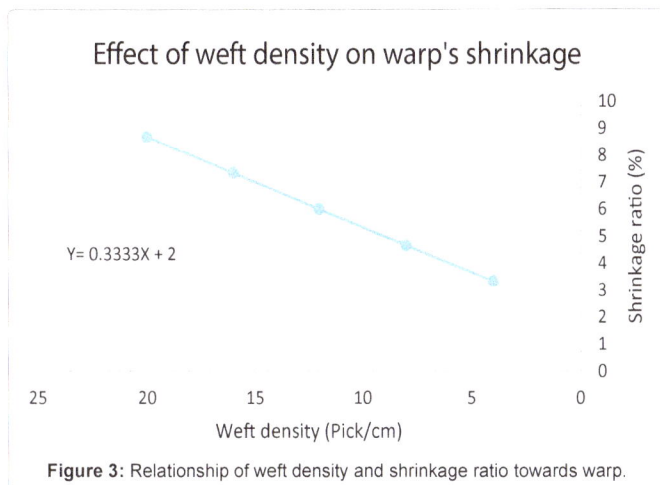

Figure 3: Relationship of weft density and shrinkage ratio towards warp.

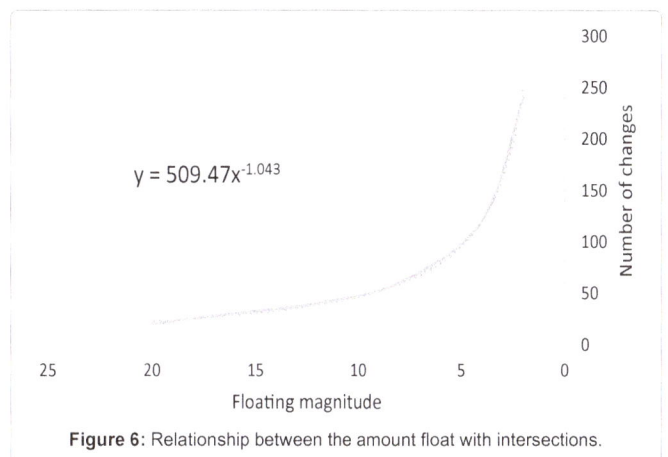

Figure 6: Relationship between the amount float with intersections.

(1/7 - 1/15), which shows that there is no effect of the type of weave whether twill or satin on the fabric shrinkage ratio (Tables 2 and 3).

On the other hand, and for Explanation of the previous result, we must study the number of intersections in a pair of weave satin and twill to one of the previous float (1-7) (Figure 6).

Can enumerate the number of intersections occurring within a weave by Ned Graphics program and note that number of intersections between each of the twill and satin is the same in one repeat.

We conclude from the foregoing that if the number of weave intersections similar must have the same effect on the percentage of shrinkage towards the warp yarns [14]. The only difference between satin and twill is a difference in the distribution of the intersection between the warp and weft yarns points.

Effect of weave floating on fabric shrinkage ratio toward warp

Amounts of floats were tested covering more weaves traded in the practical field, namely (2, 4, 8, 12, 16, 20).

The figure shown that the relationship between the percentage of shrinkage towards warp yarns and the amount of float is inverse (Table 4).

This result did not match with the result of the researcher Milašius and Vytautas in the relationship between the type of weave and the fabric shrinkage, so the reason for this is that weaves tested by the researcher is of derivatives plain, this weaves don't effect on float amount, while the method of studying the effect of the amount of float on the shrinkage ratio is comprehensive.

To explain the previous result we must enumerate the number of intersections between the warp and weft in every pick in one repeat, as the woven fabric caused by friction between the warp and weft yarns linked directly with the number of intersections, and you have to enumerate the number of intersections occurring in each of the previous weaves within a unified measure paper squares, the suitable measurements for (2, 4, 8, 12, 16, 20) weaves is 240×240, results of the number of intersections shown in Table 5.

The relationship between the intersections and the amount of float in weaves is an inverse relationship, which means that by increasing

the float amount will decrease the number of intersections so would weaken the fabric structure woven due to lack of friction between the warp and weft yarns, but when increasing the number of intersections amount required from the warp yarn length will increase in order to achieve the thread intersections required of it, and this explains the results obtained, that the greater amount of float in weave gives larger shrinkage ratio.

Effect of warp's yarn tension on fabric shrinkage toward warp

The range of warp's yarn tension tested followed to the range of tensile could be applied on warp yarns, depends on the allowable area of warps is (120 - 200) Kg and due to that the total number of warps are 10000 thread, the range of study is (12 - 20) cN/Tex, it divided into five sections (20, 18, 16, 14, 12) cN/Tex (Table 6).

In definite from the Figure 7 that the inverse relationship between the percentage of shrink warp direction and value of warp's yarn tension.

sample	Length lost (%)	Length after test (cm)	Length before test (cm)	Warp tension (cN/Tex)
D 001	8	13.8	51	12
D 002	6.67	14	51	14
D 003	5.33	14.2	15	16
D 004	4.67	14.3	15	18
D 005	4	14.4	15	20

Table 6: Shrinkage ratios toward warp according to warp's Yarn tension.

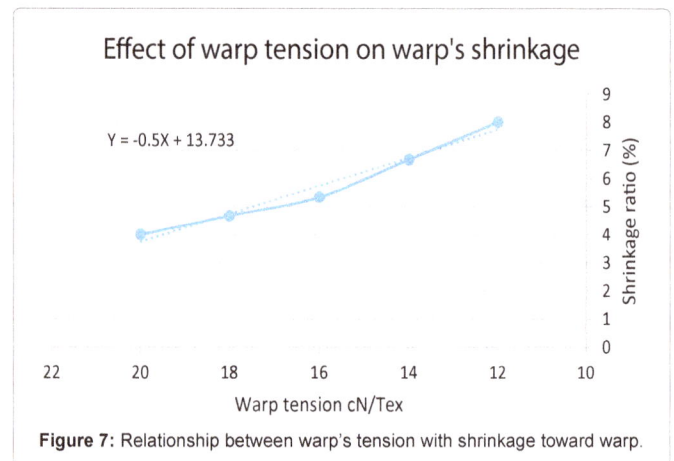

Figure 7: Relationship between warp's tension with shrinkage toward warp.

sample	Length lost (%)	Length after test (cm)	Length before test (cm)	Length of floating (pick)
C 001	10.67	13.4	51	2
C 002	8	13.8	51	4
C 003	6	14.5	15	8
C 004	4.67	14.3	15	12
C 005	3.33	14.5	15	16
C 006	1.33	14.8	15	20

Table 4: Shrinkage ratios for each of the amounts floats tested.

Intersections	Floats
240	2
120	4
60	8
40	12
30	16
20	20

Table 5: Intersections for each amount float.

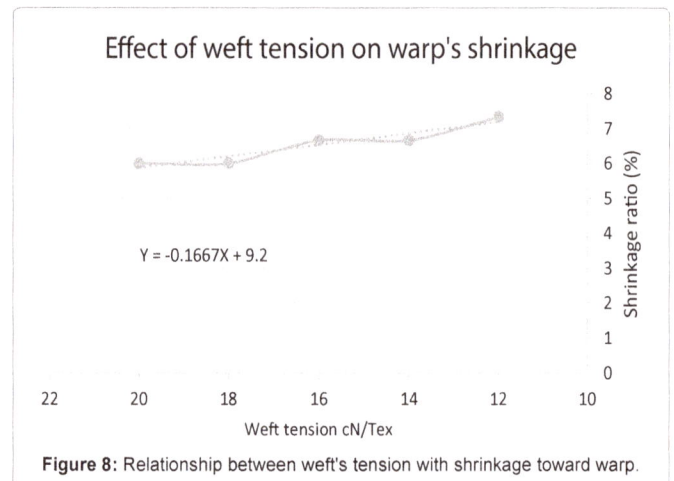

Figure 8: Relationship between weft's tension with shrinkage toward warp.

When warp's yarn tension increase will lead to increase pressure on the warp thread not to allow it to form excess pile during interlace with the weft thread, and thus decrease the length of the consumer warp thread, this explains the previous result.

Effect of weft's yarn tension on fabric shrinkage toward warp

Range of weft's tension tested are (20, 18, 16, 14, 12) cN/Tex. Definite from the Figure 8 that the inverse relationship between percentage of shrinkage towards warp and weft's tension. However, the effect of weft's tension on the fabric shrinkage toward the warp is a slight effect, where the difference between the highest percentage of shrinkage and the lowest is up 1%, and the reason that the increased weft's tension will press the warp thread just in weave point, on the other hand weft's tension does not effect on the amount of used warp thread (Table 7).

Effect of weft's type and count on fabric shrinkage toward warp

Types of weft yarn tested are (Chenille, Poly Propylene with

sample	Length lost (%)	Length after test (cm)	Length before test (cm)	Warp tension (cN/Tex)
E 001	7.33	13.9	51	12
E 002	6.67	14	51	14
E 003	6.67	14	15	16
E 004	6	14.1	15	18
E 005	6	14.1	15	20

Table 7: Shrinkage ratios toward warp according to weft's Yarn tension.

sample	Length lost (%)	Length after test (cm)	Length before test (cm)	Warp tension (cN/Tex)
F 001	14.67	12.8	51	5512
F 002	12.67	13.1	51	1500
F 003	10.67	13.4	15	1125

Table 8: Shrinkage ratios toward the warp for each weft count (Chenille).

sample	Length lost (%)	Length after test (cm)	Length before test (cm)	Warp tension (cN/Tex)
F 004	6	14.1	51	150
F 005	7.33	13.9	51	300
F 006	8.67	13.7	15	500
F 007	10.67	13.4	15	1000

Table 9: Shrinkage ratios toward warp for each weft count (Poly Propylene).

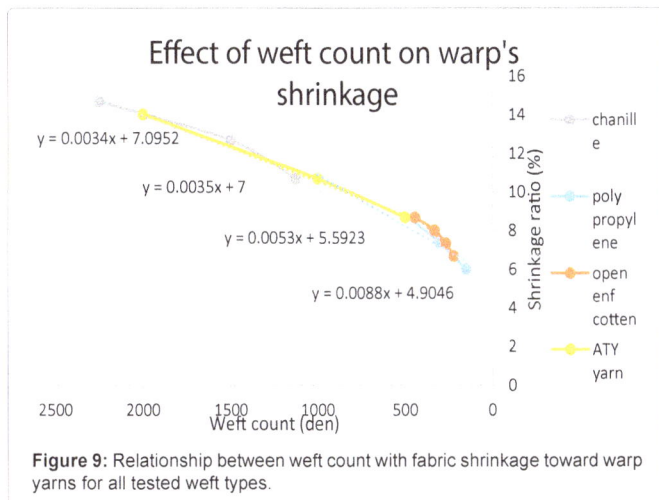

Figure 9: Relationship between weft count with fabric shrinkage toward warp yarns for all tested weft types.

continuous filaments, spinning turbine cotton, threaded inflated by air ATY), and because of the large difference in thread count tested it is difficult to compare, so it will be compared to each type of weft yarn (Tables 8 and 9).

By studying the charts in Figure 9 it is clear that a positive relationship between fabric shrinkage towards warp and weft count when count are directly (den). When thread diameter increases the fabric shrinkage will increase thereby increasing the length of thread required to complete the interlacement between warp and weft (weave structure), as shown in the Figure 10, which is a fabric cross section showing the three weft count for same weave, figure shows that warp length needed for complete the weaving increases when weft diameter increased, on other hand when weft count decreases in metric system [15] (Tables 10 and 11).

Effect of warp density on fabric shrinkage toward weft

Warp densities are selected (66, 45, 33) end/cm, because the warp density related to harness density, because of the tested fabrics are upholstery fabrics and curtains, harness density are (66, 45, 33) end/cm.

Definite from the Figure 11 that when warp density increases fabric shrinkage toward weft increases, the reason for this result is when the number of warp yarns per length unit increases the number of weft intersections increases, thus the shrinkage ratio towards weft increases (Table 12).

Effect of weave type on fabric shrinkage ratio toward weft

Chosen weaves are more useful in the practical field, namely: (a Satin 7/1, (b twill 7/1, (c Satin 15/1, d twill 15/1).

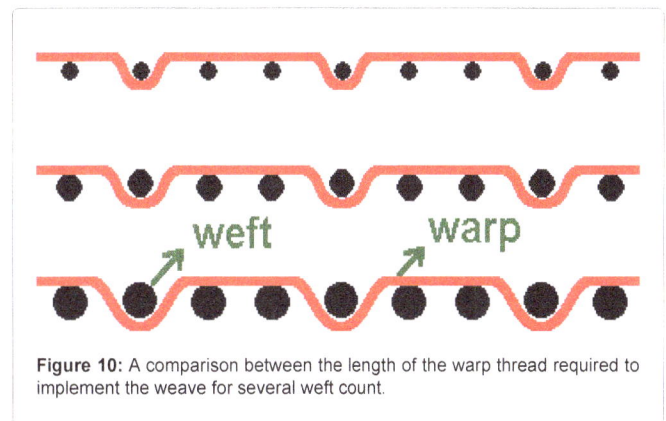

Figure 10: A comparison between the length of the warp thread required to implement the weave for several weft count.

sample	Length lost (%)	Length after test (cm)	Length before test (cm)	Warp tension (cN/Tex)
F 008	8.666667	13.7	51	442.5
F 009	8	13.8	51	331.87
F 010	7.333333	13.9	15	265.5
F 011	6.666667	14	15	221.25

Table 10: Shrinkage ratios toward the warp for each weft count (spinning turbine cotton).

sample	Length lost (%)	Length after test (cm)	Length before test (cm)	Warp tension (cN/Tex)
F 012	8.67	13.7	51	500
F 013	10.67	13.4	51	1000
F 014	14	12.9	15	2000

Table 11: Shrinkage ratios toward the warp for each weft count (ATY).

Effect of warp density on weft's shrinkage

$Y = 0.3193X + 2.4729$

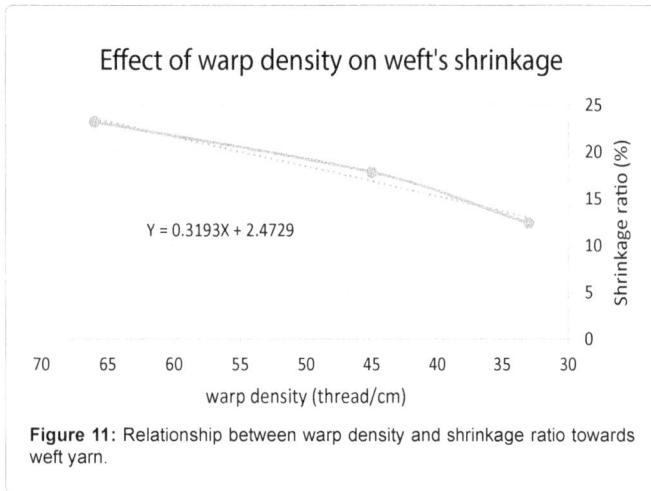

Figure 11: Relationship between warp density and shrinkage ratio towards weft yarn.

sample	Length lost (%)	Length after test (cm)	Length before test (cm)	Warp tension (cN/Tex)
G 001	12.40	14.6	16.667	33
G 002	17.80	13.7	16.667	45
G 003	23.20	12.8	16.667	66

Table 12: Shrinkage ratios toward weft for each warp density.

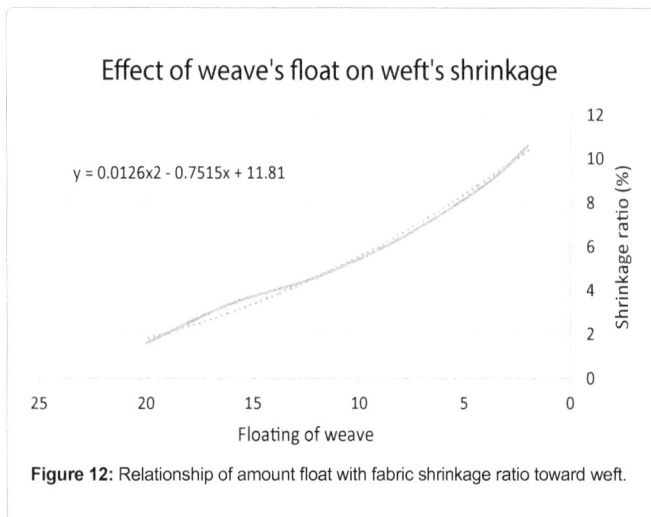

Effect of weave's float on weft's shrinkage

$y = 0.0126x2 - 0.7515x + 11.81$

Figure 12: Relationship of amount float with fabric shrinkage ratio toward weft.

sample	Length lost (%)	Length after test (cm)	Length before test (cm)	Warp tension (cN/Tex)
H 001	6.40	15.6	16.667	Satin
H 002	6.40	15.6	16.667	Twill

Table 13: Shrinkage ratios toward weft for each of weave satin and twill 7/1.

sample	Length lost (%)	Length after test (cm)	Length before test (cm)	Warp tension (cN/Tex)
H 003	3.40	16.1	16.667	Satin
H 004	3.40	16.1	16.667	Twill

Table 14: Shrinkage ratios toward weft for each of weave satin and twill 15/1.

As the only difference between satin and twill is the method of point distribution of weave Figure 12, because that there isn't any difference between fabric shrinkage towards weft for each of satin twill and this explains that there is no effect on the weave type whether satin or twill

(Tables 13 and 14).

Effect of weave floating on fabric shrinkage ratio toward weft

Amount of weave float has been selected are (2, 4, 8, 12, 16, 20).

Equation shows that the inverse relationship between the percentage of shrinkage towards weft and the amount of float, as the number of intersections increases the required weft length increases, thus the shrinkage ratio increases (Table 15).

Effect of weft's tension on fabric shrinkage toward weft

Definite from Figure 13 that inverse relationship between fabric shrinkage towards weft and value of weft's tension, it explains that when weft's tension increases it will prevent the shrinkage of fabric

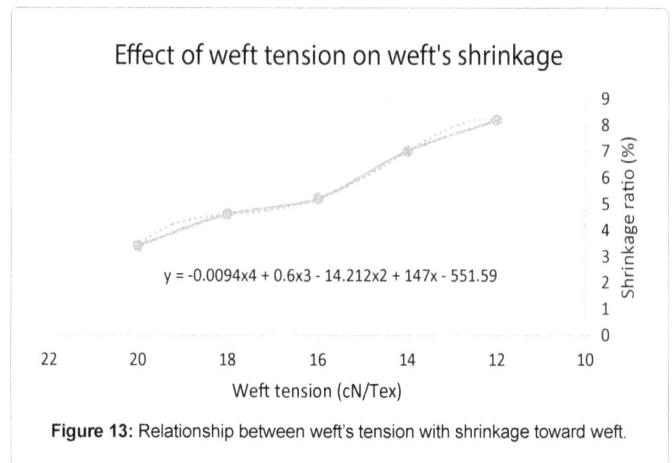

Effect of weft tension on weft's shrinkage

$y = -0.0094x4 + 0.6x3 - 14.212x2 + 147x - 551.59$

Figure 13: Relationship between weft's tension with shrinkage toward weft.

sample	Length lost (%)	Length after test (cm)	Length before test (cm)	Warp tension (cN/Tex)
I 001	10.60	14.9	16.667	2
I 002	8.80	15.2	16.667	4
I 003	6.40	15.6	16.667	8
I 004	4.60	15.9	16.667	12
I 005	3.40	16.1	16.667	16
I 006	1.60	16.4	16.667	20

Table 15: Shrinkage ratios for each of the tested amounts float.

sample	Length lost (%)	Length after test (cm)	Length before test (cm)	Warp tension (cN/Tex)
J 001	8.20	15.3	16.667	12
J 002	7.	15.5	16.667	14
J 003	5.20	15.8	16.667	16
J 004	4.60	15.9	16.667	18
J 005	3.40	16.1	16.667	20

Table 16: Shrinkage ratios toward the weft for each of tested weft's tension.

sample	Length lost (%)	Length after test (cm)	Length before test (cm)	Warp tension (cN/Tex)
K 001	7.33	13.9	51	12
K 002	6.67	14	51	14
K 003	6.67	14	15	16
K 004	6	14.1	15	18
K 005	6	14.1	15	20

Table 17: Shrinkage ratios toward the weft for each of tested warp's tension.

towards weft depends on the friction between the warp and weft (Table 16).

Effect of warp's tension on fabric shrinkage toward weft

Definite of Figure 14 that there an inverse relationship between fabric shrinkage and value of warp's tension, the reason is that increase in warp's tension would increase the pressure on the contact points

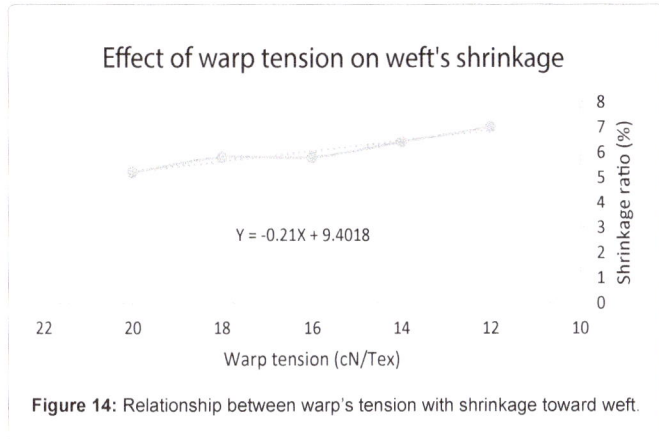

Figure 14: Relationship between warp's tension with shrinkage toward weft.

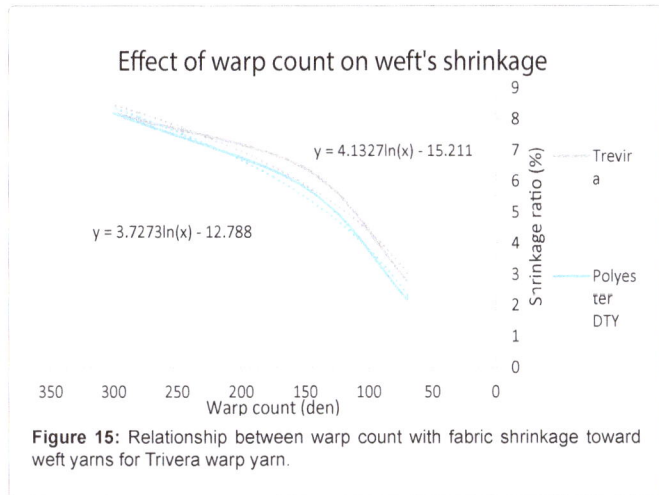

Figure 15: Relationship between warp count with fabric shrinkage toward weft yarns for Trivera warp yarn.

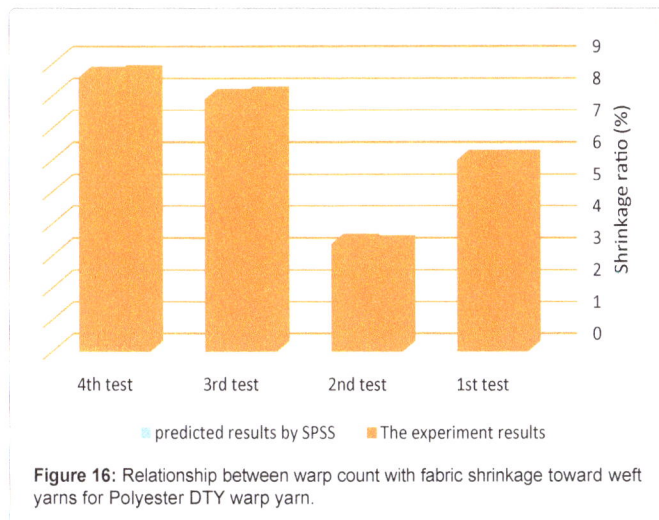

Figure 16: Relationship between warp count with fabric shrinkage toward weft yarns for Polyester DTY warp yarn.

between the warp and weft and reduce the shrinkage, but by light effect (Table 17).

Effect of warp's type and count on shrinkage toward weft

Figures 15 and 16 show great similarities between the planners path for each of the polyester and trivera, and both indicate that the relationship between the fabric shrinkage ratio toward weft and warp count is a positive relationship. This result does not matched with the result of researcher [11] conducted on warp count (70-100-150) den and densities (10-30) thread/cm, where the percentage of shrinkage for count 150 den is 5.80%, while a researcher [11] is greater than that where it is 9.11% and the reason for this is that researcher tested weave is plain 1/1 while our tested weave is Satin 1/7, and as we saw previously that when the amount of float less the shrinkage ratio be greater and this explains the difference between the two results (Tables 18 and 19).

Calculate fabric shrinkage ratio by deriving an equation

Using SPSS program (Statistical Package for the Social Sciences) from the IBM company to conclude the shrinkage ratio equation, we get equations one to calculate the percentage of shrinkage towards warp numbered (1) and another to calculate the percentage of shrinkage towards weft numbered (2).

- Shwr: shrinkage ratio towards warp (%)
- Shwf: shrinkage ratio towards weft (%)
- Dwf: weft density (pick/cm)
- Dwr: warp density (thread/cm)
- Fl: weave float
- Twr: warp's Yarn tension (cN/Tex)
- Twf: weft's Yarn tension (cN/Tex)
- Cwf: weft count (den)
- Cwr: warp count (den)

To ensure the validity of equations (1) and (2), four experiments are performed on random values of variables of fabric samples and

sample	Length lost (%)	Length after test (cm)	Length before test (cm)	Warp tension (cN/Tex)
L 001	2.80	16.2	16.667	70
L 002	6.40	15.6	16.667	150
L 003	8.20	15.3	16.667	300

Table 18: Shrinkage ratios toward weft for each weft yarn (Trivera) count.

sample	Length lost (%)	Length after test (cm)	Length before test (cm)	Warp tension (cN/Tex)
L 004	2.20	16.3	16.667	70
L 005	5.80	15.7	16.667	150
L 006	8.20	15.3	16.667	300

Table 19: Shrinkage ratios toward weft for each weft count (Polyester DTY).

Weft type	α	β	ξ	η	φ	λ
Chenille	+0.191	-0.017	-0.104	-0.018	+0.001	6.187
Poly propylene	+0.168	-0.040	-0.151	-0.023	+0.001	7.387
Turbine spinning cotton	+0.194	-0.016	-0.143	-0.018	+0.006	6.777
ATY yarn	+0.184	-0.024	-0.093	-0.016	+0.003	6.011

Table 20: Constants of the equation (1).

Warp type	α	β	ξ	η	φ	
polyester	+0.291	-0.008	-0.050	-0.008	+0.001	3.324
trivera	+0.257	-0.006	-0.077	-0.014	+0.001	3.109

Table 21: Constants of the equation (2).

sample	Dwf	Fl	Twr	Twf	Cwf	Actual Shwr	Predict Shwr
1st test	12	8	15	15	150	6	6.002
2nd test	12	16	15	15	150	3.33	3.38
3rd test	12	8	12	15	150	8	7.93
4th test	12	8	15	15	500	8.66	8.6

Table 22: A comparison of the actual samples results and predictive results.

compare it with the theory of equations results, and found that the error rate does not exceed 0.006 (Tables 19-22).

Results

➢ Variables that effect on fabric shrinkage ratio towards warp are: weft density, weft count, weft type, weft's Yarn tension, warp's Yarn tension and weave float.

➢ Variables that effect on fabric shrinkage ratio towards weft are: warp density, warp count, warp type, weft's Yarn tension, warp's Yarn tension and weave float.

➢ Relationship between weft density with fabric shrinkage ratio toward warp is a positive correlation. It matches with the result of the researcher Edita M.

➢ Relationship between weft count with fabric shrinkage ratio toward warp is a positive correlation.

➢ Relationship between weave float with fabric shrinkage ratio toward warp is an inverse correlation. This result does not matches with the result of the researcher Milašius and Vytautas.

➢ Relationship of each warp and weft Yarn tension with fabric shrinkage ratio toward warp is an inverse correlation.

➢ Relationship between warp density with fabric shrinkage ratio toward weft is a positive relationship.

➢ Relationship between warp count with fabric shrinkage ratio toward weft is a positive relationship. This result does not matches with the result of the researcher Rukuižienė and Milašius.

➢ Relationship of each warp and weft Yarn tension with fabric shrinkage ratio toward weft is an inverse correlation.

➢ There isn't effect of weave type whether Satin or Twill on fabric shrinkage ratio toward both warp and weft direction.

References

1. Lord M (1982) Weaving conversion of yarn to fabric.
2. Watson W (1955) Advanced textile design.
3. Nisbet H (1980) Grammar of textile design.
4. Moghe A (2002) Study and characterization of small diameter woven tubular fabrics.
5. Edita M (2008) Investigation of Linen Honeycomb Weave Fabric Shrinkage After Laundering in Pure Water, Kaunas.
6. Çeven E, Özdemir Ö (2006) Using Fuzzy Logic to Evaluate and Predict Chenille Yarn's Shrinkage Behavior, Bursa.
7. Sirková B, Vyšanská M (2012) Methodology for Evaluation of Fabric Geometry on the Basis of the Fabric Cross-Section, Liberec.
8. Milašius A, Milašius V (2008) New Representation of the Fabric Weave Factor, Kaunas.
9. Gabrijelčič H, Černoša E, Dimitrovski K (2007) Influence of Weave and Weft Characteristics on Tensile Properties of Fabrics, Ljubljana.
10. Malčiauskienė E, Rukuižienė Z, Milašius R (2008) Investigation of Linen Honeycomb Weave Fabric Shrinkage After Laundering in Pure Water, Kaunas.
11. Rukuižienė Z, Milašius R (2006) Influence of Reed on Fabric Inequality in Width, Kaunas.
12. Kumpikaitė E (2007) Influence of Fabric Structure on the Character of Fabric Breakage, Kaunas.
13. Anderson K (2004) Seamless textiles with inherent shape.
14. Alderman S (2004) Mastering weave structures.
15. Vander H (1993) The complete book of drafting for handweavers.

Simulation and optimisation of warp tension in the weaving process

Gloy YS [1]*, **Renkens W**[2], **Herty M**[3] **and Gries T**[4]

[1]Institut für Textiltechnik der RWTH Aachen University, Aachen, Germany
[2]Renkens Consulting, Aachen, Germany
[3]Mathematics (IGPM), RWTH Aachen University, Aachen, Germany

Abstract

Warp tension is a major parameter of the weaving process. The system analysis of a weaving machine leads to a simulation model for calculating the warp yarn tension. Validation of the simulation has demonstrated that the results correspond well with the reality. In a second step, an improved model of this simulation was used in combination with a genetic algorithm and a gradient based method to calculate optimised setting parameters for the weaving process. A cost function was defined taken into account a desired course of the warp tension. It is known, that a low and constant warp tension course is suitable for weaving. Using the genetic algorithm or the gradient based method leads to optimised weaving machine parameters. Applying the optimised setting parameters on a loom demonstrated that the quality of the produced fabrics can be improved. Further analysis of produced fabrics did not show an influence of optimised weaving machine parameters on the mechanical properties or productivity of the weaving process.

Keywords: Weaving; Optimisation; Simulation; Warp tension

Introduction

Development of weaving mills in high-wage countries aims at rising productivity and flexibility. The weaving mills are confronted by customer wishes to more and more product variety and shorten article length. Therefore productivity describes the capability to produce more meters of woven fabrics in a shorter period of time. It is important to reduce setup times in order to raise productivity. Setup time in weaving mills occurs mainly because of the change of the warp beam and adaptation of the weaving machine like changing working width or reeds finesse. In general, the mill workers have the necessary know-how to set-up a loom or the machine setting is stored in databases of the weaving mill [1].

In order to find optimised setting parameters for unknown articles, it is generally necessary to conduct experiments within the weaving mill. In order to raise productivity it is also important to reduce down-time of weaving machines. Down-time normally occurs through wrong weft insertion or broken warp yarns. While the removal of incorrect inserted weft is already automated on modern air jet looms, the repair of a broken warp has to be done manually. Exceptions are of course high strength warp materials like aramid or carbon. Breakage of warp yarns occurs if the warp tension is too high. In contrast, if the warp tension is too low, warp yarns attend to jam and then break. Furthermore, a too low warp tension leads to an unclean shed formation. A clear shed is needed in order to have less problem of weft insertion.

As described, knowledge on the course of the warp tension during weaving is essential in order to perfectly setup a weaving machine. Keller [2] and Schlichter [3] described methods for measuring warp tension during weaving. They presented an adequate measurement system that measures the dynamic warp tension during the weft insertion. Based on the results, De Weldige [4] presents an algorithm that can calculate the warp tension based on the examination of equilibrium of torque around the axis of a mass-spring backrest system. Chen [5], and Mirjalili [6] described a simulation of warp tension. Another approach can be obtained by using a model of the behaviour of the weaving machine, as described by Beitelschmidt [7] and yet another approach by Grobmann et al. is using transient computation of dynamic interactions between weaving machine and the fabric take-off system [8].

Klöppels et al. developed a motor driven backrest system to an air jet loom in industry [9]. They also calculated motion curves for the backrest taking into account parameters like fabric construction, warp yarn properties, warp drawing in and weaving machine settings. For the production of a 6- pick twill under industrial conditions, the motor driven backrest system reduced warp breaks around 61% compared to weaving machines with negative backrest and 74% compared to identical weaving machines.

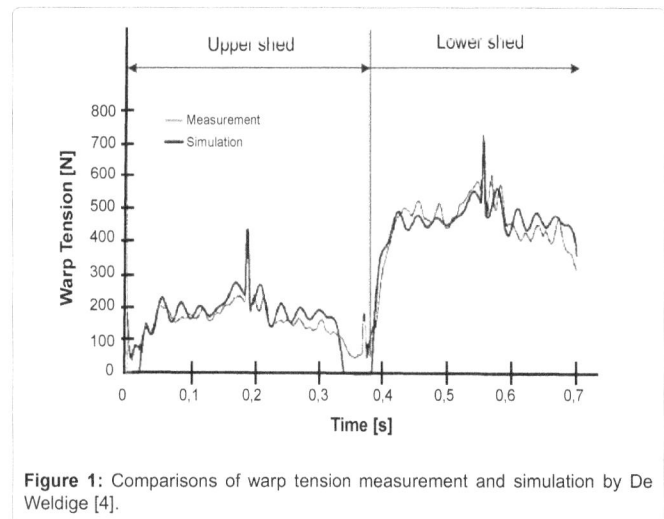

Figure 1: Comparisons of warp tension measurement and simulation by De Weldige [4].

***Corresponding author:** Gloy YS, Institut für Textiltechnik der RWTH Aachen University, Aachen, Germany
E-mail: yves.gloy@ita.rwth-aachen.de

Çelik and Eren present a mathematical analysis of warp elongation in weaving machines with positive backrest system [10]. Influences of backrest swinging angle, shed closing angle, backrest orientation angle and backrest phase difference on warp elongation are investigated. Results shown that especially shed closing angle and backrest phase difference have an important effect on warp elongation curve during one weft insertion. They concluded that by producing different motion curves with motor driven positive backrest, optimum warp elongation or warp tension curves can be obtained for weaving of different type of fabrics.

The research of Wolters aimed at an optimisation of the warp tension [11]. He trains neural networks with data obtained from well performing weaving machines. These neural nets are used in combination with the warp tension from De Weldige and a genetic algorithm and additional quality criteria for the course of the warp tension. His results demonstrates that such optimisation of warp tension results in less machine stops but the fabric quality could not be improved significantly. Furthermore, it is very time consuming to obtain data from well running weaving machines for the training of the neural networks. The difference in warp tension in the upper and lower shed position is caused by the shed geometry.

Likewise research was conducted for the knitting process. Recently Metzkes presented a continuum model with spatial dynamic of thread and axial transport movement [12]. He describes also the need of a constant and low yarn tension in order to obtain a flawless knitting process.

Thus, an alternative modelling of the weaving process will be presented in this paper in order to simulate the warp tension. In order to obtain optimised setting parameters for a weaving machine, by a given quality function, the simulation is combined with a genetic algorithm. In a test series, simulation and optimisation are validated regarding their accuracy and the effect of the optimisation on the fabric properties.

Simulation of Warp Tension

To introduce the principle of the simulation of warp yarn tension, some basic information about the weaving process needs to be given. A woven fabric is formed by crossing longitudinal yarns with transversal yarns. The longitudinal yarns are warped on the warp beam. Each warp yarn is fed through the eye of a heddle and then stretched to the fabric beam. Producing a woven fabric (by inserting the transversal weft yarns between the warp yarns) involves the following basic actions (Figure 1).

♦   Let off of yarn from the warp beam via the backrest system

♦   Forming of the shed by movement of the heddles

♦   Insertion of the weft yarn

♦   Beat up by the reed movement

♦   Taking up of the fabric by the fabric beam [13]

The shed is formed by moving some warp yarns in the upper position of the shed and the others in the lower position. Then, the weft yarn is inserted in the shed, using a method that varies depending on the type of loom: with a rapier, by air, by water, with a needle, with a projectile or with a shuttle. Subsequently, the weft is moved by the reed to the fabric edge. Next, another shed is formed and the next weft is inserted, beaten up and so on. Ultimately, by the shed movement, the warp yarns form a quad limited by the fabric edge, the upper and lower rest positions of the eye of the heddles and the warp stop system.

The mean tension exerted on the warps is dependent mainly on

the difference between the circumference speeds of the warp beam and the fabric beam. Superimposed on that mean tension, there are tension changes caused by movements of the backrest system, the heddles, and the reed.

The system analysis shows that the four points of the shed form a quad. The displacement of the four edges of this quad is the basis for the warp tension model. This displacement can be determined in an analytical way by calculation, or by measuring of the movement on an actual machine. First of all, it is necessary to further clarify what is happening in these four points with respect to warp yarn. The warp is fixed at two of these points: one at the rim of the backrest system (P_1) and the other at the fabric edge (P_4) (Figure 3). The warp yarn glides at the following two points: at the warp stop (P_2) and in the eye of the heddles (P_3) [14]. The loom model used to calculate warp tension is based on the movement of these four points, which will be described in the following paragraphs.

P_1 belongs to the backrest system. The main tasks of the backrest system are as follows:

♦   Direction change of warps into the production direction

♦   Compensation of changes in warp tension

♦   Measurement of elements for the control of warp tension

The backrest system can be designed in different ways. Two main variants of this system can be distinguished: passive and active controlled backrest systems. In the passive variant, the roller of the backrest system is connected by a spring to the machine frame. Formerly, the deviation of the backrest-system was used to govern the mean yarn tension. If a passive backrest system is used, the movement of P_1 depends on the running speed of the loom and the transfer function of the mass/spring/damping parameters [15]. From this transfer function (amplitude and phase shift), the amplitude and the angle of the movement of P_1 can be determined.

In modern machines, there is no need to use the backrest system to measure the actual value of the total yarn tension since this can be measured easily, e.g. by yarn tension sensors. Thus, the backrest system can be controlled by an actor to compensate for yarn tension changes. In general, this actor is realized by an eccentric. The absolute position of P_1 caused by the rotation motion of an active backrest system can be calculated by the following formula:

$$s_{br} = s_{b0} . \sin(\phi - \phi_{b0})\qquad (1)$$

where $s_{br} \triangleq$ movement of P_1, $s_{b0} \triangleq$ amplitude of the movement, $\phi \triangleq$ loom crank angle and $\phi_{b0} \triangleq$ Phase shift of the loom crank angle. Figure 2 shows the main elements of the active backrest system.

The position of P_2 is fixed, so there is no movement of this point. P_3 is located in the eye of heddles inside the healds. As described, the heald movement is necessary to form the shed. To obtain fast running looms, this movement is realized by cam systems. The movement of the heddle in such a case can be described analytically, e.g. by a polynomial function. An easier method is to measure the movement and present the obtained data in a look-up Table. Because of the movement of the heddles to form the shedding quad, we identify P_{3o} as the upper position and P_{3u} as the lower position (Figure 3). A measurement of the heald movement is shown in Figure 4. With the help of the table the movements of the point P_{3o} and P_{3u} during the weaving process can be calculated. The reference point or zero point of the model is defined as the position P_2 of the warp stop motion in case of a symmetrical shed

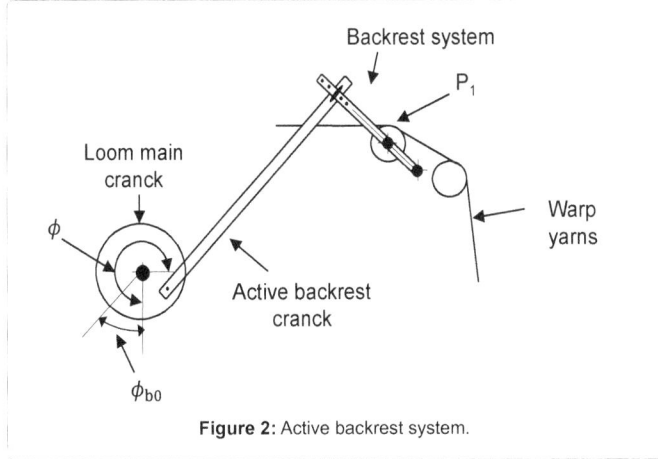

Figure 2: Active backrest system.

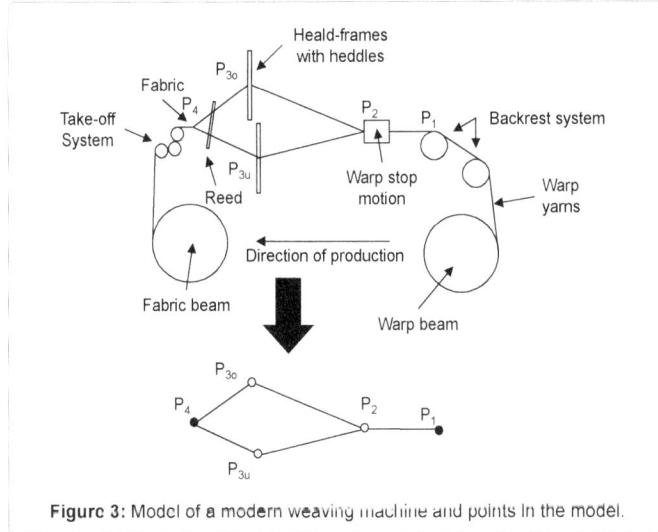

Figure 3: Model of a modern weaving machine and points in the model.

and no initial movement of the warp stop motion.

The heald movement can also be calculated by the following formula in this case

$$h_{heald}(\phi) = 2 . \hat{h}_{heald} . \left\{ \frac{\phi}{\phi_T} - \frac{1}{2\pi} \left[\sin\left(2\pi \frac{\phi}{\phi_T}\right) \right] \right\} \tag{2}$$

Where $h_{heald}(\phi) \triangleq$ heald movement during the weaving process, $\hat{h}_{heald} \triangleq$ maximum heald movement, $\phi \triangleq$ loom crank angle and $\phi_T \triangleq$ loom crank angle during change of shed [16]. In Figure 4 the formula applies only for the movement between the rest positions. Position 1 indicates the upper heald position and Position 0 the lower heald position.

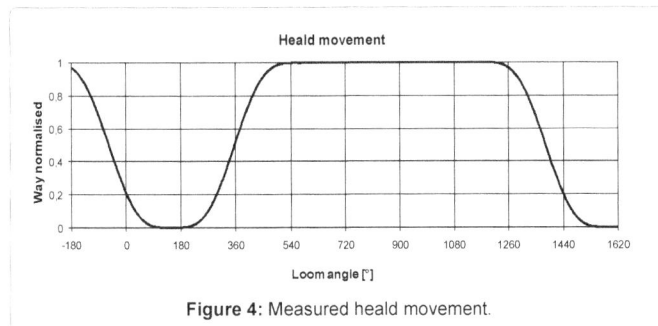

Figure 4: Measured heald movement.

The reed movement is needed to push the weft to the fabric edge, where P_4 is located. The impact of the reed movement has an influence on the warp tension, since the edge of the fabric is moved and the warps are lengthened. A good approach is to use the following formula (depicted in Figure 5). In reality the reed movement can be more complex.

$$s_{fe} = \min\left\{ 0, \frac{\left[s_{r0} - \sin(\phi - \phi_{r0}) - s_{r1} \right]}{s_{r0} + s_{r1}} \right\} \tag{3}$$

Where $s_{fe} \triangleq$ movement of the fabric edge, $s_{r0} \triangleq$ amplitude of the reed movement, $s_{r1} \triangleq$ position for touching the fabric edge, $\phi \triangleq$ loom crank angle and $\phi_{r0} \triangleq$ displacement angle of the reed.

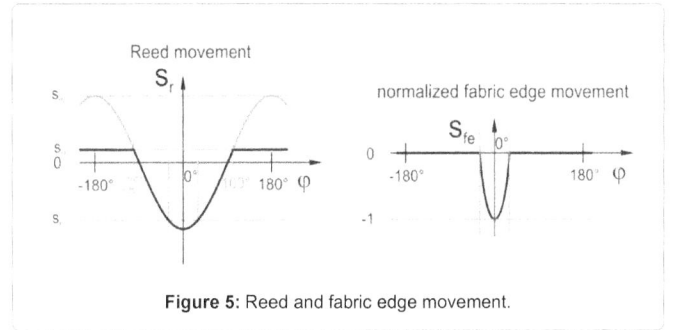

Figure 5: Reed and fabric edge movement.

The tension of the yarn is described by of the following function:

$$\sigma = E . \varepsilon \tag{4}$$

Where $\sigma \triangleq$ tension, $E \triangleq$ Young Modulus and $\varepsilon \triangleq$ elongation. In the textbooks, the Young Modulus is normally presented as a constant value, which is correct if the elongation ε is small and the material is a metal or something similar. In this model, the force/strain relationship is important; in general it is represented as follows:

$$F = \frac{EA}{s} . ds \tag{5}$$

Where $A \triangleq$ area, $s \triangleq$ yarn length and $ds \triangleq$ change of yarn length. It is known that in real yarns the factor $\frac{EA}{s}$ is nonlinear [17]. In a future model of the warp tension, this nonlinearity can be taken into account by using a look-up Table.

By determining the movement of P_1 to P_4 it allows the whole model to be described and the warp tension simulation to be realized. The simulation is programmed using the MatLab software from MathWorks, Natick (MA), USA [18]. First of all, the position of the four points is calculated for each degree of the main-shaft revolution represented by φ of the loom. The yarn tension $F(\varphi)$ is then calculated as follows:

$$F(\phi) = EA . \varepsilon(\phi) \tag{6}$$

Where

$$\varepsilon(\phi) = \frac{l_\phi - l_0}{l_0} \tag{7}$$

and

$$l_\varphi = s_{br} + s_h + s_{fe} + l_0 \tag{8}$$

With $1_\varphi \triangleq$ length of warp during the weaving process, $s_{br} \triangleq$ change of warp length caused by back-rest movement, $s_h \triangleq$ change of warp length caused by heddle-frame movement, $s_{fe} \triangleq$ change of warp length caused by reed movement and $1_0 \triangleq$ the initial (closed shed) distance of the points P_1 to P_4.

Therefore, another way to describe the movement of a point (e.g., for P_1) can be obtained by the following equation:

$$P_1(\varphi) + dP_1(\varphi) \qquad (9)$$

Where $P_1(\varphi)$ is the movement during the weaving process, P_{1in} is the initial position and dP_1 is the change of the point position.

Thus far, there have been no calculations of the forces caused by the friction in the heddles or the warp stop. Also, the forces caused by the yarn regarded as swinging string have not been taken into account. Furthermore in this paper, the simulation of the eccentric controlled backrest system is realized. The algorithm for the movement was realized as an analytic function within Matlab.

By using Matlab, the simulation can be used as a subroutine for a general user interface (GUI). Therefore a GUI was provided for the convenience of the user. The GUI is separated into three main areas: input data, visualization of the model, and simulated warp tension. In the input area, the user can define the initial points P_{1i} to P_{4i} material data such as fineness and Young-Modulus, amplitude of backrest movement, the maximum value of the elongation of warp caused by reed movement, and mean warp tension. Hence, the simulation needs 19 input parameters and the program calculates the warp tension for four weft insertions, which is equal to 1440° machine angle.

Optimisation and Cost Function of Warp Tension

The general objective of an optimisation is the search for the minimum of a target function. The target function is sometimes called cost function. One of the most challenging problems is the formulation of this cost function, in such a way, that the result of the mathematical calculation meets the intentions of the user of the optimisation. The choice of this cost function is essential for the calculating time and the quality of the result [19,20]. In the present case the cost function CF is defined as the difference between a desired warp tension $F_{desired}(\varphi)$ and the simulated warp tension $F_{Sim}(\varphi)$. The cost function is calculated for four loom crank angle turns, equal to 8π, since the used weave pattern for this research is a 3/1 satin structure.

$$CF = \int_0^{8\pi} (F_{desired}(\varphi) - F_{sim}(P_1,...,P_n,\varphi))^2 d\varphi \qquad (10)$$

Criteria for an optimal warp tension course, meaning $F_{desired}(\varphi)$ are described by Wolters [21]. In general, these criteria aim at having a low and constant warp tension in the lower shed. In the upper shed a sufficient tension is needed in order to have a clear shed. By combining the desired cost function within the warp tension simulation and an appreciated optimisation algorithm it is therefore possible to receive an optimised loom setting parameter. In the already described GUI, it is possible for the user to integrate the necessary $F_{desired}(\varphi)$.

After having the right cost function, it is necessary to choose an optimisation algorithm in order to minimize the cost function. There are many different algorithms proposed in literature to reach objective of the optimisation, such as Newton's method, gradient methods, hill climbing, artificial bee colony optimisation, genetic algorithm etc. All optimisation methods can be described by the following three characteristics. First, they need a set of starting values for the calculation. Second, the more information about the optimisation problem will be integrated in the algorithm, the faster the optimisation will be conducted. Third, it needs to be decided whether the result of the optimisation is a local or global optimum [19,20]. In this presented paper an evolutionary algorithm and a gradient based method is used.

termination criterion is not straightforward From the optimisation point of view using the gradient based method, the problem is reduced to a quadratic minimization problem with nonlinear equality and box constraints on the parameters. Due to the small number of parameters subject to optimisation a discrete gradients using finite differences is easily obtained and also computationally efficient. Then; an interior point method is used to solve the constrained minimization problem. The method converges with less than 100 iteration steps and up to a tolerance of $1 \cdot e^{-6}$ in all tested cases [22,23].

One disadvantage of gradient-based methods using adjoints is the development cost. In addition the method is relatively intolerant of difficulties like noisy objective function spaces or categorical variables. Also the gradient-based methods find a local rather than a global optimum. Rapid convergence is the main advantage of a gradient-based method [24].

With evolutionary algorithms an arbitrary starting combination of the genotypes will be generated and from them the cost function will be calculated. In each generation those individuals will survive, who will best fit the minimization of the cost function. These individuals will reproduce by recombination and mutation the genotypes of the next generation. This circuit will continue, until either the maximal amount of generations is reached or the cost function of subsequent generations will be less than a minimum [20,21].

Evolutionary algorithms treat the function evaluation as a "black box". Consequently, development cost is minimal. Evolutionary algorithms are tolerant of noise and have no difficulty with categorical variables or topology changes. In general evolutionary algorithms find a global optimum. The key disadvantage is they are very slowly, especially near an optimum. A second weakness is that determining a termination criterion is not straightforward [24].

Validation of warp tension simulation and optimisation

Validation in general contains two steps. First, the simulation of warp tension will be validated, since the results of the simulation have to be close to the measurement of the warp tension on a real weaving machine. Only in the case that the analogy between simulation and measurement is adequate, the use of a genetic algorithm in order to improve the weaving process makes sense. Hence, in a second step it will be checked, if the weaving process and quality of the fabrics does improve using new calculated weaving machine parameters.

For the validation of the warp tension simulation, measurements of the warp tension on an OmniPlus 800 air jet weaving machine from Picanol nv, Ieper, Belgium with an active backrest system were conducted. The measurements were done using an analogue yarn tension sensor form BTSR International S.p.A. Partita, Olgiate Olona, Italy (Figure 6). The sensor is installed between back rest system and warp stop motion of the weaving machine. Setup of the weaving machine is done according to Table 1. These parameters were also used as input parameters for the simulation.

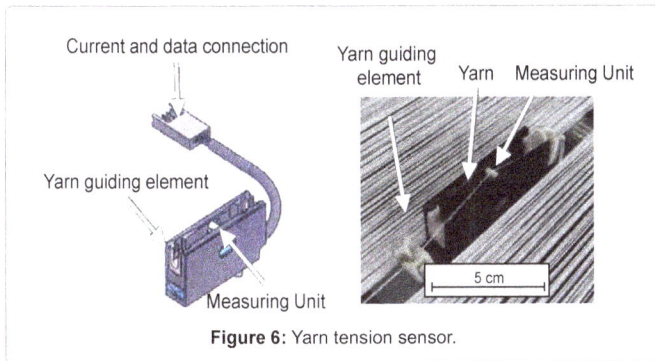

Figure 6: Yarn tension sensor.

Parameter	Value	Unit
General weaving machine settings		
Speed	800	rpm
Warp / Weft material	PES	
Warp count	740	dtex
Weft count	740	dtex
Warp density	20	yarns/cm
Weft density	14	yarns/cm
Weaving pattern	twill 1/3	
Closed heald angle	340	°
Warp mean force (single yarn)	150	cN
Active backrest position P1		
Backrest X-position	560	mm
Backrest Y-position	30	mm
Warp stop position P2		
Warp stop X-position	225	mm
Warp stop Y-position	20	mm
Upper heald position P3o		
Upper Heald X-position	-365	mm
Upper Heald Y-position	37	mm
Upper heald position P3u		
Lower Heald X-position	-365	mm
Lower Heald Y-position	-28	mm
Fabric Edge position P4		
Fabric edge X-position	-552	mm
Fabric edge Y-position	0	mm
Active back rest system		
Amplitude of movement	0,9	mm
Phase shift to loom crank angle	20	°

Table 1: Setup of loom before optimisation.

In Figure 7 the comparison with the measurements shows a good similarity with the simulation, especially the change of tension caused by the movement of the heald frames. In addition, changes in warp tension caused by the reed movement are simulated in a sufficient way.

There of course still are some discrepancies. These discrepancies result from simplifications of the model. Especially the position of point P_2 is not constant during weaving. Additionally there are dynamic effects: vibrations of the warp yarns, jamming of the warp yarns at the point of collision during up and down movement, oscillating heddles in the shed, the movement of the warp stop drop wires etc. Because of the complexity of the effects they are not taken into account for the presented simulation.

In a second step, the optimisation was investigated. The necessary cost function is chosen in such a way, that the warp tension is reduced and in the same time much more constant compared to the first warp tension measurements (Figure 8). Four weaving machine parameters

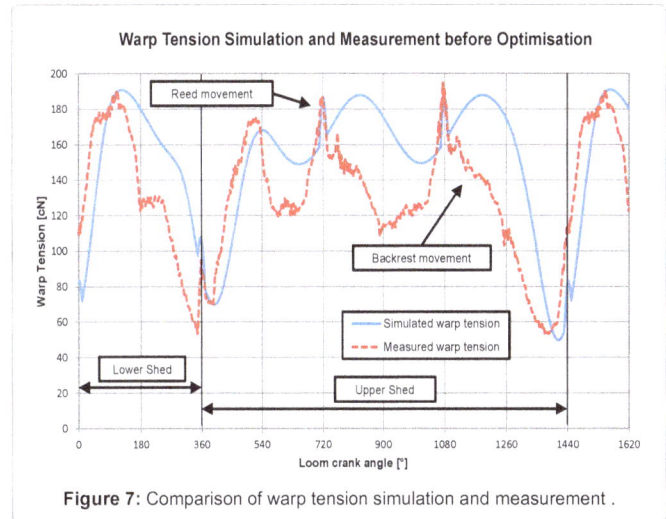

Figure 7: Comparison of warp tension simulation and measurement .

were chosen to be optimised: vertical warp stop motion position, closed heald angle, backrest angle phase shift to loom crank angle and the warp mean force. These parameters can be adapted on the loom with less effort of time compared to the other weaving machine parameters used in the simulation.

Changes in the warp mean force have an influence on the general level of the warp tension during weaving, changes in the horizontal warp stop motion position influences mainly the relation of warp tension in upper and lower shed. Changes in the backrest phase shift to the loom crank angle mainly influence the timing of distribution in warp tension due to the warp beam movement. Therefor it is for example possible to reduce the tension peak occurred by the reed movement, when the reed peak takes place in the minimum of the back rest movement. Changes in the closed heald angle influences the 'position' of the reed-peak relative to the closed shed minimum.

The genetic algorithm was using 200 population, 50 generations and 20 Iterations and needed 300 seconds on a PC with an intel core i3 CPU at 3.3 GHz. The gradient based method took less than 10 seconds on the same computer. Both optimisations methods nearly calculated the same optimised settings points for the weaving machine (Table 2).

Figure 8 shows the simulation of the warp tension with weaving machine parameters before and after optimisation. The optimised weaving machine parameter creates a warp tension course, which is much closer to the desired warp tension course. However, the new calculated weaving machine parameters will not lead to a constant desired warp tension course, since the movement of the back rest system and reed always disturb a constant warp tension course in the simulation.

The optimised machine settings were used on the weaving machine

Parameter	Value before optimisation	Value after optimisation with genetic algorithm	Value after optimisation with gradient method	Unit
Warp stop Y-position	20	50,7	50,9	mm
Closed heald angle	340	345	340	°
Warp mean force (single yarn)	150	97,1	97,1	cN
Backrest phase shift to loom crank angle	20	-10	20	°

Table 2: Comparison of weaving machine setting before and after optimisation.

in order to analyse the quality of the produced fabric and productivity of the weaving process. Analysis of the produced fabric before and after the optimisation of weaving machine settings was conducted accordingly:

♦ Breaking force and breaking elongation in warp and weft direction according to ISO 13934-1:2013[25]

♦ Fabric weight according to EN 12127:1997[26]

♦ Warp and weft crimp according to DIN 53852:1991-09[27]

♦ Air permeability according to EN ISO 9237:1995[28]

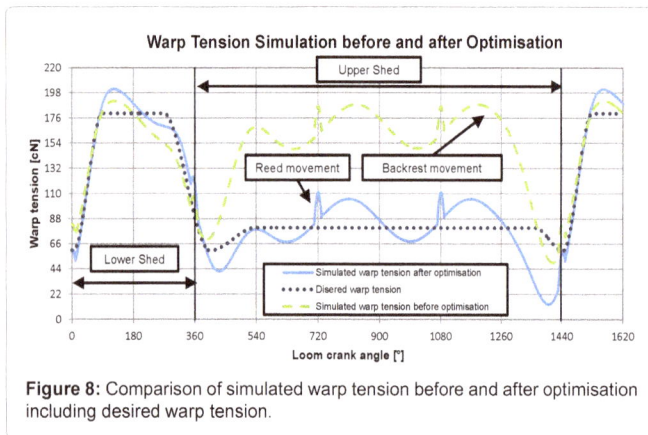

Figure 8: Comparison of simulated warp tension before and after optimisation including desired warp tension.

It is visible that the changes of weaving machine parameters closely do not influence the woven fabric quality, also the medium warp tension is reduced from 132,04 cN to 106,32 cN. Only the breaking force in warp direction was improved by around 0,3 kN. It is well known that the warp tension influence fabric quality, for the present case warp tension reduction was probably too low in order to have a stronger impact. Table 3 shows the results of these measurements and Figure 9a comparison of measured warp tension before and after optimisation.

In a last step, the long-time running behaviour of the loom before and after optimisation was analysed. In order to do so, the machine was kept running for three hours with the initial setting and an additional three hours with the optimised setting parameters. In both cases 144.000 woven picks were produced. The amount of weaving machine stop related to warp breakage or non-insertion of the weft was recorded. In summary it is shown, that by using the present simulation, optimisation

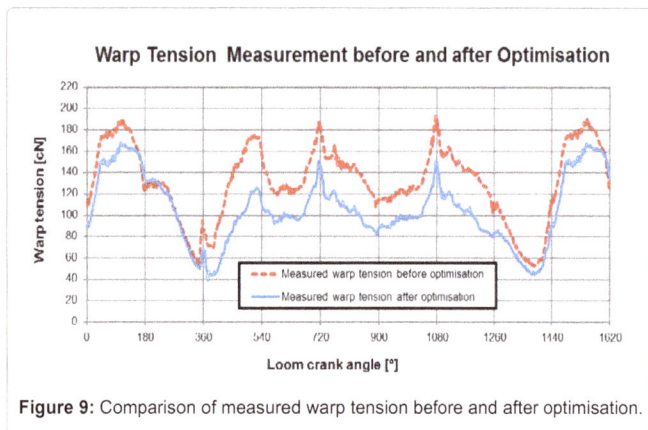

Figure 9: Comparison of measured warp tension before and after optimisation.

algorithms and the right cost function, there was no notable change in the weaving machine productivity. The machine was running without any stops using the initial or the optimised setting parameters (0 warp stops per 100000 picks). Thus, fabric defects occurred irregularly in the middle of the fabric using the initial weaving machine settings (Figure 10). Filaments of the warp threads break in the eye of the heddles. These broken filaments did form a fabric defect. One reason for filament breakage could be because the warp tension was too high.

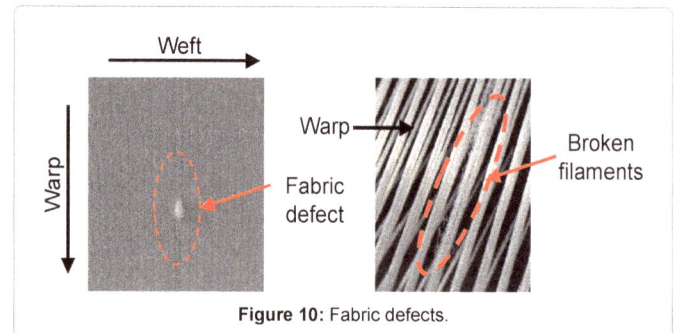

Figure 10: Fabric defects.

Also the fact that the defects just occurred in the middle of the weaving machine can prove this theory. Warp tension is highest in the middle of a weaving machine due to the warp tension bow [29]. The fabric defects vanished using the optimised machine settings. A major explanation of these results is that the choice of desired warp tension course $F_{desired}(\varphi)$. The course was chosen in such a way, that stress on the warp is lower and more constant compared to the initial settings. In addition, warp tension in the upper shed of $F_{desired}(\varphi)$ was still high enough choose to produce a clear shed.

Conclusion and outlook

In this paper, the development of a warp tension simulation was presented. A system analysis of modern weaving machines lead to suitable simulation model to calculate the warp tension. The validation of the simulation demonstrates that the results correspond well in reality. In a second step, an improved model of this simulation was used in combination with a genetic algorithm and a gradient based method to calculate optimised setting parameters for the weaving process. In order to do so, a cost function was defined taking into account a desired course of the warp tension. By literature it is known, that a low and constant warp tension course is suitable for weaving.

Using the genetic algorithm or the gradient based method leads to optimised weaving machine parameters. Both algorithms do get nearly the same results for the optimised weaving machine setting. Applying the optimised setting parameters on a loom did not demonstrate that the productivity of a weaving machine can be raised. Analysis of the produced fabrics did not show an influence of optimisation on the fabric quality. The reduction of warp tension was not sufficient in order to have an impact on the mechanical properties of the fabric. Thus, fabric defects could be eliminated using the optimised weaving machine settings.

In the future, adaption of the simulation using a spring-mass backrest system can be done. Thereby the movement of the backrest system is described by a differential equation of second order. Other aspects like the dynamic behaviour of a yarn during the beat up process were not taken into account for the simulation. In order to improve the simulation, the influence of the movement of the heddles, the friction caused by the yarn movement in the heddles, and the crossing of warps could be added. In real yarns, the factor EA/s is quite nonlinear [6]. In a future model, this can be taken into account by using a look-up Table.

Also a combination of the simulation and optimisation system with the automatic position system of back rest system and warp stop motion, as presented by Osthus [30] could further fasten the setup process of the weaving machine. In addition use of other models like regression models and aspects of self-optimisation could be of interest in order to further improve the weaving process.

Acknowledgement

The authors would like to thank the German Research Foundation DFG for their support of the depicted research within the Cluster of Excellence "Integrative Production Technology for High-Wage Countries".

References

1. Schmitt R (2001) Self-optimising Production Systems.

2. Keller H (1943) Messung der Kettspannung beim Weben. Universität Zürich, Switzerland.

3. Schlichter S (1987) Der Einfluß der einzelnen Maschinenelemente auf die Bewegungs- und Kraftverläufe in Kette und Schuß an Hochleistungswebmaschinen. Techn Hochsch Germany.

4. De Weldige E (1996) Prozeßsimulation der Kettfadenzugkräfte in Webmaschinen. Techn Hochsch, Germany.

5. Chen M (1998) Computergestützte Optimierung des Webprozesses bezüglich Kettfadenbeanspruchung und Kettlaufverhalten. Stuttgart Univ, Germany.

6. Mirjalili SA (2003) Computer Simulation of Warp Tension on a Weaving Machine. J Text Eng 49: 7-13.

7. Beitelschmidt M (2000) Simulation of warp and cloth forces in weaving machines. Melliand Textilberichte 81: 45-48.

8. Großmann K, Mühl A, Löser M (2007) Integrated take-up system for weaving of space preforms for textile-reinforced composite strucutres. ZWF Zeitschrift für wirtschaftlichen Fabrikbetrieb 102: 216-221.

9. Klöppels M, Gries T, Bösing T, Pothoff PJ (2002) Practical Trial of the Freely Programmable Active Back Rest Roller System. Melliand International 8: 115-116.

10. Çelik Ö, Eren R (2014) Mathematical analysis of warp elongation in weaving machines with positive backrest system. TEKSTİL ve KONFEKSİYON 24: 56-65.

11. Wolters T (2003) Verbesserte Webmaschineneinstellungen mittels Simulationsrechnung. Techn. Hochsch, Diss, Germany.

12. Karoline M, Rolf S, Jan M, Gerald H, Chokri C (2013) Simulation of the yarn transportation dynamics in a warp knitting machine. Textile Research J 83: 1251-1262.

13. Adanur S (2001) Handbook of weaving. Taylor and Francis, Boca Raton, London, New York.

14. Gloy YS, Renkens W, Kato S, Gries T (2012) Simulation of warp tension for power looms.

15. Milkov M, Kyosev Y (2000) Mechano-techological modelling of system "warp-fabric" on the weaving looms. Application of Mathematics in Engineering and Economics 25: 137-140.

16. Wulfhorst B, Schneemelcher S (1990) Analyse der dynamischen Belastung von Schaftmaschinen. Textil Praxis International 8: 831-838.

17. Hättenschwiler P, Pfeiffer R, Schaufelberger J (1984) Die Zugfestigkeit von Garnen.

18. www.mathworks.com.

19. Adami J (2007) Fuzzy Logik, Neuronale Netze und Evolutionäre Algorithmen.

20. Veit D (2010) Simulation in textile technology: Theory and applications. Woodhead Publishing, Cambridge, UK.

21. Wolters T, Wulfhorst B (2000) Ermittlung praxisrelevanter Gütekriterien für intelligente Einstellhilfen für Webmaschinen.

22. Byrd RH, Mary EH, Nocedal J (1999) An Interior Point Algorithm for Large-Scale Nonlinear Programming. SIAM Journal on Optimization 4: 877-900.

23. Coleman TF, Li Y (1996) An Interior, Trust Region Approach for Nonlinear Minimization Subject to Bounds. SIAM Journal on Optimization 6: 418-445.

24. Zingg D, Nemec M, Pulliam P (2008) A comparative evaluation of genetic and gradient-based algorithms applied to aerodynamic optimization .Revue Européenne de Mécanique Numérique 17: 103-136.

25. Textiles - Tensile properties of fabrics - Part 1: Determination of maximum force and elongation at maximum force using the strip method. ISO 13934-1:2013.

26. Textiles - Fabrics - Determination of mass per unit area using small samples. ISO 13934-1:2013.

27. Testing of textiles, determination of yarn length ratios in woven and knitted fabrics. ISO 13934-1:2013.

28. Textiles - Determination of permeability of fabrics to air. EN ISO 9237:1995.

29. Neumann, Florian; Ein Verfahren zur Reduzierung der Welligkeit von Gewebekanten Aachen, Techn. Hochsch., Diss., 2013 ; Zugl. Aachen : Shaker, 2013.

30. Osthus T, De Weldige E, Wulfhorst B (1995) Reducing set-up times and optimizing processes by the automation of setting procedures on looms. Mechatronics 2-3: 147-163.

4

Tencel Process Optimization in Conventional Cotton Processing Machineries and a Quality Comparison with Similar Cotton Yarn Count

Hasan NB[1]*, Begum AR[2], Islam A[3] and Parvez M[4]

[1]Senior Lecturer, Green University of Bangladesh, Private University in Dhaka, Bangladesh
[2]Professor and Head, Bangladesh University of Textiles, University in Dhaka, Bangladesh
[3]Executive, Quality Assurance Department, Delta Spinning Ltd, Bangladesh
[4]Senior Lecturer, Atish Dipankar University of Science and Technology, Bangladesh

Abstract

Tencel a regenerated fiber manufactured from cellulose derivatives, is getting popularity nowadays due to its biodegradability and environment amenity depending on the less land and water consumption relative to organic and conventional cotton. Spinning process optimization of tencel is performed and reviewed in this article in comparison with conventional cotton process. Wider setting with low production speed is observed for tencel processing with superior yarn quality. Quality parameter of different yarn count evaluated and compared. Mass uniformity and its cv is significantly higher in tencel yarn; breaking strength is 50% greater in tencel along with breaking elongation; tencel yarn possess more twist (>15%) while twist cv is slightly lower; yarn cuts in autoconer machine is lower in tencel yarn.

Keywords: Tencel; Organic cotton; Unevenness; Co-efficient of variation (CV); Yarn; Spinning

Introduction

Lyocell is a manmade fiber derived from cellulose [1], better known in the United States under the brand name Tencel. Tencel fiber, though related to rayon, is a kind of regenerated cellulose fiber produced in a way called "solvent spinning method", the production mainly uses coniferous wood pulp as raw materials. Tencel fiber has the advantages of both naturalfiber and synthetic fiber. Such as cellulose fiber, tencel fiber has good hygroscopic, permeability and performance, its wearing comfort is much better than polyester, and its feel, gloss, drape are all good. The strength of tencel fiber is higher than that of cotton fiber and viscose fiber; it also has advantages such as warm and soft and comfortable; in addition, tencel has good performance and dimensional stability; it can be blended with othernatural fiber and synthetic fiber [2]. The source ingredients of tencel fiber mainly come from green cellulose fiber in nature, chemical solvent used in production can be recycled, so it won't cause any damage to human andthe environment. After using, tencel fiber can be decomposed completely in soil, which can greatly reduce the environmental destruction, therefore, tencel fiber is also known as "green fiber and eco-friendly fiber". Tencel fiber, as a kind of biodegradable fiber, will become the mainstream in the future [3]. Because, throughout the world modern civilization's greater inclination and attention towards environmental protection, social development towards the direction of environmental protection and sustainability, processing and manufacturing green textiles. It is an extremely strong fabric with industrial uses such as in automotive filters, ropes, abrasive materials, bandages and protective suiting material. It is primarily found in the garment industry, particularly in women's clothing (Table 1).

Historical background and production

Whatever it's called, Tencel or Lyocell is a sustainable fibre, regenerated from wood cellulose. It is similar in hand to rayon and bamboo, both regenerated fibres. However, Tencel is one of the most environmentally friendly regenerated fibre for several reasons. Tencel fibres are grown sustainably. Tencel has earned Forest Stewardship Council (FSC) certification that the products come from environmentally responsible forests. Tencel eliminates the negative environmental impacts of traditional fibre processing, using new

sustainable technologies. As a new kind of cellulose fiber, tencel fiber has been published for many years. It's now popular in UnitedStates, Japan and other countries mainly for the production of high fashion. 1939 Patent appears describing the dissolution of cellulose in amine oxide. Kaoerzi Company had started researching cellulose fiber production with NM-MO solvent method since 1978. 1969-1979 American Enka/Akzona Inc. work on spinning fiber from a solution of cellulose in amine oxide but did not scale up. In 1981, the company used amine oxide as a new solvent to spin; the experiment showed that the spinning method is feasible. In 1983, the company set up experimental factory; in 1987, the continuity of large scale plant was established. In 1989, it's named tencel fiber after recognized by the International Bureau of Artificial and Synthetic Fiber Standards, then

	Organic Cotton	Tencel
Synthetic pesticides and fertilizers	Not allowed	None used
Water use	As low as about 10.6 gallons/lb for rain-fed from Brazil and as high as 782 gallons/lb for CA organic cotton	154.7 gallons/lb fiber
Land use	3.5 acres/ton	0.52 acres/ton
Heat-trapping gases	3.5 pounds/lb	Less than 1 lb/lb fiber
3rd party certification	USDA organic, GOTS in some cases	Oeko Tex 100, FSC, PEFC-accredited certifiers

Table 1: Comparison of tencel *vs.* cotton in environmental perspective.

*Corresponding author: Hasan NB, Senior Lecturer, Green University of Bangladesh, Private university in Dhaka, Bangladesh
E-mail: nakib.tex@green.edu.bd

American factories began commercial production of tencel and its production capacity had reached 55 million kilograms. Tencel has some distinct advantages over traditional fibres in terms of chemical processing, which can often be extensive and toxic. For example, Rayon manufacturing generates highly polluting air and water emissions, uses catalytic agents containing cobalt or manganese, and creates a strong unpleasant odor. Tencel products can be decomposed, solvent of which is non-toxic, the celllose used for Tencel is treated in what is known as a closed loop process in which the solvents are recycled with a recovery rate of 99.55%. The tiny amount of remaining emissions is decomposed in biological purification plants. Because of the natures of the material, the processing never requires bleach. This methos of manufacturing of fibre was awarded the "European Award for the Environment" by the European Union. So it is regarded as the third generation of regenerated cellulose fiber. Because of its unique properties, it had been applied into a variety of goods in many fields. The whole Asia has also appeared upsurge of tencel development [4]. Japan is currently the largest developers of tencel textiles, lots of men products in department stores were tencel goods. It can be said that tencel fiber goods in Japan entered a new stage. Textile software and hardware for tencel, especially the commercialization of tencel commodity, were also under development in Japan. This had helped Japan to maintain a leading position in the manufacturing of tencel fiber. In 2000, the British tencel fiber output reached 6,140,000 tons. Tencel fiber with its incomparable superiority is in rapid development in international and domestic market. Tencel is used in a variety of applications, including men's wear, sheets, and blankets. Since it's absorbent and dries quickly, it is also suitable for towels. Clothes made from this material are often recommended for traveling because they are light and keep their shape well. Tencel® is also available as fabric for sewing, as yarn for knitting or crocheting, and as fiber for spinning. Besides it use as a cloth, it is also used in making bandages, baby wipes, oil filters and carpeting for cars, as well as conveyor belts and plastic parts. In powder or fiber form, this material is used in making specialized papers, as an additive for building materials, and in making foam mattresses [5].

The purpose of this study is to analyze the suitability of conventional cotton processing machineries in tencel processing and establish a successful benchmark and provide necessary guidelines for yarn manufacturers.

Tencel in comparison with organic and conventional cotton fibers

Bearing in mind that the figures below include wide room for variation that makes any strict head-to-head comparison impossible, we can distill the discussion down to some basics (Table 1).

Lyocell fiber spinning process

This section provides a description of the process steps required for making lyocell [6]. A diagram of the process is shown in Figure 1. The principles are simple. Firstly, the pulp is wetted out with dilute aqueous amine oxide to fully penetrate the pulp fibers. The subsequent removal of the excess water under heat and vacuum is a very effective way of making a homogenous solution with a minimum of undissolved pulp particles and air bubbles. The solution is highly viscous at its operating temperature (90 to 120°C) and must be processed in similar high pressure equipment to that used in melt polymer systems. The fibres are formed by spinning into an air gap and then coagulating in a water/amine oxide bath. They are then washed and dried and cut. The wash liquors are recovered, purified, concentrated then recycled. The process description below applies to the two commercial-scale operations of

Tencel® and Lenzing. Variations in detail have been cited in patent applications and the literature but these are at a much smaller scale of operation (Figure 1).

Methodology

Fiber selection

Before start of the production following fiber characteristics were measured and bale laydown was done according to the quality parameters (Table 2).

Yarn Spinning Process

Spinning is the twisting together of drawn out strands of fibers to form yarn, and is a major part of the textile industry (Figure 2).

Testing of machine gauge and yarn characteristics

Following tests were carried out on the Square Texcom ltd. Situated at kathali, valuka, Mymenshing (Table 3).

Results and Discussion

Process optimization and benchmarking for Tencel and Cotton fiber:

Opening and cleaning

Fiber length uniformity is the main reason for closer setting in Mixing Bale Opener (MBO). In addition to the length uniformity tencel fiber is free from foreign matter. Which lead to narrower setting of grid bar angle as 2 to 3. Another important aspect of tencel processing is the bypass of two saw tooth beater and it is advised to use only one pin beater. More beater exploits fiber to pass through more beating action and which was highly responsible for curling of fibers and finally generation of fiber to fiber entanglement neps. Some of the fabrics are

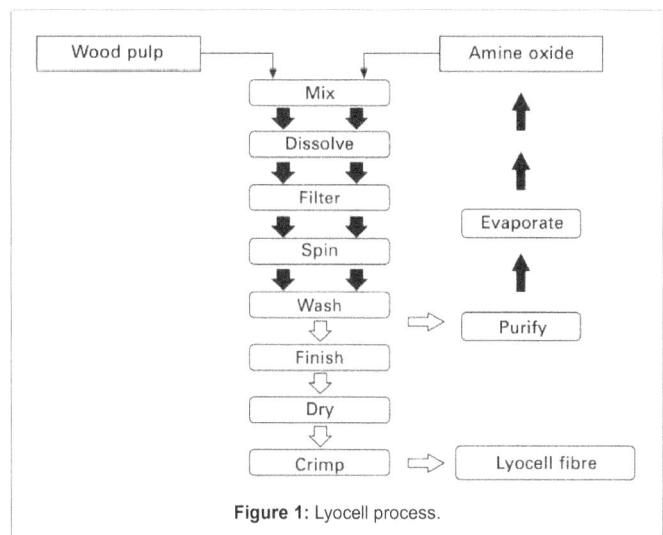

Figure 1: Lyocell process.

Fiber specifications	Tencel	Cotton (CO)
Country of origin	Austria	Australia, Brazil, Turkmenistan, USA, Uzbekistan
Color	Bright natural white	Natural yellow white
Fiber count	1.3 dtex	4.3 Mic.
Commercial staple	35 mm	28.75 mm
Tenacity	30 g/tex	29.4 g/tex
Elongation	7.90%	7.40%

Table 2: Properties of both Tencel and cotton fiber.

Figure 2

TENCEL	COTTON
Mixing Bale Opener (MBO)	Mixing Bale Opener (MBO)
RSK Pin Beater	Saw tooth beater
Multimixture	RSK Pin Beater
Condenser	Multimixture
Carding	Saw Tooth Beater
Breaker Drawing	Condenser
Finisher Drawing	Carding
Simplex	Breaker Drawing
Ring Spinning	Finisher Drawing
Autoconer	Simplex
	Ring Spinning
	Autoconer

Figure 2: Flow chart of tencel and cotton processing.

Test name	Test method (ASTM standard)	Testing instrument
Different machine gauge		Supplied by machine manufacturers
Yarn count	D1059-97	Wrap reel, balance
Tensile strength, breaking elongation	D2256-02	USTER Tensorapid
Twist/unit length, twist CV	D1422-99	Single yarn twist tester
Hairiness	D5647-01	USTER Tester 5 (UT5)
Yarn fault classifying system		USTER CLASSIMAT QUAMTUM
Yarn evenness	D1425-96	USTER Tester 5 (UT5)
Thin/km, thick/km, or Nep/km	D1425-96	USTER Tester 5 (UT5)

Table 3: Test method and testing instrument.

Machine name	Process parameter	100% Tencel	100% Cotton
MBO	MBO grid bar angle:	1	4
Pin Biter	Grid bar setting:	1	4
	Condenser-III:	1	3.5

Table 4: Setting of Mixing Bale Opener and other opening and cleaning machineries.

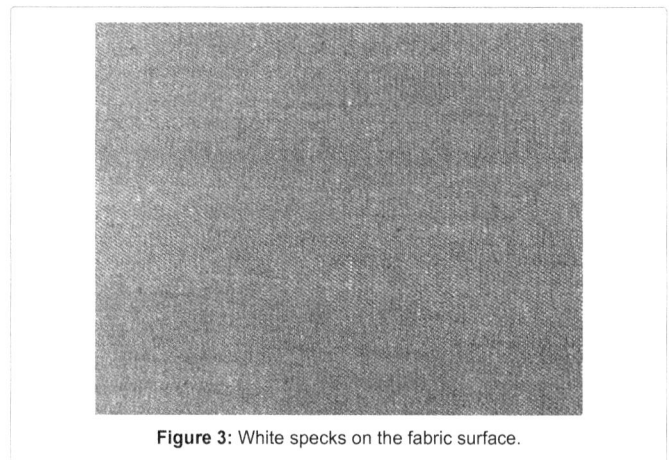

Figure 3: White specks on the fabric surface.

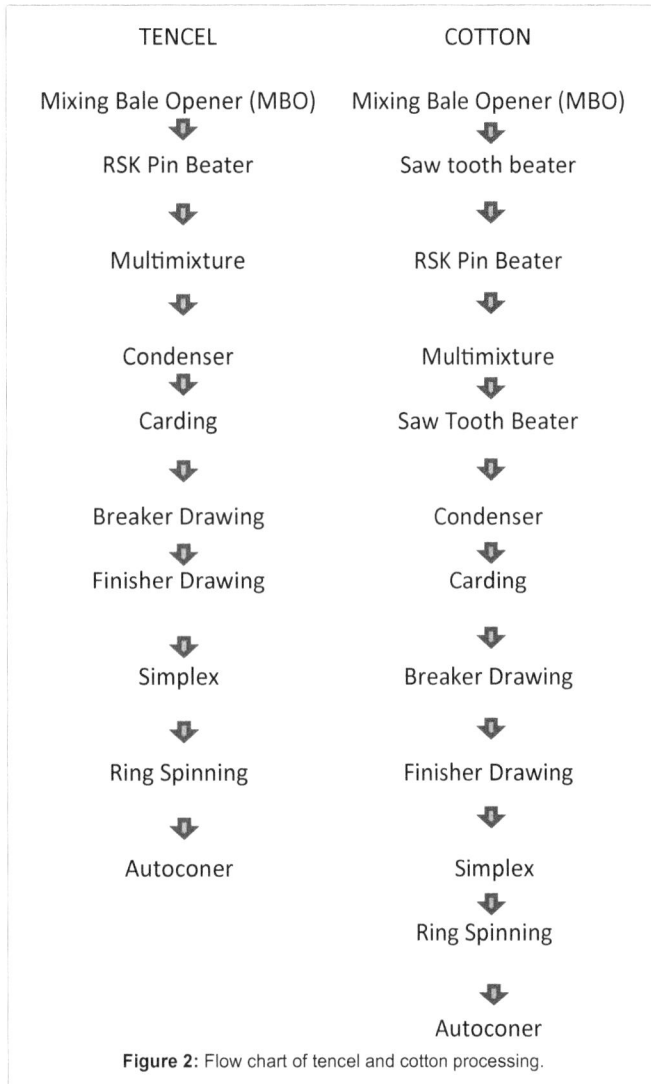

observed to be white specks on the surface. It is advised to set larger beater grid angle and more beating sections for cotton than that of tencel processing due to less length uniformity and large number of foreign matter (Table 4).

Carding

As tencel fiber processing was performed in the carding, usually designed for cotton processing, with higher point density more than 960 points per square inch does not allow higher speed of production. Higher jamming of cylinder-flat region and cylinder-doffer region was highly noticeable and less nep generation was observed in carding machine. Ultimately numerous complaints of nappy yarn and white specks (Figure 3) on fabric surface were done by the buyers of the yarn. Finally, in optimization of carding setting at a speed of 45-50 kg per hour at a delivery speed of 250 m/min was maintained successfully in comparison to cotton processing speed, which was significantly higher 65-70 kg per hour in the same machine. In order to improve web quality; licker-in speed, cylinder and flat speed was reduced as well as gauge of those elements were amplified. On the other hand for cotton setting optimized narrower with higher processing speed. Tencel fiber is with larger fiber volume quality is smooth and fluffy, its moisture regain rate is higher but its cohesion is not good, so the carding should

Machine name	Process parameter	100% Tencel	100% Cotton
Carding Machine	Sliver weight:	6.5 Ktex (91.7 gr/yd)	5.2 Ktex (73.4 gr/yd)
	Speed:	45-50 kg/hr	70kg/hr
	Flat to cylinder gauge:	0.30/0.30/0.25/0.25/0.25	0.225/0.20/0.20/0.175/0.17
	BSF to Cylinder:	0.45/0.40/0.45/0.40	0.40/0.40/0.40/0.40
	FSF to Cylinder:	0.40/0.35/0.40	0.35/0.35/0.35
	Licker in speed:	1200 rpm	1200 rpm
	Cylinder speed:	550 rpm	850 rpm
	Flat speed:	0.26 m/min	0.35 m/min
	Cylinder to Doffer	0.175	0.175

Table 5: Carding machine setting for both tencel and cotton processing.

choose with less number of card clothing 800-850 points per square inch clothing similar to polyester or polyester blend in order to reduce yarn neps and improve cotton web quality [7,8] (Table 5).

Drawing (Breaker and finishing)

Higher staple length and length uniformity are the primary reason for wider setting in tencel processing compare to the cotton processing in drawing frame machine. Hygroscopicity, higher moisture regain rate, is highly responsible for generation of roller lapping and jamming of the drawing frame machines. It is optimized as 47/49 roller setting for front to back zone whereas in cotton 37/42. Higher top roller weighting 380-380-320 (front top roller to back roller) along with larger amount of break draft 1.6-1.9 was optimized for tencel and on the other hand for cotton those were 320-320-320 and 1.15, respectively [9]. Second drawing is mainly in order to make the fiber straight and to make the weight unevenness reduced to a certain extent. Draft configuration may be the most important aspect for tencel processing as well as lower production speed to improve straightening of back hook and enhancement of web quality (Table 6).

Simplex

Number of twist per unit length of tencel roving was carefully observed and was substantially lower than that of cotton; twist value was appropriately increased which allows lower appearance of spinning head. Beside this precise measurement of spinning tension and its optimization was done at 1.033 both for cotton and tencel. Optimization of the above mentioned due to less roller lapping and unevenness of yarn. may be slipping tendency due to lack of fiber cohesion and spin finish on the tencel fiber creates layer on the roller surface and roller slip occurred thus higher break draft was advised to intensify desired total draft [10,11] (Table 7).

Ring frame

Roller gauge both for tencel and cotton processing was

Machine name	Process parameter	100% Tencel	100% Cotton
Breaker Draw Frame	Sliver weight:	4.9 Ktex (69gr/yd)	4.9 Ktex (69gr/yd)
	Breaker draft:	1.4	1.3
	R/R setting	47/49	37/42
	Condenser size:	4.6	4.2
	Funnel no:	8	10
	Doubling	6	6
	Speed:	450 m/min	650 m/min

Table 6: Drawing frame machine setting for cotton and tencel fiber.

Machine name	Process parameter	100% Tencel	100% Cotton
Simplex	Breaker draft:	1.206	1.1206
	Tension:	1.033	1.033
	TPI:	0.78	1.05
	Roving hank:	0.9	0.75
	Spacer color:	Blue	Green
	Flyer speed:	1050	1200
	Roller gauge:	40/49.5/60	40/49.5/60

Table 7: Simplex machine setting and quality parameters to be maintained for tencel and cotton.

60/42.5 (back to front roller) maintained and yarn count variation was noticeable for tencel. Suitable drafting force in optimizes the yarn quality; exact roller pressure on pneumatic drafting with apron guidance and proper spacing with 70/42.5 (back to front roller) for tencel fiber. Higher break draft 1.72 facilitates achieving tencel yarn of less thick, thin places and minimum end breakage. Lighter traveller with flat cross-sectional shape optimizes the yarn quality profile for tencel yarn (Table 8).

Analysis of yarn quality parameters

Comparison of yarn properties of Tencel and Cotton yarn, greater cohesion between the fibers and length uniformity may be the significant contributor on higher tensile strength of tencel yarn. Fiber migration, the displacement of fibers along the yarn axis during spinning, is dependent on the amount of twist inserted on the yarn. Higher amount of twist per unit length higher the packing density thus cotton yarn is firm than compare to tencel yarn and exhibited lower elongation percentage than compare to that of tencel yarn (Tables 9 and 10).

By lubricating the fiber surfaces, the fibers move smoothly against each other and machine parts like the yarn guide, trumpets and aprons. This prevents friction and potential fiber breakage and protects the fibers from the static electricity charges that are generated due to inter-fiber friction, friction between the fibers and machines parts, and the low electrical conductivity of the fiber. A spin finish can also prevent fibers from splitting from each other, thus leading to less hairiness and less fiber lapping in the final yarn. By promoting greater adherence between fibers, a spin finish can improve the efficiency of bale opening. The spin finish increases cohesion between the fibers and the machine parts, reducing fly which otherwise introduces defects into the yarn [12].

Uniform draft distribution after several trial and errors allows much uniformity of tencel yarn along with less thick, thin and neps which finally exploits less yarn cut in autoconer machine than compare to that of the cotton yarn. This is because of less thick, thin, and abnormal neps in the tencel yarn (Tables 11-14).

Conclusion

Tencel yarn, comprise of high strength fibers, exhibits higher breaking strength over conventional 100% cotton yarn. Other yarn parameters like breaking elongation, mass uniformity, and hairiness of tencel yarn are significantly better in comparison with that of cotton yarn. Another important aspect of tencel yarn is Eco friendliness which overweighed lower production speed. Ultimate quality profile of tencel yarn facilitates its uprising popularity. Eco friendliness will enhance the use of tencel on basic garments manufacturing as well as

Machine name	Process parameter	100% Tencel			100% Cotton		
		Knitted yarn			Knitted yarn		
		20	30	40	20	30	40
	TPI:	15.12	18.21	21.7	16.52	20.55	23.92
	TM:	3.25	3.325	3.45	3.34	3.46	3.54
	Spacer:	Yellow (2.9 mm)	White (2.8 mm)	White (2.56 mm)	Yellow (3.0 mm)	White (3.0 mm)	White (2.8 mm)
Ring Frame	Traveller size:	1/0	3/0	4/0	1/0	3/0	4/0
		Weaving			Weaving		
	TPI:	16.9	20.55	24.51	20.28	25.07	29.41
	TM:	3.55	3.6	3.7	3.78	3.75	3.74
	Spacer:	Yellow (3.25 mm)	White (3.0 mm)	White (2.8 mm)	Yellow (3.25 mm)	White (3.0 mm)	White (2.8 mm)
	Traveller size	1/0	3/0	4/0	1/0	3/0	4/0

Table 8: Ring frame machine setting parameters for cotton and tencel of different count.

Yarn Count (Ne)	Woven yarn		Knitted yarn		Woven yarn		Knitted yarn	
	Uniformity (U%)	CVm	Uniformity (U%)	CVm	Uniformity (U%)	CVm	Uniformity (U%)	CVm
20	8.36	2.2	8.27	2	9.84	3	10.62	1.7
30	10.56	1.4	10.56	1.4	12.03	2.1	12.11	2.8
40	11.58	1.8	12.16	2.2	12.97	2.23	13.62	1.7

Table 9: Mass Uniformity of tencel and cotton yarn.

Yarn Count (Ne)	Tencel Yarn				Cotton Yarn			
	Woven yarn		Knitted yarn		Woven yarn		Knitted yarn	
	Breaking Strength (CSP)	Breaking Elongation	Breaking Strength (gm/tex)	Breaking Elongation	Breaking Strength (CSP)	Breaking Elongation	Breaking Strength (CSP)	Breaking Elongation
20	4777	12	3449	11	3197	10	2740	10
30	4339	11	3773	10	2712	9	2522	9
40	4080	10	3172	9	2521	8	2119	9

Table 10: Breaking strength and elongation of tencel and cotton yarn.

	Knitted (Tencel)				Knitted (Cotton)			
Yarn count (Ne)	20	30	40	50	20	30	40	50
Hairiness	6.06	5.59	5.2	4.9	6.51	6.46	5.96	5.65
Evenness	8.36	10.56	11.55	13.13	10.62	12.11	13.26	13.42
Thin (-50%)/km	0	2	17	76	6	4	71	73
Thick (+50%)/km	8	82	211	375	126	200	520	553
Nep (200%)/km	20	166	281	794	155	266	673	600
IPI	28	250	511	1245	387	469	1263	1226

Table 11: Imperfection Index, Hairiness of tencel/cotton yarn.

Yarn Count (Ne)	Tencel Yarn				Cotton Yarn			
	Woven yarn		Knitted yarn		Woven yarn		Knitted yarn	
	Twist	Twist CV	Twist	Twist CV	Twist	Twist CV	Twist	Twist CV
20	16.9	1.04	15.12	1.02	20.28	0.8	16.52	1.01
30	20.55	0.89	18.68	0.85	25.07	1.08	20.55	1
40	24.51	1.01	21.7	1.03	29.41	1.15	23.92	0.81

Table 12: Tencel and cotton yarn twist and twist cv% comparison.

	Tencel			Cotton		
	20 Ne	30 Ne	40 Ne	20 Ne	30 Ne	40 Ne
Nep cuts	10	8	10	10	8	13
Short cuts	60	75	85	70	65	88
Long cuts	10	15	10	15	25	12
Thin cuts	8	10	10	10	13	11
Short off count cuts	2	1	1	5	1	2
Off count cuts	1	1	3	1	1	4
Splice cuts	2	4	3	2	7	3
Total	93	114	122	113	120	133

Table 13: Autoconer cut of tencel and cotton yarn comparison for knitted yarn.

	Tencel			Cotton		
	20 Ne	30 Ne	40 Ne	20 Ne	30 Ne	40 Ne
Nep cuts	8	8	10	20	25	25
Short cuts	50	65	70	70	65	78
Long cuts	6	10	10	15	18	12
Thin cuts	7	8	10	15	13	18
Short off count cuts	2	1	1	5	5	2
Off count cuts	1	1	2	3	3	4
Splice cuts	2	2	2	3	7	3
Total	74	95	105	131	136	142

Table 14: Autoconer cut of tencel and cotton yarn comparison for Woven yarn.

intensify use of tencel as blend with cotton, jute etc. which will solve the problem of dependency on synthetic fiber to a large extent and sustain environment [13].

References

1. Mortimer SA, Peguy A (1995) Spinning of fibres through the N-methylmorpholine-Noxide process, 8th International cellucon conference, Cellulose and cellulose derivatives: physico-chemical aspects and industrial applications. Woodhead Publishing: 561-567.

2. Wen HY, Yang XJ (2007) The Spinning Process of Tencel Pure and Blended Yarn. Progress in Textile Science and Technology 1: 44-45.

3. Xu YJ, Wang JH (2006) A New Generation of Cellulose Fibers-Tencel and its Analysis. China Fiber Inspection 1: 43-45.

4. Xu YJ, Wang JH (1999) Taiwan Textile Industry Development Fever Caused by Tencel. Foreign Textile Technology 4: 45-46.

5. Murray S (2016) The overseas textile mills that make our clothes are incredibly wasteful and polluting. Through NRDC's Clean by Design program. Fixing the Fashion Industry.

6. Blackburn R (2005) Biodegradable and sustainable fibres, Wood-head Publishing Limited. Cambridge 188: 158-159.

7. Morley RJ, Taylor JM (2002) Easy Care TENCEL: Best Practice in Fabric Construction and Finishing. International Dyer 4: 17-22.

8. Tu ZX, Fu BK, Zhou GQ (2001) Spinning Practice of Tencel Fiber. Cotton Textile Technology 29: 45-46.

9. Zhang ZP, Weng Y (2003) The Spinning Process of Tencel Fiber. Shandong Textile Science & Technology 5: 24-25.

10. Diao WL (2006) Properties of Tencel Fiber and Its Spinning Practice. Progress in Textile Science and Technology 3: 70-71.

11. Gupta BS (2008) Friction in textiles, Woodhead Publishing Limited, Cambridge: 65-67.

12. Lawrence CA (2010) Advances in yarn spinning technology, Woodhead Publishing Limited, Cambridge: 425-426.

13. Gupta VB, Kothari VK (1997) Manufactured Fiber Technology, Chapman and Hall, London: 978-994.

5

Pistachio Hulls, A New Source of Fruit Waste for Wool Dyeing

Amir Kiumarsi[1]* and Mazeyar Parvinzadeh Gashti[2]

[1]*Department of Chemistry and Biology, Ryerson University, Toronto, Canada*
[2]*Department of Textile, College of Engineering, Yadegar-e-Imam Khomeini (RAH) Branch, Islamic Azad University, Tehran, Iran*

Abstract

Natural dyes have been employed in dyeing Persian carpet piles for many years. In this study, the dried pistachio hulls were powdered and used for dyeing wool yarns. The Iranian wool yarn was first scoured with nonionic detergent and mordanted using some metal salts including Cu, Cr and Al. It was then dyed with different amounts of dried pistachio hull powder. Taguchi statistical method was employed to find the effective factors and results of the planned experiments, in order to optimize the dyeing factors. A L$_{18}$ orthogonal array (seven factors in three levels) was employed to evaluate the effects of different parameters in dyeing process. The colorimetric properties of the dyed yarns were evaluated in CIELAB system. Pistachio hulls found to have good agronomic potential as a natural dye in Iran. Metal mordants when used in conjunction with pistachio hulls were found to enhance the dyeability and its fastness properties. The stepwise process of dyeing with pre-mordanting showed to be energy and time saving and found to achieve high dye retention. Therefore this natural dye has good scope in the commercial dyeing of wool yarns used as Persian carpet piles.

Keywords: Natural dye; Wool dyeing; Mordant; Pistachio hull; Taguchi

Introduction

Today, natural colorants are emerging globally due to the fact that are safer and environment-friendly and thus the application of natural dyes should be considered as a better alternative to synthetic dyes. Using natural dyes contributes to the added value of textiles and also responses to the increasing demand of compatibility with the environment [1-7]. From the point of color chemist view, the actual coloring matters used and the methods by which they were so skillfully applied are of considerable interest. So, it is necessary to study and modify the ways of using natural materials in textiles [5,7]. Many studies have been done on natural dyes covering such areas as: variation in the quantity of dyes concerning natural sources, combination of dyes, properties of natural dyes, effects of mordant and auxiliaries on different properties of dyed samples, light fastness behavior of natural dyes, improvement in natural dyes production and discovering other natural dye sources [8-16]. The colorimetric properties of natural yellow dyes including henna (*lawsonia inermis*), dolu (*rheum emodi*), kamala (*mallotus phillipinensis*), onion (*allium cepa*) and turmeric (*curcuma tinctoria*) with different mordants have been studied by Gulrajani et al. [11-13]. Tsatsaroni and Eleftheriadis also discussed the color and fastness of natural saffron [17]. Many research papers have been published on natural dyes and mordanting but a few papers have been published on the alteration of all different factors in one approach. *Pistacia vera L.*, from family *Anacardiaceae*, genus *Pistacia* is a small tree up to 10 meters tall, with deciduous pinnate leaves and edible delicious nuts. It is native to southwestern Asia and Asia Minor, from Syria to the Caucasus, Iran, Turkey and Afghanistan. There are archaeological evidences of pistachio nuts were used for food as early as 7,000 B.C. The plant was then introduced to Italy from Syria early in the first century A.D. Subsequently, its cultivation spread over other Mediterranean countries, USA, Australia and China. Botanically a drupe consists of Pericarp (ovary wall) (hull), Endocarp (shell) and ripened ovule (seed) that is the edible portion. Pericarp, in turn, consists of Exocarp and Mesocarp. Pistachio hulls are the byproduct of nut production, which is usually useless. Iran is the world's leading pistachio producer and produces over 350 thousands metric tons every year. There are thousands tons of pistachio hull wastes which can be used as a valuable source of natural dye for carpet piles. To optimize the design of an existing process, it is necessary to identify which factors have the greatest influences and which appropriate values produce the most consistent performance. Experimenting with the design variables, one at a time or by trial and error, until a first feasible design found would be a common approach to process optimization. However, this approach could lead one to a very long and expensive time span in completing the design process. A technique for laying out the experiments when multiple factors involved is popularly known as the "factorial design of experiments". This method helps researchers to determine possible combinations of factors and to identify the best combination. Since it is extremely costly to run a number of experiments to test all combinations, application of a full factorial design of experiments is restricted when many factors and levels are studied. The statistical experimental analysis is proposed to investigate the significances of systematic effects. The application of this kind of experiments requires careful planning, prudent layout of experiment, and expert analysis of results. A commonly applied statistical method, Taguchi experimental design and analysis of variance (ANOVA), could be used to analyze results of the experiments on the response and to determine how much variation quality influencing factors contribute [18-25]. Orthogonal arrays can be used to assign factors to a series of experimental combinations, in which results could then be analyzed by using a common mathematical procedure. The variables can generally be grouped into two major classifications: (a) independent variables or factors, and (b) dependent variables or responses. In this factorial design, the main effects of independent variables and the interactions between dependent variables can be studied. The latter would be the

*Corresponding author: Amir Kiumarsi, Department of Chemistry and Biology, Ryerson University, Toronto, Canada
E-mail: akiumarsi@ryerson.ca

major advantage of this technique while a major disadvantage for one-at-a-time variable testing method.

In this research, the usage of the dried pistachio hulls for dyeing wool yarns was studied. To achieve the maximum color strength of wool samples dyed with the pistachio hulls with a minimum number of experiments, a Taguchi experimental design L_{18} orthogonal array (seven factors in three levels and one factor in 2 levels) was employed to evaluate the effect of different factors in the dyeing process. The Iranian wool yarn was first scoured with a nonionic detergent and mordanted with some metal salts including: $KAl(SO_4)_2$, $CuSO_4$, and $K_2Cr_2O_7$. The yarn was then dyed with different amounts of dried pistachio hull powder. The colorimetric properties of the dyed yarns were evaluated in CIELAB system.

Experimental

Materials

The following materials were used:

• Iranian wool yarns of 432/2 Tex with 144 twists per meter.

• Nonionic detergent (Shirley Development Limited) for scouring of wool yarns.

• Aluminum potassium sulfate $(KAl(SO_4)_2)$, copper sulfate $(CuSO_4)$ and potassium dichromate $(K_2Cr_2O_7)$ from Merck for mordanting process.

• Hydrochloric acid and sodium hydroxide for controlling the pH.

• Dried powdered pistachio hulls from genus *Pistacia vera L.* collected from Kerman province, Iran.

Procedure

The wool yarns were treated in four steps as follows:

Scouring: Wool yarns were scoured with 0.5% nonionic detergents for 30 minutes at 50°C. The L:G (Liquor to Good ratio) of the scouring bath was kept at 40:1. The scoured material was thoroughly washed with tap water and dried at room temperature. The scoured material was soaked in clean water for 30 minutes prior to dyeing or mordanting.

Mordanting: The scoured wool yarns were mordanted with $KAl(SO_4)_2$, $CuSO_4$, and $K_2Cr_2O_7$. The L:G of the mordanting bath was kept at 40:1. Hydrochloric acid was used in the mordanting bath for adjusting the pH at 5. The mordanting temperature was started at 40°C and then gradually raised to the required temperature during 20 min and kept at this temperature for 1 hr. The mordanted material was then rinsed with water thoroughly, squeezed and dried.

Dyeing: First the dye solution was prepared by pouring the appropriate amount of dye powder in water for 24 hours before dyeing. Then, the dyeing process was carried out. Dyeing started at 40°C and temperature was raised to required temperature in 20 minutes and resumed at this temperature 1 hr. The pH of the dyebath was kept at

pH=5 using dilute acid or base. The dyed material was then rinsed with water thoroughly, squeezed and dried.

Measuring of color strength: To investigate the effects of different parameters on the reflectance spectra of dyed samples, a GretagMacbeth spectrophotometer model 7000A computer integrated, was used. CIELAB color co-ordinates (L*, a*, b*, C) and color strength values (K/S) were calculated from the reflectance data (R) of dyed samples for 10° observer and D65 illuminant based on the Kubelka–Munk equation:

$$K/S = \frac{(1-R)^2}{2R}$$

In which, K, is the sorption coefficient, R is the reflectance of the dyed sample and S is the scattering coefficient.

Results and Discussion

In this research work, the optimization of the dyeing factors in applying the pistachio hulls as a new source of natural dye for wool yarns was carried out employing Taguchi method for statistical design of experiments. According to Taguchi parameter design methodology, one experimental design should be selected for the controllable factors. A L_{18} orthogonal array (that accommodates seven factors in three levels and one factor in two levels each in 18 runs) was employed to identify the optimum conditions for dyeing wool yarns with pistachio hulls [25]. Seven influencing factors were taken into account and one column was set for the determination of errors as following:

• Amount of colorant (A)

• Amount of mordant (B)

• Type of mordant (C)

• Temperature of mordanting bath (D)

• Temperature of dyeing bath (E)

• Duration of mordanting (F)

• Duration of dyeing (G)

The three levels of factors were selected in accordance with the preliminary test and the previous author's experience [26,27]. The factors and levels considered for this experiment are shown in Table 1. The K/S ratios for all samples from different conditions was calculated and shown in Table 2. In these experiments, the system was optimized according to the maximum response value of (K/S). The analysis of the variance (ANOVA) was employed to determine the factors influencing the average response (K/S) ratios. Table 3 presents the degree of freedom (df), the sum of squares (SS), the mean square (variance, V), and the F ratio of variances (F). The insignificant factors are pooled to reduce the chance of making alpha mistakes. As a role, the factor, which influence is 10% or lower than the most influential factor is pooled [25]. The F-Value implies that the model is significant. One column was set for the determination of errors to consider the effects

Levels	Amount of colorant, A (g)	Amount of mordant, B (g)	Type of mordant, C	Temp. of mordanting bath, D (°C)	Temp. of dyeing bath, E (°C)	Duration of mordanting, F min	Duration of dyeing, G min
1	0.4	2	Cr	50	60	30	60
2	0.1	5	Al	70	80	60	90
3	2	10	Cu	Boil	Boil	120	120
Optimum	0.4	2	Cu	50	60	120	120

Table 1: Studied levels of different factors and the optimum level.

Factors Exp. No	Error	A	B	C	D	E	F	G	K/S
1	1	1	1	1	1	1	1	1	0.06077
2	1	1	2	2	2	2	2	2	0.05804
3	1	1	3	3	3	3	3	3	0.06073
4	1	2	1	1	2	2	3	3	0.05729
5	1	2	2	2	3	3	1	1	0.04999
6	1	2	3	3	1	1	2	2	0.05730
7	1	3	1	2	1	3	2	3	0.05911
8	1	3	2	3	2	1	3	1	0.05747
9	1	3	3	1	3	2	1	2	0.04033
10	2	1	1	3	3	2	2	1	0.06026
11	2	1	2	1	1	3	3	2	0.05847
12	2	1	3	2	2	1	1	3	0.05652
13	2	2	1	2	3	1	3	2	0.05753
14	2	2	2	3	1	2	1	3	0.06043
15	2	2	3	1	2	3	2	1	0.05233
16	2	3	1	3	2	3	1	2	0.05796
17	2	3	2	1	3	1	2	3	0.05720
18	2	3	3	2	1	2	3	1	0.05714

Table 2: Design of experiments (DOE) and the calculated K/S ratios for different conditions.

	df	SS	V	F
Error	1	1.56968E-05	1.56968E-05	1.742903
A	2	6.01627E-05	3.00813E-05	3.340098
B	2	6.90913E-05	3.45457E-05	3.835796
C	2	6.45276E-05	3.22638E-05	3.582426
D	2	6.15764E-05	3.07882E-05	3.418584
E	2	1.49631E-05	7.48157E-06	0.83072
F	2	4.80049E-05	2.40024E-05	2.665124
G	2	3.9758E-05		
e	2	3.2291E-05		
(e)	4	3.60245E-05	9.00613E-06	
Sum	17	4.06072E-04		

Table 3: The analysis of the variance (ANOVA) for samples dyed with pistachio hulls.

of factors excluded from the experiment and/or Uncontrollable factors (beta mistakes).

The variation of K/S against the different levels of factors is shown in Figure 1.

The response of software calculation shows that the optimum levels of each factor are as follows:

- Amount of colorant at level 1 (0.4 g).

- Amount of mordant at level 1 (2.0 g).

- Type of mordant at level 3 ($CuSO_4$).

- Temperature of mordanting bath at level 1 (50°C).

- Temperature of dyeing bath at level 1 (60°C).

- Duration of mordanting at level 3 (120 min).

- Duration of dyeing at level 3 (120 min).

The ANOVA predicts the maximum value of K/S=0.7 for the above optimum conditions. It also suggests that the factors influencing the dye absorption of the wool yarn are of different importance and that the most important factor is the amount of mordant (F=3.835796)

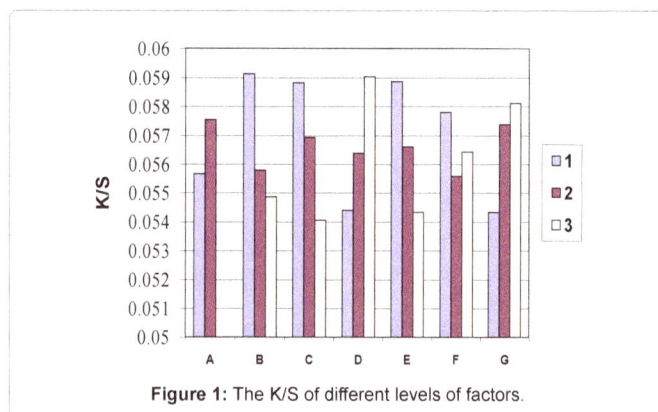

Figure 1: The K/S of different levels of factors.

followed by the type of mordant (F=3.582426), and temperature of mordanting bath (F=3.418584). The amount of colorant is the forth most important factor presenting the F ratio of 3.340098. The higher the percent influence of a factor, the tighter the tolerance, and vice versa.

Effect of mordant

Most of the natural dyes have poor affinity for natural fibers. Moreover, their fastness is often enhanced by metal mordants, which form an insoluble complex with the dye molecules. It is known that flavonols have greater tendency towards chelate formation due to the presence of hydroxyl–keto functionality [28,29]. The probable way of chelation depends on the nature of the mordant–dye complex which has been discussed earlier [27]. Mordant makes strong coordination bonds with the wool yarn on one side and the colorant molecule on the other side. The dye molecules are capable of forming five- and six-membered chelate rings with different metal ions. These chelates utilize the ortho-dihydroxy structure and/or carbonyl ortho to the hydroxyl group in the flavonol dye molecules, which in turn greatly enhance the affinity of natural dye and fiber. Therefore the type and amount of mordant is of the great importance to make such chelation [27].

Conclusion

The use of the experimental design is described for optimizing analytical methods. Many factors can be studied, so the interactions can be determined. In addition to the factors identified as being significant, they have more credibility since they are studied several times. The Taguchi's method simplifies both the assignments of factors and the calculations.

The designs described in this work include three level factors. The results and analysis of the variance show that the above-mentioned conditions were considered as optimum. Experiments were carried out at the stated levels, and the optimum conditions were achieved. Confirming experiments were also carried out. Limitations to the experimental design may be seemed obvious, but they are worth being stated, neglecting them often leads to the failing of this approach. The variance observed for a factor is only valid over the range studied for that factor. Pistachio hulls found to have good agronomic potential as a natural dye in Iran. Metal mordants when used in conjunction with pistachio hulls were found to enhance the dyeability and its fastness properties. The color shades on wool yarns ranges from light orange to dark orange. The stepwise process of dyeing with pre-mordanting showed to be energy and time saving and found to achieve high dye retention. Therefore this natural dye has good scope in the commercial dyeing of wool yarns used as Persian carpet piles.

Acknowledgement

The authors express their sincere thanks to Institute for Color Science and Technology of Iran (ICST) for financial support.

References

1. Kirk Othmer (1998) Encyclopedia of Chemical Technology. John Wiley and Sons, New york, USA.

2. Parvinzadeh Gashti M, Katozian B, Shaver M, Kiumarsi A (2014) Clay nanoadsorbent as an environmentally friendly substitute for mordants in the natural dyeing of carpet piles. Color Technol 130: 54-61.

3. Parvinzadeh M (2007) Effect of proteolytic enzyme on dyeing of wool with madder. Enzym Microb Technol 40: 1719-1722.

4. Anavian R, Anavian G (1975) Royal Persian and Kashmir brocades. Senshoku to Seikatsusha Ltd, Japan 5-24.

5. Parvinzadeh M (2009) An Environmentally Method for Dyeing Rug Pile using Fruit Waste Colorant. Res J Chem Environ 13: 49-53.

6. Montazer M, Parvinzadeh M, Kiumarsi A (2004) Colorimetric properties of wool dyed with natural dyes after treatment with ammonia. Color Technol 120: 161-166.

7. Montazer M, Parvinzadeh M (2004) Effect of ammonia on madder-dyed natural protein fiber. J Appl Polym Sci 93: 2704-2710.

8. Perkin AG, Everest AG (1918) The natural organic coloring matters. Longman publications.

9. Moir M, Thomson RH (1973) Naphthaquinones in Lomatia species. Phytochem 12: 1351-1353.

10. Takeda N, Seo S, Ogihara Y, Sankawa U, Itaka I, et al. (1973) Studies on fungal metabolites-XXXI: Anthraquinonoid colouring matters of Penicillium islandicum sopp and some other fungi (−)luteoskyrin, (−)rubroskyrin, (+) rugulosin and their related compounds. Tetrahedron 29: 3703-3719.

11. Gulrajani ML, Gupta D, Agrawal V, Jain M (1992) Indian Text J 102: 50.

12. Gulrajani ML, Gupta D, Agrawal V (1992) Indian Text J 102: 85.

13. Gulrajani ML, Gupta D, Agrawal V (1992) Indian Text J 102: 76.

14. Montazer M, Parvinzadeh M (2007) Dyeing of wool with Marigold and its properties. Fibers Polym 8: 181-186.

15. Nishiba K, Kobayashi K (1999) Dyeing properties of natural dyes from natural sources. Am Dyestuff Rep 81: 44.

16. Balazsy AT, Eastop D (1998) Chemical principles of textile conservation, John Wiley Ltd.

17. Tsatsaroni EG, Eleftheriadis IC (1994) The colour and fastness of natural Saffron. J. S. D. C. 110: 313.

18. Hicks CR (1982) Fundamental concept in the Design of Experiments. Holt, Rinehart and Winston, Inc. New York.

19. Fisher RA, Yates F (1953) Statistical Tables for Biological, Agricultural, and Medical Research. Oliver and Boyd, Edinburgh.

20. Bayne C, Rubin I (1979) In practical Experimental Designs and Optimization Methods for Chemists. VCH Publisher Inc., New York.

21. Duckworth WE (1968) Statistical Techniques in Technological Research: An aid to research productivity. Methuen, London.

22. Taguchi G (1987) System of Experimental Design. Vol.1, KRAUS International publisher.

23. Taguchi G (1987) System of Experimental Design. Vol.2, KRAUS International publisher.

24. Montgomery Doglas C (1991) Design and Analysis of Experiments. John Willy and Sons, USA.

25. Roy RK (2001) Design of Experiments Using the Taguchi Approach. Wiley, USA.

26. Parvinzadeh M, Moradian S, Rashidi A, Yazdanshenas ME (2010) Effect of the Addition of Modified Nanoclays on the Surface Properties of the Resultant Polyethylene Terephthalate/Clay Nanocomposites. Polym-Plast Technol 49: 874-884.

27. Parvinzadeh M, Kiumarsi A (2008) Using Eggplant Skin as a Source of Fruit Waste Colorant for Dyeing Wool Fibers. Prog Color, Colorants and Coat 1: 37-43.

28. Bayer E, Egeter H, Fink A, Nether K, Wegmann K (1966) Complex Formation and Flower Colors. Angew Chem Int Ed 5: 791-798.

29. Ferreira ESB, Hulme AN, McNab H, Quye A (2004) The natural constituents of historical textile dyes. Chem Soc Rev 33: 320-336.

6

Designing an Adaptive 3D Body Model Suitable for People with Limited Body Abilities

Kozar T*, Rudolf A, Cupar A, Jevšnik S and Stjepanović Z

Faculty of Mechanical Engineering Department of Textile materials and Design, University of Maribor, Slovenija

Abstract

The purpose of this research was to develop a generalized adaptive 3D body model the posture of which could be adapted to different positions in order to perform a virtual prototyping of garments for people with limited body abilities. The digital data of a tested person's body was acquired using the GOM Atos scanning system. Careful surface reconstruction was performed in order to provide an adequate mesh for further posture adaptation. The experimental part of this article presents the usages of a variety of graphic programs in order to provide an adaptive 3D body model through kinematic skeleton construction. The mesh-deformation during posture adaptation was improved using advanced tools of applied software packages. The usefulness of the gained 3D body model was determined by calculating the differences between the real and digital 3D body model measurements.

Keywords: Scanning technology; 3D body model; 3D body model posture adaptation; People with limited body abilities

Introduction

The utilities of three-dimensional body models, also called avatars or humanoids, are unlimited just as human activities are unlimited. The book "Virtual Humans" clearly demonstrates the scope of their applicability from virtual presenters for Television (TV) and World Wide Web (WEB), to virtual assistants for training in the case of emergency, virtual workers in industrial applications, virtual actors in computer-generated movies to virtual characters for the garment industry [1]. The complexity of today's garment industry has led to unit advanced computer-aided (CAD) technologies and 3D graphic software for assisting designers' creativities, reducing garments' manufacturing costs and more importantly to serve customers' needs and increase their satisfaction percentage [2]. Recently, an obvious trend in garment manufacturing processes has become epitomized by individualized garment production supported by CAD systems for garment pattern construction [3-6] and their appearances on parametric body models in standing positions, the body dimensions of which could be adjusted to specific customers' body measurements. It refers to garment virtual prototyping based on a made-to-measure concept for ensuring garments' pattern adaptation by taking individual requirements for a garment's style and functionality into consideration [7]. It is well-known that the use of 3D CAD systems for garment pattern design has decreased the production of real garments' prototypes by 20% and the amount of manufacturing waste [8]. A major issue regarding garment virtual prototyping is when evaluating a garment's fit appearance and the fabric's draping behavior on virtual body models are limited to a standing posture with standard body shape characteristics. These body shapes may be adaptable for healthy individual body measurements but useless if there is a need to construct individualized garments for people with physical disabilities.

Based on the United Nation's definition, persons with physical disabilities include those who have long-term physical, mental, intellectual, or sensory impairments, which during interactions with various barriers may hinder their full and effective participation in society on an equal basis with others [9]. In our presented contribution we focused on people with limited body abilities, caused by several forms of injuries, diseases, and amputations. This group of people within the world's population have different kinds of limitations and health conditions and therefore these people with limited body abilities are sensitive about their garments and general appearances. Therefore, greater attention should be focused on constructing special needs garments, a category of functional garments for persons with physical limitations, whose body shapes, sizes, mobility or dexterity are significantly different from people with average body shapes [10]. While the application field of 3D body models is unlimited and improvements and progressions are focused on imitating real humans' behavior and appearance within virtual environments, there is need to observe an adaptive 3D body model and thereby adaptive garment prototyping for improving the qualities of life of this special group of people.

To date several techniques for human body modeling have been devoted to automatic modifications resulting in models of different shapes with the help of example models [11,12]. One researcher developed a method for fitting high-resolution template meshes to detailed human body range scans from the CEASAR dataset. In order to find a consistent shape representation of non-structured data, this researcher observed two surface-matching algorithms for creating a shape model that could generate different kinds of body shapes in different poses but with limitations to the skinning framework, showing a non-linear relation to the proportion of the body at the joint angles [13]. The parameterization approach synthesizes sample models to a new body model according to the input sizing dimensions. Along with synthesizing the system is characterized by feature wireframes of scanned human bodies for enabling garment design [14]. In addition, there are several methods for fitting deformable body models to body shapes, as parameters to ensure a variety of body shapes [15,16]. Other methods have used data-driven models where the variation of human body shapes was based on a training set of 3D body shapes

*Corresponding author: Kozar Tatjana, Faculty of Mechanical Engineering Department of Textile materials and Design, University of Maribor, Slovenia E-mail: tatjanka.kozar@gmail.com

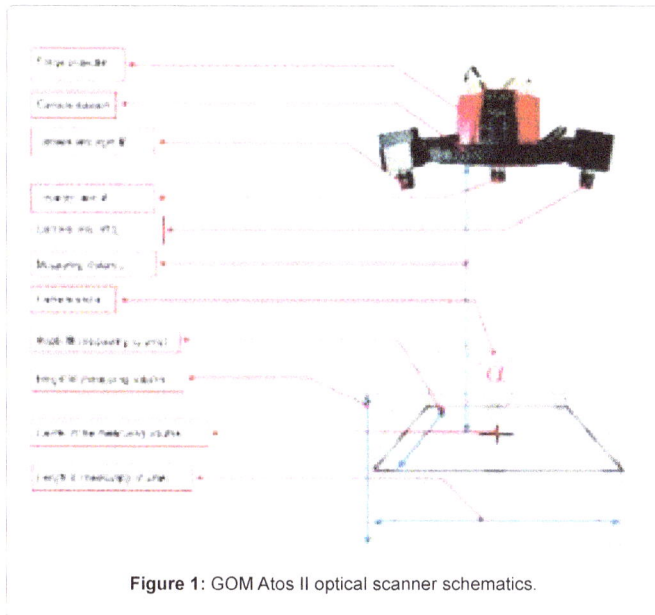

Figure 1: GOM Atos II optical scanner schematics.

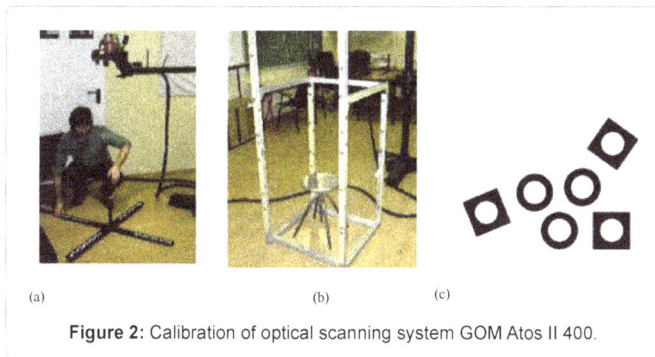

Figure 2: Calibration of optical scanning system GOM Atos II 400.

[17]. Obviously, the mentioned human body modelling techniques are focused on a standard standing posture, involving healthy individuals and exclude persons with physical limitations.

Over recent years, certain research has been devoted to acquiring an accurate 3D body model in a sitting posture for garments' virtual prototyping. Two types of scanning technologies have been applied for obtaining a suitable three-dimensional image of a human body in a sitting posture [18,19]. Firstly, the Vitus Smart scanner was introduced for acquiring a polygonal meshed surface of a test person in a sitting posture in order to achieve a garment's virtual prototyping and its fit appearance. Its technical features, operation principle and mesh-processing are clearly described in [18]. Next, scanning of a human body in a sitting position by using a GOM ATOS scanner was introduced and performed with the help of a specially developed frame with hand rails [19]. Finally, by analyzing a 3D body model's measurements in both standing and sitting positions, the researcher assessed the fit of the garments intended for paraplegics and took terms of wearing comfort on specific body parts into consideration [20]. The GOM Atos scanning system's primary use refers to the exploration of an object's digital data within the field of mechanical and civil engineering. Its accuracy was analyzed in a study [21], where five digitization techniques like laser scanners, one fringe projection system and one x-ray system on calibrated objects, with dimensions of less than 300 × 500 mm were compared. Another research was oriented towards finding a suitable 3D scanner for the digitization of human body parts within the field of

prosthetics and orthotics. Due to higher resolution features and higher requirements for preparing the scanning process, the scanning systems ATOS I and ATOS Triple Scan predominate among other tested scanners when scanning a plaster cast among other tested scanners, while the tested hand scanner Artec MH showed a better alternative for a direct scanning of a patient's body parts [22].

In order to increase the interest of people with limited body abilities within the garment manufacturing process, the aim of this study was to develop a generalized adaptive 3D body model within the following activities:

- Scanning the surface of a test person's body in standing and sitting positions.

- Reconstruction of observed polygonal meshed surfaces.

- Kinematic skeleton construction inside a watertight polygonal mesh in a standing position.

- Adjusting the 3D body model from a standing to a sitting position.

- Comparing the real body measures with a 3D body model's measures in standing-, sitting- and adjusted sitting positions.

Experimental

Scanning of a human body with a GOM Atos II 400 3D optical scanner

In the field of computer graphics for human body modeling, scanning technology represents an effective technique for obtaining the polygonal meshed surface of a human body in order to avoid manual sculpting techniques of meshed surfaces. In this research, the surface of a human body in both standing and sitting positions was obtained by using a three-dimensional laser scanner.

The GOM Atos™ II 400 [23], is installed at the Faculty of Mechanical Engineering in Maribor and used for different research purposes. Figure 1 shows a schematic representation of a GOM Atos scanning system, which operates using a triangulation principle [20,21], for converting the obtained point clouds to three- dimensional surfaces. Its configuration includes the following technical features: measuring volume of 1200 × 960 × 960 mm³ (L × W × H), measuring point distance of 0.94 mm, 6 mm projector lens and 8 mm camera lens.

The GOM Atos scanning system uses white light for obtaining polygonal meshed surfaces, when different fringe patterns, recorded by two cameras, are projected onto the scanned object – in our case on a human body through the sensor units. Garment pattern construction of a tight suit was established for observing accurate shape information from a tested person's body. Generally, such a scanning method is an accepted procedure for scanning human bodies which does not particularly differ from scanning human bodies wearing underwear. The digital body information of a scanned human body was obtained in three steps. First, the measuring system was calibrated with a calibration cross, (Figure 2a) to ensure the system's dimensional consistency. Next, the test person was scanned in both standing and sitting postures from different heights and angles, while each measurement generated up to 4 million data points. In order to avoid incorrect alignment of individual scans in a sitting posture, a frame with hand rails was used to improve the static position of a human body. The frame was marked with reference points or circular markers in order to improve the process of alignment (Figure 2b). Self-adhesive markers had a defined geometry (white Circle on a black background) (Figure 2c), and served

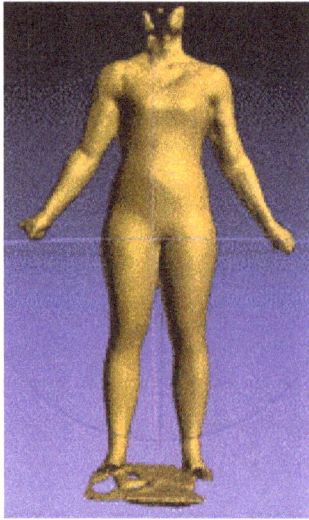

Figure 3: Aligned individual scans in a global coordinate system using Atos V6.0.2 software.

as connection points for the individual measurements to provide their transformation into a common coordinate system. Therefore, we finally used the Atos V6.0.2-6 software, where each scan was polygonized into one independent mesh in order to perform further fit registration (Figure 3).

Scanned surface processing

The polygonal meshed surfaces, obtained by general 3D laser scanners, represent in many cases starting points for further computer modelling and visualization of physical objects. In order to ensure the body model's appropriate fitting by taking into account the real subjects, we obtained an optimal surface description through pre-processing and post-processing procedures, which undoubtedly stood out as the more significant and challenging tasks in obtaining the digital model from the physical object [24]. Table 1 clearly demonstrates a short description of applied programs and indicates the applied tools for performing surface-processing procedures.

The pre-processing procedure was carried out by using the GOM Inspect program. GOM Inspect is a 3D inspection and mesh processing software for three dimensional analysis of 3D point clouds, obtained from white light scanners, laser scanners, CT's and other sources. The program is independently tested and certified by national measurement laboratories, such as German PTB (Physikalisch-Technische Bundesanstalt) and American NIST (National Institute of Standards and Technology) [25]. We performed the noise reduction through cleansing and removing those points which did not correspond to the body surface, like the digital data of a chair from the point data set shown in Figure 4a. The resulting holes are shown in Figure 4b, which were repaired manually through modeling and adding new surfaces, Figure 4. The obstructed areas were repaired as also were the scanner had failed to capture any data, especially at the thigh- and calf areas. Consequently, the tool "Mesh Bridge" was used to reduce mesh errors.

In addition, pre-processed polygonal meshes usually need higher refinements and the surface was reconstructed through a post-processing procedure which was also done manually. In addition, graphic programs like MeshLab and Blender were used to correct imperfections and errors on the surface. Meshlab is an open source,

portable and extensible system for the processing and editing of unstructured 3D triangular meshes, providing a set of tools for editing, cleaning, healing, inspecting, rendering and converting of this kind of meshes. The system is based on the VCG Library (The Visualization and Computer Graphics Library) developed at the Visual Computing Lab of ISTI-CR (Instituto di Scienza e Technologie dell' Informazione) in Italy [26]. During the first step we created a new average mesh over the existing scan by using the tool "Select All-Filter Selection" from the program MeshLab. Afterwards, the surface was reconstructed using the tool "Poisson", Figure 6.

Partially reconstructed polygonal meshes were imported in the form of .stl files into the Graphic Program Blender for smoothing the remaining irregularities on the meshed surface. The tool "Sculpt Mode-Smooth" was used to provide an entirely watertight mesh (Figure 7).

Although it was necessary to reconstruct the mesh at some areas, especially at the region after removing the scanned data of a chair, the corrections do not affect the actual shape of the body. It is evident that the body shape of the reconstructed 3D body model in

Applied program	Features Description	Applied Tools
GOM Inspect	3D inspection and mesh processing software for editing 3D point clouds, containing evaluation and editing tools for an extensive analysis of parts and components.	"Delete Neighborhood", "Mesh Bridge"
MeshLab	3D mesh processing software for editing unstructured 3D triangular meshes, containing editing filters for cleaning meshes, remeshing and mesh inspections.	"Select All-Filter Selection". "Poisson"
Blender 2.71	Open source 3D graphics and animation software, enabling 3D modeling, rigging, sculpting, rendering and interactive creation of 3D models.	"Sculpt Mode-Smooth"

Table 1: Applied programs and tools for surface reconstruction.

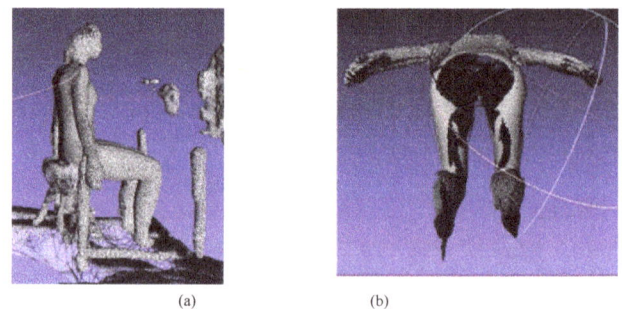

(a)　　　　　(b)

Figure 4: Mesh errors after removing the digital data of a chair by using the tool "Delete Neighborhood".

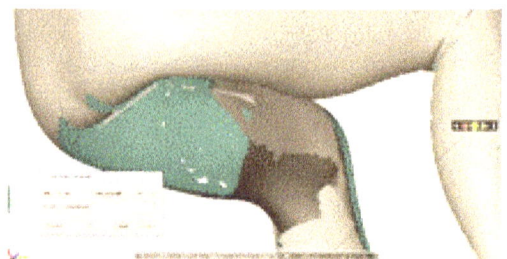

Figure 5: Applying the "Mesh Bridge" tool for surface modeling.

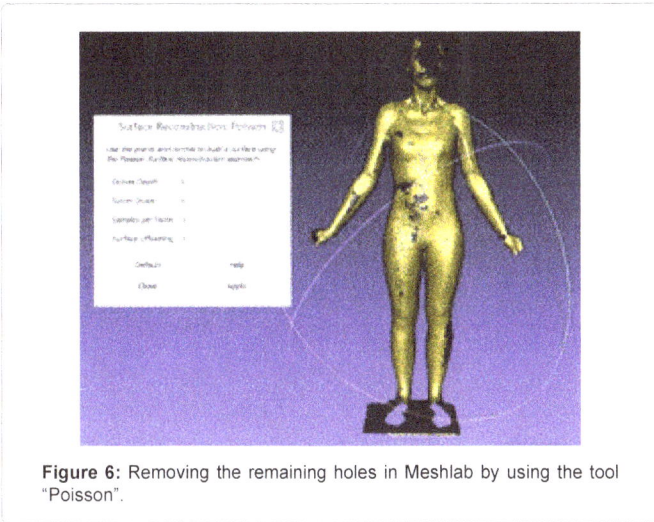

Figure 6: Removing the remaining holes in Meshlab by using the tool "Poisson".

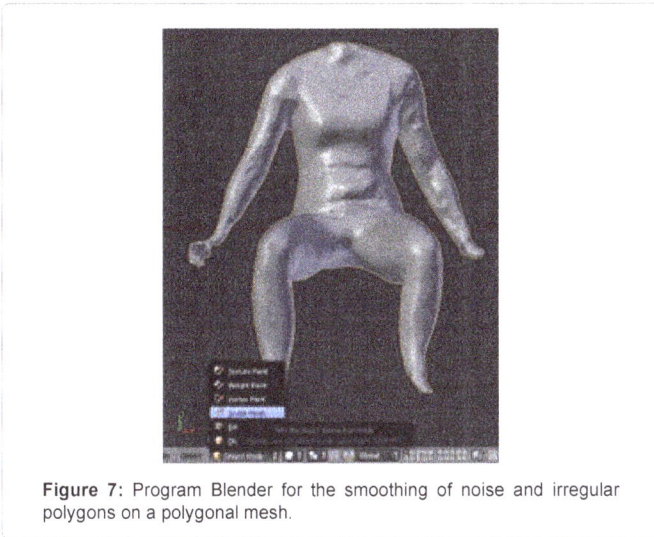

Figure 7: Program Blender for the smoothing of noise and irregular polygons on a polygonal mesh.

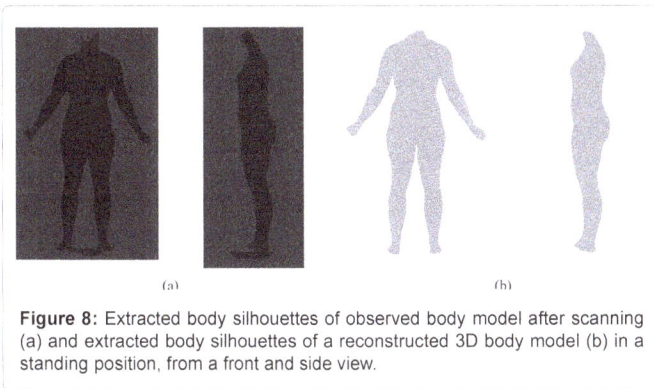

Figure 8: Extracted body silhouettes of observed body model after scanning (a) and extracted body silhouettes of a reconstructed 3D body model (b) in a standing position, from a front and side view.

a standing position; (Figure 8a) does not significantly differ from the observed body model after scanning procedure (Figure 8b). The same applies to the visual assessment of differences between body shapes of the reconstructed 3D body model and the observed body model after scanning procedure (Figures 9a and Figure 9b).

Kinematic skeleton construction

A related study discussed 3D body models' posture adaptations through the kinematic skeleton construction of 19 bones and 15 joints [27]. Pants- and T- shirt patterns construction was carried out in order to simulate a garment's behavior and virtual fitting using the 3D body model in a sitting position. Its accuracy was tested by measuring the virtual body measurements and comparing them with a real person's body measurements in a sitting position. An additional indicator when calculating surface areas in square centimeters was used after adapting a 3D body model to images taken from two different angles. Differences between the body measurements were encountered at the hips, and under the knee and calf circumferences. Even then there were no larger differences between the silhouettes extracted from the images, the visual assessment displayed clear differences between the extracted silhouettes. In order to provide a generalized adaptive 3D body model, which should imitate the real human's behavior and ensure accurate anthropometric measurement of those three-dimensional body measurements in a sitting position, the kinematic skeleton construction inside the observed polygonal mesh was improved by taking specific anatomical landmarks into consideration. Anthropometry, as one of the methods within anthropology, studies human body measurements and the proportions between different parts of the body [28] beginning with accurate placements of anatomical landmarks or feature point on the human body's surface that indicate the locations of body organs or components [1]. In order to ensure accurate anthropometric measurements of the three-dimensional body measurements, we learned from the H- Anim standard [29] and used the H-Anim feature point's definition as a basic guide line for constructing the kinematic skeleton. In regard to the skeleton construction inside the watertight mesh of a scanned body in a standing position, we used the program Blender 2.68. Blender 2.68 is open-source 3D graphic software, as maintained by the Blender foundation [30] which assists designers with similar functions and animation tools as commercial software. Its suitability for creating virtual humans with realistic human animations was proved during the study [31] where the researcher extended the Blender's functionality for computer vision applications. In our contribution, we built an articulated 3D body model consisting of a polygonal mesh and a kinematic skeleton. The observed polygonal mesh of vertices, edges, and faces or polygons, describes the highly complex geometric structure of a scanned human. Each polygon is defined by 3 or more vertices with x, y, and z coordinates within a 3D space. For the skeleton construction we imported the 3D body model in a standing posture in the form of a .stl file into the program Blender. The "Armature Modifier" was used to construct a hierarchical skeleton of 20 bones and 15 joints (Figure 10).

In Blender the term "Armature" refers to an object that deforms a mesh model and it borrows many ideas from real life skeletons. In

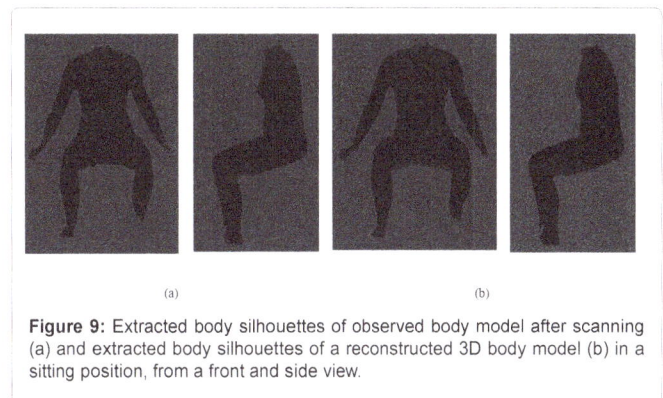

Figure 9: Extracted body silhouettes of observed body model after scanning (a) and extracted body silhouettes of a reconstructed 3D body model (b) in a sitting position, from a front and side view.

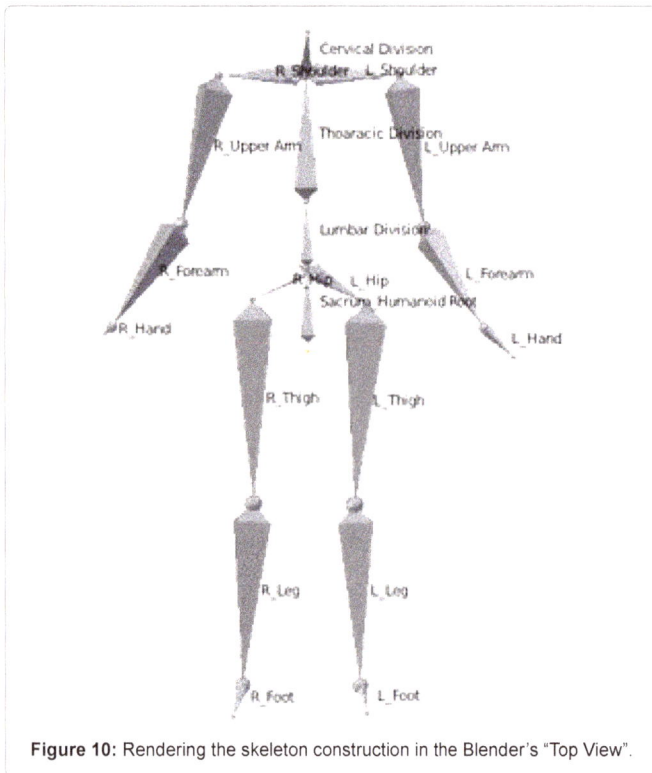

Figure 10: Rendering the skeleton construction in the Blender's "Top View".

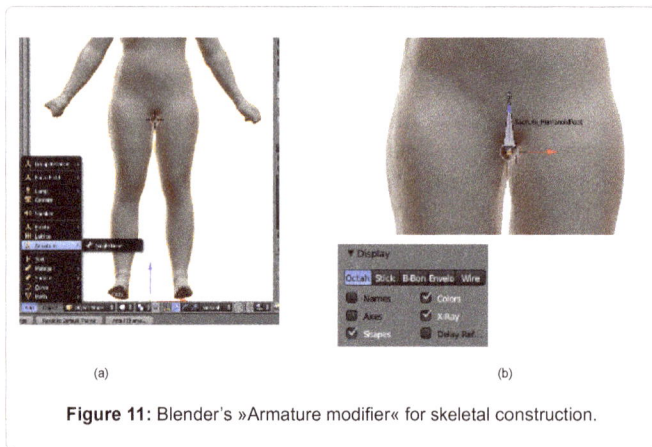

Figure 11: Blender's »Armature modifier« for skeletal construction.

"object mode" with a selected mesh of a 3D body model the 3D Cursor was placed at the crotch point. By adding the "Armature Modifier", the skeleton construction started up with a single bone, called the "Sacrum_ Humanoid Root" or the base- bone of a skeleton's hierarchical structure (Figure 11). In the "Properties Window" an "Object button" was used which makes the bone visible using the option X-Ray (Figure 11b).

Firstly, in "Edit Mode" the spinal cord construction was carried out by selecting the tail of each particular constructed bone through extruding it to a new bone (Figure 12a). The skeleton construction continued for the shoulder and upper limbs at the left and right sides shown in Figure 12b, and completed at the lower limb for the Left and Right sides in Figure 12c.

In order to provide effective posture adaptation, it was necessary to establish a "Parent-Child" relationship between the constructed bones. For example, in "Edit Mode" with the bone head and the bone

tail selected, the "Lumbar Division" bone was added by extruding its "Parent bone", which parents it to the constructed bones automatically, (Figure 13).

For example, all the bones of the spinal cord, from the "Lumbar Division" bone to the "Cervical Division" bone, have their rotational centers at their parent bones. This means that the "Cervical Division" bone is the child of the lower bone, called the "Thoracic Division" bone, which is further the child of the "Lumbar Division" bone, which is further the child of the "Sacrum Humanoid Root" base-bone. This means if we rotate or move the lower bone, its child will also be affected. However, if we rotate or move a child bone, its parent bone will not be affected.

Further, the head or the tail of each particular bone represents the location of an anatomical feature, the position of which has to be located by visual assessment of extreme feature points from the side view of the body and by taking specific anthropometric rules into consideration. The anatomical feature points are summarized and defined in Table 2.

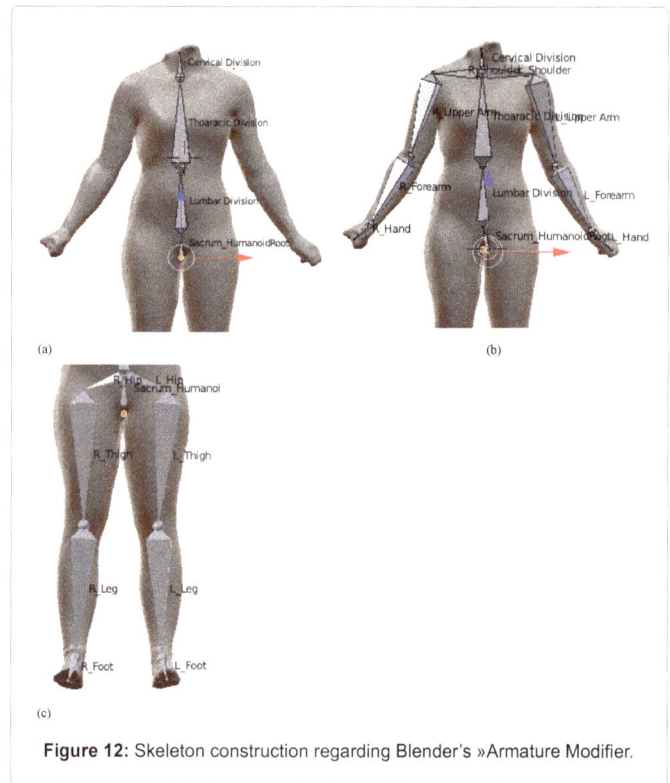

Figure 12: Skeleton construction regarding Blender's »Armature Modifier.

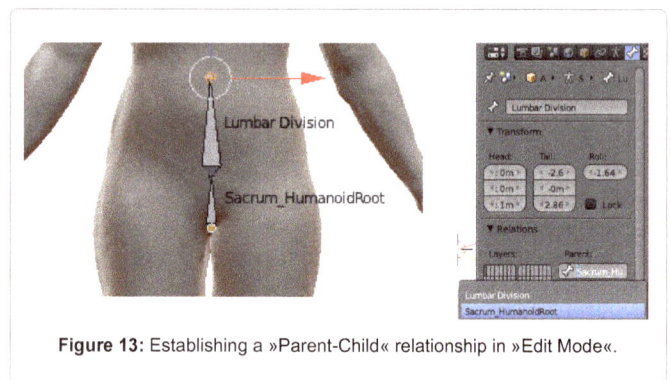

Figure 13: Establishing a »Parent-Child« relationship in »Edit Mode«.

Anatomical feature point	Definition
Crotch	The lowest part of the trunk, located on the head of the Sacrum_humanodid Root, between the Left- and Right legs.
Navel (Belly Button Point)	Located at the abdomen area, between the Abdominal extension and the Tenth Rib. It is placed on the bone tail of the Lumbar Division bone
Cervical Vertebrae (Vertebra Prominous)	At the base of the neck portion of the spine and located at the tip of the spinous process of the 7th cervical vertebra determined by palpitation, often found by bending the neck or head forward.
Acromion (Shoulder Point)	The most prominent point on the upper edge of the acromial process of the shoulder blade (scapula) as determined by palpitation (feeling). It is placed on the bone tails of Left- and Right shoulders.
Iliocristal	Highest palpable point of the iliac crest of the pelvis, half the distance between the front (anterior) and back (posterior) upper (superior) iliac spine. It is placed on the tails of the left- and right hip bones.
Olecranon (Elbow)	When an arm is bent, the furthermost (lateral) point of the olecranon which is the projection of the end of the innermost bone in the lower arm (ulna); the joint between the upper and lower arm. It is placed on the tails of the Left and Right upper arm bones
Carpus (Wrist)	Joint between the lower arm and hand; Distal ends (toward the fingers) of the ulna (the innermost bone) and radius (the outermost bone) of the lower arm. It is placed on the tails of the Left and Right Forearms.

Table 2: Anatomical features' locations and definitions [1, 29]

Bone part	Bone head location			Bone tail location			
Bone name	x	y	z	x	y	z	Parent bone
Sacrum_Humanoid Root	0	23.2	18	0	23.2	1.18	HumanoidRoot
Lumbar Division	0	23.2	1.18	-2.5	38.7	2.93	Sacrum_Humanoid Root
Thoaracic Division	2.58	38.7	2.93	-2.5	62.1	6.41	Lumbar Division
Cervical Division	2.58	62.1	6.41	0.42	-32.3	7.28	Thoaracic Division
L*_Hip	0	23.2	1.18	1.39	23.2	71.5	Sacrum_Humanoid Root
L_Thigh	1.39	23.2	71.5	1.11	1.55	-3.5	L_Hip
L_Leg	1.11	1.55	3.5	1.21	1.55	-7.6	L_Thigh
L_Foot	1.21	1.55	7.60	1.3	-6.67	-8.2	L_Foot
R**_Hip	0	23.2	1.18	-1.2	23.2	74.1	Sacrum_HumanoidRoot
R_Thigh	1.29	23.2	74.1	-1.2	23.2	-3.4	R_Hip
R_Leg	1.26	23.2	3.47	-1.5	23.2	-7.6	R_Thigh
R_Foot	1.52	23.2	7.6	-1.7	54.2	-8.30	R_Leg
L_Shoulder	2.58	62.1	6.41	2.05	29.7	6.19	Thoaracic Division
L_Upper Arm	2.05	29.7	6.19	2.74	29.7	2.81	L_Shoulder
L_Forearm	2.74	29.7	2.81	4.18	3.88	54.1	L_Upper Arm
L_Hand	4.18	3.88	54.1	4.88	3.99	-13.0	l_Forearm
R_Shoulder	2.58	62.1	6.41	-1.9	19,0	6.23	Thoaracic Division
R_Upper Arm	1.96	19.0	6.23	-2.9	14.7	2.91	R_Shoulder
R_Rorearm	2.94	14.7	2.91	-4.5	-41.2	56.0	R_Upperarm
R_Hand	4.54	41.2	56.0	-4.8	-36.9	28.8	R_Forearm

Table 3: The bone locations in a standing position.
L* - left
R** - Right

The bone locations in a standing position are summarized in Table 2 with a given "Parent bone" relationship.

The Blender also supports an "Inverse Kinematics" tool for enabling advanced animation and rendering. The "Bone Constrain" modifier was used to set the position of the last bone within the bone-chain, where other bones are positioned automatically. In order to connect the mesh to the skeleton, we applied the tool "Rotation and Scale" and then created a "Parent-Child relationship" between the mesh and the skeleton using the tool "With automatic weights". After the mentioned procedure, the prepared 3D body model was ready to perform posture animation for different poses.

3D body measurement analyses

The objective of the research work was to test the accuracies and reliabilities of the 3D body models in both standing and sitting positions, as obtained by the scanning system GOM Atos II 400. In addition, we also tested the accuracy of a generalized adaptive 3D body

model, developed from the reconstructed 3D body model in a standing position. An investigation of anthropometric measurements was carried out in order to test the 3D body model's anthropometry in comparison with the real human's body measurements. Great importance was placed upon the adaptive 3D body model and its anthropometric behavior during posture adaptation. Therefore, we investigated 11 body measurements, which are defined in the standard ISO 8559 for garment construction and anthropometric surveys [32]. The definitions and locations of the body measurements are described in Table 3.

Traditional body measuring and virtual extraction of body measurements were adequately performed by taking important body landmarks into consideration. After measuring the human body in both standing and sitting postures, we performed three-dimensional body measurements' analysis, using the Rhinoceros 3D software program. Rhinoceros is 3D NURBS (Non-Uniform Rational B-Splines) curve, surface and solid modeler for Windows. It represents the primary modeling tool for designers of free-form physical shapes whose primary

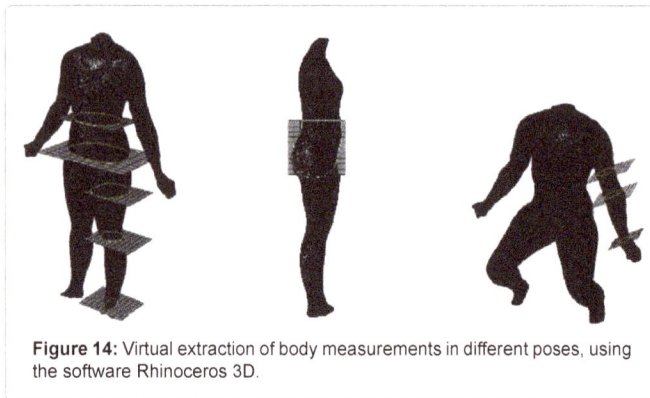

Figure 14: Virtual extraction of body measurements in different poses, using the software Rhinoceros 3D.

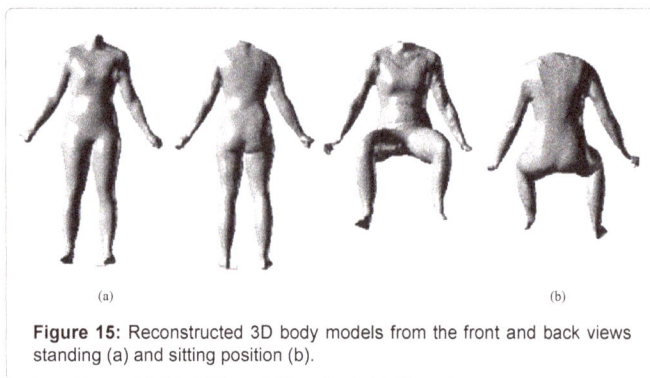

(a)   (b)

Figure 15: Reconstructed 3D body models from the front and back views standing (a) and sitting position (b).

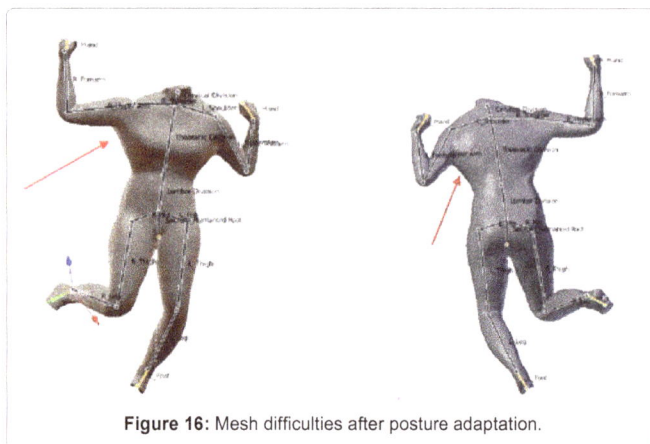

Figure 16: Mesh difficulties after posture adaptation.

goal is to design a human experience. It offers virtual measuring tools that provide accurate information about the objects [33]. In our study we used different analysis commands like the "Length" command for measuring different circumferences and lengths/distances of a 3D body model. By creating several rectangular planar surfaces at landmark locations of the body, we intersected meshes in order to extract the 3D body circumferences. Figure 14 shows the 3D body models, imported in the form of .stl files into the modeler for extracting the body measurements.

Results and Discussion

For achieving applicable 3D body models in both standing and sitting positions, we performed several activities within the course of this research: digitizing the surface of a tested person's body, using the optical scanning system GOM Atos II 3D, accurate surface

reconstruction procedures for obtaining an optimal surface description of a scanned body and developing a generalized adaptive 3D body model for the purpose of posture adaptation.

3D body scanning and reconstructions of body models

The scanning procedure was performed to digitize the entire body for each particular position. Difficulties did occur at overlapped regions which were difficult to capture, for example at regions belong to the armpits, knees and crotch. Therefore, careful surface processing was required for improving the surface descriptions of the meshes, which refers to section 4.2. Figure 15 shows the resulted 3D body models, scanned in both standing and sitting positions, imported into the program Rhinoceros 3D to determine 3D body measurements.

Design of a kinematic skeleton

For the purpose of this study we developed a kinematic skeleton consisting of 20 bones and 15 joints. The skeleton can be used for creating the digital 3D body models representing those people with limited body abilities. It was firstly evident that the mesh did not deform adequately and proportionally. Although each vertex group of the mesh were attached to the skeleton bones previously, difficulties regarding mesh stretching appeared, for example, at the vertex groups between the upper arms and the torso area, Figure 16. Blender's "Weight modifier" was used to increase the bone influence on the corresponding mesh-vertices and finally their influence on body measurements during posture adaptation

We altered the effect of the bones on vertices using a method called "Weight Painting". This method sustains a large amount of weight information in a very intuitive way. It was also used for rigging the meshes, where the vertex groups define the relative bone influence on the mesh. The selected mesh is shown slightly shaded with a rainbow color spectrum, where each color visualizes the weights associated to each vertex within the active vertex group. The weight is literally painted on the mesh surface by using a set of "Weight brushes", where the weights are visualized by a cold-hot color system [34]. By marking a bone in the "Weight modifier", the influence of a bone on the corresponding vertices can be concluded through the color-gradients. Areas with low influence are displayed in blue colors at cold areas, with weights close to the rate 0.0. Areas with high influence are displayed in red colors as hot areas, with weights up to the rate 1.0. For example, the bone "Sacrum_HumanoidRoot" in default at the crotch area exhibits low influence on the corresponding vertices with a blue-green color-spectrum and weights' ratings from 0.1 to 0.3, Figures 17a and 17b show an average influence of the left forearm on the vertices at the elbow area. A green color-spectrum depicts weights ranging up to 0.5, while a red color-spectrum shows high influence on the vertices of the left forearm. Vertices of the mesh at the torso area depict also a blue-green color-spectrum, when marking the "Thoracic Division" bone to display its influence on the correspondent vertices (Figure 17c).

In order to improve an adequate mesh deformation we chose different weights of the vertices to affect them regarding a particular bone. Their influence ranged from 0.5 to over 0.8 to avoid the robustness of mesh-deforming and the resulting difficulties of mesh-stretching and its sharp bending. Figure 18a shows the difficulties of mesh stretching at the breast circumference. Therefore, we altered the effect of the "Thoracic Division" bone on the related vertex group by using a weight influence of 0.8 and acquired a slight mesh deformation in Figure 18b. The same procedure remained ineffective for the mesh at the right side shown in Figure 18c. Namely, the mesh in this area

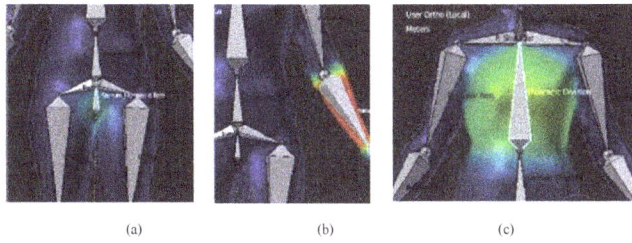

Figure 17: Color-spectra of different vertex groups for displaying the bone influence. Color-spectra of different vertex groups for displaying the bone influence.

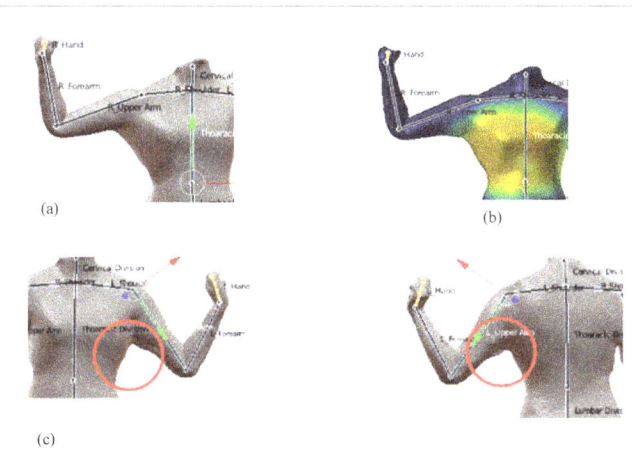

Figure 18: Using Blender's "Weight modifier" tool to improve adequate mesh deformation at The armpits.

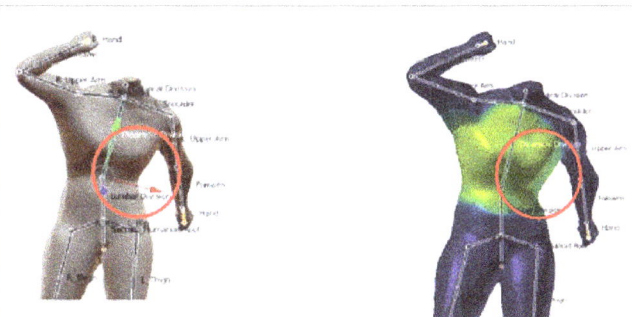

Figure 19: Using Blender's "Weight modifier" tool to improve adequate mesh deformation at the waist circumference.

was not credible for the real surface of the body, caused by the complex unification of individual scans into a global co-ordinate system and further incomplete surface reconstruction in this area.

Towards a generalized adaptive 3D body model

During the course of this research we developed a generalized adaptive 3D body model based on the previously designed kinematic skeleton. The purpose of an adaptive 3D body model is to enable a reliable and fast adaptation to the body measurements and posture specificities of real persons – people with limited body abilities. For this reason, some further adaptations of the developed body model were needed.

In order to avoid sharp bending of the mesh at the waist

circumference, we increased the influence of the "Thoracic Division" and "Lumbar Division" bones to 0.5 weight in order to create a gently mesh-deformation after adapting the 3D body model to different positions (Figure 19).

Figure 20 displays the altered effect of the "Left Thigh" bone on the related vertex group at this area. We used weight influences ranging from 0.6 to 0.8 to create a smooth mesh-deformation and consequently to improve the thigh circumference measurement

Analysis of the differences of body measurements caused by posture change

One of the goals of this study was to test the 3D body model's accuracy regarding both standing and sitting positions, and the suitability of the developed 3D body model the posture adaptation of which had been improved by upgrading the kinematic skeleton structure inside the watertight mesh of a 3D body model in a standing position. The adapted 3D body model in a sitting position, which is equal to the sitting posture of a tested person during the scanning procedure, is shown in Figure 21.

We compared the body measurements of a real person with the corresponding 3D digitized body model and determined their differences in square centimeters and percentages. In general, the 3D

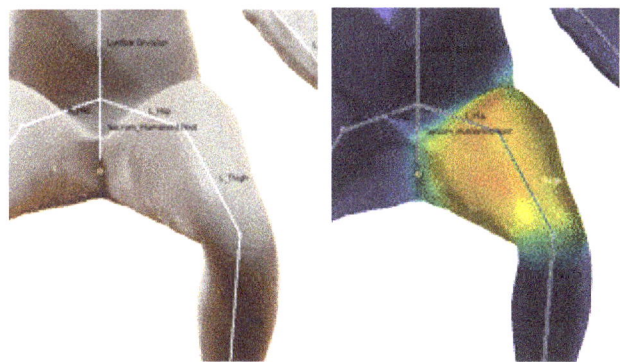

Figure 20: Using Blender's "Weight modifier" tool to improve adequate mesh deformation at the thigh circumferences.

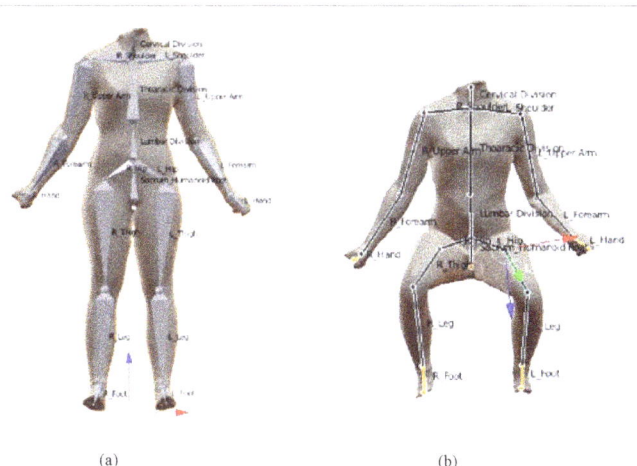

Figure 21: 3D body model posture adaptation in a standing position with an octahedral skeleton presentation (a) and adaptation to a sitting position with a stick representation of a skeleton (b).

Horizontal measurements	Body measurements' definitions
Breast circumference	The maximum horizontal circumference measured during normal breathing with the subject standing upright and the tape-measure passed over the shoulder blades (scapulae), under the armpits (axillae), and across the breast nipples.
Under breast circumference	The horizontal girth of the body just below the breasts.
Waist circumference	The circumference of the natural waistline between the top of the hip bones (iliac crests) and the lower ribs, measured with the subject breathing normally and standing upright
Hips circumference	The horizontal circumference measured round the buttocks at the level of the greatest lateral trochanteric projections, with the subject standing upright.
Thigh circumference	The horizontal circumference measured without constriction, at the highest thigh position, with the subject standing upright.
Knee circumference	The circumference of the knee measured with the subject standing upright and the upper border of the tape-measure at the tibia level.
Calf circumference	The maximum circumference of the calf measured with the subject standing upright, legs slightly apart, and with the body mass equally distributed on both legs.
Upper arm circumference	The maximum circumference of the upper arm at lowest scye level, measured with the subject standing upright with arms hanging naturally.
Vertical measurements	Body measurements' definitions
Front waist length	The distance from the neck shoulder point, over the nipple, then vertically straight to the front waist.
Cervical to breast point	The distance from the 7th cervical vertebra, round the base of the neck, to the nipple.
Total crotch length	The distance, measured using the tape-measure, from the center of the natural waist level at the front of the body, over the crotch, to the center of the waist level.
Back waist length (Cervical to waist)	The distance, measured using the tape-measure, from the 7th cervical vertebra, following the contour of the spinal column, to the waist.
Arm length /shoulder to wrist)	The distance, measured using the tape-measure, from the armscye/shoulder line intersection (acromion) to the elbow
Outside leg length	The distance from the waist to the ground measured using the tape-measure following the contour of the hip, then vertically down.

Table 4: Body measurements' definitions and measuring locations [32].

Body measures	Test person/ standing position	3D body model/ standing position	Difference between body measurements (cm)	Percentage Difference (%)
Breasts circ.	88.50	88.78	0.28	0.24
Under breasts circ.	79.80	80.28	0.48	0.38
Waist circ.	73.20	73.51	0.31	0.26
Hips circ. above thigh	102.20	103.38	1.18	1.21
Thigh circ.	61.30	61.62	0.32	0.19
Knee circ.	41.30	41.65	0.35	0.14
Calf circ.	39.29	38.78	-0.51	-0.20
Upper arm circ.	27.76	29.20	1.44	0.4
Front waist Length	46.50	46.95	0.45	0.21
Cervical to Breast point	26.50	26.68	0.18	0.05
Total crotch length	72.80	73.22	0.42	0.31
Back waist length	41.76	41.95	0.19	0.08
Arm length	57.00	57.60	0.60	0.35
Outside leg length	103.80	104.10	0.30	0.31

Table 5: Differences in body measurements at a standing position.

body model exhibited primarily increased circumferences and body-lengths. The measured differences between standing posture bodies (Table 4) ranged from -0.2% to 1.21%. Hips' circumferences showed the greatest difference, namely 1.18cm (1.21%), while calf circumferences showed the lowest differences, decreasing for 0.51cm (-0.2%) due to difficult processes of alignment for individually observed meshes into one independent mesh when observing 3D body meshes using the GOM Atos optical scanning system. These were followed by breasts, under the breasts, and waist circumferences which showed differences of not even 0.5cm. For body lengths we obtained the best result for the length at cervical to breast point, with a difference for not even 0.2cm, followed by back waist length with a difference of 0.19cm and outside leg length with a difference of 0.3cm.

Three-dimensional body measurement differences of a 3D body

model in a sitting position shown in Table 5 is ranged from -0.43% to 2.03%. Even in this case, hips-circumference showed the greatest difference of 1.89cm (2.03%), while the calf circumference decreased by 1.1cm (-0.43%), as a result of aligning individual scans into one global coordinate system. We obtained slightly increased results for breast and waist circumference than in the previous case but without 0.5cm to achieve. Total crotch length differed by 1.17cm (0.85%) due to difficult scanning of overlapped regions and further surface reconstruction. Back waist length showed the best result, namely a difference of 0.21cm, followed by outside leg length with a difference of 0.21cm when comparing the two types of measurements.

The differences in the 3D skeletal body model's measurements (Table 6), ranged from -0.73% to 0.78%. It was evident that the waist circumference had decreased by 0.99cm (0.73%) while the hip

Body measurements	Test person/sitting position	3D body model/ sitting position	3D adaptive body model	Differences between body measurements (cm)		Percentage difference (%)	
		1.	2.	1.	2.	1.	2.
Breasts circ.	88.2	88.57	88.72	0.37	0.52	0.33	0.46
Under breasts circ.	80.56	80.71	80.3	0.15	-0.24	0.12	-0.19
Waist circ.	74.5	75.10	73.51	0.60	-0.99	0.45	-0.73
Hips circ. above thigh	107.49	109.38	108.22	1.89	0.73	2.03	0.78
Thigh circ.	60.20	60.40	60.54	0.20	0.34	0.12	0.2
Knee circ.	40.40	42.20	41.84	1.8	1.44	0.72	0.58
Calf circ.	39.29	38.2	38.78	-1.1	-0.51	-0.43	-0.2
Upper arm circ.	27.76	28.75	28.63	0.9	0.87	0.27	0.24
Front waist Length	45.2	45.57	46.50	0.37	1.3	0.17	0.58
Cervical to Breast point	26.34	26.67	26.6	0.33	0.26	0.08	0.06
Total crotch length	73.33	74.50	73.22	1.17	-0.11	0.85	-0.08
Back waist length	41.26	41.47	41.64	0.21	0.38	0.08	0.16
Arm length	57.00	57.43	57.60	0.43	0.6	0.25	0.35
Outside leg length	104.48	104.70	104.28	0.22	-0.2	0.23	-0.21

Table 6: Differences in body measurements in a sitting position.

Body measurements	TP	SD	3D BM$_A$	SD	F_c	F_t	$H_0: \sigma_1^2 = \sigma_2^2$	t_c	t_t	S.D.
Breasts circ.	88.2	2.39	88.72	1.51	1.58	6.451	Confirmed	0.064	2.878	No
Under breasts circ.	80.56	0.04	80.3	0.06	1.5	6.451	Confirmed	2.26	2.878	No
Waist circ.	74.5	0.438	73.51	0.0297	14.7	6.451	Rejected	0.05	2.878	No
Hips circ. above thigh	107.49	0.287	108.22	0.35	1.21	6.451	Confirmed	7.00	2.878	Yes
Thigh circ.	60.20	0.074	60.54	0.18	2.4	6.451	Confirmed	2.23	2.878	No
Knee circ.	40.40	0.11	41.84	0.174	1.58	6.451	Confirmed	0.96	2.878	No
Calf circ.	39.29	0.019	38.78	0.206	10.8	6.451	Rejected	0.2	2.878	No
Upper arm circ.	27.76	0.085	28.63	0.015	5.6	6.451	Confirmed	1.77	2.878	No
Front waist Length	45.2	0.033	46.50	0.067	2.0	6.451	Confirmed	1.87	2.878	No
Cervical to Breast point	26.34	0.057	26.6	0.085	1.49	6.451	Confirmed	0.85	2.878	No
Total crotch length	73.33	0.052	73.22	0.015	3.5	6.451	Confirmed	1.18	2.878	No
Back waist length	41.26	0.033	41.64	0.012	2.75	6.451	Confirmed	2.08	2.878	No
Arm length	57.00	0.028	57.60	0.0179	4.36	6.451	Confirmed	2.02	2.878	No
Outside leg length	104.48	0.31	104.28	0.038	8.15	6.451	Rejected	1.67	2.878	No

Table 7: Descriptive statistics to determine significant differences between the observed measurements.
TP: test person
3D BM$_A$: 3D adaptive body model
SD: Standard deviation
Fc: F-value calculated
Ft: F- tabular value
σ_1^2, σ_2^2 : sums of square deviations of the two groups of measurements
$H_0: \sigma_1^2 = \sigma_2^2$: Hypothesis
t_c : t-values calculated
t_t : t-tabular value
S.D.: Significant difference

circumference had increased by about 0.73cm (0.78%), and thus confirmed a better result than in the previous case. The obtained measurements of the breasts' circumferences exhibited a slightly higher result of 0.52cm (0.46%), while the results for under the breast, knee and upper arm circumferences confirmed an improvement in the 3D skeletal model's measurements in comparison with traditional ones. The body length at the cervical to the breast point increased by just 0.26cm (0.06%) and confirmed a better result than in the previous case. A small decreased difference appeared for the total crotch length, namely 0.11cm (0.08%), which confirmed a better result as well. Outside leg length decreased for 0.2cm which is comparable to previous result of a 3D body model in a sitting position. We obtained also slightly increased differences for back waist length, namely a difference of 0.38cm, followed by arm length with a difference of 0.6cm, which is slightly higher as in previous case.

In order to test the suitability of the generalized adaptive 3D body model developed from the reconstructed 3D body model in a standing position, descriptive statistics of real human's measurements in comparison with 3D body model's measurements were analyzed to determine significant differences between the observed measurements by F- and t- test (Table 7).

Analysis of the difference between real human's measurements in comparison with 3D body model's measurements using tests, such as F-test and t-test, revealed a significant difference in hips circumference above thigh only.

It can be concluded that the improved body measurements of a generalized adaptive 3D body model resulted from the upgrading its kinematic skeleton and the use of virtual mesh-deformation modeling tools like Blender's "Mesh modifier" tool. It is possible to adapt the

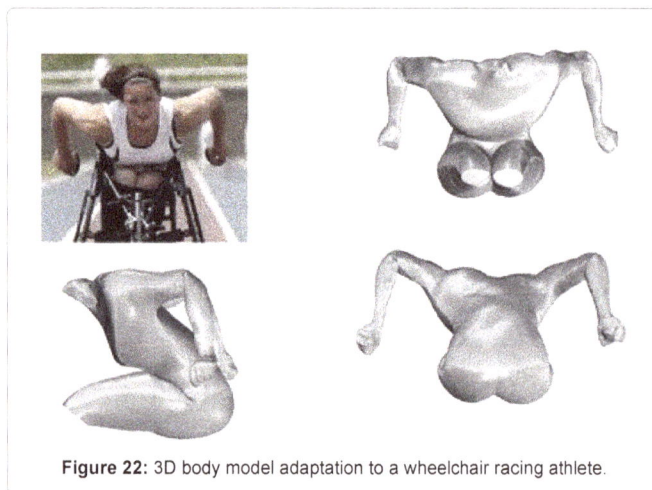

Figure 22: 3D body model adaptation to a wheelchair racing athlete.

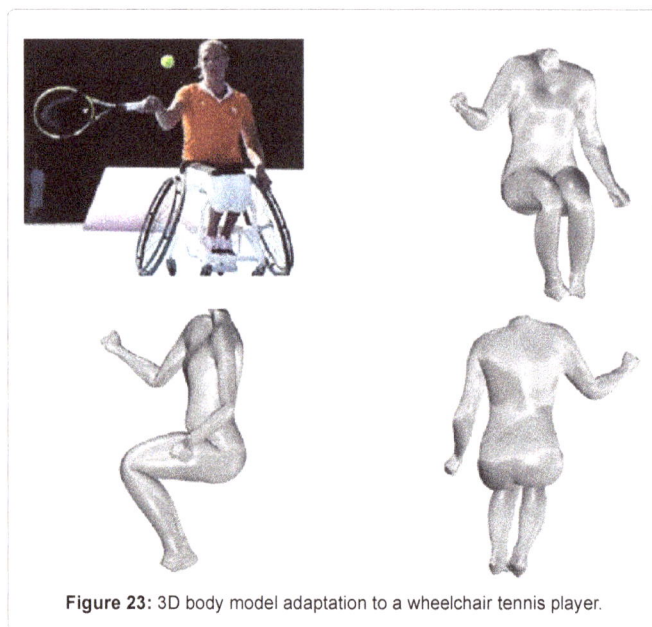

Figure 23: 3D body model adaptation to a wheelchair tennis player.

observed 3D body model to different positions while sustaining body measurements as in real-life.

It is well-known that people with limited body abilities achieve greater results in sports and in other social activities. In sport especially they compete professionally at the highest levels at international competitions supported by organizations like the International Wheelchair and Amputee Sports Federation (IWAS) in the United Kingdom [35] or the Association for disability sports in Slovenia [36]. In order to provide comfortable and aesthetic sportswear for this group of people of the world's population, there is a demand for supplying garments virtually prototyped using a general 3D body model for non-standard body shape characteristics, adaptable for different bodies' positions?

Further, the efficiency of mesh-deforming was determined after adapting a 3D body model to images of people with limited body abilities occupying positions while playing sports games. Figure 22 shows an adapted 3D body model to a female athlete with amputated legs during wheelchair racing, which represents one of the more exciting Paralympics disciplines [37]. The 3D body model was completely

adapted to the desired position. Incorrect mesh-deforming occurred in the elbow area and could not be corrected with the "Mesh modifier" tool.

Figure 23 shows a female player during wheelchair tennis at the Paralympics Games of the year 2012 [38]. Again, the 3D body model was completely adapted to the desired position but again with incorrect mesh-deformation occurring in the elbow area. These problems can be solved by manual correction procedures with the majority of 3D modelers.

Figures 24a and 24b show two additional examples of adapting the generalized 3D body model to different positions, for example to the paraplegic female basketball player [39], (Figure 24a) or even to the athlete with an amputated leg [40], (Figure 24b).

It is evident, that in spite of carefully carried out scanning and mesh processing procedures, difficulties from incorrect mesh-deformations remained during posture adaptation. Some of them could be corrected with the "Mesh modifier" tool through altering the effect of the bones on the correspondence mesh vertices. Other difficulties remain less consistent with real humans' movements, especially in the elbow area, where the bone of the kinematic skeleton was imprecisely attached to the feature points. Additional adjustments to kinematic skeleton constructions at exact feature points' areas are therefore necessary.

In order to show the usefulness of the observed 3D body models, we performed garments' pattern designing using two different systems for this purpose, namely Marvelous designer 3 and Optitex PDS. The observed 3D body model in a standing position was imported into a form of .dae file (Collada file) and then into the Marvelous designer program for performing tights and tunic pattern designing (Figure 25a). The Optitex Pattern Design System was used to construct T-Shirt and trouser patterns. The generalized adaptive 3D body model was imported in the form of a .stl (Stereo Litography file) to simulate garments' appearances (Figure 25b).

It is evident that in spite of carefully carried out scanning and mesh processing procedures, difficulties of incorrect mesh-deformations remained during posture adaptations. Some of them could be corrected using the "Mesh modifier" tool through altering the effects of the bones on the correspondence mesh vertices. Other difficulties remain less consistent to real humans' movements, especially in the elbow area, where the bone of the kinematic skeleton was imprecisely attached to the feature points. Additional adjustments of a kinematic skeleton construction at exact feature points' areas are therefore necessary.

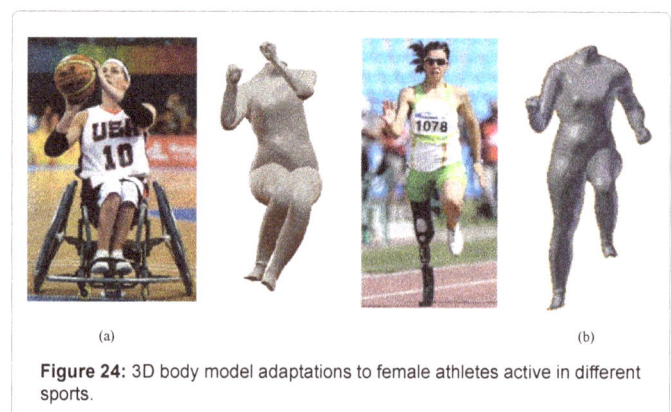

(a)   (b)

Figure 24: 3D body model adaptations to female athletes active in different sports.

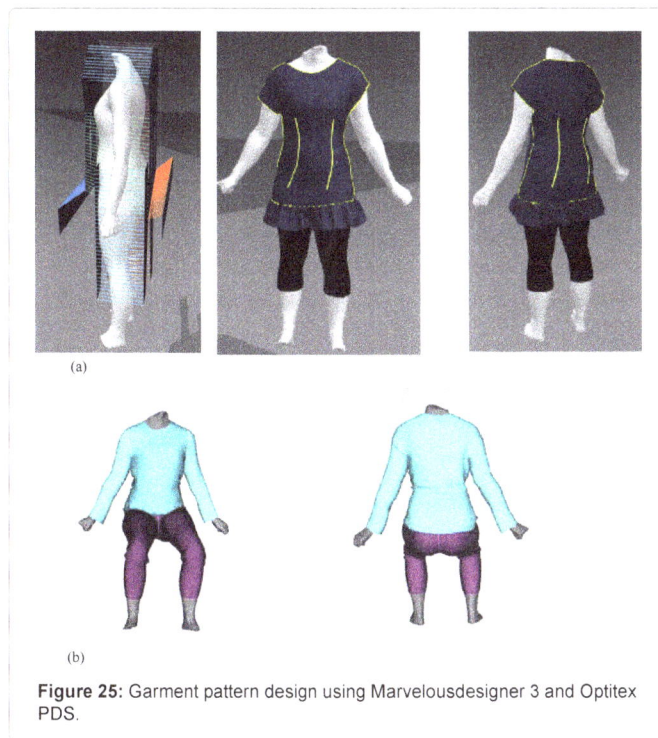

(a)

(b)

Figure 25: Garment pattern design using Marvelousdesigner 3 and Optitex PDS.

Conclusions

We have described an approach for developing a generalized adaptive 3D body model, suitable for people with limited body abilities, the positions of which could be adapted to different positions. An adequate mesh of a tested person's body in both standing and sitting positions was achieved using a general-purpose scanner GOM Atos II 400. Through pre-processing and post-processing procedures, the mesh was reconstructed using several graphic programs, such as GOM Inspect, MeshLab and Blender. Kinematic skeleton construction was carried out in order to provide posture adaptations to different positions. Further, incorrect mesh-deformations were improved through altering the effect of the bones on the corresponding mesh-vertices by using Blender's "Mesh modifier" tool. While the differences in body measurements between real persons and the adaptive 3D body model in a sitting posture confirmed the best results, consequently the "Mesh modifier" tool was ineffective when correcting mesh imperfections in the elbow areas, after adapting the 3D body model to images of people with limited body abilities. Due to the high densities of the observed polygonal meshed surfaces, their processing is difficult to approximate regarding real humans' behavior/movements. Future work will be focused on developing a meshed 3D body model, with lower densities of polygons in order to facilitate their processing for the purpose of posture adaptation. In the future, our research will focus on 3D body model parametrization to ensure 3D body model's fitting to body shapes, taken from the images. In this way, it will be possible to design appropriate parametric 3D body models, which will be digitized reflections of real persons with all individual characteristics required for virtual prototyping and virtual fitting of garments.

Within the course of this research we have developed a generalized adaptive 3D body model based on the previously designed kinematic skeleton. This developed generalized adaptive 3D body model will be used for a reliable and fast adaptation to the body measurements and posture specifics of real persons – people with limited body abilities.

Adapted body models, maximally personalized to mimic the real persons, will be imported into commercial CAD/PDS program packages and further tested for efficient garment prototyping and visualizations of two types of garments: special sports clothing and elegant dresses for people with limited body abilities. Both garment types are needed and desired by this special group of people. Our intention is to help them reach this goal.

References

1. Magnenat- Thalmann N, Thalmann D (2004) in "Handbook of Virtual Humans", John Wiley & Sons Ltd., The Atrium, Southern Gate, Chichester, UK.

2. Mpampa ML, Azriadis NP, Sapidis NS (2010) A new methodology for the development of sizing systems for the mass customization of garments. International Journal of Clothing science and Technology 22:49-68.

3. Lectra (2012)Smart design, Design room,Lectra, France.

4. Marvelous Designer (2012) Marvelous designer, Korea.

5. Gerber Technology (2012) Product design, Apparel retail, Gerbertechnology, USA.

6. OptiTex (2012) Apparel, Optitex,USA.

7. Pilar T (2012) Development of 3D prototypes of women's clothing. Master's Thesis, University of Maribor, Faculty of Mechanical Engineering.

8. Lectra (2014) Lectra 3D technology is the centrepiece of the end-to-end development process, Press Release, France.

9. United Nations (2012) What is disability and who are persons with disabilities, USA.

10. Gupta D (2011) Functional Clothing- Definition and Classification. Indian Journal of Fibre and Textile research 36: 321-326.

11. Seo H, Magnenat-Thalmann N (2004) An example- based approach to human body manipulation. Graphical Models 66: 1-23.

12. Zhengdong L, Shuyuan S (2010) Free-form deformation algorithm of human body model for garment. The 2010 International Conference on Computer Application and System Modelling 11: 602-605.

13. Allen B (2005) Learning body shape models from real-world data. Ph. D. Dissertation, University of Washington, USA.

14. Wang CCL (2005) Parameterization and parametric design of mannequins. Computer-Aided Design 37: 83-98.

15. Hilton A, Beresford D, Gentils T, Smith R, Sun W (1999) Virtual People: Capturing human models to populate virtual worlds. Proc.Computer Animation 174–185.

16. Lee W, Gu J, Magnenat-Thalmann N (2000) Generating animatable 3D virtual humans from photographs. Proc. Eurographics ,2000, Computer Graphics Forum 19: 1-10.

17. Bǎlan A (2010) Detailed human shape and Pose from images, Ph. D. Dissertation, Brown University, USA.

18. Kozar T, Rudolf A, Jevšnik S, Stjepanović Z (2012) Developing the accurate sitting position 3D body model for garment prototyping. Proceedings of 4th International Scientific Professional Conference Textile Science and Economy, Zrenjanin 133-138.

19. Rudolf A, Kozar T, Jevšnik S, Cupar A, Drstvenšek I (2013) Research on 3D body model in a sitting position obtained with different 3D scanners. Istanbul Technical University, Faculty of Textile Technologies and Design 6.

20. Rudolf A, Kozar T, Cupar A, Jevšnik S, Stjepanović Z (2013) Development of appropriate garment pattern designs for a sitting position 3D body model. The 10th Conference of Chemists, Technologists and Environmentalists of Republic of Srpska, University of Banja Luka, Faculty of technology 380.

21. Barbero BR, Ureta Santos E (2011) Comparative study of different digitization techniques and their accuracy. Computer-Aided Design 43: 188-206.

22. Brajlih T, Tasić T, Drstvenšek I, Valentan B, Hadžistević M, et al. (2011) Possibilities of Using Three-Dimensional Optical Scanning in Complex Geometrical Inspection. Strojniški vestnik - Journal of Mechanical Engineering 57: 826-833.

23. GOM (2014) Optical Measuring Techniques, GOM, USA.

24. Ma J (2011) Surface reconstruction from unorganized point cloud data via progressive local mesh matching. Ph. D .Dissertation, The University of Western Ontario, Canada.

25. GOM (2014) GOM Inspect software, GOM, USA.

26. MeshLab (2014) 3D – Coform project.

27. Kozar T, Rudolf A, Jevšnik S, Cupar A, Priniotakis G, et al. (2014) Accuracy evaluation of a sitting 3D body model for adaptive garment prototyping. Proceedings of AUTEX 2014 Conference, Bursa.

28. Petrak S (2007) The method of 3D garment construction and cutting pattern transformation models. Doctoral Dissertation, Faculty of Textile Technology, Zagreb.

29. H-Anim (2014) Humanoid animation ISO/IEC/FCD 19774:200x Specification.

30. Blender (2014) 3D animation suite, Blender, Netherlands.

31. Mazany O (2007) Articulated 3D human model and its animation for testing and learning algorithms of multi-camera systems. Maters's Thesis, Czech Technical University, Faculty of Electrical Engineering, Czech Republic.

32. International Standard (1989) Garment construction and anthropometric surveys- Body dimensions, International Standard ISO 8559 , Switzerland.

33. McNeel R (2002) Rhinoceros Version 3.0.: Nurbs modeling for Windows.

34. Weight Paint mode

35. Weight paint mode (2014) Meshses, Modelling manual.

36. IWAS (2014) International Wheelchair and Amputee Sports Federation, UK.

37. NSIOS (2014) Association for the disability sports, Slovenia.

38. Paralympic Disciplines (2014) Paralympic organization, Germany.

39. Paralympics (2012) Wheelchair tennis winner Esther Vergeer does it again, Germany.

40. Paralympic (2014) Taxonomy, London.

41. ABC radio national (2014) Bionic athlete, Body sphere, ABC radionational, Australia.

7

Antimicrobial Studies of Knitted Fabrics from Bamboo, Soybean and Flax Fibers at Various Blends

Muhammad Qamar Tusief[1,2]**, Nabeel Amin**[1]**, Nasir Mahmood**[2]**, Israr Ahmad**[2] **and Mudassar Abbas**[1]*

[1]*School of Textile and Design, University of Management and Technology, C-II Johar Town Lahore54770, Pakistan*
[2]*Department of Fiber and Textile Technology, University of Agriculture, Faisalabad, Pakistan*

Abstract

In the current study, the single jersey knitted fabric from natural fibers bamboo (*Dendrocalamus strictus*), soybean (*Glycine max*) and flax (*Linum usitatissimum*) at various blend ratios are prepared for comparison of physical strength and antimicrobial properties. For the general characteristic evaluations it is deduced that the fiber strength of pure flax is comparatively stronger than bamboo and soybean, whereas the antimicrobial properties of the bamboo fibers are the highest. Therefore, the blended fabric with multiple compositions for these three fibers is prepared and compared with the fabric made from individual pure fibers. In sum, the blended fabrics showed enhanced results for both antibacterial activity and strength, where the strength of flax/bamboo blended fabric with equal ratios (50/50) at higher twists and antibacterial activity of soybean/bamboo blended fabric with blend ratio 10/90 was found the best as compared to other combinations.

Keywords: Flax; Bamboo; Soybean; Fiber blend ratios; Antimicrobial activity; Fabric strength

Introduction

In recent years, the demands of textiles have changed with the development in technology and the raised living standards. Along-with style and durability, the clothing-comfort that includes thermo-physiological comfort. It is evident that fiber type, yarn properties, fabric structure, finishing treatments and clothing conditions are the main factors affecting the clothing comfort [1]. Over the last few years, there has been growing interest in knitted fabrics due to comfortable stretch, freedom of movement and good air-, and water vapor-permeability characteristics. Therefore, knitted fabrics are also preferred for sportswear, casual wear and all types of intimate apparel like hosiery, bathing suits, fit-in t-shirts and socks applications [2].

The development of bioactive textile materials and clothing loaded with antimicrobial properties has become the pioneer demand of the smart textiles. Wool, silk, and cellulosic material such as cotton, jute, and flax in contact with body provide ideal environment for growth and sustainability of pathogenic microbes [3]. The persistence of such microorganism in textile materials results in alteration of final structure of the fabric causing decoloration, fetid odor generation, skin infections, product deterioration and allergic responses. To avoid such counter effects, the textile goods loaded with antimicrobial agents, being the special type of medical textiles have got attention of consumers and manufacturers in all over the world [4]. Specifically, a broader market for anti-microbial fibers has been developed in outdoor textiles, air filters, automotive textiles and domestic home furnishings.

Antimicrobial properties can be achieved by the application of diversified chemicals and finishes but negative effects of such chemicals on skin and on overall health of the wearer cannot completely be ignored [5]. Hence, the use of natural fibers in fabric manufacturing as value addition is preferred in order to make the product more environmentally friendly.

Until lately, the bamboo is being consumed as a raw material for textiles mainly due to its renewability, biodegradability and carbon sequestering abilities [6]. Similarly, flax, due to similar properties as that of cotton and comparatively inexpensive in price has become the good substitute for garments with subsidized costs. The use of flax in blends helps in reducing the shrinkage, absorbing of dye, good wicking properties and shortening of drying times for the fabric [7]. In the same context, soybean fibers due to luster, similar to silk have excellent drape properties, are soft to touch, carry light weight and have good moisture transmission values. Additionally, the fiber contains the properties of anti-creasing, ease at washing and quick drying [8]. Keeping in mind the above highlighted specialties of bamboo, flax and soybean fibers, the present research study was conducted to compare the fiber strength and antimicrobial properties of knitted fabric composed of three fibers individually and at a blend ratio of 10-50% with respect to each other.

Materials and Methods

The proposed research was initiated with evaluation of fiber characteristics of bamboo (*Dendrocalamus strictus*), soybean (*Glycine max*) and flax (*Linum usitatissimum*), fibers to assess their potential according to International standards of ASTM D-5867 [9]. The physical characteristics of bamboo like staple length, fiber bundle strength, denier and elongation to break are comparable to soybean. However, the soybean protein fiber due to small fineness of single fiber has low specific gravity and tensile elongation. The moisture absorption and discharge activity, permeability properties, heat-retaining ability and spinning performance of soybean is also intermediate [10]. The physical strength of flax fiber is higher and the moisture absorbent properties for flax are lesser in this comparison. Therefore, the pre-treatment process is essentially conducted for flax in order to remove the hydrophobic impurities and to increase the water absorption of the fibers [11].

Spinning process

For the present research the blending of Bamboo, Soybean and

*Corresponding author: Mudassar Abbas, School of Textile and Design, University of Management and Technology, C-II Johar Town Lahore54770, Pakistan
E-mail: mudassirabbas@yahoo.com

Flax fibers and the manufacturing of blended yarn as per the decided variables were made in a small spinning laboratory that contains unique miniature textile machinery. This facility provides a quick approach for evaluation of the fibers and their blends. The flax, bamboo and soybean fibers were blended and processed at mini spinning system for making 30S yarn samples. Control specimens were composed of 100 percent material. Following variables were trialed in the research work, keeping hundred percent flax, soybean and bamboo as under control treatment (F4, F5 and F6) and spun into yarn of 30S at different levels as given in Tables 1 and 2.

Knitting process

Knitting of the fabric from the yarns was made according to the selected variables was accomplished in a local textile mill. The knitting was processed in the machine of 33/4 inch gauge, 2 feeders, 33/4 inch cylinder diameter and dial height of 4 inch was engaged for the preparation of knitted fabric samples from all the yarn samples of different blends.

Knitted fabric evaluation

The knitted fabric samples were placed on the flat surface for 24 hours at 65 ± 2% relative humidity and 20 ± 2% temperature for conditioning purpose to maintain the characteristics of the samples at standard test methods.

Fabric strength

Bursting strength is the force in kilopascals (kPa) was determined according to the ASTM standard method **D-3786** [12]. Mullen Burst Strength Tester was used for measuring bursting strength.

Antimicrobial activity (%)

Antimicrobial activity of the fabric samples was estimated using AATCC standard method **AATCC 147-2011** [13]. According to this disc diffusion method with some minor modification was used for screening the fabric samples for antimicrobial activity. The dispensed, sterilized and autoclaved nutrient agar was poured into flat bottomed Petri dish. Then, nutrient agar was allowed to gel firmly before inoculating. The nutrient agar plates were inoculated with 0.1 ml of an appropriate dilution of tested culture. The fabric samples of 1 cm diameter were placed on the surface of inoculated plates. The plates were incubated at appropriate temperature for 24 hours. Then the diameter of inhibition zone (mm) including the disc diameter was measured for each treatment and the antibacterial activity against gram positive bacteria *Staphylococcus aureus* was calculated and microorganism inhibition is reported as relative percentage.

Statistical evaluation

The results were statistically analyzed by factorial design using Statistical Package for Social Sciences (SPSS) as suggested by Montgomery [14]. The factorial design was preferably selected because the interaction effects can only be studied in factorial experiments. Moreover, all factors involved in this study are studied at the same of level of precision. The small letters a-e used in Tables 3 and 4, reflect the significance or non-significance difference among the means at α = 0.05. The same letters mean non-significant difference among the mean values at 0.05 level of probability.

Results and Discussion

The comparative analysis of three fibers concludes that bamboo fiber has excellent antimicrobial properties that make it ideal for processing into textile. The bactericide characteristics is mainly due to presence of a bio-agent "bamboo kun" in the fiber. In contrast, the staple length, fiber bundle strength and denier for flax fiber are higher as compared to bamboo and soybean [15]. The effects of different blend ratios of Bamboo, Soybean and Flax fibers at various twist levels on the strength and antimicrobial activity of the knitted fabric were calculated in order to investigate the performance of the fabric with special reference to its strength and Antibacterial activity [16].

Fabric bursting strength

The statistical comparison of individual means for Fabric bursting strength measured in kPa is presented in Table 3, The statistical distribution clearly indicates that the maximum strength was attained for F2 (Flax/Bamboo fabric) at B5 (50/50) blend ratio having high twist level T3 (4.2) with the mean values 409.82, 438.92 and 440.78 kPa. The results revealed that the use of Flax fiber in the blend improved the strength of the fabric, whereas the share of bamboo fiber if increased in the blend results for the decrease in fabric strength. The studies confirmed that the bursting strength of the fabric reduces with increasing bamboo content in the blend. The elongation at break of the bamboo fiber is lower which may result in decreasing elongation of yarn samples with increasing bamboo content in the blend and this could be the reason for lower bursting strength of bamboo rich fabrics. It may also be attributed to the lower strength of bamboo fibers [17]. These findings also get support from the results of hundred percent pure bamboo, soybean and flax knitted fabric as given in Table 4 where the strength of the fabrics made from pure bamboo and soybean fibers were lower than that of the fabric made from pure flax fiber.

The fabric bursting of variously blended fabrics is presented in Figure 1. The results clearly present the fabric strength at different blend ratio and twist count. The regular pattern was observed for the fabric strength as the twist count and / or the ratio of flax fiber was increased

Fiber Types	Staple Length (mm)	Fiber Bundle Strength (g/tex)	Fiber Fineness (dtex)	Elongation (%)	Moisture Regain (%)
Bamboo	38	31.41	1.5	23.80	13.03
Soybean	38	26.04	1.80	18.40	8.60
Flax	40.83	43.4	2.5	5.6	7

Table 1: Physical characteristics of Bamboo, Soybean and Flax fibers.

Fiber Type (F)	Blend Ratios (B)	Twist Multiplier (T)
F_1 = Flax/Soybean	B_1 = 10:90	T_1 = 3.8
F_2 = Flax/Bamboo	B_2 = 20:80	T_2 = 4.0
F_3 = Soybean/Bamboo	B_3 = 30:70	T_3 = 4.2
F_4 = 100% Bamboo	B_4 = 40:60	
F_5 = 100% Soybean	B_5 = 50:50	
F_6 = 100% Flax		

Table 2: Variables selected for the study.

Fiber Type	Blend Ratio	Twist Multiplier
F_1 = 399.48a	B_1 = 416.72e	T_1 = 414.10c
F_2 = 409.82c	B_2 = 421.34d	T_2 = 427.88b
F_3 = 378.04b	B_3 = 428.23c	T_3 = 440.78a
	B_4 = 432.78b	
	B_5 = 438.92a	

Table 3: Comparison of individual treatments means for Fabric Strength (kPa).

in the fabric. Therefore, minimum fabric bursting strength was observed for soybean/bamboo with the ratio of 10/90 at 3.8 yarn twist multiplier, whereas maximum strength was attained for Flax/Bamboo fabric at 50/50 blend ratio having high twist level of 4.2. The results revealed that the use of Flax fiber in the blend improved the strength of the fabric and the share of bamboo fiber, if increased in the blend, decreases the strength of the fabric. The results were in accordance with the reported data in literature [18]. The elongation at break of the bamboo fiber is lower which may result in decreasing elongation of yarn samples with increasing bamboo content in the blend and hence can be the reason for lower bursting strength for bamboo rich fabrics. Finally, the fabric bursting strength tests for pure bamboo, soybean and flax knitted fabric with yarn twist fixed at 4 were evaluated. The strength of fabrics from 100% bamboo and soybean fibers were lower than that of the fabric made from pure flax fiber. Twist multiplier factor plays a significant role on the strength of the fabric. The increase in twist results in increase of fabric strength (Figure 1).

Fabric antimicrobial activity

The antibacterial activity for blended fabrics is presented is in Figure 2, which indicates that the maximum antimicrobial activity was achieved from soybean/Bamboo blended fabric at the blend ratio of 10/90. It is clear from these results that both soybean and bamboo fibers put a decisive impact on the antimicrobial activity of the fabric. More influentially, the bamboos share in the blend, the more it plays a significant role to improve the antibacterial properties of the fabric. These findings were verified by the exploration of previous studies which revealed that the soybean protein fabric had good capacity to retain warmth, had good biocompatibility and were also beneficial to human health [19]. Furthermore, it was evaluated that twist yarn count does not contribute significantly for the variations in results against microorganism growth.

From the results in Table 4, the antimicrobial activity made from pure bamboo, soybean and flax fibers, it was evident from the results that bamboo fibers improve the antibacterial activity of the fabric.

Fiber Type	Fabric bursting strength (kPa)	Fabric Antimicrobial activity (%)
F_4	403.6	78
F_5	383.7	59
F_6	446.62	29

Table 4: Fabric bursting strength and antimicrobial activity of 100 % pure yarn fabric.

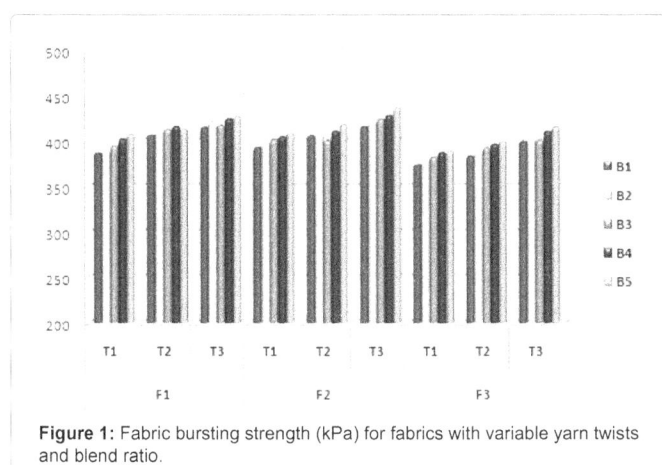

Figure 1: Fabric bursting strength (kPa) for fabrics with variable yarn twists and blend ratio.

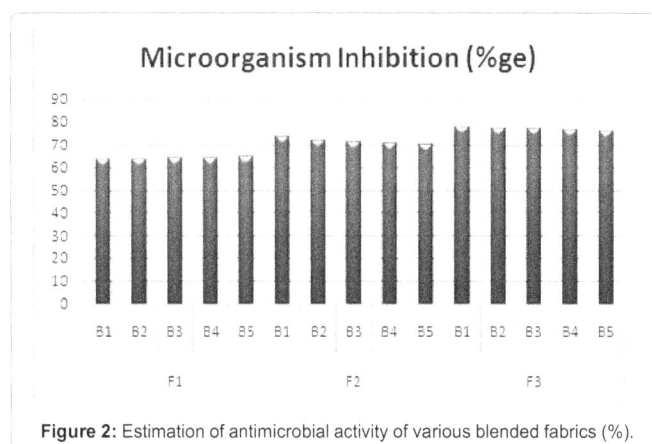

Figure 2: Estimation of antimicrobial activity of various blended fabrics (%).

Conclusions

In the light of above observations, it can be deduced that

- The use of Bamboo fibers in the blend increases the antimicrobial activity of the knitted fabric significantly.

- In addition to improve the antibacterial properties of the fabric the use of bamboo/ soybean and flax fiber in blends had good impact on the strength of the knitted fabric.

- The natural fibers can be the better choice in order to create antibacterial properties in the fabric along-with improved strength and can be good substitute for the chemical or biofinishes.

References

1. Sampath M, Aruputharaj A, Mani S, Nalanki G (2012) Analysis of thermal comfort characteristics of moisture management finished knitted fabrics made from different yarns. J Indus Text 42: 19-33.

2. Firgo H, Suchomel F, Burrow T (2006) Tencel High Performance Sportswear. Lenzinger Bericht 84: 44-50.

3. Gao Y, Cranston R (2008) Recent advances in antimicrobial treatments of textiles. Text Res J 78: 68-72.

4. Kramer A, Guggenbichler P, Heldt P, Jünger M, Ladwing A, et al. (2006) Hygienic Relevance and Risk Assessment of Antimicrobial-Impregnated Textiles. Skin and Biofunctional Textiles. Current Problems in Dermatology 33: 78-109.

5. Haug S, Rolla A, Schmid-Grendelmeier P, Johansem P, Wüthrich B, et al. (2006) Coated Textiles in the Treatment of Atopic Dermatitis. Skin and Biofunctional Textiles. Currents Problems of Dermatology 33: 144-151.

6. Waite M (2009) Sustainable Textiles: the role of bamboo and a comparison of bamboo textile properties. J Text Apparel Technology and Management 6: 1-22.

7. Rodie JB (2011) Flax Unshackled.

8. Zhao Q, Feng H, Wang L (2014) Dyeing properties and color fastness of cellulase -treated flax fabric with extractives from chestnut shell. Journal of Cleaner Production 80: 197- 203.

9. ASTM (2012) Standard Test Methods for Measurement of Physical Properties of Raw Cotton by Cotton Classification Instruments. ASTM Designation: D5867-12. American Society for Test and Materials, Philadelphia, U.S.A.

10. Yi-you L (2004) The soybean fiber- A healthy & comfortable fiber for the 21st century. Fibers and Textiles in Eastern Europe 12: 8-9.

11. Fakin D, Golob V, Kleinschek KS, Marechal AML (2006) Sorption properties of flax fibers depending on pretreatment processes and their environmental Impact. Text Res J 76: 448-454.

12. ASTM (2013) Standard Test Method for Bursting Strength of Textile Fabrics—Diaphragm Bursting Strength Tester Method. ASTM Designation: D3786. American Society for Test and Materials, Philadelphia, USA.

13. AATC (2011) Antibacterial Activity Assessment of Textile Materials: Parallel Streak Method. AATC Designation: 147-2011. American Association of Textile Chemists and Colorists, Research Triangle Park, N.C., USA.

14. Montgomery DC (2009) Design and analysis of experiments.

15. Gericke A, Van der Poll J (2011) A comparitive study of regenerated bamboo, cotton and viscose rayon fabrics Part 2: anti-microbial properties. Journal of Family Ecology and Consumer Science 39.

16. Kathirvela KP, Ramachandranb T (2014) Development of Antimicrobial Feminine Hygiene Products Using Bamboo and Aloevera Fibers. J Nat Fibers 11: 242-255.

17. Mahish SS, Patra AK, Thakur R (2012) Functional properties of bamboo/polyester blended knitted apparel fabrics. Ind J Fib Text Res 37: 231-237.

18. Arif M (2008) Studies on knitting performance under the interaction of twist, twist type and some technical variables of the knitting machine.

19. Askew PD (2009) Measuring activity in antimicrobial textiles. Chemistry Today 27: 16-20.

8

Impacts of Air Pollution on Colour Fading and Physical Properties of Wool Yarns Dyed with Some Natural Dyes in Residential Site

Helmy HM*[1], Shakour AA[2], Kamel MM[1] and Rashed SS[3]

[1]National research centre, Textile research division, Dokki, Giza, Egypt
[2]National research centre, Air pollution department, Dokki, Giza, Egypt
[3]The Islamic museum, Cairo, Egypt

Abstract

Wool yarns are dyed with natural colouring matter extracted from Cochineal, Turmeric and Madder using exhaustion method. These dyed yarns are pre-mordanted with different types of mordants. Different measurements have been carried out for the dyed wool yarns after exposure to air pollution in residential site (Kalla region) for one year. These measurements values include colour data (K/S, L*, a*, b*, ΔE, hue and chroma) and physical properties (tensile strength, tenacity and elongisty). Also, the present work studied the air pollution in (Kalla region), through determination of the suspended and deposited particulate matter and sulpher dioxide concentrations for different periods of time in one year.

Keywords: Wool Yarns; Cochineal; Turmeric; Madder; Mordants; Air Pollution; Colour Fading

Introduction

During a few last decades, revival and using of natural dyes has gained a great deal of attention and care. A renewed international interest has arisen in natural dyes due to remarkable increase of the environmental and health hazards associated with the awareness of synthesis, processing and use of synthetic dyes. Cochineal (red dye), Turmeric (yellow dye) and Madder (red dye) are extracted from (Daclylopius coccus) bug, (Curcuma longa. L, rotunda. L) plant and (Rubia tinctorum) plant respectively. When using these natural dyes for dyeing wool yarns, the light and washing fastness were very poor. Enhancement of fastness properties of wool yarns can be done by the help of mordants, usually metallic salts. Mordants have an affinity for both natural colouring matter and the fibre. After wool yarns being impregnated with such mordants, they were subjected to dyeing with different natural dyes (in case of pre-mordanting). In case of post mordanting method wool yarns were treated with different mordants after dyeing procedure. These metallic salts (mordants) after combining with dye in the fibre, they form an insoluble precipitate and thus both the dye and the mordant get fixed which improve light fastness to some extent [1-14]. Generally, concentrating residential and commercial activities without air quality management policy, led to complex mixtures of all types and sizes of uncontrollable air pollution sources.

Experimental

Materials

Natural colouring matter: Colouring substance used in this work was extracted from Cochineal as an animal source, Turmeric and Madder plants as a planted source.

Fabrics: Wool yarns were kindly supplied by El Mehalla spinning and weaving Company, Egypt.

Mordents: The following mordants were used: alum salt, copper salt, tin salt and iron salt. They were of pure grade chemicals.

Methods

Extraction of natural coloring matter: Cochineal dye (10 g/L) was immersed in water for 12 hours and then boiled for 15 minutes. Turmeric dye (powder form) was immersed in water with 15 g/L concentration for

12 hours and then boiled for 60 minutes. Madder dye (powder form) was immersed in water with 17 g/L concentration for 12 hours and then boiled for 60 minutes. At the end, the solution was filtered off and left to cool down [15-18].

Dyeing methods

Dyeing of wool yarns using traditional method: Wool yarns samples (10 gm each) were dyed with the dye extracted from Cochineal, Turmeric and Madder at liquor ratio 1:50. Dyeing was carried out at pH (4-5). Yarns samples were entered to the dyeing solution in a water bath at 70°C for 15 minutes. Then yarns were dyed for one hour (in case of Cochineal dye) and for 30 minutes (in case of Turmeric and Madder) and the dyed samples were rinsed with cold water and washed for 30 minutes in a bath containing 3 g/L of non-ionic detergent at 45°C. Finally, the yarns were rinsed and air dried [19-21].

Pre-mordanting of wool yarns: The pre–mordanting method was used in case of dyeing wool yarns with dyes extracted from Cochineal, Turmeric and Madder using alum salt, copper salt, and chromium salt. The mordant was dissolved and added to the bath with liquor ratio 1:50. Then wetted wool samples were added to the mordanting bath and the whole brought slowly to 90°C for one hour. It was then allowed to cool at room temperature and the wool samples were removed and squeezed. Mordanted wool should be used at once because some mordants are very sensitive to light [22].

Post-mordanting method was used in case of dyeing wool yarns with dyes extracted from Cochineal, Turmeric and Madder using iron salt and tin salt. In this method the mordant was added to the bath for the final ten minutes of simmering. The fabrics were lifted out while the

*Corresponding author: Hany M Helmy, National Research Centre, Textile Research Division, Dokki, Giza, Egypt
E-mail: hany_helmy2001@yahoo.com

dissolved salts are thoroughly mixed into the dye liquor. The wool yarns were interred into the bath and the dyeing process was continued for 30 minutes at 90°C, then rinsed and washed [23-25].

Preparation and exposure of samples to the ambient atmosphere: The dyed wool samples were placed over roofs of building in the investigated site. Wool samples were exposed for a period of one year. Five samples of each type were removed from each site after exposure of three months interval, and were taken off to the laboratory for the measurements. Unexposed samples were used as control.

Air pollution determination: The Area under investigation is residential site (Kalla region). It is located in the middle of Cairo city (residential area). It is characterized with heavy population and mixed activities commercial and residential beside the heavy traffic. The present investigation was undertaken to study the air pollution in residential site (Kalla region), through determination of the suspended and deposited particulate matter and sulpher dioxide concentrations.

Determination of deposited particulate matter: Deposition rate values for settled particulate matter were determined according to standard methods [26]. Dust fall collectors were used for collecting dust fall samples as previously used in Egypt [27]. The collectors consist of cylindrical glass beakers 17 cm in height and 8 to 9.5 cm diameter. The cylindrical glass beaker was half filled with distilled water to avoid re-entrainment of the collected dust and mounted on iron tripods at a height of 50 cm above roof level to avoid the collection of surface dust. Monthly collected samples were transferred quantitatively carefully to a dry, clean weighted beaker using successive washing with distilled water and a policeman until the inside of the jar became clean. Successive drying and weighing of the beaker was made until constant weight. The differences in weight represent the amount of deposit dust during the corresponding month at each site. Particulate deposition were calculated and expressed as gm/m².30 days.

Determination of suspended particulate matter: The filtration technique for collecting atmospheric suspended particulate matter [28]. Determination of sulpher dioxide: West and Gaeke method was used for the determination of SO_2 [26,28]. Air was aspirated (one liter / minute) through a glass bubbler sampler containing 50 ml of absorbing solution (0.1 M sodium tetrachloromercurate). Non-volatile dichlorosulfito mercurate ion was formed when the sulphur dioxide in the ambient air is absorbed in 0.1 M sodium tetrachloro mercurate. Addition of acid bleached pararosaniline and formaldehyde to the complex ion produces red-purple pararosaniline methyl sulphuric acid, which is determined spectrophotometrically at a wavelength of 560 mu.

Testing

Color measurements of the dyed fabrics: Colour-difference formula ΔE CIE (L*, a*, b*)

The total difference ΔE CIE (L*, a*, b*) was measured using the Hunter-Lab spectrophotometer (model: Hunter Lab DP-9000).

The total difference ΔE CIE (L*, a*, b*) between two colours each given in terms of L*, a*, b* is calculated from:

$$\Delta E^* = [(\Delta L^*)^2 + (\Delta a^*)^2 + (\Delta b^*)^2]^{1/2}$$

Where:

ΔE* value: is a measure of the perceived colour size of the colour difference between standard and sample and cannot indicate the nature of that difference.

ΔL* value: indicates any difference in lightness, (+) if sample is lighter than standard, (-) if darker.

Δa* and Δb* values: indicate the relative positions in CIELAB space of the sample and the standard, from which some indication of the nature of the difference can be seen.

ΔH= arktan b*/a* Where (ΔH) indicates any difference in hue of colour

and ΔC= (a²+b²)^{1/2} Where (ΔC) indicates any difference in chroma (saturation of colour)

Physical measurement: The physical properties (tensile strength-tenacity- elongisty) are measured for the mordanted dyed yarn samples with Cochineal, Turmeric and Madder using (Uster tensor apid) device.

Results and Discussion

Air pollution: Both the gaseous and particulate components of an atmospheric aerosol contribute to deterioration in air quality. Dust-fall samples are generally indication of atmospheric particulate concentration. Particles of size larger than 20 μm has appreciable settling velocities and relatively short atmospheric residence time. The annual mean rate of deposited particulate matter over Kalla region during the year of study was illustrated in Table 1.

Table 1 shows that Kalla region (residential site) is highly polluted with a suspended dust comparing to Helwan city (industrial area). While Helwan city is highly polluted with fine particles and SO_2 more than Kalla region. Also, Table 1 indicated that the annual mean rate of deposited dust was 16.93 g/m².month. According to Pennsylvania guidelines for dust-fall, these values are considered a heavy deposition rates.

Pennsylvania guidelines for dust-fall [29].

Class Dust-fall (g/m².month)

Slight--------------------- (0-7)

Moderate---------------- (7-14)

Heavy-------------------- (14-35)

Very Heavy ------------ (>35)

The annual average concentration of suspended particulate in residential site (Kalla region) atmosphere was 951.04 μg/m³ this concentration is about 19 times higher than the value of 50 μg/m³ concentration limit of US National Ambient Air Quality Standards [28], which is the same value recommended by UK. Expert Panel on Air Quality Standards [30]. It is also about 13 times higher than the maximum allowable concentration that is given by the Egyptian Environmental Law No. 4, 1994 (70 μg/m³) [31]. The reactive acids such as sulfuric acid cause the melting of wool especially at high temperatures and transformation it into amino acids and peptides. In addition wool fibres release sulfur gases.

It was noticed that concentrations of suspended particulate matter varied from one season to another during the period of the study and the maximum concentration recorded during autumn. The amounts of deposited and suspended dust were also affected by the location of the study "Kasr El-Jawahara Museum" in Kalla region as it is near to both Moqattam Mountain and the downtown. The percentage of the acidic particles resulting from the exhaust and the large particles coming from Moqattam Mountain was carried by the wind. Although there is a certain level of dust in the air at all times. The amount and

Season	rates of deposition (g/m².month)		Suspended dust concentration(μg/m³)		SO₂ (μg/m³)	
	residential site (Kalla region)	urban area (Helwan city)	residential site (Kalla region)	urban area (Helwan city)	Residential site (Kalla region)	urban area (Helwan city)
Winter	15.32	19.18	922.45	328.6	32.6	111.38
Spring	13.73	19.88	900.59	381.0	36	60.14
Summer	12.76	19.31	637.28	267.3	7.02	44.6
Autumn	11.92	18.90	1343.84	362.3	3.20	74.6
Annual mean	13.43	19.32	951.04	334.8	19.71	72.68

Table 1: Seasonal variation the rate of deposition of dust-fall (g/m².month), Suspended dust concentration (μg/m³) and SO₂ (μg/m³) over a residential site (Kalla region) and urban area (Helwan city).

type of dust varies considerably and depends on many factors including source, climate, wind direction, and traffic. Dust is generated from man-made and natural sources and may be made up of soil, pollen, volcanic emissions, vehicle exhaust, smoke or any other particles small enough to be suspended or carried by wind. The stronger the wind the larger the particles lifted and the more dust carried. Sulpher dioxide: SO_2 is a prominent anthropogenic pollutant and contributes to the formation of sulphuric acid, the formation of sulphate aerosols, and the deposition of sulphate and SO_2 at the ground surface. Seasonal and annual concentrations of Sulpher dioxide in the atmosphere of the Kalla region are given in Table 1. From this table it can be noted that sulpher dioxide concentrations greatly varied from one season to the other maximum concentration of 32.6 was recorded during winter. While annual mean concentrations of sulpher dioxide, reaching. 19.71 $μg/m^3$ These concentrations were lower than the value of 60 $μg/m^3$ set by the US Ambient Air Quality Standard, and also the Egyptian limit for the annual concentration of SO_2 [31]. Also, It was less than the primary US National Ambient Air Quality Standard (80 $μg/m^3$) for SO_2 [32,33]. Sulpher in the atmosphere originate either from natural processes or anthropogenic activity [34]. Fuel combustion as well as metal production is the dominant source for SO_2 emissions into the atmosphere [35].

Measurements of colorimetric data (CIE L*, a*, b*) in residential site (Kalla region)

Effect of exposure to air and light on the Change of colour (ΔE) values) for mordanted wool yarns dyed with natural colouring matter extracted from Cochineal, Turmeric and Madder using different salts in Kalla region: Tables 2-4 give values of change of colour (ΔE) for the three plants used. The change of colour (ΔE) increases with increasing the period of exposure to air and light. Good light resistance was observed in fabrics dyed with natural colouring matter extracted from Cochineal, Turmeric and Madder using different salts. This is due to the formation of complex with the metal which protects the chromatophore from photolytic degradation.

From Table 2 it can be observed that the colour change (ΔE) for the mordanted wool samples dyed with natural colouring matter extracted from Cochineal using different kinds of mordants follows the order: alum salt> tin salt> iron salt> copper salt. While, Table 3 shows that change of colour (ΔE) for mordanted wool samples dyed with Turmeric using different kinds of mordants follows the order: alum salt> tin salt> iron salt> copper salt. Finally, Table 4 indicated that change of colour (ΔE) for mordanted wool samples dyed with Madder using different kinds of mordants follows the order: tin salt> alum salt> copper salt> iron salt.

Effect of exposure to air and light on the lightness (L* values) for mordanted wool yarns dyed with natural colouring matter extracted from Cochineal, Turmeric and Madder using different salts in Kalla region: Tables 2-4 gives the (L* values) for the three plants used.

From Table 2, it can be concluded that lightness (L*values) for the mordanted wool samples dyed with natural colouring matter extracted from Cochineal, become lighter comparing to standard sample in case of using Iron or copper salts, but in case of using alum or tin salts , the colour become darker after time of exposure to air and light. From Table 3, it can be observed that the lightness (L*values) for the mordanted wool samples dyed with natural colouring matter extracted from Turmeric, become darker comparing to standard sample in case of using Iron, copper, alum or tin salts after time of exposure to air and light. It is clear from Table 4, that the lightness (L*values) for the mordanted wool samples dyed with natural colouring matter extracted from Madder, has no change of colour comparing to standard sample in case of using Iron or copper salts, but in case of using alum or tin salts, the colour become slightly darker after time of exposure to air and light.

Effect of exposure to air and light on the nature of colour (a* values) for mordanted wool yarns dyed with natural colouring matter extracted from Cochineal, Turmeric and Madder using different salts in Kalla region: Table 2-4 gives the (a*) values for the three plants used. Table 2 shows (a*) values for all wool samples dyed with natural colouring matter extracted from Cochineal using Iron salts which indicated that there is a slight change of colour in the direction of green region in the [CIE L*, a*, b*] zone, but in case of using copper salts the green colour concentration increases and in case of using alum and tin salts, a decrease in red colour concentration is happened. From Table 3 it can be observed that (a*) values for wool samples dyed with natural colouring matter extracted from Turmeric using Iron salt, copper salt, alum salt and tin salt show no change of colour after time of exposure (12 months). It is observed from Table (4) that (a*) values for wool samples dyed with natural colouring matter extracted from Madder using Iron salt or copper salts show a change of colour in the direction of green region in the [CIE L*, a*, b*] zone, but in case of using alum or tin salts a decrease in red colour concentration is happened.

Effect of exposure to air and light on the nature of colour (b* values) for mordanted wool yarns dyed with natural colouring matter extracted from Cochineal, Turmeric and Madder using different salts in Kalla region: Tables 2-4 gives the (b*) values for the three plants used. It is observed from Table 2 that (b*) values for all wool samples dyed with natural colouring matter extracted from Cochineal using iron or copper salts show a decrease in blue colour concentration, but in case of using tin salt an increasing of blue colour concentration is happened, and there is no change in colour happened in case of using alum salt. It can be seen from Table 3 that (b*) values for wool samples dyed with Turmeric using iron salts, copper salts, alum salts and tin salts show a change of colour in the direction of blue region in the [CIE L*, a*, b*] zone after exposure to air and light comparing to the blank. Table 4 shows that, (b*) values for wool samples dyed with Madder using tin salt show a change of colour from yellow region to blue region in the [CIE L*, a*, b*] zone. Also, in case of using alum salt, the blue

Type of salt	Colour data	Blank (without exposure)	After 3 months	After 6 months	After 9 months	After 12 months
Iron salt	L*	28	28	31	31	31
	a*	0.4	-3	-4	-5	-5
	b*	-39	-33	-32	-30	-30
	ΔE	0	6.9	8.8	11	11.3
	H	89.41	84.81	82.87	80.54	80.54
	C	39	33.13	32.25	30.41	30.41
Copper salt	L*	37	37	38	38	40
	a*	-4	-7	-8	-8	-10
	b*	-32	-29	-26	-26	-23
	ΔE	0	4.4	7.2	7.3	11
	H	82.87	76.43	72.90	72.90	66.50
	C	32.25	29.83	27.20	27.20	25.08
Alum salt	L*	40	37	36	33	33
	a*	27	17	14	12	9
	b*	-34	-33	-31	-30	-28
	ΔE	0	10.4	13.2	14.3	18.1
	H	50.71	62.74	66.62	67.37	73.81
	C	42.63	37.12	36.77	34.93	32.28
Tin salt	L*	43	40	39	37	34
	a*	27	19	17	15	12
	b*	-16	-21	-23	-25	-28
	ΔE	0	10.4	11.5	13.2	14.8
	H	30.65	49.18	50.71	53.97	54.46
	C	31.38	29.07	28.43	27.20	25.81

Table 2: Values of L*, a*, b*, ΔE, H and C for the mordanted wool yarns samples dyed with dyes extracted from **Cochineal** using different salts in Kalla region.

Type of salt	Colour data	Blank (without exposure)	After 3 months	After 6 months	After 9 months	After 12 months
Iron salt	L*	60	47	46	46	45
	a*	-10	-9	-9	-9	-9
	b*	-22	-30	-29	-30	-29
	ΔE	0	15.3	16.7	16.2	16.6
	H	65.56	73.3	72.76	73.3	72.76
	C	24.17	31.32	30.36	31.32	30.36
Copper salt	L*	49	45	44	43	41
	a*	-12	-14	-15	-15	-16
	b*	-7	-15	-16	-17	-19
	ΔE	0	9	9.9	10.3	13.6
	H	23.63	45	46.85	47.85	53.62
	C	17.46	19.21	21.93	21.93	23.6
Alum salt	L*	67	56	53	52	50
	a*	-10	-11	-12	-12	-13
	b*	-16	-19	-22	-29	-29
	ΔE	0	39.6	40.2	47.5	49.5
	H	57.99	59.72	61.43	68.23	69.23
	C	18.87	24.6	25.60	31.02	32.02
Tin salt	L*	62	57	51	51	50
	a*	-13	-12	-11	-10	-10
	b*	-15	-28	-30	-30	-31
	ΔE	0	15.8	20.3	20.4	21.4
	H	52.43	70.34	71.12	71.57	72.4
	C	18.40	29.73	31.20	31.62	31.62

Table 3: Values of L*, a*, b*, ΔE, H and C for the mordanted wool yarns samples dyed with dyes extracted from **Turmeric** using different salts in Kalla region.

Type of salt	Colour data	Blank (without exposure)	After 3 months	After 6 months	After 9 months	After 12 months
Iron salt	L*	32	31	31	31	32
	a*	2	-0.3	-2	-2	-3
	b*	-21	-22	-23	-23	-23
	ΔE	0	2.7	4.6	5.6	5.4
	H	84.56	89.22	85.03	85.03	82.57
	C	21.10	22	23.08	23.09	23.19
Copper salt	L*	35	35	35	35	36
	a*	3	-1	-1	-2	-3
	b*	-20	-22	-21	-20	-20
	ΔE	0	4.1	4.4	6.5	6.9
	H	88.47	87.40	87.27	84.29	81.87
	C	23.22	22.02	21.02	21.10	20.21
Alum salt	L*	40	37	37	37	36
	a*	20	12	10	9	7
	b*	-7	-15	-16	-17	-18
	ΔE	0	11.7	13.7	15.2	17.5
	H	19.29	51.34	57.99	62.10	68.75
	C	21.19	19.21	18.87	17.94	17.31
Tin salt	L*	41	40	40	38	38
	a*	19	13	10	9	7
	b*	7	-1	-2	-3	-7
	ΔE	0	10	12.8	14.5	18.6
	H	20.22	18.4	17.31	16.44	15
	C	20.25	13.04	10.20	9.89	9.20

Table 4: Values of L*, a*, b*, ΔE, H and C for the mordanted wool yarns samples dyed with dyes extracted from **Madder** using different salts in Kalla region.

colour concentration increases. But in case of using iron salt and copper salt, there is no change of colour after exposure to air and light.

Effect of exposure to air and light on the hue of colour (H values) for mordanted wool yarns dyed with natural colouring matter extracted from Cochineal, Turmeric and Madder using different salts in Kalla region: Table 2 shows that the hue of the colour values (H) for wool samples dyed with natural colouring matter extracted from cochineal and mordanted with iron salt is almost the same during all periods of exposure to air and light. However, when mordanted with copper salt, the H values shift towards the green axis. This may be because copper ions cause a bathochromic shift of the long wave length absorption bands of cochineal. When mordanted with alum or tin salt, the H values shift towards the blue axis. From Table 3, it can be observed that the H values for wool samples dyed with natural colouring matter extracted from turmeric and mordanted with iron or alum show no changes after exposure. On the other hand, when copper salt or tin salt is used, the H values shift towards the blue axis. This may be due to the fact that the ultraviolet (UV) visible spectra of turmeric show significant changes that have occurred in the band absorbance at the longest wave length in the presence of copper or tin ions, and these changes are characteristic of copper ions or tin ions. Copper ions and tin ions also cause a bathochromic shift of the long wave length absorption bands of turmeric. It is observed from Table 4 that the H values for wool samples dyed with natural colouring matter extracted from madder and mordanted with alum show a shift in the direction of the blue axis. The reason could be that alum cause a bathochromic shift of the long wave length absorption bands of madder. However, when iron, copper or tin salt is used, H values are almost the same during all periods of exposure to air and light.

Effect of exposure to air and light on the saturation of colour (chroma- C values) for mordanted wool yarns dyed with natural colouring matter extracted from Cochineal, Turmeric and Madder using different salts in Kalla region: It is observed from Table 2 that the chroma values (C) for all wool samples dyed with natural colouring matter extracted from cochineal and mordanted with iron, copper, alum and tin salts slightly decrease when exposure time to air and light increases. It can be seen from Table 3 that the C values for wool samples dyed with natural colouring matter extracted from turmeric and mordanted with iron, copper alum and tin salts slightly increase when the exposure time to air and light increases. Moreover, for almost all samples, the concentration of the chromophores decreased as the colour faded. Table 4 shows that the C values for wool samples dyed with natural colouring matter extracted from madder when tin salt is used decrease when the exposure time to air and light is increased. However, when mordanted with iron, copper or alum salt, the C values are almost the same during all periods of exposure to air and light.

Physical Measurement: The Physical measurements which have done on the dyed wool samples indicated that a severe decline in the tensile strength for the mordanted dyed wool sample with cochineal, turmeric and madder increases with increasing time of exposure to air and light.

Effect of time of exposure to air and light on the tensile strength (B-Force) for mordanted wool yarns dyed with natural colouring matter extracted from Cochineal, Turmeric and Madder using different salts in Kalla region: From Figure 1, it can be seen that, almost full damage of tensile strength occurred with copper salt after 9 months from exposure to air and light. Also, a complete damage happened in the tensile strength for the mordanted dyed samples with cochineal using iron, alum and tin salts after 12 months from the exposure to air and light compared to the standard.

It is observed from Figure 2, that, a severe decline of tensile strength occurred with all mordants used (iron, copper, alum and tin salts) after exposure to air and light for 12 months. Also, a complete decline happened in the tensile strength for the mordanted dyed samples with turmeric using iron and alum salts after exposure to air and light for 9 months.

Figure 3 shows that a severe decline of tensile strength occurred with copper salt for mordanted wool samples dyed with madder after

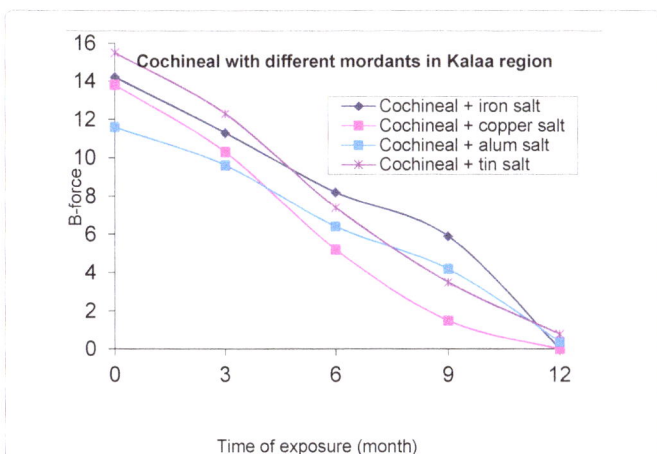

Figure 2: Effect of time of exposure to light on the tensile strength (B-Force) of the dyed mordanted wool yarns with Turmeric.

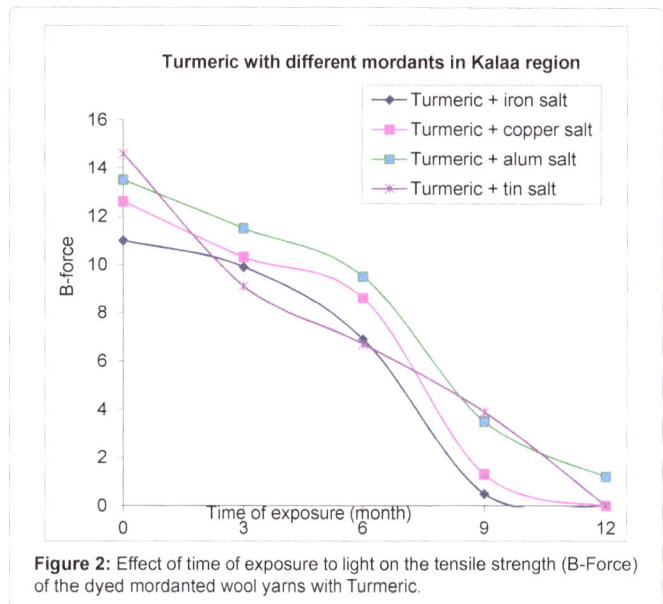

Figure 3: Effect of time of exposure to light on the tensile strength (B-Force) of the dyed mordanted wool yarns with Madder.

full time of exposure (12 months). Also, Figure 3 shows that good results of the tensile strength were obtained when using iron, alum and tin salts for mordanted wool samples dyed with madder after exposure to air and light for 12 months. Also, a gradual decline happened in the tensile strength for the mordanted dyed samples with Madder using iron, copper, alum and tin salts after exposure to air and light for 12 months compared to the standard.

However, the decline of tensile strength which occurred with mordanted wool samples dyed with Madder is better than the decline of tensile strength which occurred with cochineal and Turmeric during the time of exposure to air and light (12 months). Finally, the tensile strength for mordanted wool samples dyed with Madder using different kinds of mordants follows the order: copper salt > tin salt > iron salt > alum salt.

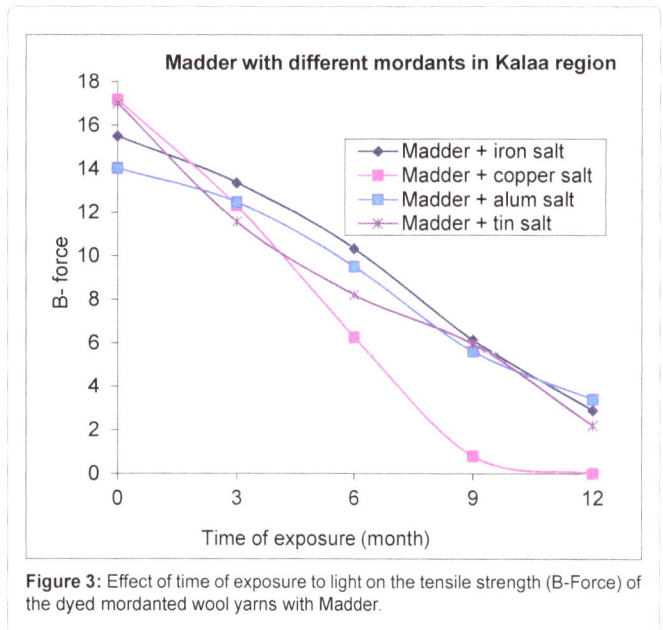

Figure 1: Effect of time of exposure to light on the tensile strength (B-Force) of the dyed mordanted wool yarns with Cochineal.

Effect of time of exposure to air and light on the Tenacity for mordanted wool yarns dyed with natural colouring matter extracted from Cochineal, Turmeric and Madder using different salts in Kalla region: Figure 4 shows that, a severe decline in the tenacity of the mordanted dyed wool yarns with cochineal using different kinds of salts (iron, copper, alum and tin salts) is occurred with increasing time of exposure. A complete damage of tenacity is occurred for the mordanted wool yarns with all mordants used during the exposure time (12 months). Also, a complete decline of tenacity is occurred for the cochineal dyed yarns mordanted with copper salts after 9 months from exposure to air and light comparing to the standard.

It can be observed from Figure 5 that, a severe decline in the tenacity of the mordanted dyed wool yarns with turmeric using different kinds of salts (iron, copper, alum and tin) is occurred with increasing time of exposure to air and light. Also, a complete damage of tenacity is occurred for the dyed mordanted wool yarns with turmeric using tin salts after 9 months of exposure time comparing to the standard.

From Figure 6, it can be concluded that, a severe decline in the tenacity of the mordanted wool samples dyed with madder using copper salt is occurred with increasing time of exposure comparing to the standard. Good results regarding the tenacity of the mordanted wool samples dyed with madder using different kinds of salts (iron, copper, alum and tin) are obtained with increasing time of exposure comparing to the standard. Also, a gradual damage of tenacity is occurred for the dyed mordanted wool yarns with different kinds of salts (iron, copper, alum and tin) during the exposure time (12 months).

Finally, the tenacity for mordanted wool samples dyed with madder using different kinds of mordants follows the order: alum salt > tin salt > iron salt > copper salt. To sum it all up, the tenacity of mordanted wool yarns dyed with madder is the best one among the tenacity of other mordanted wool yarns dyed with Cochineal or Turmeric, due to its small decline during all periods of exposure time to air and light.

Effect of time of exposure to air and light on the Elongisity for mordanted wool yarns dyed with natural colouring matter extracted from Cochineal, Turmeric and Madder using different salts in Kalla region: It can be observed from Figure 7 that, a severe decline

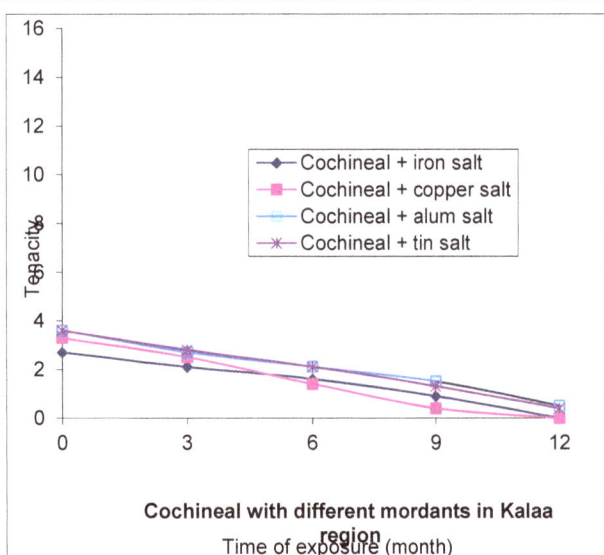

Figure 4: Effect of time of exposure to light on the Tenacity of the dyed mordanted wool yarns with Cochineal

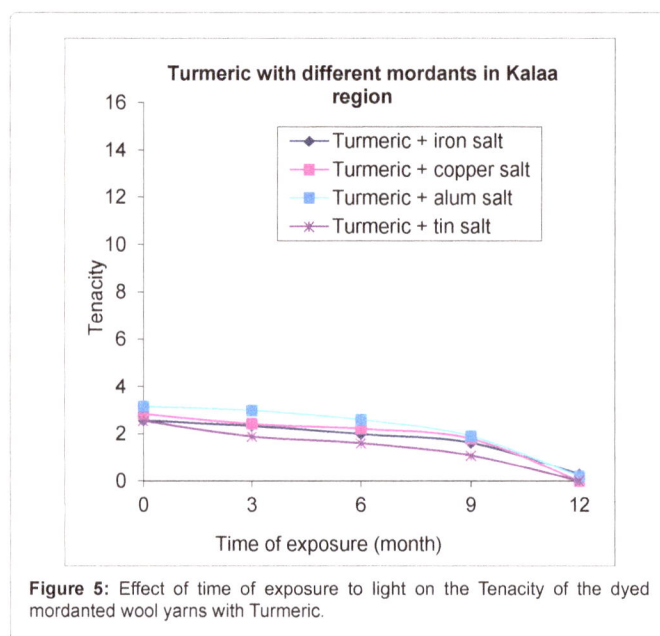

Figure 5: Effect of time of exposure to light on the Tenacity of the dyed mordanted wool yarns with Turmeric.

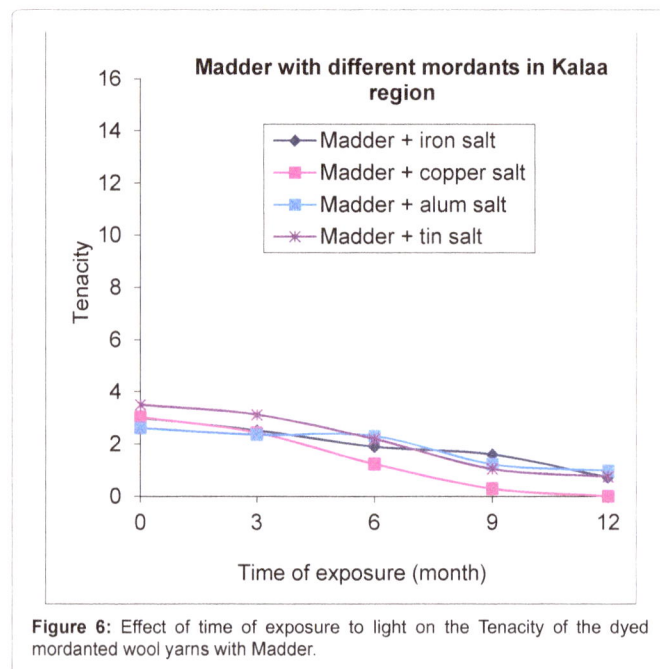

Figure 6: Effect of time of exposure to light on the Tenacity of the dyed mordanted wool yarns with Madder.

in the elongisity of the mordanted dyed wool samples with cochineal using different mordants (iron, copper, alum and tin) salts occurs with increasing time of exposure to air and light comparing to the standard. Also, a complete decline of elongisity is happened for the mordanted wool yarns dyed with cochineal using iron and copper salts after 9 months from beginning of exposure to air and light comparing to the standard. Moreover, the elongisity for mordanted wool yarns dyed with cochineal using different mordants follows the order: alum salt > tin salt > iron salt > copper salt during all periods of exposure time to air and light.

Figure 8 shows that, a severe decline in the elongisity of the mordanted dyed wool yarns with turmeric using different mordants (iron, copper, alum and tin) salts occurs with increasing time of exposure to air and light comparing to the standard. Also, a complete

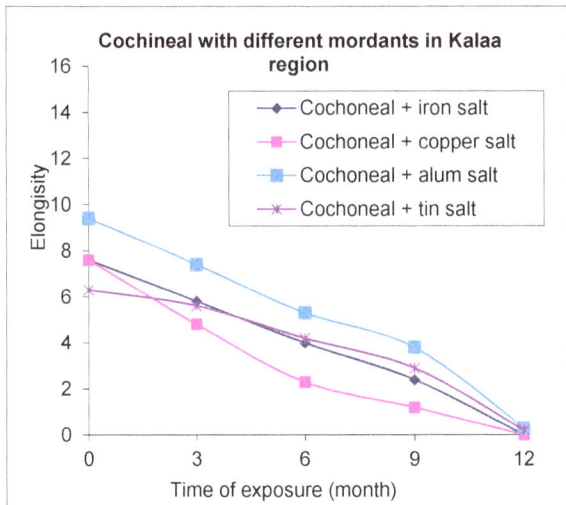

Figure 7: Effect of time of exposure to light on the Elongisity of the dyed mordanted wool yarns with Cochineal.

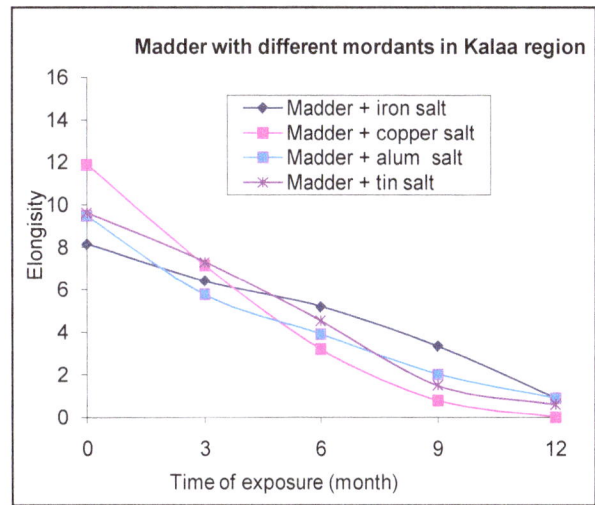

Figure 9: Effect of time of exposure to light on the Elongisity of the dyed mordanted wool yarns with Madder.

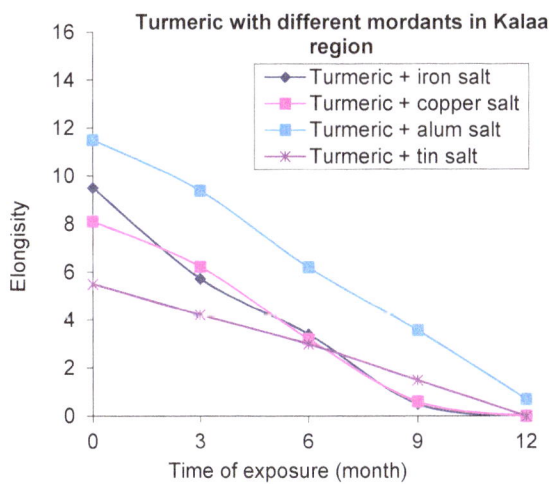

Figure 8: Effect of time of exposure to light on the Elongisity of the dyed mordanted wool yarns with Turmeric.

decline of elongisity is happened for the mordanted wool samples dyed with turmeric using copper and iron salts after 9 months from beginning of exposure to air and light comparing to the standard. In addition, the highest elongisity of mordanted wool samples dyed with turmeric salts using different mordants is alum salt then the rest of all salts after full time of exposure to air and light (after 12 months).

From Figure 9, it can be seen that, a severe decline in the elongisity of the mordanted dyed wool yarns with madder using different mordants (iron, copper, alum and tin) salts occurs with increasing time of exposure to air and light comparing to the standard. Moreover, a complete decline of elongisity is happened for the mordanted wool samples dyed with madder using copper salt after 12 months from the beginning of exposure to air and light comparing to the standard. Also, a gradual damage of elongisity is happened until it reach the complete damage of elongisity for the mordanted wool yarns dyed with madder

using tin and copper salts after full time of exposure to air and light (12 months). Moreover, the elongisity for mordanted wool samples dyed with madder using different kinds of mordants follows the order: iron salt > alum salt > tin salt > copper salt during all periods of exposure time to air and light.

Finally, the elongisity of mordanted wool yarns dyed with madder is the best one among the elongisity of other mordanted wool yarns dyed with cochineal or turmeric, due to its gradual decline during all periods of exposure time to air and light.

Conclusion

Kalla region (residential site) is highly polluted with a suspended dust comparing to Helwan city(industrial area) [36]. While Helwan city (industrial area) is highly polluted with fine particles and SO_2 more than Kalla region (residential site). The values of change of colour (ΔE) of mordanted wool yarns dyed with madder are the lowest one among other yarns dyed with cochineal or turmeric. From these results, it can be concluded that the light fastness for wool yarns dyed with madder is the best one among other wool yarns dyed with cochineal or turmeric. Using different mordants as well as different kinds of sources of dyes for mordanted dyed wool yarns gives beautiful colorful wide range of hues.

References

1. Patel BH (2011) 11-Natural dyes. In: Clark M (eds.) Handbook of Textile and Industrial Dyeing. B Woodhead Publishing 1: 395-424.

2. Chan-Bacab MJ, Sanmartín P, Camacho-Chab JC, Palomo-Ascanio KB, Huitz-Quimé HE, et al. (2015) Characterization and dyeing potential of colorant-bearing plants of the Mayan area in Yucatan Peninsula, Mexico. Journal of Cleaner Production 91: 191-200.

3. Błyskal B (2015) Fungal deterioration of a woollen textile dyed with cochineal. Journal of Cultural Heritage 16: 32-39.

4. Haddar W, Ben Ticha M, Guesmi A, Khoffi F, Durand B (2014) A novel approach for a natural dyeing process of cotton fabric with Hibiscus mutabilis (Gulzuba): Process development and optimization using statistical analysis. Journal of Cleaner Production 68: 114-120.

5. Har Bhajan S, Bharati KA (2014) 6 - Enumeration of dyes. In: Singh HB, Bharati KA (eds.) Handbook of Natural Dyes and Pigments. Woodhead Publishing India: 33-260.

6. Sinha K, Chowdhury S, Saha PD, Datta S (2013) Modeling of microwave-assisted extraction of natural dye from seeds of Bixa orellana (Annatto) using response surface methodology (RSM) and artificial neural network (ANN). Industrial Crops and Products 41: 165-171.

7. Kamel MM, El-Hossamy MM, Helmy HM, El-Hawary NS (2008) Some studies on dyeing properties of cotton fabrics with Curcuma Longa (turmeric)(roots) using ultrasonic method. Polish journal of applied chemistry.

8. Bechtold T, Turcanu A, Ganglberger E, Geissler S (2003) Natural dyes in modern textile dyehouses — how to combine experiences of two centuries to meet the demands of the future? Journal of Cleaner Production 11: 499-509.

9. Han S, Yang Y (2005) Antimicrobial activity of wool fabric treated with curcumin. Dyes and Pigments 64: 157-161.

10. Shahid M, Shahid ul I, Mohammad F (2013) Recent advancements in natural dye applications: A review. Journal of Cleaner Production 53: 310-331.

11. Abel A (2012) The history of dyes and pigments: From natural dyes to high performance pigments. In: Best J (eds.) Colour Design. Woodhead Publishing 433-470.

12. Har Bhajan S, Bharati KA (2014) History of natural dyes. In: Singh HB, Bharati KA (eds.) Handbook of Natural Dyes and Pigments. Woodhead Publishing India 4-8.

13. Grifoni D, Bacci L, Di Lonardo S, Pinelli P, Scardigli A, et al. (2014) UV protective properties of cotton and flax fabrics dyed with multifunctional plant extracts. Dyes and Pigments 105: 89-96.

14. Khan MI, Ahmad A, Khan SA, Yusuf M, Shahid M, et al. (2011) Assessment of antimicrobial activity of Catechu and its dyed substrate. Journal of Cleaner Production 19: 1385-1394.

15. Har Bhajan S, Bharati KA (2014) Introduction. In: Singh HB, Bharati KA (eds.) Handbook of Natural Dyes and Pigments. Woodhead Publishing India, pp. 1-3.

16. Punrattanasin N, Nakpathom M, Somboon B, Narumol N, Rungruangkitkrai N, et al. (2013) Silk fabric dyeing with natural dye from mangrove bark (Rhizophora apiculata Blume) extract. Industrial Crops and Products 49: 122-129.

17. Baliarsingh S, Panda AK, Jena J, Das T, Das NB (2012) Exploring sustainable technique on natural dye extraction from native plants for textile: Identification of colourants, colourimetric analysis of dyed yarns and their antimicrobial evaluation. Journal of Cleaner Production 37: 257-264.

18. Khan AA, Iqbal N, Adeel S, Azeem M, Batool F, Bhatti IA (2014) Extraction of natural dye from red calico leaves: Gamma ray assisted improvements in colour strength and fastness properties. Dyes and Pigments 103: 50-54.

19. Janhom S, Griffiths P, Watanesk R, Watanesk S (2004) Enhancement of lac dye adsorption on cotton fibres by poly(ethyleneimine). Dyes and Pigments 63: 231-237.

20. Prabhu KH, Teli MD (2014) Eco-dyeing using Tamarindus indica L. seed coat tannin as a natural mordant for textiles with antibacterial activity. Journal of Saudi Chemical Society 18: 864-872.

21. Bhajan Singh H, Bharati KA (2014) Preface. In: Singh HB, Bharati KA (eds.) Handbook of Natural Dyes and Pigments. Woodhead Publishing India ix-x.

22. Feng XX, Zhang LL, Chen JY, Zhang JC (2007) New insights into solar UV-protective properties of natural dye. Journal of Cleaner Production 15: 366-372.

23. Cristea D, Vilarem G (2006) Improving light fastness of natural dyes on cotton yarn. Dyes and Pigments 70: 238-245.

24. Ali NF, El-Mohamedy RSR (2011) Eco-friendly and protective natural dye from red prickly pear (Opuntia Lasiacantha Pfeiffer) plant. Journal of Saudi Chemical Society 15: 257-261.

25. selvam RM, Singh AJAR, Kalirajan K (2012) Antifungal activity of different natural dyes against traditional products affected fungal pathogens. Asian Pacific Journal of Tropical Biomedicine 2: S1461-S1465.

26. Stern AC (1986) Air Pollution (3rd Eds.) vol 111. Academic Press. Inc., New York.

27. Shakour AA (1992) Evaluation of dust deposited over urban and rural areas in Egypt. TESCE 18: 186-196.

28. Harrison RM, Perry R (eds.) (1986) Handbook of Air Pollution Analysis (2ndedn.) Chapman and Hall Ltd., London.

29. Stern AC (1976) Air Pollution (3rd Eds.) vol 11. Academic Press. Inc, New York.

30. EPAQS (1995) Department of the Environment Expert Panel on Air Quality Standards. HMSO, London.

31. Egyptian Environmental Affair Agency (1994) Environmental Protection Law No. 4. Ministry of State for Environmental Affairs, Cairo.

32. Colls J (1997) Air pollution, an introduction. London: E & FN Spon.

33. Heinsohn RJ, Kabel LR (1999) In: Stenquist B, Horton M (eds.). Sources and control of air pollution. Prentice-Hall, Inc., Upper Saddle River, New Jersey.

34. Aneja VP, Agarwal A, Roelle PA, Phillips SB, Tong Q, et al. (2001) Measurements and analysis of criteria pollutants in New Delhi, India. Environment International 27: 35-42.

35. Isakson J, Selin Lindgren EA, Foltescu VL, Pacyna JM, Tørseth K (1995) Behaviour of sulphur and nitrogen compounds measured at marine stations Lista and Säby in Scandinavia. Water, Air, Soil Pollut 85: 2039-2044.

36. Kamel MM, Helmy HM, Shakour AA, Rashed SS (2012) The Effects of Industrial Environment on Colour Fastness to Light of Mordanted Wool Yarns Dyed with Natural Dyes. Research Journal of Textile and Apparel 16: 46-57.

9

Geometrical Analysis of Warp Knit Auxetic Fabrics

Samuel C Ugbolue[1]*, Olena Kyzymchuk[2] and Yong K Kim[1]

[1]*Department of Bioengineering, University of Massachusetts, Dartmouth, USA*
[2]*Kyiv National University of Technology and Design, Ukraine*

Abstract

Despite considerable interests that have been shown on the formation, properties and characteristics of auxetic knit structures there remains a dearth of information about the fundamental geometrical analytics of warp knit auxetic fabrics. This paper examines the geometrical model of auxetic warp knit structure and validates its characteristics with data obtained from experimental analysis of nine recently produced warp knit auxetic fabrics.

Keywords: Auxetic structure; In-laying yarn; Fabric geometry; Pre-knitting

Introduction

In recent years, considerable interests have been shown on the formation, properties and characteristics of auxetic knit structures [1-7]. Auxetic structures can enable an article to exhibit an expansion in a lateral direction, upon subjecting the article to a longitudinal stress or strain. Conversely, auxetic structures also exhibit a contraction in the lateral direction upon subjecting such an article to longitudinal compression. Such materials are understood to exhibit a negative Poisson's ratio. However, very limited studies have been produced on the fundamental geometrical analytics of warp knit auxetic fabrics. It is recognized that the geometry of fabrics has significant effects on their behavior. Hence Ugbolue et al. [4] focused on the geometry and structural properties of warp knit auxetic fabrics in one of their papers. Generally, the complexity of warp knitted structures have posed major challenges to researchers but several studies on the geometry of warp knitted structures have been published in the literature [8]. Early attempts on the study of the subject were experimental studies in which the dimensional properties of the warp-knitted structure were investigated by Fletcher and Roberts [9,10] and by Stimmel [11,12]. Later, a geometrical model for warp-knitted structures was developed by Allison [13] and then Grosberg [14] presented a full picture of the warp-knitted-loop configuration for two-bar warp-knitted fabrics. Subsequently, Grosberg [15] improved the previously suggested geometrical model wherein he assumed that the root end of the loop laid at the widest section of the previous loop and the underlap was part of a circle. The third geometrical model was developed by Shinn and El-Araf [16]. Wheatley [17] and Jacobsen [18] also proposed some geometrical loop models for warp-knitted structures produced from wool yarns. These researchers were keenly interested in finding some relationships between fabric parameters and fabric dimensions that were akin to those found for weft-knitted structures in order to facilitate some production calculations in industrial settings and encourage efforts in predicting some fabric properties after [19]. Our thrust in prior research combined our knowledge of geometry and fabric structural characteristics to engineer auxetic textiles and to determine the properties of such auxetic textile fabrics. Auxetic warp knit fabrics have great potentials in many areas of application such as protective clothing, blast resistant uniforms and other industrial textiles.

Our continued efforts to produce auxetic knit structures from non-auxetic yarns are described here. Specifically, this paper examines our geometrical model of auxetic warp knit structure and validates its characteristics with data obtained from experimental analysis of nine recently produced warp knit auxetic fabrics.

Theoretical analysis

A typical warp knit structure under strain is shown in Figure 1. The length of interlooping repeat that is formed by alternation of tricot (n_t) and chain (n_c) courses during drawing off at knitting machine (Figure 1) could be calculated as follows:

$$L^s_R = 2n_t B_t \max + 2n_c B_c \max \qquad (1)$$

Where $B_{t\ max}$ and $B_{c\ max}$ are maximum height of tricot and chain courses under a strain respectively mm:

$$B_{t\max} = \frac{l_t - 2.5\pi d_{\min}}{3} \quad and \quad B_{c\max} = \frac{l_c - (3 + 2.5\pi)d_{\min}}{3}$$

Where lt and lc are loop length of tricot and chain, mm respectively, dmin is the diameter of yarn under strain, mm;

After relaxation as shown in Figure 2,

$$L_R = a_1 + h \qquad (2)$$

Where a_1 is length of vertical rib, mm

h is distance between two vertical ribs in wale wise direction, mm.

It is obviously clear that:

$$a_1 = (n_t - 1)B_t \qquad (3)$$

Where B_t is height of tricot course mm,

$$B_t = 0.25l_t - 0.03\sqrt{T}$$

Where T is the linear density of yarn, tex

For conventional warp knit structure with hexagonal net (Figure 2a) distance h is calculated by:

$$h = (n_t - 1)B_t + 2(B_t + n_c B_c)\cos\alpha \qquad (4)$$

Where B_c – height of chain courses, mm

$$B_c = \frac{l_c - 3(\pi - 1)d}{3}$$

*Corresponding author: Ugbolue SC, Lecturer, Department of Bioengineering, University of Massachusetts, Dartmouth, USA, E-mail: sugbolue@umassd.edu

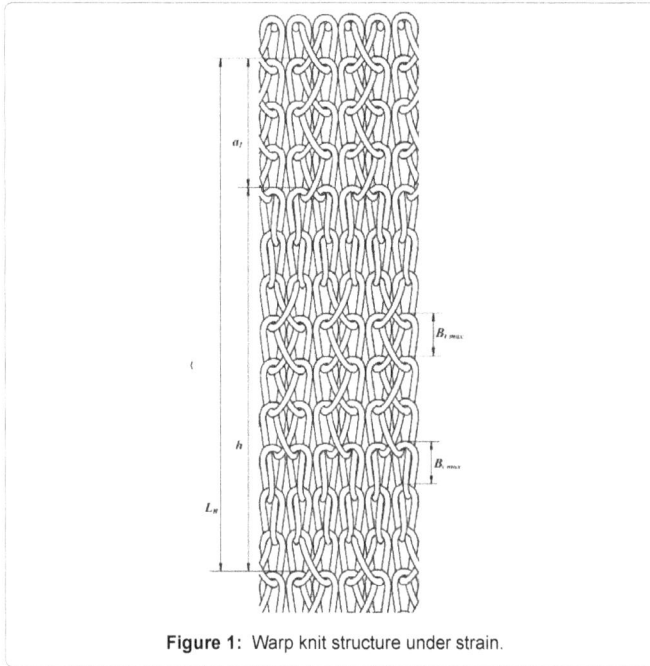

Figure 1: Warp knit structure under strain.

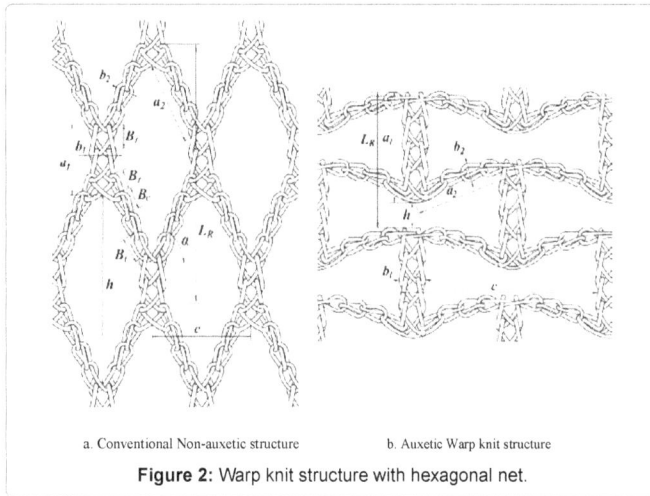

a. Conventional Non-auxetic structure   b. Auxetic Warp knit structure

Figure 2: Warp knit structure with hexagonal net.

Where d is the diameter of yarn, mm, and α is angle between diagonal rib and horizontal.

The geometrical model depicted in Figure 2b shows the configuration of our Auxetic warp knit structure. Thus, the length of interlooping repeat of conventional warp knit structure is given by:

$$L_R = a_1 + h = 2(n_t - 1)B_t + 2(B_t + n_c B_c)\cos\alpha \qquad (5)$$

Also, Figure 2b depicts the auxetic warp knit geometrical structure with hexagonal mesh and its auxetic configuration is realized when the following condition is satisfied:

$$h < a_1 - 2b_2 \quad or \quad h < (n_t - 1)B_t - 2b_2 \qquad (6)$$

Therefore, the length of interlooping repeat of auxetic warp knit structure is:

$$L_R = (n_t - 1)B_t + h \quad or \quad L_R < 2(n_t - 1)B_t - 2b_2 \qquad (7)$$

As was proposed by Ugbolue et al. [3,4] to achieve auxetic property,

it is required to employ a high elastic yarn in the basic filet warp knit structure. This yarn must to be placed between the stitch wale in the knitting direction to insure that the fabric structure will retain necessary configuration after relaxation. The filling yarn makes contact with the ground yarn just at the vertical ribs from tricot courses and is inlaid between vertical and diagonal ribs inside the hole at the other part of net. Also in their previous paper, Kyzymchuk and Ugbolue [20] affirmed that the positioning of inlaid yarns in filet warp knit structure is determined by the amount and direction of inlaying and ground guide bars' shifting behind the needles as well as disposition of guide bars with ground and filling yarns. It was determined that the in-laid yarn could be positioned inside the structure in one, two, three or five courses of vertical rib and might also be laid between the tricot's junctures from different ground yarns or wrapped junctures from one or both ground guide bars. Thus, two variants of inlaid yarn positioning are showed at Figure 3.

Consequently, the length of interloping repeat of such structure could be calculated as:

$$L_R = L_i + L_f \qquad (8)$$

Where L_i is length of elastomeric yarn that is fixed in the knitted structure,

L_f is length of elastomeric yarn that is not fixed in the knitted structure

For our modelling we adopt the following assumptions:

(i) The relaxation of elastomeric yarn, that is fixed in the structure, is insignificant and does not affect the size and shape of the loops;

(ii) A full relaxation of elastomeric yarn occurs only in the area where the yarn is not fixed in the structure.

Thus, lengths of elastomeric yarn that are fixed in knitted structure are determined as follows:

-      during drawing off at knitting machine:

$$L_i^s = (i - 0.5)B_{t\max} \qquad (9) \qquad\qquad (9)$$

-      after relaxation:

$$L_i = (i - 0.5)B_t \qquad (10)$$

Where i is the number of tricot courses in which the elastomeric yarn is fixed.

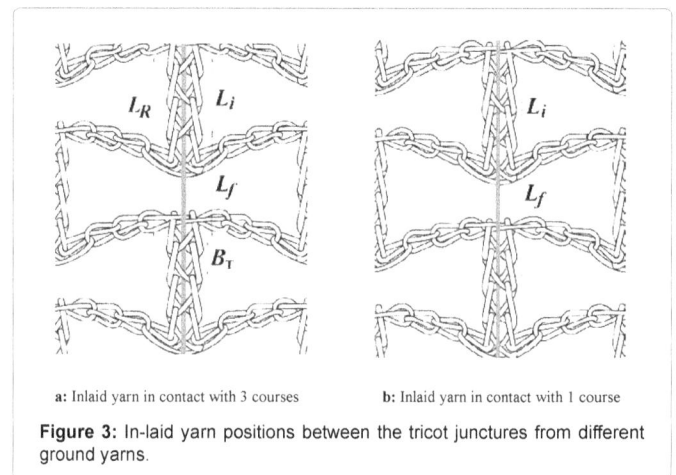

a: Inlaid yarn in contact with 3 courses   b: Inlaid yarn in contact with 1 course

Figure 3: In-laid yarn positions between the tricot junctures from different ground yarns.

Based on prior presented dependences, the lengths of elastomeric yarn, which are not fixed in the structure, are given as follows:

- length the knitting machine:

$$L^s_{if} = (2n_t - i + 0.5)B_{tmax} + 2n_c B_{cmax} \qquad (11)$$

- length needed to achieve auxetic effect after relaxation:

$$Li_f < (2n_t - 1.5 - i)B_t - 2b_2 \qquad (12)$$

Thus, to achieve the auxetic effect in warp knitted filet structure the elastomeric in-laid yarn must be fed with the prior stretching, , which is determined as follows:

$$\varepsilon > \left[\frac{(2n_t - i + 0.5)B_{tmax} + 2n_c B_{cmax}}{(2n_t - 1.5 - i)B_t - 2b_2} - 1 \right] 100\% \qquad (13)$$

It is evident that the strain value decreases with increasing number of tricot courses and decreasing number of chain courses within a repeat of filet interlooping as well as the decreasing number of courses in which the in-laid yarn is fixed within the warp knit structure.

Typical results of calculated theoretical strain value of elastomeric yarn that predict auxetic property of filet net structure are shown at Table 1. The theoretical Poisson's ratio has been calculated with reference to the relationships derived from Figure 2 as follows:

$$V_{xy} = -\frac{(2a_2 - c)}{(a_1 - 2b_2 - h)} \frac{h}{c} \qquad (14)$$

It is observed that the value of negative Poisson's ratio increases with the number of tricot courses as well as chain courses of the interlooping repeat. Thus, the best auxetic property of the filet warp knit is expected for the structure formed by alternation of 7 tricot and 3 chain courses

at repeat, but prior stretching of elastomeric yarn should be increased.

Experimental

Production of Auxetic Fabrics

In order to verify the auxetic behavior of the model, nine different filet warp knit fabrics with in-lay yarn that is fixed in two courses of repeat as shown in Table 2 were produced. These fabrics were made on a 10 gauge Jakob Muller crochet knitting machine with one needle bed. The warp knit fabrics were made from 250 denier polyester yarn as ground. The 150 denier (96 filaments) polyester yarn covered with one end of 70 denier spandex was used to achieve a high elastic in-lay component. Elastomeric yarn was fed into the knitting zone with elongation of 150% [21]. This elastomeric yarn is in-laid between the stitch wale in the knitting direction to ensure that the fabric structure retains the necessary configuration after relaxation.

Measurement of some fabric physical properties

A description of each measured property is given below:

Courses per unit length and Wales per unit length: A course is a predominantly horizontal row of loops produced by adjacent needles during the same knitting cycle. In warp knitting each loop in a course normally is composed of a separate yarn. Courses per inch (cpi) or courses per cm are used to evaluate the loops along the axial or walewise direction. A wale is a predominantly vertical column of needle loops produced by the same needle knitting at successive knitting cycles and thus intermeshing each new loop through the previous loop. In warp knitting a wale can be produced from the same yarn. Wales per inch (wpi) or wales per cm are used to evaluate the loops along the transverse or coursewise direction. The number of courses and the

Sample Designation	Number of courses at interlooping repeat		Minimum strain value ε for different positions of in-laid yarn, %			Poisson's ratio, vxy
	Tricot, n_t	chain n_c	i=1	i=2	i=3	
AX3-1	3	1	483	947	-	-0.031
AX3-2	3	2	614	1222	-	-0.130
AX3-3	3	3	746	1497	-	-0.196
AX5-1	5	1	198	226	270	-0.144
AX5-2	5	2	240	278	334	-0.698
AX5-3	5	3	283	329	399	-1.070
AX7-1	7	1	143	152	164	-0.195
AX7-2	7	2	168	181	196	-1.179
AX7-3	7	3	194	209	228	-1.846

Table 1: Theoretical strain values of elastomeric yarn.

Sample Designation	Number of courses at interlooping repeat		Number of loops per cm		Stitch	Basis weight, g/sq.m	Thickness mm
	tricot, n_t	chain, n_c	Wales N_w	Courses N_c	Loops per cm²		
AX3-1	3	1	3.9	21.2	82.7	283.2	0.97
AX3-2	3	2	3.5	23.0	80.5	254.7	1.06
AX3-3	3	3	3.3	27.0	89.1	244.9	1.09
AX5-1	5	1	4.0	23.2	92.8	289.6	1.08
AX5-2	5	2	3.7	24.0	88.8	261.6	1.16
AX5-3	5	3	3.3	27.7	91.4	253.1	1.07
AX7-1	7	1	4.0	23.7	94.8	293.4	1.15
AX7-2	7	2	3.7	25.5	94.4	285.7	1.32
AX7-3	7	3	3.3	29.1	96.0	264.8	1.26

Table 2: Structural parameters of auxetic warp knit fabrics

Sample Designation	Unit size, mm				Poisson's ratio v_{xy} at various strain levels	
	Length of vertical rib a_1	Length of diagonal rib a_2	Distance between two vertical ribs		10%	30%
			in walewise direction h	in coursewise direction c		
AX3-1	3.76	1.94	2.01	6.31	0.003	0.052
AX3-2	4.07	2.52	2.12	7.49	- 0.022	0.038
AX3-3	3.13	4.80	4.53	8.57	- 0.062	0.011
AX5-1	5.26	1.71	3.69	5.94	- 0.034	0.068
AX5-2	5.19	2.56	5.28	6.54	- 0.076	-0.016
AX5-3	5.48	4.29	5.26	8.14	- 0.082	-0.004
AX7-1	5.46	1.90	5.73	7.02	- 0.108	-0.094
AX7-2	5.26	2.41	6.29	7.75	- 0.143	-0.073
AX7-3	6.09	3.47	5.82	8.86	- 0.162	-0.024

Table 3: Analysis of unit cell geometry and Poisson's Ratio of fabrics.

number of wales per unit length are obtained by using a counting glass. The results are reported using SI units namely, number of courses or wales in 10 mm. The mean of ten readings is recorded for each direction of the fabric.

Stitch density, S: The term loop or stitch density, S, is the total number of needle loops in a square area measurement. It is obtained by multiplying, for instance, the number of courses and wales, per square unit length together. The unit is loops/in^2 or loops per cm^2.

Thickness and basis weight: The thickness (with the unit of mm) of each sample is tested using a Thickness Testing Instrument according to ASTM D1777-64.8. Each sample is also weighed on an electronic balance to determine its basis weight or areal density (g/m^2).

Measurement of Poisson's ratio: To measure the Poisson's ratio of the identified fabrics, video-extensometry along with micro-tensile testing techniques are employed by using Instron 5569 Mechanical Tester ASTMD5034-95 (2001). All samples are tested by straining the entire fabric strip evenly and each duly marked 2 cm × 2 cm square of the sample is measured to obtain the Poisson's ratio by using the equation, $v_{xy} = -\varepsilon_x/\varepsilon_y$ where ε_x is the strain in the x-direction, or transverse strain, and ε_y is the strain in the y-direction, or the axial strain. Initially all samples of 10 cm long are strained at a rate of 5.08 cm/min, in the walewise and coursewise directions. The test process is observed with a Canon Pro 300 camera. The strain of the sample is measured using the camera to capture an image of the sample at different strain levels, totaling 16 pictures per sample. The width of each sample is measured in three locations to insure that the measured Poisson's ratio is as accurate of measurement as possible [22]. Each fabric structure is tested three times, using different samples of the fabric structure each test. The Poisson's ratio is obtained after all samples had been photographed and strain values obtained using appropriate image analysis software. It should be noted in the photographs that the relationship of interest (for determining if auxetic) is not necessarily the relationship of the strain in the y direction to the x direction. The axial strain can still be greater than the transverse strain (as pictured), yet the auxetic nature is a reflection of the transverse strain increasing under axial load (as measured, the width of the box increased relative to the initial width). All the samples were tested along the wale direction. Then the Poisson's ratio was calculated using the equation:

$$V_{xy} = -\varepsilon_x / \varepsilon_y \qquad (15)$$

where vxy is Poison's ratio, ε_x is transverse strain, and ε_y is axial strain.

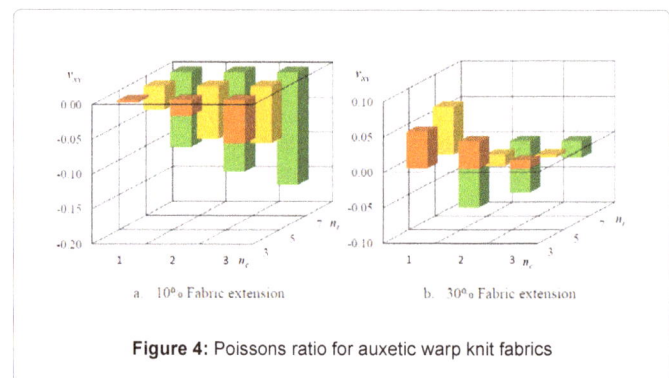

a. 10% Fabric extension   b. 30% Fabric extension

Figure 4: Poissons ratio for auxetic warp knit fabrics

Poisson's ratio is the ratio of transverse contraction strain to longitudinal extension strain in the direction of stretching force.

Results and Discussion

The structural parameters of the auxetic warp knit fabrics are shown in Table 2. Details of the experimental measurements of unit cell geometry and Poisson's ratio are presented in Table 3. Also, the values of the Poisson's ratio at different strain levels are presented in Figure 4. The measurement of Poisson's ratio is an important fundamental tool for determining the auxetic property of materials. If the Poisson's ratio is negative then the structure is auxetic. Ugbolue et al. [4] have shown that the factor which influences the Poisson's ratio is identified as intrinsic unit size displacement, which depends on chain course numbers. It is surmised that the Poisson's ratio values decrease as the number of tricot courses increase. Indeed, the auxetic properties depend on the interaction of vertical and horizontal ribs in the knitted structure as shown in the model.

As shown in Figure 4, the Poison's ratio of warp knit structures is increasing with increasing number of tricot and chain courses per interlooping repeat. However, such tendency is observed at low strain levels, up to 10% of fabric stretching. At 30% fabric extension, only the knit structure with 7 tricot courses as its interlooping repeat has negative Poisson's ratio whose value is shown to decrease with increasing number of chain courses at repeat. The observed low Poisson's ratio effect is attributed to the poor prior (pre-knitting) stretching of the elastomeric in-laid yarn.

Conclusion

A geometrical analysis of warp knit auxetic fabrics has been presented in which theoretical computations of the auxetic model were undertaken. Experimental results validate the theoretical computation. Also, it has been established that the experimental values produced lower negative

Poisson's ratios primarily because of the low pre-tensioning of the inlay spandex yarns employed. On the basis of these results, future fabrication of auxetic warp knit structures must employ higher pre-tensioned 100% elastomeric inlay yarn in order to improve on the auxetic properties and characteristics of the warp knit auxetic fabrics.

Acknowledgement

One of the authors, Olena Kyzymchuk, is grateful for the funds provided by the Council for International Exchange of Scholars (CIES) and Institute of International Education (IIE) under the Fulbright Visiting Scholar Program at the Department of Bioengineering, University of Massachusetts, Dartmouth, MA 02747, USA.

References

1. Alderson A, Alderson K (2005) Expanding materials and applications: exploiting auxetic textiles. Text Int 14: 29-34.

2. Ugbolue SC, Warner SB, Kim YK, Fan Q, Chen Lu, et al. (2010) The Formation and Performance of Auxetic Textiles, Part 1: Theoretical and Technical Considerations, J Text Inst 101: 660-667.

3. Ugbolue SC, Warner SB, Kim YK, Fan Q, Chen Lu, et al. (2011) The Formation and Performance of Auxetic Textiles, Part 1I: Geometry and Structural Properties. J Text Inst 102: 424-433.

4. Liu YP, Hu H, Lam JKC, Su Liu (2010) Negative Poisson's ratio weft-knitted fabrics. Text Res J 80: 856-863.

5. Hu H, Wang ZY, Liu S (2011) Development of auxetic fabrics using flat knitting technology. Text Res J 81: 1493-1502.

6. Alderson K, Alderson A, Anand S, Simkins V, Nazare S, et al. (2012) Auxetic warp knit textile structures, physica status solidi(b) 249: 1322-1329

7. Fletcher HM, Roberts SH (1956) The Geometry and Properties of Two-bar Tricot Fabrics of Acetate, Viscose, and Cotton. Text Res J 31: 151-159.

8. Siimmel R (1951) An Approach to Tricot Fabric Engineering. Text World 101: 130.

9. Stimmel R (1952) Standardisation of Tricot Construction. Knit. Outers Times 20: 71-73.

10. Allison GL (1958) Warp-knitting Calculations Made Easy, Skinner's Silk Rayon Rec 32: 28i-283

11. Grosberg P (I960) The Geometry of Warp-knitted Fabrics. J Text Inst 51: T39-T48.

12. Grosberg P (1964) The Geometrical Properties of Simple Warp-knit Fabrics. J Text Inst 55: T18-T30.

13. Shinn WE, El-Araf M (1966) The Geometry of Warp-knit Structures, Knit. Outerwears Times 35: 24-27.

14. Wheatley B (1973) The Dimensional Properties of Raschel-knitted Fabrics Constructed from Wool Yarns.

15. Goktepe O, Harlock SC (2002) A 3D Loop Model for Visual Simulation of Warp-knitted Structures. J Text Inst 93:11-28

16. Kyzymchuk O, Ugbolue SC (2012) The effect of positioning of inlaid yarns in fillet warp knit structures.

17. Dalidovich AS (1970) Basics of knitting technology.

18. Goktepe O, Harlock SC (2002) 3D Computer Modelling of Warp-knitted Structures. Text Res J 72: 266-272

19. Ugbolue SC, Warner SB, Kim YK, Fan Q, Chen Lu, et al. (2006) The Formation and Performance of Auxetic Textiles.

20. Ugbolue SC, Warner SB, Kim YK, Fan Q, Chen Lu, et al. (2007) The Formation and Performance of Auxetic Textiles.

21. Ugbolue SC, Warner SB, Kim YK, Fan Q, Chen Lu, et al. (2008) The Formation and Performance of Auxetic Textiles.

22. Ugbolue SC, Warner SB, Kim YK, Fan Q, Chen Lu, et al. (2009) The Formation and Performance of Auxetic Textiles.

10

Synthesis of a Novel Nanoencapsulated n-Eicosane Phase Change Material with Inorganic Silica Shell Material for Enhanced Thermal Properties through Sol-Gel Route

Mohy uddin H.G, Jin Z and Qufu W*

Key Laboratory of Eco-textiles, Jiangnan University, Wuxi, Jiangsu, China

Abstract

A novel nanoencapsulated phase change material (PCM) based on an n-eicosane core and an inorganic silica shell was synthesized through sol-gel route by using tetraethyl-orthosilicate (TEOS) and sodium silicate as an inorganic silica-precursor at different conditions for enhanced thermal stability and phase change properties. Fourier transform infrared spectra confirm the chemical composition of synthesized nanocapsules. Scanning electronic microscopic images show that the nanocapsules consist of spherical morphology. Furthermore, nanocapsules present different particle size range between 250-550 nm with respect to their pH values; the optimum pH for n-eicosane/TEOS nanocapsules is 2.20, and 2.94 for n-eicosane/sodium silicate, respectively. In addition to this, capsules synthesized by using TEOS show small particle size distributions as compared to the ones integrated by using sodium silicate as a silica-precursor. Differential scanning calorimetry suggests that by controlling the acidity of the reaction solution nanoencapsulated n-eicosane/silica can achieve good phase change properties and high encapsulation efficiency. Thermogravimetric analysis (TGA) shows that silica-nanocapsules have good thermal stability and phase change performance. Synthesis of nanoencapsulated n-eicosane (PCM) with the silica shell material through sol-gel process can be a perspective technique to prepare the nano-PCMs with enhanced thermal transfer and phase change properties for potential applications to thermal-regulating textiles and fibers.

Keywords: N-eicosane; Sol-gel process; Nanoencapsulation; Silica-precursors; Thermal and phase-change performance

Introduction

In the past two centuries the industrial development has produced remarkable wealth and brought prosperity to the whole world. However, it has also triggered a series of environmental issues [1]. Therefore, the development of new energy storage materials has been drawing huge attention in both academic and industrial communities all around the globe [2]. Phase change materials (PCMs) are considered as a good candidate with efficient utilization of energy to reduce the dependence on traditional fossil energy sources [3]. PCMs can also be used for thermal-regulating fiber and textiles [4]. When PCMs are integrated into fibers or textile materials, they can change their phase from solid to liquid to absorb the access heat as the temperature increases above the human body temperature. Otherwise, when the temperature decreases below the human body comfortable temperature these PCMs change their phase from liquid to solid by desorbing heat [5]. During this phase-change process these PCMs can actually bring the human body to the comfortable feel [6].

The organic n-alkanes (paraffin waxes), fatty alcohols, fatty acids, neopentyl glycol, eutectic mixtures, and some inorganic substances like salt hydrates are most usually applied PCMs to the thermal regulating fiber and textiles, which have high phase change enthalpies between 150 and 240 J/g [7]. Especially paraffin waxes (n-alkanes), are one of the most promising organic PCMs, can be generally applied in textiles like garments and home furnishing products [8,9]. These linear hydrocarbons have suitable melting–freezing temperature range between 18°C and 36°C, making humans feel comfortable [10,11].

However, these paraffin PCMs are mobile when molten to low viscous liquids and may diffuse throughout other materials which leads to a decrease in their lifespan [12]. Hence, the use of pristine PCMs is usually not endorsed in most cases. Therefore the encapsulation of PCMs has been considered as a vital solution for the problems mentioned above. The encapsulated PCMs cannot only maintain their microscopic solid form during the phase change processes to improve the ease of handling but also provide a large heat transfer area [13]. Most of the literatures indicated that both organic polymers and inorganic materials could be employed as shell materials to encapsulate PCMs through chemical processes; such as suspension polymerization [14], interfacial polycondensation [15], *in situ* polycondensation [16], *in situ* precipitation [17] and other special *in situ* processes [18]. These shell materials covered polyurea-formaldehyde resin [19], melamine-formaldehyde resin, poly(methyl methacrylate) (PMMA) [20], polystyrene [21], $CaCO_3$ [22], SiO_2 [23], TiO_2 [24], Al_2O_3 [25], and so more.

Though, these polymeric wall materials have some drawbacks, such as flammability, low mechanical strength and poor thermal and chemical stability [26,27]. Most of inorganic materials show a better rigidity and strength than the polymeric ones, and therefore, a high-strength inorganic shell material not only improves the thermal transfer performance of a PCM system but also increases the durability and working reliability of encapsulated PCMs [27,28]. Recently, silica materials as carriers in controlled drug release, has gained growing interest due to their several attractive features such as stable structure, high surface area, tunable pore sizes, well-defined surface properties,

*Corresponding author: Wei Q, Key Laboratory of Eco-textiles, Jiangnan University, Wuxi 214122, China
E-mail: qfwei@jiangnan.edu.cn

nontoxic nature [29], and good biocompatibility [29]. Moreover, these silica spheres expedite a high storage capacity, chemical, thermal, excellent thermal conductivity, and environmentally inert characteristics [30].

As cited in many articles, the inorganic silica was most employed as a wall material for the encapsulation of PCMs. Wang et al. [31] first reported the encapsulation mechanism of PCMs with silica through *in situ* polycondensation in an oil-in-water emulsion. Li et al. [32] and Fang et al. [33] reported the synthetic method of the silica/paraffin phase-change composites through interfacial polycondensation and found that these phase change composites exhibited a good heat storage performance and also had a form-stable feature. Jin et al. [34] developed a one-step synthetic method for the microencapsulated PCMs with the silica wall in dispersant free condition.

In this article, we attempted to nano-encapsulate the n-eicosane PCM with silica wall, n-eicosane is an important member of the paraffin family, it has been used as a thermal-regulating functional PCM for clothing application due to its appropriate phase-change temperature around 37°C, comfortable for the human body as well as its high latent heat of around 248 J/g. Nanocapsules were prepared by sol-gel method through an oil-in-water (O/W) emulsion route due to its easy control over size and morphology of the capsules and size distributions. The PCM material first dispersed in an aqueous solution to form droplets with the aid of surfactant, which resulted in a stable O/W emulsion. Low-temperature hydrolysis and condensation reactions occurred to accomplish the gelation of the solution. In this study we extract silica wall material by using two different precursors; Tetraethyl-orthosilicate (TEOS) and sodium silicate, though both are inorganic materials with their own advantages and disadvantages, generally TEOS is more often used in laboratory-scale work due to its high cost as a raw material on other hand sodium silicate consider to be easy available and of low cost as compared to TEOS. The aim of this work to develop a novel inorganic encapsulation technique for PCMs to enhance their performance in heat energy storage and thermal regulation, and to investigate the formation mechanism of these silica nanocapsules, influence of pH range and comparison of two different precursors.

Experimental section

Materials

Tetraethyl orthosilicate (TEOS), Sodium silicate and hydrochloric acid (HCl 36.46 wt%) were purchased from Sinopharm Chemical Reagents Company, China. N-eicosane with the purity of (99 wt%) and Poly (ethylene oxide-b-propylene oxide-b-ethylene oxide) triblock copolymer (PEO-PPO-PEO, Pluronic P123) were commercially obtained from Shanghai Vita chemical Reagents Company, China. All chemicals were of reagent quality and used without further purification.

Synthesis of capsules

A series of nanoencapsulated n-eicosane with silica shell were synthesized at various pH values through a sol-gel route by using two different silica-precursors. The nanoencapsulation of n-eicosane was derived from the hydrolysis and condensation reactions of TEOS/sodium silicate in an O/W emulsion and HCL was used as a catalyst. A synthetic procedure for TEOS derived n-eicosane/silica capsules is described as follows: In one beaker, PEO-PPO-PEO (0.25 g) was dissolved in 150 ml deionized water at 55°C. When the emulsifier completely agitates in water and the temperature become stable then, n-eicosane (15.0 g) was added into this solution and continuously stirred for 3 h to form a stable emulsion. In another beaker, the

mixture of TEOS (15.0 g) was taken and to get the desired pH value, HCl aqueous solution was added dropwised and stirred at 35°C until a homogeneous solution was obtained, indicating that the hydrolysis of the TEOS was completed and the silica-sol solution as encapsulation precursors was formed. Subsequently, the silica-sol solution was added drop wise into the prepared emulsion and kept it stirring for 24 h to complete the gel process.

Procedure for sodium silicate derived n-eicosane/silica capsules is described as follows: PEO-PPO-PEO (0.5 g) was dissolved in one beaker with 250 mL of deionized water at 70°C. Then, n-octadecane (10.0 g) was added into this solution, and to get a stable emulsion continually stirred it for 1 h to form a stable emulsion. In another beaker, sodium silicate (5.0 g) was dissolved in 50 mL of deionized water at 35°C, and the HCl aqueous solution was added to obtain a desired pH value under a vigorous stirring until a homogeneous solution was obtained. Which indicated that the silica-sol solution was formed as silica monomers and oligomers and hydrolysis of sodium silicate was completed. Afterwards, this silica sol solution was added dropwise into the prepared n-eicosane emulsion under continuous stirring. This mixture was heated to 70°C with agitation for 24 h to complete the silicate condensation.

After filtration and washed with deionized water some white powders were collected. Finally, the silica-nanoencapsulated n-eicosane was obtained after being washed with ethanol to effectively remove the residual surfactants and drying at 50°C overnight.

Characterization

Morphologies of nanocapsules were obtained by using scanning electron microscope (SEM, SU1510). During the synthetic process before filtration, the status of the nanocapsules was monitored by Moritex ML-Z07545D optical microscope equipped with JVC TK-C9201EC digital camera. The size of nanocapsules was detected using NanoBrook Omni (280039) a laser particle size analyzer, and the mean diameters of the silica-nanoencapsulated n-eicosane were determined.

Fourier transform infrared (FTIR) spectra of the sample were obtained by using a Nicolet iS10 FT-IR spectrometer (Thermo Fisher Scientific), which were recorded in the range of 400-4000 cm^{-1}.

PCMs were measured through differential scanning calorimetry instrument (DSC-Q200) with the scanning rate of 10°C/min in nitrogen atmosphere. DSC curves were recorded with the temperature ranging from 5 to 50°C with a scanning rate of 10°C/min. The precisions of measurements were ±2.0% and ±2.0°C for calorimeter and temperature, respectively.

Thermogravimetric analysis (TGA) for the nanocapsules was carried out at a heating rate of 10°C/min under a nitrogen atmosphere from room temperature to 550°C by using a TA thermal gravimetric analyzer.

Results and Discussions

Synthesis mechanism of nanocapsules with silica wall

Nano-encapsulation of n-eicosane PCM with inorganic silica was achieved via sol-gel process through *in situ* condensation. Figure 1 shows the schematic route of this sol-gel synthesis. This encapsulation mechanism is described as follows: Firstly the oily n-eicosane was dispersed in an aqueous solution including a nonionic emulsifier. To obtain the stable O/W emulsion, vigorous agitation was carried out. During this period, the hydrophilic segments of the emulsifier alternatively arrange along its hydrophobic chains, and thus are linked with the water molecules and trimly cover the surface of the oil droplets of n-eicosane with hydrophobic chains oriented into the oil droplets and hydrophilic groups out of the oil droplets. One the other hand, the silica sol was prepared by dissolving TEOS/sodium silicate in water under acidic conditions. The reaction can be explained as the hydrolysis and condensation processes of the silica source. In case of TEOS, hydrolysis reaction occurs in the presence of an acidic catalyst with a controlled pH around 1.9-2.5. The hydrolysis rate of TEOS is significantly higher than the condensation rate of the silica precursors at pH around 2.0. When the silica-sol solution is added drop wise into the emulsion containing n-eicosane micelles, these silica precursors are attracted onto the surfaces of the micelles through a hydrogen-bonding interaction between the silanol groups and the hydrophilic PEO segments of the surfactant, and by mixing the silica-sol solution with the emulsion reduces the concentration of hydrogen ions, which results in a rise in pH values. This leads to an acceleration of the condensation rate of the silica precursors, while an opposite trend occurs for the hydrolysis rate as long as the pH value is below 7. Here, the alcoholysis and oxolation reactions become dominant, and consequently, though this sol-gel process a silica shell is successfully fabricated onto the surface of the n-eicosane droplets.

Microstructure and particle size and distribution of nanocapsules

Images were taken by using optical microscope, when the reaction was completed in the solution before the filtration of nanocapsules. Figure 2a and 2b displays the formation of n-eicosane/TEOS nanocapsules in the solution and Figure 2b and 2d shows n-eicosane/sodium silicate, respectively. Growth of nanocapsules can be seen in all images, which illustrates the successful encapsulation of PCM in silica wall material.

Figure 3 shows the SEM images of the nanoencapsulated n-eicosane/silica synthesized at different pH values and silica-precursors. It can be observed from these micrographs that the nanocapsules obtained by n-eicosane/TEOS at pH 2.20 demonstrate some regular spheres with the diameter of about 330 nm, and these well-defined nanocapsules have very compact and even surfaces (Figure 3a and 3b). However, in case of TEOS when the pH was under 2 (Figure 3c and 3d), it is observed that the condensation rate of the silica precursors is so slow

in the solution at pH 1.88 in this case, the shell is so thin that it can easily be cracked by brisk agitation, and then the subsequent silica oligomers continuously accumulate on the surface of the nanocapsules. As a result, the nanocapsules synthesized at pH below 2 have a thin and porous silica shell with a rough surface. However, when the acidity of the reaction solution increases the surface structure becomes finer. Figure 3e and 3f shows the morphologies of n-eicosane/sodium silicate nanocapsules, these capsules also shows spherical morphologies but it is found that they have bigger diameters about 528 nm at pH 2.94. Thus the most important factor along with temperature and agitation is pH, the capsules synthesized at pH values (2.9-3) have shown good spheres for sodium silicate but in case of TEOS best morphologies were obtained in the range of (pH 1.9-2.3).

The particle size and distribution of the silica nanocapsules synthesized at different pH values and silica-precursors were determined by using a laser particle size analyzer. The obtained mean diameters are listed in Table 1, and the size distribution plots are displayed in Figure 4, which indicates a multidispersed particle size distribution. Though there is a uniform Gaussian distribution for these nanocapsules, it can also be noted that there are small differences in the mean particle sizes and size distributions of the nanocapsules obtained under different conditions. It is obvious that the acidity of the reaction affects the particle size and size distribution with great impression. For the sample (a) synthesized with n-eicosane/TEOS at pH 2.20 exhibit a greater mean particle size and broader distributions than sample (b and c) which were synthesis under different pH values, whereas (b and c) showed wider size distributions than sample (a). The nanocapsules synthesized with n-eicosane/sodium silicate at pH 2.94 (Figure 4e) exhibit large mean particle size of 528 nm compared to the sample obtained at pH 2.50 (Figure 4d).

Evidently, the acidity plays an important role in the formation of the shell materials, and silica condensation can be conducted continuously and steadily with the proper acidity of the reaction solution. Figure 4d and 4e, when the acidity of the reaction increases to pH 2.94 from pH 2.50 the silica condensation also accelerates and more original n-eicosane micelles are encapsulated and we can see the wider particle size distribution. The effect of different silica-precursors on particle size and size distribution of nanocapsules is also illustrated in Figure

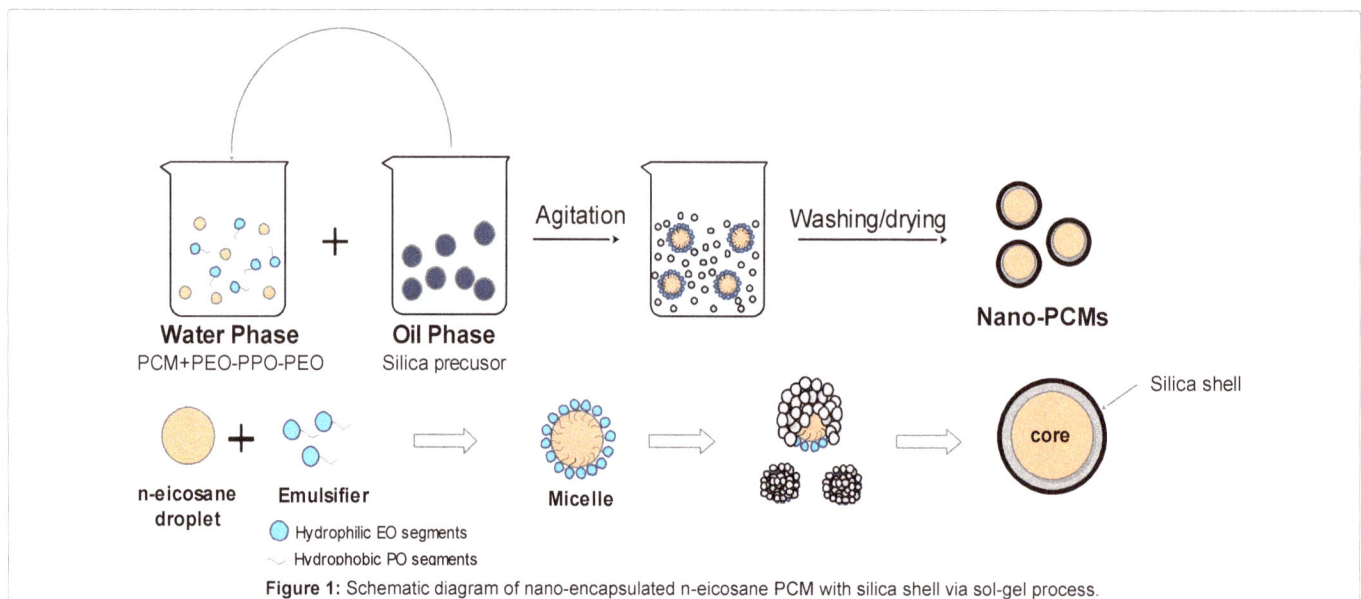

Figure 1: Schematic diagram of nano-encapsulated n-eicosane PCM with silica shell via sol-gel process.

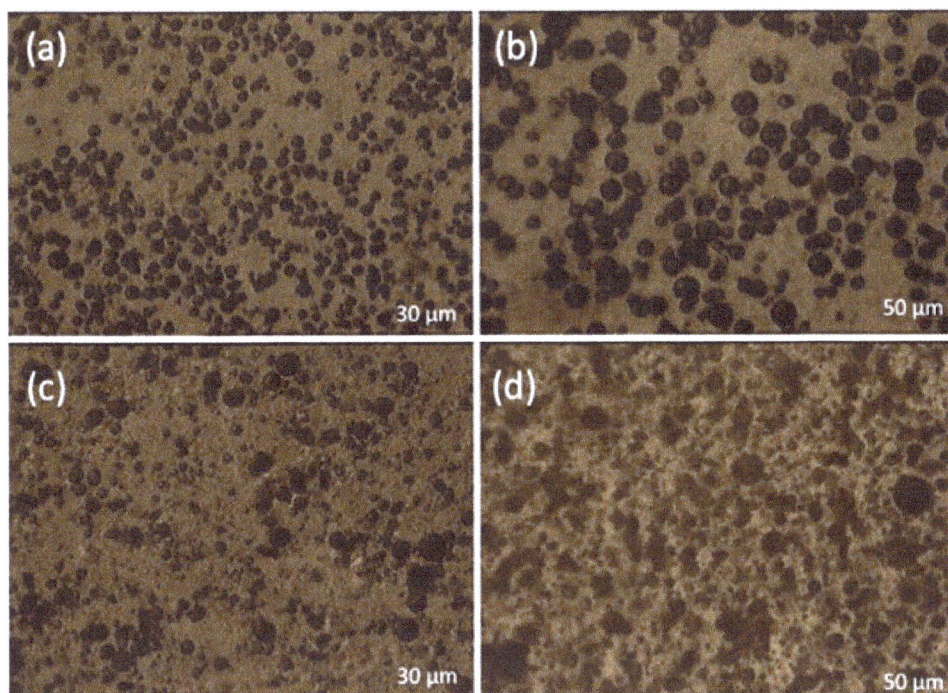

Figure 2: Optical microscope images of the n-eicosane nanocapsules after completeion of 24 h reaction in solution before filtration and washing; (a and b) n-eicosane/TEOS at pH 2.20, (c and d) n-eicosane/sodium silicate at pH 2.94.

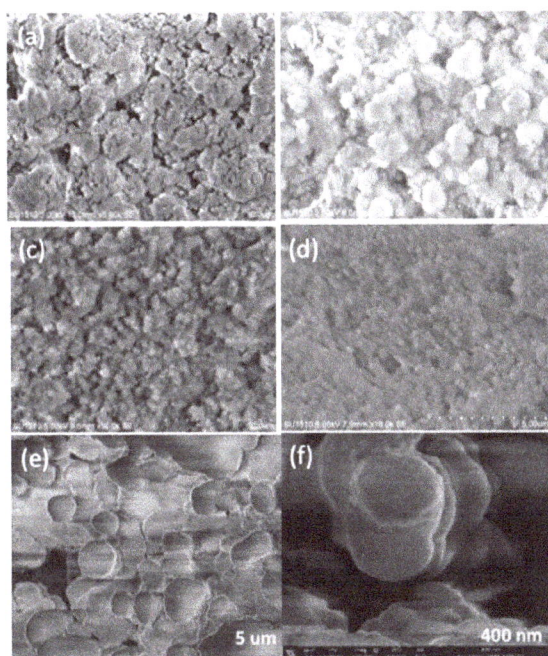

Figure 3: SEM images of nanocapsules synthesized at different pH values and by using TEOS/sodium silicate as a silica-precursors: (a and b) n-eicosane/TEOS at pH 2.20; (c and d) n-eicosane/TEOS at pH 1.88; (e and f) n-eicosane/sodium silicate at pH2.94.

4 and Table 1. It is found that the nanocapsules synthesized with n-eicosane/TEOS have a small mean particle size as compared to the n-eicosane/sodium silicate. As it can be seen, the mean particle size of the sample (b) is 274 nm and when the pH value decreases to pH 1.88,

the mean particle size also increases. And the samples (d, e) prepared with n-eicosane/sodium silicate showed large mean particle sizes. This result indicates that the size of silica-encapsulated n-eicosane is strictly dependent on the pH values, which determines the ultimate particles size of the nanocapsules.

Chemical composition

In Figure 5a displays the FTIR spectra of the bulk n-eicosane and n-eicosane/TEOS nanocapsules synthesis at different pH values and Figure 5b shows the FTIR spectra of bulk n-eicosane and n-eicosane/ sodium silicate nanocapsules prepared at different pH values respectively. In the spectrum of pure n-eicosane and n-eicosane/ TEOS the C-H bonding vibrations at 1471 and 1376 cm^{-1} are due to methylene brides and the alkyl C-H stretching vibrations of methyl and methylene groups can be seen at 2921 and 2851 cm-1, respectively. (Figure 5a). It is found that the spectra of all the silica nanocapsules prepared by TEOS are quite similar, in which the peaks at 1082 and 457cm^{-1} are respectively attributed to asymmetric and symmetric Si-O-Si stretching vibrations of the silica shells. However, n-eicosane/sodium silicate nanocapsules displayed only one Si-O-Si vibration at 1064 cm^{-1}(Figure 2a). In Figure 2a an intense absorption band at 954 cm-1 and a weak one is at 3385 cm-1 assigned to Si–OH bending and stretching vibrations as in Figure 5b at 958 cm-1 and at 3390 cm-1, respectively. It is noteworthy that all the characteristic peaks of n-eicosane can be distinguished in the spectra of the nanocapsules, which confirms the successful encapsulation of n-eicosane within the silica shell. As the nanocapsules were synthesized through the condensation of the silica precursors via a so-gel route, it is expected that a large number of silanol groups could be detected on their silica shells.

Phase-change performance

Dynamic DSC scans were performed to examine the phase

Sample code	n-Eicosane/Silica precursor ratio (wt/wt)	Acidity (pH)	Mean diameter (nm)	T_c(°C)	T_m(°C)	Encapsulation ratio (%)	Encapsulation efficiency (%)	Char yield at 500°C (wt%)
1	100/0	-	-	33.6	38.6	-	-	-
2	50/50 (TEOS).	2.25	298	32.1	39.8	79.5	77.8	14.45
3	50/50 (TEOS)	1.88	274	33.1	40.8	70.2	70.5	17.42
4	50/50 (TEOS)	2.2	348	33.3	39.8	74	72.4	13.1
5	50/50 (S.S)	2.94	528	30.5	38.5	56.6	60.6	14.1
6	50/50 (S.S)	2.5	472	29.4	41	21.3	21.5	23.41

Table 1: The mean particle sizes, melting and crystallization temperatures and thermal performances of nanoencapsulated n-eicosane/silica.

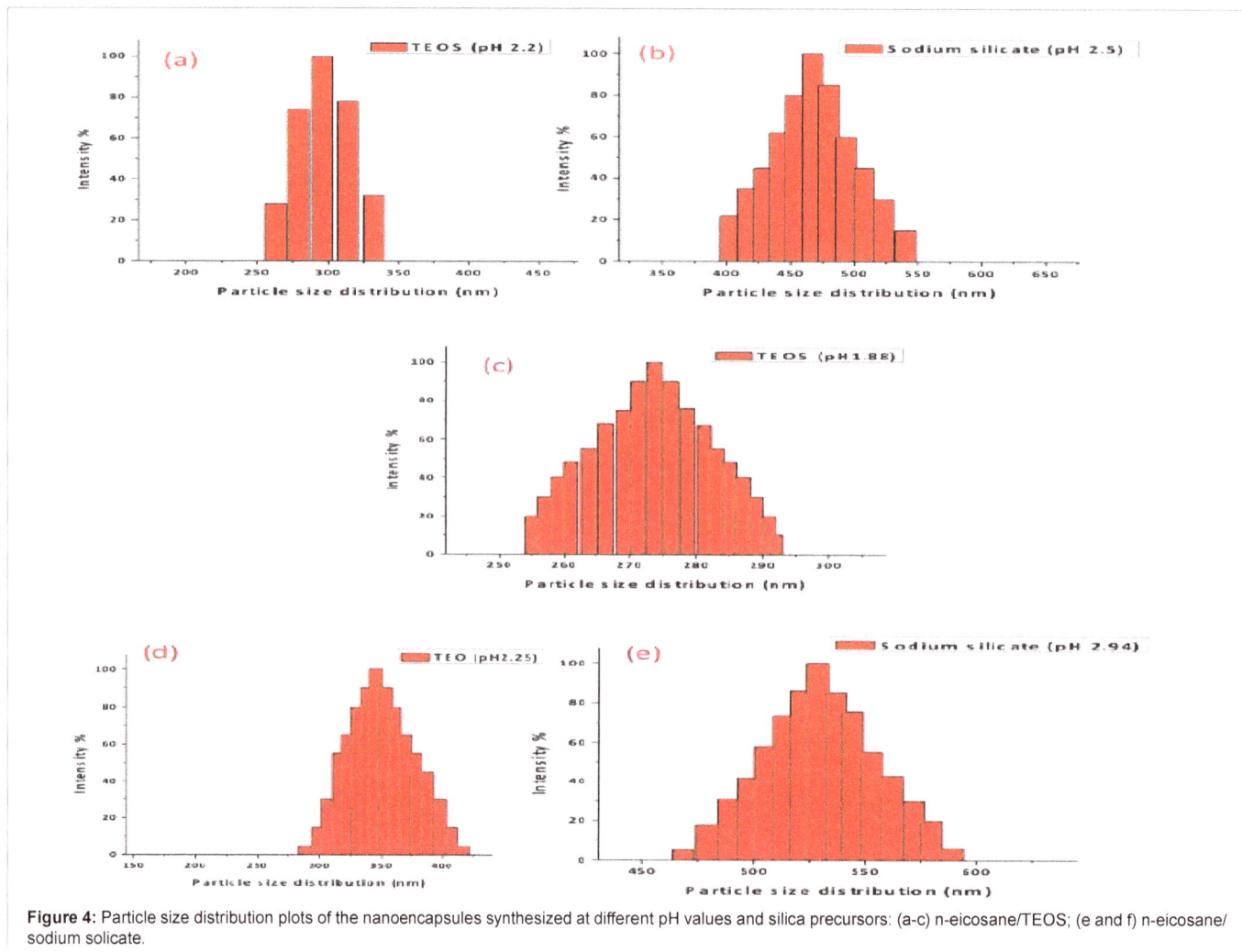

Figure 4: Particle size distribution plots of the nanoencapsules synthesized at different pH values and silica precursors: (a-c) n-eicosane/TEOS; (e and f) n-eicosane/sodium solicate.

change behavior and latent heat storage performance of nanocapsules synthesized at different pH values and by using two different silica-precursors. The DSC curves and phase change temperatures is presented in Figure 6 and Table 1, respectively. It is noteworthy in Figure 6a that pure n-eicosane exhibits a bimodal phase transition behavior with two exothermic peaks during its crystallization process. A number of studies have indicated that normal alkanes like n-eicosane usually present a rotator phase above the bulk crystallization temperature during the phase transition from liquid to solid [35]. Such a metastable rotator phase is considered as the orthorhombic rotator phase with molecules untitled with respect to the layers. As a result, pure n-eicosane has to undergo twice phase transitions between the isotropic liquid and stable orthorhombic phases. The first transition is from the homogeneously nucleated liquid to the rotator phase, and the second one is from the

heterogeneously nucleated rotator phase to the crystalline phase. This leads to the presence of rotator-phase-transition peak at a temperature higher than the crystallization temperature in the DSC thermogram. However, pure n-eicosane only exhibits a single endothermic peak at 38.6°C in the melting process. It is noteworthy in Figure 6a and 6b that pure n-eicosane generates phase change enthalpies of 240 and 235 J/g during the crystallization and melting processes, respectively. This indicates the excellent latent-heat storage–release performance for pure n-eicosane as an organic PCM. It can be found that the melting curves of nano-encpsulated n-eicosane are become wider which shows their melting points are improved.

It can also be observed from Figure 6a and 6b that the encapsulation of n-eicosane into silica wall significantly influences the phase change performance of the n-eicosane core, although the nanocapsule samples synthesized by using TEOS as a silica-precursor demonstrate a phase change behavior similar to pure n-eicosane at different pH values, But sample (6) which was prepared by using sodium silicate as a silica precursor at pH 2.50 found to have the lowest enthalpies, probably due to the worst encapsulation ratio, though sample (5) at pH 2.94 showed quite superior phase change properties and enthalpies.

Encapsulation ratio and encapsulation efficiencies are two important parameters in describing the phase-change properties of the nanoencapsulated n-eicosane, and can be determined by the results from the DSC measurements. As one of the most important phase-change performance affecting the working effect of the nano-PCM, the phase-change enthalpy strongly depends on the encapsulation ratio and encapsulation efficiency of the silica nanoencapsulated n-eicosane. The values can be found in Table 1. The actual encapsulation efficiency (E) of nanocapsules was calculated by the following equation 1.

$$\mathcal{R} = \frac{nano\ (\Delta Hm)}{PCM\ (\Delta Hm)} \times 100\%$$

Nano ($\Delta H_m + \Delta H_c$) in the equation represented the plus of melting and crystallizing enthalpy from nanocapsules. PCM ($H_m + H_c$) was the plus of melting and crystallizing enthalpy from PCM. The encapsulation ratio (R) can be calculated by following equation 2.

$$\mathcal{R} = \frac{nano\ (\Delta Hm)}{PCM\ (\Delta Hm)} \times 100\%$$

Where ΔHm is the fusion heat of nano-PCM and bulk n-eicosanen respectively. Encapsulation ratio describes the effective encapsulation of n-eicosane within the nanocapsules while the loading content is considered as dry weight percent of the core material. Melting and crystallization enthalpies of silica-nanocapsules are shown in Figure 7.

Thermal stability

Figure 8a shows the digital photographs of the n-eicosane/silica nanocapsules and Figure 8b of pure n-eicosane respectively, while heating at 60°C on a hot stage. These photographs were taken at different melting stages so as to record the change of shape for two specimens; n-eicosane/silica nanocapsules and pure n-eicosane, during melting process. It is witnessed that pure n-eicosane gradually lost its original shape due to a transformation from solid to liquid state while melting, and it cannot be recovered to its original shape when cooling. However, the n-eicosane/silica nanocapsules were kept its original triangle shape during long-term heating at a temperature above the melting point of n-eicosane. There is no difference in shape between the original sample and the recovered one, and it appears that no liquid n-eicosane is found to leak out of the capsules. These results indicate that the n-eicosane core is well encapsulated within the silica material and isolated it from the outside environment. In this case, the silica layer could not only maintain the original shape of PCMs but also could provide a good protection for the encapsulated PCMs, preventing the molten n-eicosane from leakage accordingly.

For the applications of nanoencapsulated n-eicosane/silica in heat energy storage and thermal regulation, thermal stability is a significant factor to be evaluated. The thermal stability of the silica nanocapsules synthesized under different pH values and silica precursors was investigated by means of TGA, which presents the mass loss of samples and corresponding temperature. The thermogravimetric analysis (TGA) and derivative thermogravimetry (DTG) curves are represented in Figure 9a and 9b, respectively and obtained char yield (wt%) is also presented in Table 1. It is observed that the weight loss profiles of all

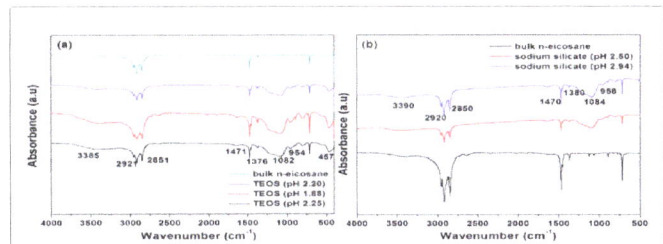

Figure 5: FTIR spectra of bulk and nanoencapsulated n-eicosane synthesized at different pH values and silica-precursors: (a) n-eicosane/TEOS; (b) n-eicosane/sodium silicate.

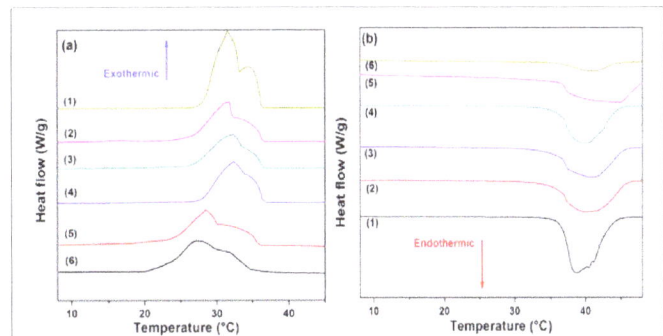

Figure 6: DSC thermograms of the bulk and nanoencapsulated n-eicosane synthesized under different conditions, the curve numbers correspond to the samples code.

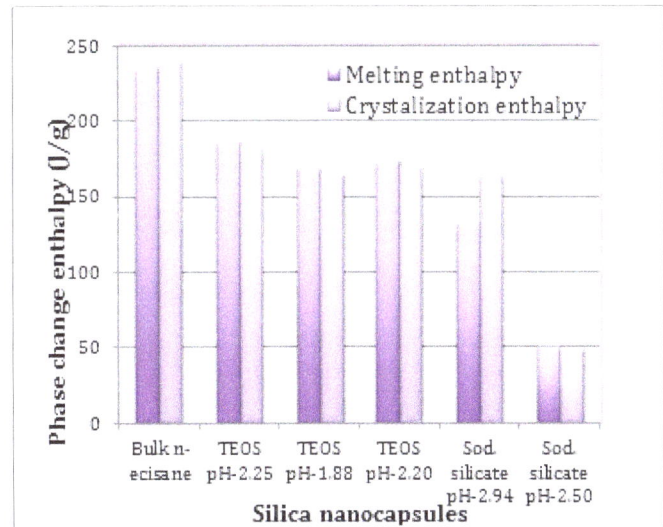

Figure 7: Phase change enthalpies of bulk n-eicosane and silica nanocapsules synthesized at different conditions.

the nanocapsules are quite similar, and their thermal decompositions performed in the temperature range off 35-500°C and occur through two-step decomposition. It can be easily found that silica nanocapsules, which were synthesized at optimum pH values, they start loosing their weight after reaching at higher temperatures, due to high encapsulation ratios and well defined thick silica walls, which indicates that they can withstand against high temperatures for more time. As Figure 9a clearly depicts that sample 2 and 5 lost their maximum weight in the range between 180-270°C, while on the other hand all other samples starts loosing their weight at around 120°C to onwards.

The DTG curves also display a narrow peak corresponding to the

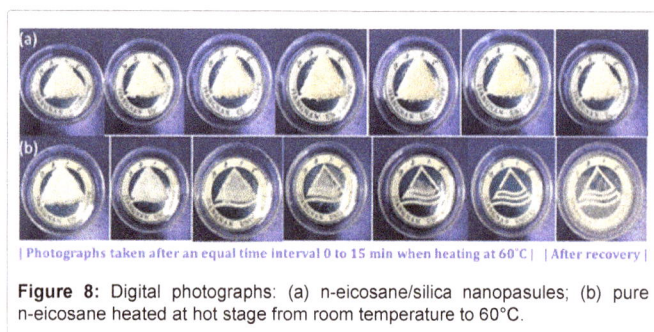

| Photographs taken after an equal time interval 0 to 15 min when heating at 60°C | | After recovery |

Figure 8: Digital photographs: (a) n-eicosane/silica nanopasules; (b) pure n-eicosane heated at hot stage from room temperature to 60°C.

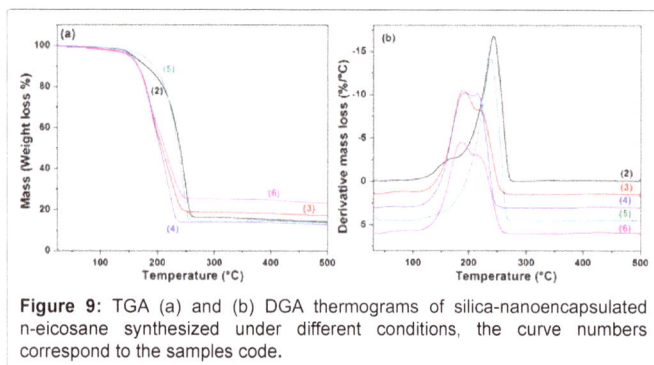

Figure 9: TGA (a) and (b) DGA thermograms of silica-nanoencapsulated n-eicosane synthesized under different conditions, the curve numbers correspond to the samples code.

temperature at the rapid weight loss of the n-eicosane when leaking from the nanocapsules. As comprehended in Figure 9b, the nanocapsules synthesized at the n-eicosane/TEOS at pH 2.25 and n-eicosane/sodium silicate at pH 2.94 resist to the highest temperature while rapid weight lose, which is attributed to the greatest thickness of their shells as a result of the highest silica loading. It is comprehensible that the compact and thick shell material prevents the leaking of the core material. However, the sample synthesized at pH 2.5 by using sodium silicate as a precursor has a loose shell due to the fast condensation rate as well as has the lowest encapsulation efficiency, resulting in a lower temperature at the rapid weight loss. Furthermore, the broad peaks of samples suggest a slower process of the further condensation of the silica shell material at high temperatures.

Conclusion

Nanoencapsulation of n-eicosane PCM with inorganic silica shell was carried out via sol-gel route by using two different inorganic silica-precursors. Capsules were synthesized at different pH values. FTIR spectra confirmed the chemical composition and crystallinity of the synthesized nanocapsules. The morphologies of the obtained nanocapsules strongly depend upon the pH value of the reaction solution. Nanoencapsulated n-eicosane derived from TEOS silica-precursor exhibited spherical morphologies when synthesized at pH 2.20~2.30, in this pH range nanocapsules also demonstrate good phase change properties and higher encapsulation rates and efficiency. Whereas, n-eicosane/sodium silicate obtained nanocapsules presented good overall properties with high encapsulation rates and thermal stability, when prepared at pH 2.90~3.00, but showed very poor encapsulation efficiency and enthalpies when prepared under pH 2.9. In addition, thermogravimetric investigation presented that n-eicosane/silica nanocapsules degrade in two steps and are resistant enough for usage at ambient temperatures. All of phase change properties with enhanced thermal performance make these nanocapsules a potential PCM for applications of thermal energy storage and thermal-regulating textiles and fibers.

Acknowledgements

This research work was financially supported by Chinese government scholarship (CSC) program.

References

1. Jiang F, Wang X, Wu D (2014) Design and synthesis of magnetic microcapsules based on n-eicosane core and Fe3O4/SiO2 hybrid shell for dual-functional phase change materials. Appl Energy 134: 456-468.

2. Chai L, Wang X, Wu D (2015) Development of bifunctional microencapsulated phase change materials with crystalline titanium dioxide shell for latent-heat storage and photocatalytic effectiveness. Appl Energy 138: 661-674.

3. Özonur Y, Mazman M, Paksoy HÖ, Evliya H (2006) Microencapsulation of coco fatty acid mixture for thermal energy storage with phase change material. Int J Energy Res 30: 741-749.

4. Mondal S (2008) Phase change materials for smart textiles - An overview. Appl Therm Eng 28: 1536-1550.

5. Kara YA (2016) Diurnal performance analysis of phase change material walls. Appl Therm Eng 102: 1-8.

6. Mohamed SA, Al-Sulaiman FA, Ibrahim NI, Zahir MH, Al-Ahmed A, et al. (2017) A review on current status and challenges of inorganic phase change materials for thermal energy storage systems 70: 1072-1089.

7. Sun Z, Zhang Y, Zheng S, Park Y, Frost RL (2013) Preparation and thermal energy storage properties of paraffin/calcined diatomite composites as form-stable phase change materials. Thermochim Acta 558: 16-21.

8. Khudhair AM, Farid MM (2004) A review on energy conservation in building applications with thermal storage by latent heat using phase change materials. Energy Convers Manag 45: 263-275.

9. Rao ZH, Zhang GQ (2017) Thermal Properties of Paraffin Wax-based Composites Containing Graphite Energy Sources Part A: Recovery 7: 587-593.

10. Kim EY, Do KH (2005) Preparation and properties of microencapsulated octadecane with waterborne polyurethane. J Appl Polym Sci 96: 1596-1604.

11. Delgado M, Lázaro A, Mazo J, Zalba B (2012) Review on phase change material emulsions and microencapsulated phase change material slurries: Materials, heat transfer studies and applications. Renew Sustain Energy Rev 16: 253-273.

12. Su W, Darkwa J, Kokogiannakis G (2015) Review of solid-liquid phase change materials and their encapsulation technologies. Renew Sustain Energy Rev 48: 373-391.

13. Borreguero AM, Carmona M, Sanchez ML, Valverde JL, Rodriguez JF (2010) Improvement of the thermal behaviour of gypsum blocks by the incorporation of microcapsules containing PCMS obtained by suspension polymerization with an optimal core/coating mass ratio. Appl Therm Eng 30: 1164-1169.

14. Zhan S, Chen S, Chen L, Hou W (2016) Preparation and characterization of polyurea microencapsulated phase change material by interfacial polycondensation method. Powder Technol 292: 217-222.

15. Zhang H, Wang X (2009) Fabrication and performances of microencapsulated phase change materials based on n-octadecane core and resorcinol-modified melamine-formaldehyde shell. Colloids Surfaces A Physicochem Eng Asp 332: 129-138.

16. Yu S, Wang X, Wu D (2014) Microencapsulation of n-octadecane phase change material with calcium carbonate shell for enhancement of thermal conductivity and serving durability: Synthesis, microstructure, and performance evaluation. Appl Energy 114: 632-643.

17. Zhao CY, Zhang GH (2011) Review on microencapsulated phase change materials (MEPCMs): Fabrication, characterization and applications. Renew Sustain Energy Rev 15: 3813-3832.

18. Zhang H, Wang X (2009) Synthesis and properties of microencapsulated n-octadecane with polyurea shells containing different soft segments for heat energy storage and thermal regulation. Sol Energy Mater Sol Cells 93: 1366-1376.

19. Sarı A, Alkan C, Karaipekli A (2010) Preparation, characterization and thermal properties of PMMA/n-heptadecane microcapsules as novel solid-liquid microPCM for thermal energy storage. Appl Energy 87: 1529-1534.

20. Fang Y, Liu X, Liang X, Liu H, Gao X, Zhang Z (2014) Ultrasonic synthesis and characterization of polystyrene/n-dotriacontane composite nanoencapsulated

phase change material for thermal energy storage. Appl Energy 132: 551-556.

21. Bayés-García L, Ventolà L, Cordobilla R, Benages R, Calvet T, et al. (2010) Phase Change Materials (PCM) microcapsules with different shell compositions: Preparation, characterization and thermal stability. Sol Energy Mater Sol Cells 94: 1235-1240.

22. Zhang X, Wang X, Wu D (2016) Design and synthesis of multifunctional microencapsulated phase change materials with silver/silica double-layered shell for thermal energy storage, electrical conduction and antimicrobial effectiveness. Energy 111: 498-512.

23. Chai L, Wang X, Wu D (2015) Development of bifunctional microencapsulated phase change materials with crystalline titanium dioxide shell for latent-heat storage and photocatalytic effectiveness. Appl Energy 138: 661-674.

24. Pan L, Tao Q, Zhang S, Wang S, Zhang J, et al. (2012) Preparation, characterization and thermal properties of micro-encapsulated phase change materials. Sol Energy Mater Sol Cells 98: 66-70.

25. Tyagi VV, Kaushik SC, Tyagi SK, Akiyama T (2011) Development of phase change materials based microencapsulated technology for buildings: A review. Renew Sustain Energy Rev 15: 1373-1391.

26. Su J-F, Wang L-X, Ren L (2007) Synthesis of polyurethane microPCMs containing n-octadecane by interfacial polycondensation: Influence of styrene-maleic anhydride as a surfactant. Colloids Surfaces A Physicochem Eng Asp 299: 268-275.

27. Sánchez-Silva L, Rodríguez JF, Romero A, Borreguero AM, Carmona M, et al. (2010) Microencapsulation of PCMs with a styrene-methyl methacrylate copolymer shell by suspension-like polymerisation. Chem Eng J 157: 216-222.

28. Tourné-Péteilh C, Lerner DA, Charnay C, Nicole L, Bégu S, Devoisselle JM (2003) The Potential of Ordered Mesoporous Silica for the Storage of Drugs: The Example of a Pentapeptide Encapsulated in a MSU-Tween 80. ChemPhysChem 4: 281-286.

29. Ozalp VC, Eyidogan F, Oktem HA (2011) Aptamer-gated nanoparticles for smart drug delivery. Pharmaceuticals 4: 1137-1157.

30. Rekuć A, Bryjak J, Szymańska K, Jarzebski AB (2009) Laccase immobilization on mesostructured cellular foams affords preparations with ultra high activity. Process Biochem 44: 191-198.

31. Wang L-Y, Tsai P-S, Yang Y-M (2006) Preparation of silica microspheres encapsulating phase-change material by sol-gel method in O/W emulsion. J Microencapsul 23: 3-14.

32. Li B, Liu T, Hu L, Wang Y, Gao L (2013) Fabrication and Properties of Microencapsulated Paraffin@SiO 2 Phase Change Composite for Thermal Energy Storage. ACS Sustain Chem Eng 1: 374-380.

33. Fang G, Chen Z, Li H (2010) Synthesis and properties of microencapsulated paraffin composites with SiO2 shell as thermal energy storage materials. Chem Eng J 163: 154-159.

34. Jin Y, Lee W, Musina Z, Ding Y (2010) A one-step method for producing microencapsulated phase change materials. Particuology 8: 588-590.

35. Wang LP, Wang TB, Gao CF, Lan X, Lan XZ (2014) Phase behavior of dodecane--hexadecane mixtures in bulk and confined in SBA-15. J Therm Anal Calorim 116: 469-476.

11

Eco-friendly Antibacterial Printing of Wool Using Natural Dyes

Teli MD*, Javed Sheikh and Pragati Shastrakar

Department of Fibres and Textile Processing Technology, Institute of Chemical Technology, Mumbai-400019, India

Abstract

Natural dyes which are known to possess antibacterial properties can be safely used for the purpose of antibacterial finishing of natural fibres like wool. Chitosan is a functional biopolymer which can be utilized for various applications in textiles. In the present work bio-waste chitosan was utilized as a mordant in the printing of wool fabric with natural dyes making it eco-friendly. The efficacy of natural mordant chitosan was studied in comparison with alum as a standard mordant. Both the mordants gave prints of similar color values. The printed fabric showed excellent antibacterial properties against both *S. aureus* and *E. coli*. This method hence can be suitable for eco-friendly printing and antibacterial finishing of wool fabrics.

Keywords: Natural dyeing; Antibacterial; Wool; Chitosan

Introduction

Natural fibres are an excellent medium for the growth of microorganisms when the basic requirements for their growth such as nutrients, moisture, oxygen, and appropriate temperature are present. The large surface area and ability to retain moisture of textiles also assist the growth of microorganisms on the fabric [1]. The growth of microorganisms on textiles inflicts a range of unwanted effects not only on the textile itself (such as tendering and its degradation) but also on the wearer (skin rashes, foul smell, etc.) [2]. In the last few decades, with the increase in new antimicrobial fibre technologies and the growing awareness about cleaner surroundings and healthy lifestyle, a range of textile products based on synthetic antimicrobial agents such as triclosan, metal and their salts, organometallics, phenols and quaternary ammonium compounds, have been developed and quite a few are also available commercially [3]. Even though the excellent antimicrobials are available, their user ecology and safety is always a question. Some of the natural dyes, on the other hand such as turmeric, catechu, marigold etc. are reported to possess antimicrobial properties. This is also substantiated by the work reported from our laboratory [4-8].

Synthetic dyes offer the flexibility of selection of proper hue and substantively with reference to the fibre. The development of synthetic dyes in the second half of nineteenth century led to a high quality dyes with better reproducible techniques of application. As a result, a distinct lowering in the dyestuff costs per kg of dyed goods was achieved [9]. Also a full gamut of color is available for application on various types of fibres. However, during the last few decades, the use of synthetic dyes is gradually receding due to an increased environmental awareness and harmful effects, of either toxic degraded products of these dyes or their non-biodegradable nature. In addition to above, some serious health hazards like allergy and, carcinogenicity are associated with some of the synthetic dyes based on azochromospheres. As a result, a ban has been imposed all over the world including European Economic Community (EEC), Germany, USA and India on the use of some synthetic dyes (e.g. azodyes) which finally triggered active research and development to revive world heritage and traditional wisdom of employing safer natural dyes [10]. Even though natural colors cannot substitute the synthetic dyes completely, there is definitely increasing market for such complete eco-friendly dyed or printed materials, if not for common market, but for those who desire to have such products.

Natural dyes with a few exceptions are non-substantive and hence must be used in conjunction with mordants. Mordant is a chemical, which can fix itself on the fibre and also combines with the dyestuff. The challenge before the natural dyers in application of natural color is thus, the necessity to use metallic mordants which themselves are pollutant and harmful. Due to the environmental hazard caused by metallic mordant while dyeing of textile fabric, dyers are always looking for safe natural mordant for natural dyes. In this paper, this issue of screening a natural mordant and exploring its potential for printing of wool with natural dyes is reported.

The main objective of printing is to produce colored patterns with sharp boundaries on textile materials without any dye spreading beyond the boundaries of the motif of design [11]. Although a lot of research has been reported in the area of natural dyeing of textiles, the research in the area of natural dye printing is reported to a limited extent. Also the subject like obtaining antibacterial finish through natural dye printing has also remained unexplored. Under this backdrop chitosan which has been utilized for various applications in textiles ranging from fibres, dyeing auxiliary, printing thickener, antimicrobial finishing, etc. [12-27], becomes an obvious choice for study. There is very limited information available in the literature with regard to use of such functional biopolymer as a mordant in natural dyeing. Application of chitosan as an eco-friendly mordant in dyeing [28] and printing [8] of cotton with natural dyes was reported from our laboratory.

In the present work, chitosan extracted from waste shrimp shells [29] was utilized as a mordant for simultaneous natural dye printing and antibacterial finishing of wool and its performance is discussed in comparison with alum as a standard mordant. The efficacy of chitosan as eco-friendly mordant and antibacterial finish against both gram positive and gram negative bacteria has been investigated.

*Corresponding author: Teli MD, Department of Fibres and Textile Processing Technology, Institute of Chemical Technology, Mumbai-400019, India
E-mail: textileudct@yahoo.com

Materials and Methods

Materials

Chitosan (mol. wt. 124999.8, degree of deacetylation, 89.06%, nitrogen content, 7.1%) was extracted from waste shrimp shells [29] and used for the study. Wool fabric as well as catechu, turmeric and marigold were purchased from the market. All other chemicals used were of laboratory grade.

Methods

Preparation of mordant: Stock solution of alum was made by dissolving 1 g alum in 100 ml double distilled water. Chitosan was dissolved in aqueous acetic acid (20 g/l). Chitosan powder 1 g, was added to 100 ml acetic acid solution followed by warming it at 60°C for 2 h with continuous stirring using mechanical stirrer.

Preparation of dye for printing: All the dyes used were first converted to powder form. For this purpose, catechu, turmeric and marigold were first dried in an oven at 50°C for 24 h and later grinded in mixer grinder. The powder so obtained was filtered through the 60 mesh nylon fabric. The fine powder of natural dye was used for printing.

Mordanting of wool fabric: Mordanting of wool was carried out with different mordants at two different concentrations on the weight of fabric (alum 10% and 20%; and chitosan 10% and 20%) at 90°C for 60 min with material to liquor ratio as 1:30. The mordanting of wool was carried out in rota dyer (Rota Dyer machine, Rossari® Labtech, Mumbai). The mordanted fabric was squeezed and dried in open width form at 100°C for 5 min and then used for printing.

Printing: Guar gum being nonionic is best suited for textile printing. The stock paste of guar gum was prepared using 2% of guar gum powder. The guar gum powder was sprinkled slowly in water with continuous stirring in order to prevent lump formation. The paste was stirred continuously at 90°C for 1 h.

The print paste was prepared using different dye concentrations (1%, 5% and 10%) on the basis of weight of print paste. The 1% of print paste was prepared using 1 g of dye powder which was first pasted with small quantity of water followed by addition of 99 g of guar gum stock paste. The print paste was continuously stirred for 30 min. The mordanted wool fabrics were then printed with two strokes of squeeze and steamed at 102°C for 10 min. The samples were washed with water and dried in air.

Analysis of printed fabrics

Color value by reflectance method: The printed samples were evaluated for the depth of color by reflectance method using 10 degree observer using Spectraflash SF 300 (Datacolor International, U.S.A.). The K/S values were determined using equation

$$\frac{K}{S} = \frac{(1-R)^2}{2R}$$

where, R is the reflectance at complete opacity, K is the Absorption coefficient and S is the Scattering coefficient.

Printed fabrics were simultaneously evaluated in terms of CIELAB color space (L*, a* and b*) values using the RayscanSpectrascan 5100+. In general, the higher the K/S value, the higher the depth of the color on the fabric. L* corresponding to the lightness (100=white, 0=black), a* to the red-green coordinate (+ve=red, -ve=green) and b* to the yellow-blue coordinate (+ve=yellow, -ve=blue). As a whole, a combination of all these parameters enables one to understand the tonal variations.

Washing fastness: The printed wool fabric was sandwiched between cotton and wool fabric and then subjected to washing as per the ISO II test method. Evaluation of color fastness to washing was carried out using ISO II method [30].

Light fastness: The light fastness was determined using artificial illumination with Xenon arc light source, Q-Sun Xenon Testing Chamber. One half of each sample was covered with an opaque black paper sheet leaving the other half exposed. The color fastness to light was evaluated as per ISO 105/B02 standards [31].

Determination of antibacterial activities of printed fabrics: The antibacterial properties of the printed fabrics against *S. aureus* and *E. coli* were estimated by AATCC Test Method 100-2004 [32]. The fabric samples were sterilized before actual antibacterial testing using autoclave followed by exposure to UV light.

Results and Discussion

Color values of natural dye printed wool

The color values of the printed fabrics using different concentrations of mordants and dyes are summarized in Tables 1-3.

When the printing was carried out without any mordant, the prints obtained were very light in shade and color bled heavily after washing. However, when alum or chitosan pre-treated fabric was printed, it gave deeper and fast prints. In case of chitosan, the color values of the prints were higher than the respective alum pretreated prints. They also were washing fast and this trend was valid for all the three dyes studied.

As the concentration of mordant increased, even with the fixed concentration of dye in the paste, the K/S values and hence depth of the prints progressively increased which may be attributed to increase in the extent of fixation of the dye. Similarly at fixed mordant concentration, with increase in dye concentration K/S values of the prints obtained increased. This clearly indicates that chitosan played significant role as a mordant and was marginally, but distinctly superior to alum with regard to its efficacy as a mordant. In case of all the three dyes studied, this trend of chitosan having a distinct edge as a mordant over alum was confirmed.

The role of chitosan as a mordant may be attributed to the presence of amino groups which get protonated and offer the sites for attachment of the dye which is mainly anionic in nature. The mechanism of dye attachment to the fibre is similar to that in case of metal mordants.

The varying combinations of mordant and dye resulted in different shades and tones of printed wool fabric. This is evident from the CIE colour co-ordinate results shown in Tables 1-3. The prints obtained using catechu dye, in case of chitosan were more reddish yellow (distinctly higher a* values for almost similar b* values) in tone as compared to the respective prints obtained using alum. In case of turmeric dye, it was alum which gave the prints more yellowish-green (distinctly higher b* values and lower a* values) as compared to those obtained using chitosan as a mordant. For marigold however, no distinct trend in the tonal variation was observed, although the K/S values were slightly but distinctly higher for the chitosan than that for alum mordanted prints.

The results in Tables 1-3 indicate the comparative fastness properties in the case of both the mordants. The printed samples without mordanting showed much inferior fastness properties to those obtained with mordanting (irrespective of the type) which indicates the positive role of a mordant in holding of the dye on the fabric. It is well

Mordant	Mordant (%)	Catechu (%)	Colour value	CIE colour coordinates			Washing fastness$	Light Fastness#
			K/S@	L*	a*	b*		
None	0	10	0.83 ± 0.04	73.35	7.52	15.39	3	3
Alum	10	1	0.96 ± 0.04	70.32	0.04	10.19	4-5	6
	10	5	1.52 ± 0.03	71.69	2.93	15.15	4-5	6
	10	10	1.63 ± 0.04	72.02	3.95	16.65	4-5	6
	20	1	1.04 ± 0.03	69.55	-0.18	8.60	4-5	6
	20	5	1.41 ± 0.04	70.22	1.96	11.92	4-5	6
	20	10	1.64 ± 0.03	72.10	3.96	16.70	4	7
Chitosan	10	1	1.20 ± 0.02	68.01	2.53	12.65	4-5	6
	10	5	1.61 ± 0.03	67.87	4.74	13.98	4-5	6
	10	10	1.75 ± 0.04	68.31	6.05	15.96	4-5	6
	20	1	1.09 ± 0.04	66.87	0.91	10.21	4-5	6
	20	5	1.68 ± 0.02	67.39	4.69	13.37	4-5	7
	20	10	1.75 ± 0.03	68.14	5.43	15.29	4-5	7

$Rating 1-5, where 1-poor, 2- fair, 3-good, 4-very good and 5-excellent.
#Ratings 1-8, where 1-poor, 2-fair, 3-moderate, 4-good, 5-better, 6-very good, 7-best and 8-excellent.
@Represents average values of three determinations.

Table 1: Effect of mordants and catechu concentration on color strength of printed wool.

Mordant	Mordant (%)	Turmeric (%)	Colour value	CIE colour coordinates			Washing fastness$	Light Fastness#
			K/S@	L*	a*	b*		
None	0	10	1.60 ± 0.02	85.14	-1.61	11.32	2	2
Alum	10	1	1.97 ± 0.03	72.39	-5.70	34.03	4-5	4
	10	5	4.45 ± 0.02	79.54	-5.41	47.18	4-5	5
	10	10	8.68 ± 0.04	84.08	-6.19	55.34	4-5	5
	20	1	2.55 ± 0.03	76.88	-7.97	42.05	4-5	4
	20	5	6.62 ± 0.02	83.12	-7.29	53.38	4-5	5
	20	10	9.32 ± 0.04	85.12	-7.17	57.05	4	5
Chitosan	10	1	2.42 ± 0.03	71.24	-3.44	34.48	5	4
	10	5	5.37 ± 0.04	77.24	-3.94	44.77	5	5
	10	10	11.31 ± 0.04	78.18	0.53	48.07	4-5	5
	20	1	2.89 ± 0.02	70.24	-1.47	33.98	4-5	4
	20	5	7.86 ± 0.03	76.59	0.30	45.11	4-5	5
	20	10	11.95 ± 0.04	82.25	-3.95	52.64	4-5	5

$Rating 1-5, where 1-poor, 2- fair, 3-good, 4-very good and 5-excellent.
#Ratings, 1-8, where 1-poor, 2-fair, 3-moderate, 4-good, 5-better, 6-very good, 7-best and 8-excellent.
@Represents average values of three determinations.

Table 2: Effect of mordants and turmeric concentration on color strength of printed wool.

Mordant	Mordant (%)	Marigold (%)	Colour value	CIE colour coordinates			Washing fastness$	Light Fastness#
			K/S@	L*	a*	b*		
None	0	10	1.8 ± 0.02	85.63	-1.57	12.2	2-3	3
Alum	10	1	2.69 ± 0.03	70.76	-3.44	33.72	4-5	5
	10	5	6.03 ± 0.02	68.78	1.83	32.49	4-5	7
	10	10	13.44 ± 0.04	73.31	4.29	39.94	4-5	7
	20	1	3.09 ± 0.03	70.24	-0.99	36.67	4-5	6
	20	5	7.59 ± 0.04	69.54	2.76	37.35	4-5	7
	20	10	11.85 ± 0.04	80.47	-0.86	50.09	4	7
Chitosan	10	1	3.52 ± 0.03	64.37	0.58	26.33	5	6
	10	5	8.67 ± 0.05	77.40	0.07	47.08	5	6
	10	10	14.0 ± 0.04	78.81	3.85	51.57	4-5	7
	20	1	3.75 ± 0.03	64.98	-0.79	25.96	4-5	7
	20	5	8.79 ± 0.01	78.27	-1.22	46.95	4-5	7
	20	10	14.09 ± 0.05	71.14	7.61	42.03	4-5	7

$Rating 1-5, where 1-poor, 2- fair, 3-good, 4-very good and 5-excellent.
#Ratings, 1-8, where 1-poor, 2-fair, 3-moderate, 4-good, 5-better, 6-very good, 7-best and 8-excellent.
@Represents average values of three determinations.

Table 3: Effect of mordants and marigold concentration on color strength of printed wool.

Mordant	Mordant conc.	Dye	Dye Conc. (in print paste)	Bacterial Reduction@ (%)	
				S. aureus	*E. coli*
None	0	None	0	0.53 ± 0.1	0.65 ± 0.1
Alum	20	-	0	74.32 ± 0.5	70.25 ± 0.6
Chitosan	20	-	0	78.10 ± 0.25	75.75 ± 0.30
-	0	Catechu	10	65.30 ± 0.3	65.80 ± 0.40
-	0	Turmeric	10	63.50 ± 0.3	61.75 ± 0.45
-	0	Marigold	10	67.75 ± 0.25	66.80 ± 0.20
Alum	20	Catechu	10	98.00 ± 0.1	97.50 ± 0.4
Chitosan	20		10	99.20 ± 0.2	99.45 ± 0.25
Alum	20	Turmeric	10	97.25 ± 0.15	97.10 ± 0.20
Chitosan	20		10	98.75 ± 0.25	99.10 ± 0.3
Alum	20	Marigold	10	98.50 ± 0.4	99.10 ± 0.25
Chitosan	20		10	99.75 ± 0.20	99.60 ± 0.3

@Represents average value of 3 determinations.

Table 4: Effect of mordant and dye combinations on antibacterial properties.

known that the use of mordants is essential to fix most of the natural dyes on the textile fabric.

The improvement in fastness properties with increase in mordant concentration clearly indicates the positive role of mordants in case of printing with natural dyes. The washing fastnesses obtained varied in the range of good to excellent grade. Light fastness was found to be improving with increase in K/S values, which is quite obvious.

Antibacterial activity of printed wool fabric

The results for quantitative antibacterial assessment of the prints are presented in Table 4. The wool fabric showed negligible antibacterial properties against both *S. aureus* and *E. coli*. The prints obtained only using the dyes such as catechu, turmeric and marigold in absence of any mordant showed significant antibacterial activity (in the range of 61-68%) indicating the inherent nature of these dyes in preventing the bacterial growth. These results are also supported in the literature [4-8]. In case of prints obtained using the mordant such as alum and chitosan, the extent of bacterial reduction was significantly increased to the range of 97-100%. This clearly indicates the positive contribution of alum and chitosan mordants in enhancing the antibacterial activity of the printed fabric. As far as antibacterial activity is concerned, the chitosan mordanted samples showed comparative antibacterial properties than that of alum in case of all the dyes studied. Chitosan is known for antibacterial activity against broad spectrum bacteria and mechanism of antibacterial activity was explained by researchers [16,33]. It is also to be noted that alum itself provides antibacterial activity which is mainly because of metal ions. The chitosan hence can be claimed as eco-friendly mordant-cum-antibacterial finishing agent for printing of wool using natural dyes.

Conclusion

The printing of wool fabric with natural dyes in fine powder form was successfully carried out using chitosan as an eco-friendly mordant and its performance was compared with alum mordant. The color values varied with the dye-mordant combination; however the chitosan mordant showed higher color values than those with alum mordant. The excellent antibacterial activity and fastness properties were displayed by the printed samples. The role of chitosan as a mordant in natural dye printing of wool fabric has been confirmed. The antibacterial printed wool fabrics can thus be obtained using eco-friendly method employing chitosan.

References

1. Su W, Wei SS, Hu SQ, Tang JX (2011) Antimicrobial finishing of cotton textile with nanosized silver colloids synthesized using polyethylene glycol. Journal of the Textile Institute 102: 150-156.

2. Gao Y, Cranston R (2008) Recent Advances in Antimicrobial Treatments of Textiles. Textile Research Journal 78: 60-72.

3. Joshi M, Ali SW, Purwar R, Rajendran S (2009) Ecofriendly antimicrobial finishing of textiles using bioactive agents based on natural products. Indian Journal of Fiber and Textile Research 34: 295-304.

4. Teli MD, Sheikh J, Kamble M (2013) Ecofriendly Dyeing and Antibacterial finishing of Soyabean Protein Fabric using Waste Flowers from Temples. Textiles and Light Industrial Science and Technology 2: 78-84.

5. Teli MD, Sheikh J, Kamble M, Trivedi R (In-press) Temple waste Marigold Dyeing and Antibacterial finishing of Bamboo rayon using Natural Mordants. International Dyer.

6. Teli MD, Sheikh J, Kamble M (2012) Simultaneous dyeing and antibacterial finishing of Soyabean protein fabric using catechu and natural mordants. Journal of Textile Association 73: 227.

7. Teli MD, Sheikh J, Valia S, Yeola P (In-press) Temple waste marigold and turmeric dyeing of milk fibre. Journal of Textile Association.

8. Teli MD, Sheikh J, Shastrakar P (2013) Exploratory investigation of Chitosan as mordant for eco-friendly antibacterial printing of Cotton with Natural dyes. Journal of Textiles 1-6.

9. Bechtold T, Turcanu A, Ganglberger E, Geissler S (2003) Natural dyes in modern textile dyehouses — how to combine experiences of two centuries to meet the demands of the future. Journal of Cleaner Products 11: 499-509.

10. Kumar JK, Sinha AK (2004) Resurgence of natural colourants: a holistic view. Natural Product Letters 18: 59-84.

11. Shenai VA (1985) Technology of printing. Sevak Publications, Mumbai, India.

12. Rippon JA (1984) Improving the Dye Coverage of Immature Cotton Fibres by Treatment with Chitosan. Journal of the Society of Dyers and Colourists 100: 298-303.

13. Lim S-H, Hudson SH (2004) Application of a fibre-reactive chitosan derivative to cotton fabric as a zero-salt dyeing auxiliary. Coloration Technology 120: 108-113.

14. Lim S-H, Hudson SH (2204) Application of a fibre-reactive chitosan derivative to cotton fabric as an antimicrobial textile finish. Carbohydrate Polymers 56: 227.

15. Tiwari SK, Gharia MM (2003) Characterization of chitosan pastes and their application in textile printing. American Association of Textile Chemists and Colorists 3: 17-19.

16. Zhang Z, Chen L, Ji J, Huang Y, Chen D (2003) Antibacterial Properties of Cotton Fabrics Treated with Chitosan. Textile Research Journal 73: 1103-1106.

17. Sharaf S, Opwis K, Knittel D, Gutmann JS (2011) Comparative investigations on the efficiency of different anchoring chemicals for the permanent finishing of cotton with chitosan. AUTEX Research Journal 11: 71-77.

18. El-tahlawy KF (2008) Chitosan phosphate: A new way for production of eco-friendly flame-retardant cotton textiles. Journal of the Textile Institute 99: 185-191.

19. El-tahlawy KF, El-bendary MA, Elhendawy AG, Hudson SM (2005) The antimicrobial activity of cotton fabrics treated with different crosslinking agents and chitosan. Carbohydrate Polymers 60: 421-430.

20. Oktem T (2003) Surface treatment of cotton fabrics with chitosan. Coloration Technology 119: 241-246.

21. Abou-Okeil A, El-Shafie A, Hebeish A (2007) Chitosan phosphate induced better thermal characteristics to cotton fabric. Journal of Applied Polymer Science 103: 2021-2026.

22. Bandyopadhyay BN, Sheth GN, Moni MM (1998) Chitosan can cut salt use in reactive dyeing. International Dyer 183: 39.

23. Davidson RS, Xue Y (1994) Improving the dyeability of wool by treatment with chitosan. Journal of the Society of Dyers and Colourists 110: 24-29.

24. Rattanaphani S, Chairat M, Bremner JB, Rattanaphani V (2007) An adsorption

and thermodynamic study of lac dyeing on cotton pretreated with chitosan. Dyes and Pigments 72: 88-96.

25. Jassal M, Chavan RB, Yadav R, Singh P (2005) Chitosan as thickner for printing of cotton with pigment colors. Chitin and Chitosan: Opportunities & Challenges, SSM International Publication, Contai, India.

26. Bahmani SA, East GC, Holme I (2000) The application of chitosan in pigment printing. Coloration Technology 116: 94-99.

27. Teli MD, Sheikh J (2012) Simultaneous pigment dyeing and antibacterial finishing of denim fabric using Chitosan as a binder. International Dyer 197: 28.

28. Trotmann ER (1984) Dyeing and Chemical Technology of Textile Fibers. (6thedn), Charles Griffin and Company ltd, London, England.

29. Teli MD, Sheikh J (2012) Extraction of chitosan from shrimp shells waste and application in antibacterial finishing of bamboo rayon. International Journal of Biological Macromolecules 50: 1195-1200.

30. ISO technical manual (2006) Geneva, Switzerland.

31. Teli MD, Sheikh J, Pradhan C Simultaneous natural dyeing and antibacterial finishing of cotton using Bio-waste chitosan. Melliand International, Communicated paper.

32. AATCC Technical Manual (2007) Assessment of Antibacterial Finishes on Textile Material. Research Triangle Park, NC, USA.

33. Lim SH, Hudson SM (2003) Review of chitosan and its derivatives as antimicrobial agents and their uses as textile chemicals. Journal of Macromolecular Science Part C: Polymer Reviews 43: 223-269.

12

Comparison Mechanical Properties for Fabric (Woven and Knitted) Supported by Composite Material

Karnoub A[2]*, Makhlouf S[2], Kadi N[2] and Azari Z[1]

[1]*Laboratory of Biomechanics, Polymers and Structures, ENIM, 57000 Metz, France*
[2]*Faculty of Mechanical Engineering, University of Aleppo, Syria*

Abstract

In many recent years the use of composite materials increases in many fields, for example agricultural uses, where these materials are characterized by good mechanical properties, tenacity and light weight. In this paper, we will shed light on the use of composite materials reinforced by knitted fabric compared with composite materials reinforced by woven fabric. Attachment materials was used in our research is resin, while supported cloth either woven or knitted were manufactured from amplified polypropylene filaments (BCF), and the testing are (tensile strength, resistance to bending, shear strength, resistance to penetration).

Comparing the results of composite material produced from knitted fabric shown better mechanical properties than woven fabric because of the knitted fabric distinct by the process of overlap between the stitches that gave better resistance.

Keywords: Composite material; Woven fabric; Knitted fabric

Introduction

In recent years, use of composite materials supported by textile fibers is increased, because the intermingling between the inorganic fibers and organic materials gives the new material with good properties. as there are materials isn't similar at all but by merging with each other produces a strong correlation materials. Some yarns are still suffering of weakness in the resistance of the surface stresses resulting from use, and by adding the appropriate materials will be improved significantly, in the case of blending polypropylene fiber with thermoplastic materials will improve the mechanical properties of polypropylene fiber and recycle damaged thermoplastic materials [1-8].

The use of products made of composite materials backed with cloth was expand, where they are used in technical applications like space and civil engineering, also entered strongly in the shipping industry because of its qualities and useful in these industries [9].

Composite material which mainly consists of resin distinct in good mechanical properties [10], but it is very smashed material [11].

Polypropylene fibers (PP) mixed with attachment material (resin) give a new material with high bending resistance and penetration resistance, followed by polyester filament and nylon yarns [12].

Researcher [13] concluded that the addition of composite materials especially resin improves the properties of knitted cloth with a single jersey knit, made of glass fibers, were tensile test results of both the parallel and vertical direction very approximate [14].

Tensile strength was improved after addition of composite materials for knitted cloth made of polyethylene or polypropylene fiber or fabric called UHMWPE (polyethylene fabric with a high specific weight), which fiber enhances the strength of the composite material [15].

The researcher [16] finds that there is a relationship between the temperatures of the composite material during pours it to knitted fabric and the amount of improvement in product properties, by increasing the temperature the product tensile strength improves and explained it to a good mix of fabric fiber with attachment material.

Research [17] compares between the two ways to add composite materials (resin) to single jersey knit fabric, first way is anoint the fabric with the composite materials, and the second way is inject composite materials to cloth by pressure. Research found that the latter way have shown better results for tensile test and shear strength resistance, because of the ability of the attachment material in the injection way by pressure to the penetration to all parts of cloth. Research found by microscopic examination of samples made by first way that cloth has air spaces within the samples lead to the formation of weak points and in turn lower product resistance.

Samples made of several layers of woven fabric have good results of tensile resistance and energy absorption characteristics by way injection composite materials, than when using the method (film-stacked) paste layers with each other using composite materials [18].

Woven fabric with composite materials shows more penetration resistant than knitted fabric. Therefore [19] compares several samples of woven cloth made of polyester filament of yarns 300 den and heavily 30 thread per cm for both warp and weft, but different types of weaves, namely, (1/1 plain, 1/2 plain, 1/4 plain, 1/4 Satin). And found that the woven fabric with weave 1/1 supported by composite materials gives a greater resistance to penetration, and the reason is that this weaves is characterized by the largest number of interlacements between the warp and weft yarns.

Bending resistance of the composite material by knit fabric has five times greater resistance than composite material by woven fabric [20].

Researcher [21] tested the behavior of two types of knit fabric made of fiberglass with composite materials under tensile test, two types are (Rib, Milano), testing was in both directions with rows and

---

*Corresponding author: Karnoub A, Faculty of Mechanical Engineering, University of Aleppo, Syria, E-mail: amerkarnoub@gmail.com

perpendicular to the rows, and found that the two types of cloth have greater resistance to perpendicular to the rows.

The researcher [22] have tested the type of composite material produced from knitted fabric (Rib 1 * 1) with resin, under two types of tests shear and tensile, in three directions (0°, 45°, 90°) and found that the product is have greater tensile resistance at 90°, either under shear test at 45° angle showed the best results.

The Aim of the Research and its Importance

The research aims to study the effect of supportive fabric type (either woven or knitted) on the mechanical properties of produced composite materials. While previous studies have not conducted comprehensive comparison between the two types of fabric, previous studies conducted one samples tested, while our research has been conducting a four tests (tensile strength, resistance to bending, shear strength, resistance to penetration), these tests which can determine the mechanical properties of the resulting material.

Materials and Methods of Search

Composite materials consist of the following basic materials

1. Polyester resin

2. Strengthening material

3. accelerated material (cobalt)

4. PP yarn (to made woven or knit fabric)

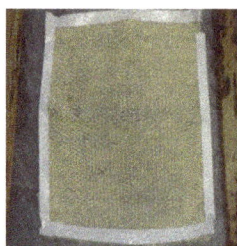

Tested fabric

➢   Woven fabric from plain weave 1/1 made of propylene filaments count 1200 den for both warp and weft, weight per square meter is 150 g/m².

➢   Knitted fabric from single jersey weave, made of propylene filaments count 1200 den, weight per square meter is 150 g/m² in Figure 1.

Preparation of the attachment materials

Attachment materials made by adding accelerated materials (cobalt + catalyst) and mix it well until it's ready in Figure 2.

Coating the metal mold and wooden textured wax

So as to ease removal of samples processed and prevent it from sticking template metal, wait about half an hour until the foam and not to touch wipe oneself and only tarnished the first layer. Anoint attachment material on all farm accurately because the presence of any part is greased makes it difficult to separate the piece from the mold in Figure 3.

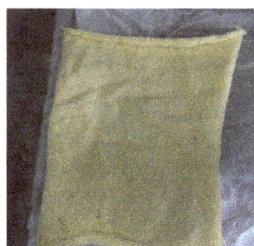

knitted fabric   woven fabric
Figure 1: Woven fabric samples.

Figure 2: Preparation Material Association.

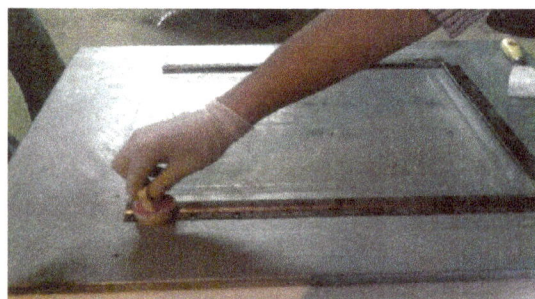

Figure 3: Mold processing.

Figure 4: Satisfy the cloth attachment material.

Adding the cloth

After making sure of the full-frame anointed completely, we put the fabric layer on the template, and then put attachment material through painted with a brush. Also using a roll pressure must be well to empty out of the air and the satisfaction of material do, and must work quickly so that material does not dry Association in an area without the other leading to conglomerate article Association and deformation piece and the loss in cost, especially when there is a metronome in Figure 4.

Retention period

After the completion of the development and cloth material piece Association leave for 24 hours to dry and hold together with each other and after this period by a screwdriver or other tool separate piece from the mold of the tip and then easily separate it by hand.

Composite materials tests

The tests that we will have are the resistance of tensile, bending, shear, and penetration, and each test we have cut composite materials to irregular forms of the test according to its ASTM.4.

Results

Three types of samples tested, namely, (resin, composite materials produced from woven fabric, composite materials produced from knitted fabric), mechanical properties are defined by four tests types, namely, (tensile test, shear test, bending test, penetration testing), and the results were as below:

Tensile test

It has been relying to conduct this test on the system: (ASTM: D 3039/D 3039M-95a).

1. The purpose of the experiment:

Determine the tensile properties and that by drawing the relationship between stress and strain or between strength and elongation.

2. The test device:

It consists of:

- Fixed jaw: Sticky key element is the center of a handle one load.

- Movable jaw: a moving element holds steady grip.

- Handles: It is in order to hold the sample and the sample placed between the handle and the handle fixed mobile

3. The test specifications:

Test speed=10 mm/min

Sample thickness=3 mm

The sample width=13 mm

The distance between the jaws=100 mm

The number of repeaters=10 samples

It determines the tensile test through three specifications, namely, (stress, strain, elongation), and with a table of the values of the arithmetic average of the results of tensile test specimens tests comes in Table 1 and Figure 5.

As it has been the comparison between the resistance woven cloth

Tensile	Stress [N/mm²]	Strain [%]	Elongation [mm]
Resin	15.1	1.491	1.491
Knitting fabric with composite material	172.114	4.298	4.298
Woven fabric with composite material	146.9	4.846	4.846
Knitting fabric	52.667	1.852	1.852
Woven fabric	77.433	2.054	2.054

Table 1: Results of tensile test of all samples.

Figure 5: Test device.

Shear	Load [N]
Resin	590.11
Knitting fabric with composite material	802.43
Woven fabric with composite material	788.97

Table 2: Results of shear test of all samples.

Figure 6: Test device.

and woven fabric under tensile test, and the results were as follows:

Shear test

It has been relying to conduct this test on a system: (ASTM D790-95a).

1. The purpose of the experiment:

Determine the shear properties and that by drawing the relationship between the force and bending.

2. The test device:

It consists of:

- Fixed jaws: Sticky key element is the center of a handle one load.

- Movable jaw: a moving element mediates jaws steadfast but from the top.

3. The test specifications:

Test speed=2 mm/d

Sample thickness=4 mm

The sample width=13 mm

The distance between the jaws=35 mm

The number of repeaters=10 samples

It determines the tensile test through three specifications, namely, (load, bending, strain, stress), and with a table of the values of the arithmetic average of the results of the shear test samples tests comes in Table 2 and Figure 6.

Bending test

It has been relying to conduct this test on a system :(ASTM D790-95a).

1. The purpose of the experiment:

- Determine the bending characteristics, and through drawing the relationship between the force and bending.

- Calculate the characteristics of a material through a curved and discuss the results.

2. The test device: consists of:

- Fixed jaws: increase the distance between the two jaws is the only difference from the shear test

- Movable jaw: a moving element mediates jaws steadfast but from the top.

3. The test specifications:

Test speed=2 mm/min

Sample thickness=3 mm

The sample width=12 mm

The distance between the jaws=80 mm

The number of repeaters=10 samples

It determines the tensile test through three specifications, namely, (load, strain, deflection, stress), and with a table of the values of the arithmetic average of the results of test samples bending tests comes in Table 3 and Figure 7.

Penetration test

It has been relying to conduct this test on a system: (ASTM D790-95a).

1. The purpose of the experiment:

Determine the bending characteristics, and through drawing the relationship between the force and bending.

2. The test device:

It consists of:

- Fixed jaw: Sticky key element is the center of a handle one load.

- Movable jaw: a moving element mediates the upper jaw circular head.

3. The test specifications:

Test speed=3 mm/min

Sample thickness=3 mm

Qatar sample=50 mm

Bending	Load [N]
Resin	47.77
Knitting fabric with composite material	178.5
Woven fabric with composite material	141.13

Table 3: Results of bending test of all samples.

Figure 7: Test device.

Penetration	Load [N]
Resin	301.3
Knitting fabric with composite material	386.167
Woven fabric with composite material	393.2

Table 4: Results of penetration test of all samples.

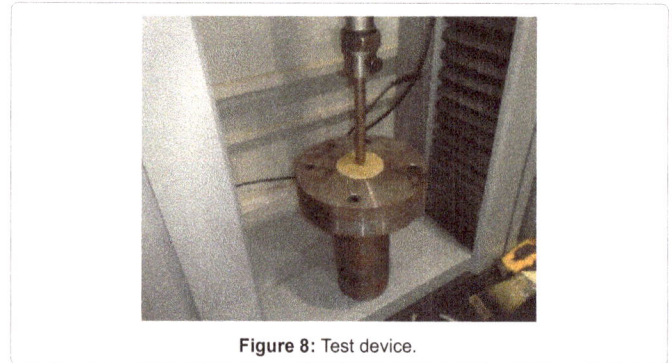

Figure 8: Test device.

Qatar test head=11 mm

The number of repeaters=10 samples

It determines the tensile test through three specifications, namely, (load, strain, bending, stress), and with a table of the values of the arithmetic average of the results of penetration testing samples tests comes in Table 4 and Figure 8.

Discussion

We will compare the results of all tests of samples consisting of resin, composite material supported by woven fabric and composite material supported by knitted fabric. To identify the samples are characterized by better specifications.

Tensile test

Woven fabric exhibits a better tensile strength compared with knitted fabric before adding the attachment material, due to the structure of woven fabric (plain 1/1) based on a friction between threads, while as the structure of knitted fabric (single jersey) based on stitches are made from one yarn only.

But after adding attachment material to knitted cloth, the tensile strength increases than woven cloth, due to fill the blanks in its structure, and on the other hand, because the test is perpendicular to the rows and not parallel to it, this gives better results depended to weave structure (single jersey) that we used. This is consistent with [14] in Figure 9.

Bending test

Attachment material falls down easily in bending test it very smashed material, on the other hand we cannot do this test for each of woven fabric and knitted fabric alone, because it are high drooping materials. But when the cloth unions with attachment material gave the new product, it have a high resistance of bending than it was for attachment material alone due to lower drooping of new material.

Resultant stress of composite material with knitted fabric is greater than the stress of woven fabric, that because of woven fabric drooping less than knitted fabric drooping, due to the structure of knitted fabric (single jersey) has a flexible behave more than the structure of

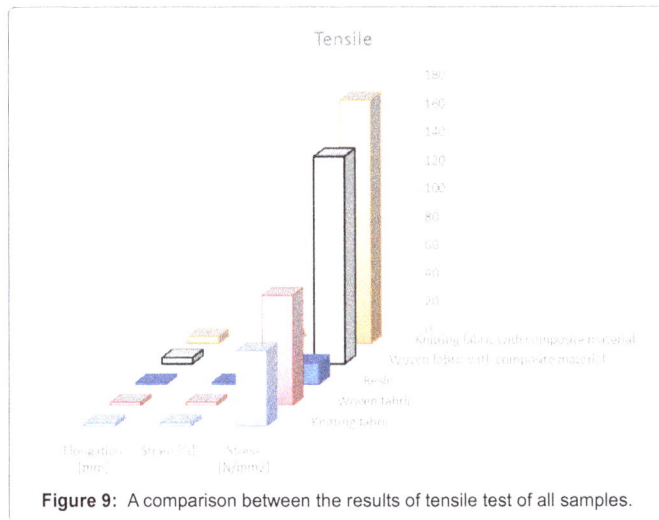

Figure 9: A comparison between the results of tensile test of all samples.

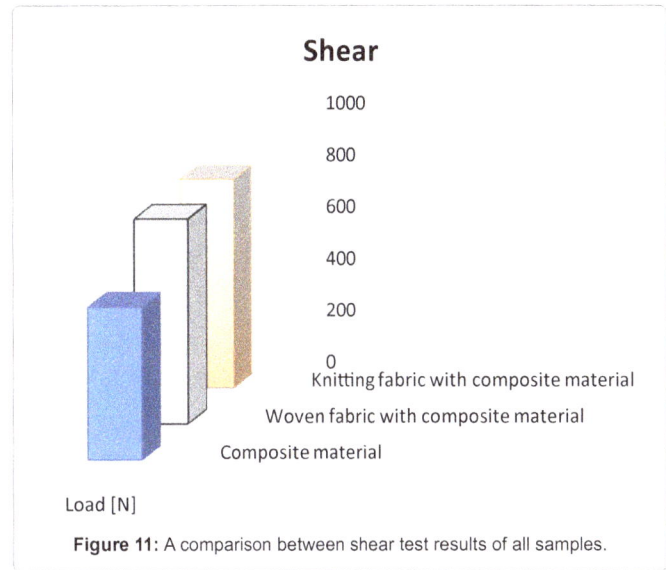

Figure 10: A comparison between bending test results of all samples.

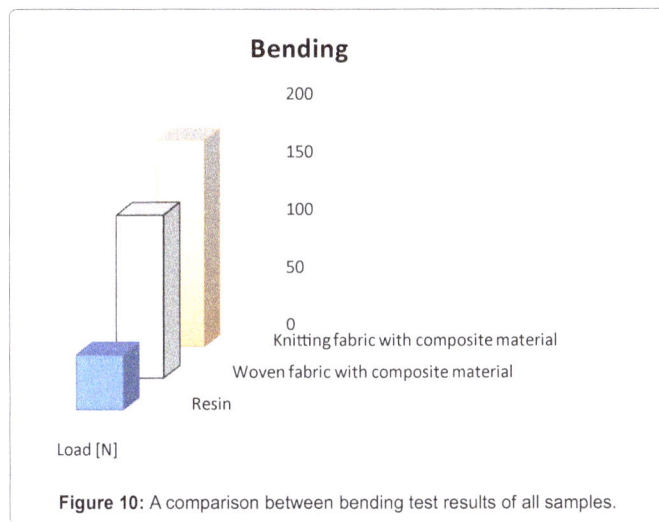

Figure 11: A comparison between shear test results of all samples.

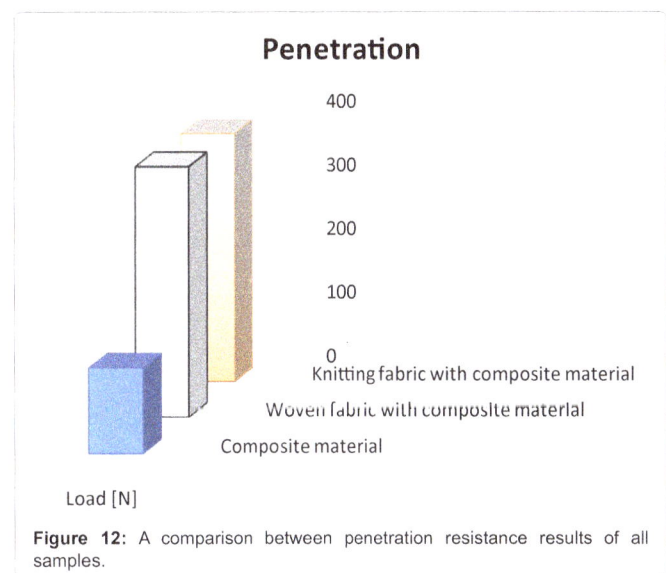

Figure 12: A comparison between penetration resistance results of all samples.

woven fabric (plain 1/1). That resulting to increase product smash and collapses under the slightest of cloth value knitted. This corresponds with the results of the researcher [20] in Figure 10.

Shear test

Attachment material has good resistance to shear its own, but after support it by either woven or knitted, stress increased in the produced material. The equal improvement between composite material from woven fabric and composite material from knitted fabric is explained of the test direction is to knitting direction leading to provide resistance knitted fabric greater than thread direction. This corresponds with the results of the researcher [12] in Figure 11.

Penetration resistance test

A good result was shown when attachment material infiltrates between knitted fabric poles and links to poles to each other, and repeated the same case with woven fabric, where that attachment material infiltrates will enter spaces occurring between the warp and weft, and the product has a greater resistance to penetration. But this result conflicted with the result researcher experiments [22]. The reason for this is that the researcher used a woven yarn density greater than the knitted fabric in Figure 12.

Conclusion

Adding attachment material (resin) to woven or knitted fabric will give a new composite material, it has mechanical properties better than the properties of attachment material alone or cloth materials alone.

Comparing the composite materials obtained from woven fabric with composite materials obtained from knitted fabric, we find the following:

• Composite materials produced from knitted fabric shows better tensile resistance than composite materials produced from woven fabric. This is consistent with [14].

• Composite materials produced from knitted fabric shows better bending resistance than composite materials produced from woven fabric. This is consistent with [20].

• Composite materials produced from knitted fabric shows similar shear resistance to composite materials produced from woven fabric. This is consistent with [12].

• Composite materials produced from knitted fabric shows similar penetration resistance to composite materials produced from woven fabric. This is inconsistent with [19].

References

1. van Wyk CM (1946) Note on compressibility of wool. J Text Inst 37: 285-292.

2. Harwood RJ, Grishanov SV, Lomov SV, Cassidy T (1997) Modelling of two component yarns Part I: The compressibility of yarns. J Text Inst 88: 373-384.

3. Gutowski TG, Dillon G (1997) the elastic deformation of fibre bundles, advanced composites manufacturing.

4. Hu J, Newton A (1997) Low-load lateral-compression behavior of woven fabrics. J Text Inst 88:242-254.

5. De Jong S, Snaith JW, Michie NA (1986) A mechanical model for the lateral compression of woven fabrics. Text Res J 57: 759-767.

6. Matsudaira M, Qin H (1995) Features and mechanical parameters of a fabric's compressional property. J Text Inst 86: 477-487.

7. Saunders RA, Lekakou L, Bader MG (1998) Compression and microstructure of fibre plain woven cloths in the processing of polymer composites.

8. Saunders RA, Lekakou L, Bader MG (1998) Compression in the processing of polymer composites - a mechanical and micro-structural study for different glass fabrics and resins. Comp Sci Technol 59: 983-993.

9. Chen B, Tsu-Wei C (1999) Compaction of woven-fabric preforms in liquid composite molding processes: single-layer deformation.Comp Sci Technol 59: 1519-1526.

10. Chen B, Tsu-Wei C (2000) Compaction of woven fabric preforms: nesting and multi-layer deformation. Comp Sci Technol 60: 2223-2231.

11. Chen B, Lang Eric J, Tsu-Wei C (2001) Experimental and theoretical studies of fabric compaction behaviour in resin transfer molding. Mater Sci Eng 317: 188-196.

12. Lomov SV, Verpoest I (2000) Compression of woven reinforcements: a mathematical model. J Reinforced Plast Comp 19: 1329-1350.

13. Young JJ, Kang TJ (2001) Analysis of compressional deformation of woven fabric using finite element method. J Text Inst 92: 1-15.

14. Ward IM (1983) Mechanical properties of solid polymers.

15. Gasser A, Boisser P, Hanklar S (1977) Mechanical behaviour of dry fabric reinforcement: 3D simulations versus biaxial tests.

16. Page J, Wang J (2002) Prediction of shear force using 3D non-linear FEM analyses for a plain weave carbon fabric in a bias extension state. Finite Elem Anal Des 38: 755-764.

17. Hearle JWS, Shanahan WJ (1978) An energy method for calculations in fabric mechanics, Part I: Principles of the method. J Text Inst 69: 81.

18. Kawabata evaluation system (KES). Kato Tekko, Kyoto, Japan.

19. Aurrekoetxea J, Sarrionandia MA, Urrutibeascoa I, Maspoch ML (2003) Effects of injection moulding induced morphology on the fracture behaviour of virgin and recycled polypropylene 44: 6959-6964.

20. Khondker OA, Leong KH, Herszberg I, Hamada H (2005) Impact and compression-after-impact performance of weft-knitted glass textile composites. Composites 36: 638-648.

21. Abrate S (1994) Impact on laminated composite: recent advances. Appl Mech Rev 47: 517-544.

22. Abrate S (1991) Impact on laminated composite materials. Appl Mech Rev 44: 155-190.

13

Linearization of Power Amplifier using the Modified Feed Forward Method

Motavalli MR[1]* and Solbach K[2]

[1]Department of Communication Technique, Faculty of Engineering, University of Qom, Qom, Iran
[2]Department of Microwave and RF-Technology, Faculty of Engineering, University of Duisburg-Essen, Duisburg, Germany

Abstract

A modified circuit for improving linearization of power amplifier based on the model of the Feed Forward circuit amplifier is proposed. With the help of mathematical model for the single power amplifier, the circuit is simulated and a demonstrator is fabricated and measured complex Taylor series are used for modeling the power amplifier by the approximation of the amplitude transfer function and the level-dependence of the transmission-phase of the power amplifier. This can be understood as a simplified form of volterra series. In our proof of concept experiment, we verified the concept but also found that the adjustment of the circuit is critically dependent on the drive conditions and linearization is achieved only for a narrow range of drive power. The proposed circuit in compared with the conventional Feed Forward amplifier in addition a significant increase in efficiency, to minimize the power of the distortion signal 3IMD-products at high drive levels.

Keywords: Volterra series; Complex taylor series; 3IMD-products, pre-distortion, Feed Forward amplifier; Power combiner

Introduction

The use of high power amplifiers with high linearity for mobile and satellite communications systems is essential. For example, in second and third generation mobile systems, GSM (Global System Mobile), 3GPP (third Generation Partnership Project), a large number of signals with different frequencies are transmitted from BS (Base Station) by high power amplifiers at the same time [1]. To avoid interference of these amplified signals, the amplifier must operate linear and that linearization is not possible by classical methods. Since the amplitude of intermodulation signals at the output of the amplifiers depends on the size of the input signals, input signals with large amplitude limit the performance of the amplifier. If the connection between an amplifier input and output signals display as transformation function that is extended in the form of a series (for example Taylor's series), we see that for the larger signal, the role of higher degrees of expression is more and more important, that is, the behavior of amplifiers is no longer linear and amplifier operates in saturation (non-linear) region. The saturation region, due to high output power and resulting high efficiency in mobile communications and satellite systems play an important role. Intermodulation signals with large amplitudes produce in this region of amplifier which leads to large distortion in output. Generally, nonlinearity in an amplifier can appear in two different forms: first production of new frequency components in the output of the amplifier and second dependence of gain amplitude and phase of the amplifiers to amplitude of input signals. If amplifiers have been multiple input signals frequency a type of distortion signals, that is, 3IMD-products (third order Intermodulation Distortion) should be considered more than other produced signals in output of amplifier, because they are near to frequency of original signals (input signals). They are in the range of useful bandwidth amplifiers and due to limitations in fabrication are not removable in practice [2].

The distortion signal of type 3IMD-products in base stations are propagated by high power amplifier in total send bandwidth and cause distortion and interference in band of inside channel as well as the neighboring channels. This problem occurs even on TV channels (by 3IMD-products and even 2IMD-products), where a large number of channels have placed at a close frequency near each other. The aim of this paper is to design a concept for a power amplifier with high linearity and high efficiency.

This paper presented the proposed circuit for improving linearization of the power amplifier based on the model of the Feed Forward circuit amplifier. In section 2, the principle of operation of the amplifier concept is discussed. A mathematical model of new amplifier concept is proposed in section 3. In section 4, a simulation model is used to investigate in detail the signals within the circuit and the performance and limitations of the amplifier. Finally in section 5, experimental proof of new amplifier concept is presented. Simulation and measurement results are compared and show good agreement.

New Amplifier Design

The classical parallel power combiner amplifier using two equal linear amplifiers have been used for many years in order to efficiently produce higher output power levels (doubles the available output power of one single amplifier) and also in order to improve the reliability and availability of the amplifier system component. However, linearity of amplification of each individual amplifier is not improved over the individual amplifiers. Power-added efficiency of the combiner amplifier circuit can be high when the amplifiers are driven close to the saturation level and consequently at high intermodulation level. Another successful concept in linearization of power amplifiers is pre-distortion, which can yield higher power efficiency, yet at lower cancellation ratios of unwanted products [3-6]. On the other hand, amplifiers for very high linearity requirements in mobile communications successfully employ the feed forward (FF) – amplifier scheme, Figure 1, which cancels the nonlinear intermodulation-products (IM) of a high power primary amplifier by superposition of signals from an auxiliary amplifier [7-11]. However, this concept suffers from a degradation of the efficiency of the amplifier which is mainly due to the linearity requirements on

*Corresponding author: Motavalli MR, Department of Communication Technique, Faculty of Engineering, University of Qom, Qom, Iran
E-mail: motavalliReza@gmail.com

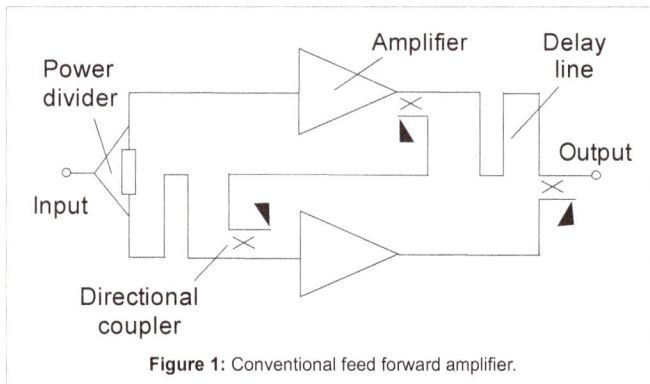

Figure 1: Conventional feed forward amplifier.

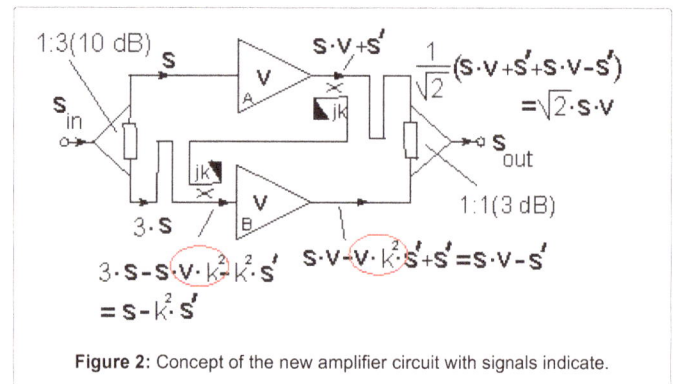

Figure 2: Concept of the new amplifier circuit with signals indicate.

the auxiliary amplifier. The FF-amplifier entails a first loop to extract the intermodulation distortion components from the power amplifier output while the second loop inverts phase and amplifies this signal in an auxiliary amplifier such that it destructively superimposes and cancels the original intermodulation signal at the output coupler. Distortions from the auxiliary amplifier have to be kept very low so that this amplifier needs to be operated far off saturation which means high dc power requirement. Since the auxiliary amplifier cannot contribute to the fundamental signal output power, its supply power degrades the power-added efficiency of the FF-amplifier [12-13] and also since the auxiliary amplifier is driven at low input power (low amplitude), a good ratio of cancelling of the 3IMD-products signals for small amplitude obtained, in other words, cancelling of the 3IMD-products signals for high output power is extremely low.

While the FF-amplifier has found wide application in communication systems due to its superior intermodulation suppression, its efficiency problem has inspired a modified concept which allows the auxiliary amplifier to contribute fundamental signal power in addition to cancelling the intermodulation products. The new circuit, Figure 2, exhibits a two-loop topology of a conventional feed forward amplifier, however, both loops are modified and the two amplifiers are assumed to be identical power amplifiers. In the first loop, the input power divider splits the input signal in a 1:3 ratio while at the output, the combiner is 1:1. The first loop acts as a pre-distortion stage while the second is a distortion cancelling and power combining loop.

For the presentation of the circuit's operation principle, we assume that the fundamental input voltage signal of the circuit is s_{in} (t). When the first power splitter divides this signal at a ratio of 1 to 3, the input signal to the upper power amplifier A is $s(t) = \frac{1}{\sqrt{10}}.s_{in}(t)$ while the signal incident to the lower power amplifier B is $3. s_{in}(t)$. The two amplifiers are assumed to be identical with equal amplification υ and distortion products s' (t) and $s''(t)$ for power amplifier A and B, respectively, under identical drive conditions.

The output voltage of power amplifier A is a superposition of the fundamental signal amplified by voltage gain v and a distortion product: $v.s(t)+s'(t)$. The upper coupler samples this combined signal and feeds it to the lower coupler in front of amplifier B. Assuming both coupling coefficients as jk (note: complex notation used to indicate a phase shift of 90°), the sampled signal offered to the input of amplifier B is $k^2.v.s(t)-k^2.s'(t)$ which is combined with the input signal $3.s(t)$ after its travel through the lower delay line to give the total voltage signal, $(3-k^2.v).s(t)-k^2.s'(t)$ Note, the coupler and delay line insertion loss have been neglected for simplification and time delay of the both power amplifiers is compensated by two delay line in two-loop).

With our aim to drive the lower amplifier at the same fundamental signal level as the upper amplifier, we choose $k^2 = \frac{2}{\upsilon}$ to achieve the total voltage signal $s(t) - \frac{2}{\upsilon}.s'(t)$.

It is seen that this lower amplifier input signal is a pre-distorted version of the input signal of the upper amplifier, with the distortion component as a replica of the upper amplifier distortion, but reversed in phase. Assuming the distortion component to be much smaller than the fundamental component, the lower amplifier will amplify the fundamental signal plus its distortion component by the voltage amplification υ. In addition, in the same way as the upper amplifier, the lower amplifier will produce a distortion component due to its fundamental signal excitation. The resulting output signal of amplifier B is then $v.s(t)-2.s'(t)+s'(t)$. Assuming perfectly equal amplifiers A and B, the distortion signal originating from the lower amplifier is equal to the one generated by the upper amplifier, such that the output signal of the amplifier B is $v.s(t)-s'(t)$.

In the upper signal path, the output signal of the power amplifier A travels through an upper delay line by neglecting the delay line's attenuation, appears at the combiner as: $v.s(t)+s'(t)$. By comparing signals in the upper and lower path, we see that the fundamental signals are equal and thus can be combined to give double power while the distortion components incident to the combiner are equal and in anti-phase and thus cancel to give the total output power: $v.s(t)+s'(t)$.

It has to be noted that this analysis is over-simplified with the assumption of perfect identity of the two amplifiers and it neglects the effect of the increased driving level of the lower amplifier due to the added pre-distortion signal. With slightly unequal amplifiers and slightly unequal driving levels, it is still possible to achieve near-perfect cancellation of intermodulation at the price of a loss in the power combining efficiency, as can be shown by simulation, section 4. However, a major performance limitation of the new circuit was found by analysis of an experimental amplifier system in section 5, to originate in the drive power level dependence of the amplifier voltage gain (magnitude and phase) and the insertion loss of the delay lines.

Mathematical Model

Characterization of single power amplifiers

To describe the behavior of the entire circuit mathematically, each power amplifier should be first characterized separately, it means that each power amplifier should be represented as a mathematical model; all other components can be described by simple mathematical models, since their behavior can be assumed to be linear in the region of interest.

For nonlinear behavior of power amplifier, the power transfer function has been measured, in the other words, the behavior of each amplifier is measured regarding 1 dB compression point and intermodulation products of orders 3, 5, 7 and 9 (3IMD-products, 5IMD-products, 7IMD-products and 9IMD-products) separately. To develop the mathematical description of this behavior, Amplifier output voltage to the input voltage is expanded in a series. At first, Taylor series is used, that due to stark non-linearity, model obtained was not fit to actual behavior of amplifiers. Next the volterra series is used that relation other series is more flexible and for description nonlinearity systems is more appropriate [14-16]. The measurements of the characteristics of the power transfer function has been done with two tone input signal. In the calculations performed, it has been demonstrated that if both amplitudes input signals are equal (such as common case in GSM), volterra series becomes Taylor series with complex coefficients [17]. Since for creating mathematical model of an amplifier with complex coefficient, complex measurements must be available, therefore, measurements of complex voltage gain in close saturation region is used. For creating mathematical model, two sinusoidal voltage signal with equal amplitudes $\hat{U}_{in}$ and Frequencies f_1 (ω_1) and f_2 (ω_2) are expended in Taylor series. The generated signals include main input frequencies and all new frequencies in output of amplifier can be summarized in a general form as follows [17]:

$$u_{out}(t) = \sum_{m=0}^{\frac{(N-1)}{2}} \left\{ \sum_{n=m}^{\frac{N-1}{2}} \binom{2n+1}{n}\binom{2n+1}{n-m}\frac{u_{in}^{2n+1}}{2^{2n}}c_{2n+1}\cos\left(((m+1)\omega_1-m\omega_2)t\right) \right\} +$$

$$\sum_{m=0}^{\frac{N-1}{2}}\sum_{n=m}^{\frac{N-1}{2}} \binom{2n+1}{n}\binom{2n+1}{n-m}\frac{u_{in}^{2n+1}}{2^{2n}}c_{2n+1}\cos\left(((m+1)\omega_2-m\omega_1)t\right) \right\} \qquad (3.1.1)$$

With:
- N as maximum Taylor series degree
- n as variable index
- m as degree of the generated intermodulation signals
- c_n as taylor series coefficients

The share of the original signal (or main frequency) and IDM-products various degrees (3 to 9) as follows:

$$\hat{u}_{out,F} = \sum_{n=0}^{\frac{N-1}{2}} \binom{2n+1}{n}\binom{2n+1}{n}\frac{\hat{u}_{in}^{2n+1}}{2^{2n}}c_{2n+1} \qquad (3.1.2)$$

$$\hat{u}_{out,3IMD} = \sum_{n=1}^{\frac{N-1}{2}} \binom{2n+1}{n}\binom{2n+1}{n-1}\frac{\hat{u}_{in}^{2n+1}}{2^{2n}}c_{2n+1} \qquad (3.1.3)$$

$$\hat{u}_{out,5IMD} = \sum_{n=2}^{\frac{N-1}{2}} \binom{2n+1}{n}\binom{2n+1}{n-2}\frac{\hat{u}_{in}^{2n+1}}{2^{2n}}c_{2n+1} \qquad (3.1.4)$$

$$\hat{u}_{out,7IMD} = \sum_{n=3}^{\frac{N-1}{2}} \binom{2n+1}{n}\binom{2n+1}{n-3}\frac{\hat{u}_{in}^{2n+1}}{2^{2n}}c_{2n+1} \qquad (3.1.5)$$

$$\hat{u}_{out,9IMD} = \sum_{n=4}^{\frac{N-1}{2}} \binom{2n+1}{n}\binom{2n+1}{n-4}\frac{\hat{u}_{in}^{2n+1}}{2^{2n}}c_{2n+1} \qquad (3.1.6)$$

To determine the coefficients series, linear equation system of considered signals using the measured values are written. The equation system for main signals (3.1.2), which is formed of the number M, the corresponding measured values $\hat{u}_{in}$ and $\hat{u}_{out}$, can be summarized as follows (Since the measured values on the left side (3.1.2) is real, the absolute terms of right side is used):

$$\begin{pmatrix} \hat{u}_{out,F,1} \\ \hat{u}_{out,F,2} \\ \vdots \\ \hat{u}_{out,F,M_1} \end{pmatrix} = \begin{vmatrix} \hat{u}_{in,1} & \frac{9}{4}\hat{u}_{in,1}^3 & \cdots & \left(\frac{N}{N-1}{2}\right)^2\frac{\hat{u}_{in,1}^N}{2^{N-1}} \\ \hat{u}_{in,2} & \frac{9}{4}\hat{u}_{in,2}^3 & \cdots & \left(\frac{N}{N-1}{2}\right)^2\frac{\hat{u}_{in,2}^N}{2^{N-1}} \\ \vdots & \vdots & \ddots & \vdots \\ \hat{u}_{in,M_1} & \frac{9}{4}\hat{u}_{in,M_1}^3 & \cdots & \left(\frac{N}{N-1}{2}\right)^2\frac{\hat{u}_{in,M_1}^N}{2^{N-1}} \end{vmatrix} \begin{pmatrix} c_1e^{j\varphi_{c_1}} \\ c_3e^{j\varphi_{c_3}} \\ \vdots \\ c_Ne^{j\varphi_{c_N}} \end{pmatrix} \qquad (3.1.7)$$

The equation system for the 3IMD-products signals is too similar to the manner as the main signals equation and the basis of (3.1.3) for the M_2 measured values can be summarized as follows:

$$\begin{pmatrix} \hat{u}_{out,3IMD,1} \\ \hat{u}_{out,3IMD,2} \\ \vdots \\ \hat{u}_{out,3IMD,M_2} \end{pmatrix} = \begin{vmatrix} \frac{3}{4}\hat{u}_{in,1}^3 & \frac{50}{16}\hat{u}_{in,1}^5 & \cdots & \left(\frac{N}{N-1}{2}\right)\left(\frac{N}{N-3}{2}\right)\frac{\hat{u}_{in,1}^N}{2^{N-1}} \\ \frac{3}{4}\hat{u}_{in,2}^3 & \frac{50}{16}\hat{u}_{in,2}^5 & \cdots & \left(\frac{N}{N-1}{2}\right)\left(\frac{N}{N-3}{2}\right)\frac{\hat{u}_{in,2}^N}{2^{N-1}} \\ \vdots & \vdots & \ddots & \vdots \\ \frac{3}{4}\hat{u}_{in,M_2}^3 & \frac{50}{16}\hat{u}_{in,M_2}^5 & \cdots & \left(\frac{N}{N-1}{2}\right)\left(\frac{N}{N-3}{2}\right)\frac{\hat{u}_{in,M_2}^N}{2^{N-1}} \end{vmatrix} \begin{pmatrix} c_1e^{j\varphi_{c_1}} \\ c_3e^{j\varphi_{c_3}} \\ \vdots \\ c_Ne^{j\varphi_{c_N}} \end{pmatrix} \qquad (3.1.8)$$

The equation system of the intermodulation signals higher order are obtained same manner.

To write equations system related to the gain amplifier, the output voltage signal according to the input voltage with amplitude $\hat{u}_{in}$ and Frequency f_1 (ω_1) is expended in Taylor series. In main frequency, output voltage can be summarized as follows:

$$\hat{u}_{out,G} = c_1\hat{u}_{in} + \frac{3}{4}c_3\hat{u}_{in}^3 + \frac{10}{16}c_5\hat{u}_{in}^5 + \ldots + \left(\frac{N}{N-1}{2}\right)\frac{\hat{u}_{in}^N}{2^{N-1}}c_N \qquad (3.1.9)$$

The linear equation system of the gain amplifier for the M_3 measured values as follows:

$$\begin{pmatrix} v_{-1} \\ v_{-2} \\ \vdots \\ v_{-M_6} \end{pmatrix} = \begin{vmatrix} 1 & \frac{3}{4}\hat{u}_{in,1}^2 & \cdots & \left(\frac{N}{N-1}{2}\right)\frac{\hat{u}_{in}^{N-1}}{2^{N-1}} \\ 1 & \frac{3}{4}\hat{u}_{in,2}^2 & \cdots & \left(\frac{N}{N-1}{2}\right)\frac{\hat{u}_{in,2}^{N-1}}{2^{N-1}} \\ \vdots & \vdots & \ddots & \vdots \\ 1 & \frac{3}{4}\hat{u}_{in,M_8}^2 & \cdots & \left(\frac{N}{N-1}{2}\right)\frac{\hat{u}_{in,M_8}^{N-1}}{2^{N-1}} \end{vmatrix} \begin{pmatrix} C_{-1} \\ C_{-3} \\ \vdots \\ C_{-N} \end{pmatrix} \qquad (3.1.10)$$

Now magnitude and phase gain of amplifier $\varphi_{S_{21}} = \angle \underline{\upsilon}$, $|\underline{\upsilon}| = |\underline{S}_{21}|$ should be regarded as real values separately. Since Taylor coefficients should satisfy all equations system, the individual linear equation system should be solved simultaneously, that is, the following linear equations system for M_1, M_2, M_3, M_4, M_5, M_6 measured values should be considered together:

$$[U_{out,3IMD}]_{M_2 \times 1} = \left| [D_{in,3IMD}]_{M_2 \times N} \times [\underline{C}_{\ N}]_{N \times 1} \right|$$

$$[U_{out,3IMD}]_{M_2 \times 1} = \left| [D_{in,3IMD}]_{M_2 \times N} \times [\underline{C}_{\ N}]_{N \times 1} \right|$$

$$[U_{out,5IMD}]_{M_3 \times 1} = \left| [D_{in,5IMD}]_{M_3 \times N} \times [\underline{C}_{\ N}]_{N \times 1} \right|$$

$$[U_{out,7IMD}]_{M_4 \times 1} = \left| [D_{in,7IMD}]_{M_4 \times N} \times [\underline{C}_{\ N}]_{N \times 1} \right|$$

$$[U_{out,9IMD}]_{M_5 \times 1} = \left| [D_{in,9IMD}]_{M_5 \times N} \times [\underline{C}_{\ N}]_{N \times 1} \right|$$

$$\left| [\underline{\upsilon}_{\ out,G}]_{M_6 \times 1} \right| = \left| [D_{in,G}]_{M_6 \times N} \times [\underline{C}_{\ N}]_{N \times 1} \right|$$

$$[\varphi_{\upsilon_{out,G}}]_{M_6 \times 1} = \angle \{ [D_{in,G}]_{M_6 \times N} \times [\underline{C}_{\ N}]_{N \times 1} \}$$

With

$U_{out,F}$, $U_{out,3IMD}$ as output voltag for main signals, 3IDM-products, ... in (3.1.2), (3.1.3), ...

$D_{in,F}$, $D_{in,3IMD}$, ... , as main matrix in (3.1.2), (3.1.3), $^{\prime\prime\prime}$

$D_{in,G}$ as main matrix in (3.1.10) and

$$\left| [\underline{\upsilon}_{\ out,G}] \right|_{M_6 \times 1} = \begin{pmatrix} \left| v_{-1} \right| \\ \left| v_{-2} \right| \\ \vdots \\ \left| v_{-M_6} \right| \end{pmatrix}, \; [\ [\ ]_{out,G} \ ]_{M_6 \times} = \begin{pmatrix} \varphi_{v_{-1}} \\ \varphi_{v_{-2}} \\ \vdots \\ \varphi_{v_{-M_6}} \end{pmatrix}, \; \underline{C}_{\ N \times} = \begin{pmatrix} C_{-1} \\ C_{-3} \\ \vdots \\ C_{-N} \end{pmatrix} = \begin{pmatrix} c_1 e^{j\varphi_{c_1}} \\ c_3 e^{j\varphi_{c_2}} \\ \vdots \\ c_N e^{j\varphi_{c_N}} \end{pmatrix}$$

The number of measurements M_1, M_2 etc. is not the same but has been selected depending on the available number of measurement points (length of the curve).

Optimization

To determine the complex coefficients, the transfer characteristics of the amplifier such as a model in the form of the mathematical approximation are presented; in other words using determine coefficients model, the difference between the model and measurements are minimized. To do this, the numerical optimization techniques are used. For optimization, the complex least square method is an appropriate technique, in which the model coefficients are determined through setting zero of the partial derivatives [18] that is:

$$e = \sum_{i=1}^{n} g_i \left(\left| y_m(C) \right| - y_i \right)^2 \tag{3.2.1}$$

And for phase relationship

$$e = \sum_{i=1}^{n} g_i \left(\angle y_m \left(\underline{C} \right) - \varphi_i \right)^2 \tag{3.2.2}$$

With

e as error function,

y_i as measured output value

y_m as model (desired) output value

φ_i as measured output value for phase gain

g_i as weighting function

n as the measured value and

C as model parameters that must be found

It is natural that all discussed equations system must be considered in error function. For this reason, cost function (CF) as a function of the total error and the sum of all the dividable functions (functions error) are included and formed. For 6 output signals, we have:

$$e_{Sum} = e_1 + e_2 + e_3 + e_4 + e_5 + e_6 = \sum_{i=1}^{n} g_i \left(\left| y_{1m}\left(\underline{C} \right) \right| - y_{1i} \right)^2$$
$$+ \cdots + \sum_{i=1}^{n} g_i \left(\left| y_{6m}\left(\underline{C} \right) \right| - y_{6i} \right)^2 \tag{3.2.3}$$

Our research show that it is not possible to determine all the coefficients of the model with the same lowest error, especially the higher order ID-products can always be modeled worse [17]. Because of that, the weighting function, $(g_1, g_2....)$ is added in the cost function, the weighting factors for each output signal also be used so that be able to output signal or output signals with changes of this weights as requires to be optimized. So (3.2.3) is written in a new form as follows:

$$e_{Sum} = G_1 \times e_1 + G_2 \times e_2 + G_3 \times e_3 + G_4 \times e_4 + G_5 \times e_5 + G_6 \times e_6 \tag{3.2.4}$$

The accuracy of the model depends on the series degree (n). A better accuracy can be achieved in principle by increasing of this value. However, it is difficult to enhance the performance for $n > 13$.

For the numerical solution of the optimization problem, the function "minsearch" to find the minimum of the cost function in MATLAB has been used. Examples of the measured results and of the model results can be seen in Figure 3(a), where the voltage gain magnitude and phase of amplifier A is plotted as a function of input power level for single-tone excitation, and in Figure 3(b), where the two-tone fundamental signal level and the intermodulation product levels are plotted versus input power level.

Figure 3 shows some deviations between model and measurements in the fundamental signal levels as well as in the intermodulation products; note that in the optimization of the model coefficients the largest weights were used for the fundamental and third-order intermodulation products.

Figure 3(a) shows a good match for the magnitude and phase of the amplifier. Error obtained for the phase just a few degrees and for the magnitude is less than 0.5 dB. Figure 3(b) shows the divergence of the model when the two-tone input power goes beyond the highest input power level that was used as measurement data in the calculation of the Taylor series coefficients (-13.8 *dB*). Less dramatic but notable is the characteristic behavior below the divergence region: Deviations appear as slight oscillations with increasing amplitude closer to the divergence limit.

Using the mathematical model, it was possible to calculate the phase variation of the fundamental output signals for two-tone excitation, Figure 4, which was not accessible when measuring with a spectrum analyzer. Again, a slight oscillatory deviation is included in the calculated variation of the phase versus input power since a smooth parabolic-shaped variation should be expected. The observed oscillatory model errors, though not large on an average over the total

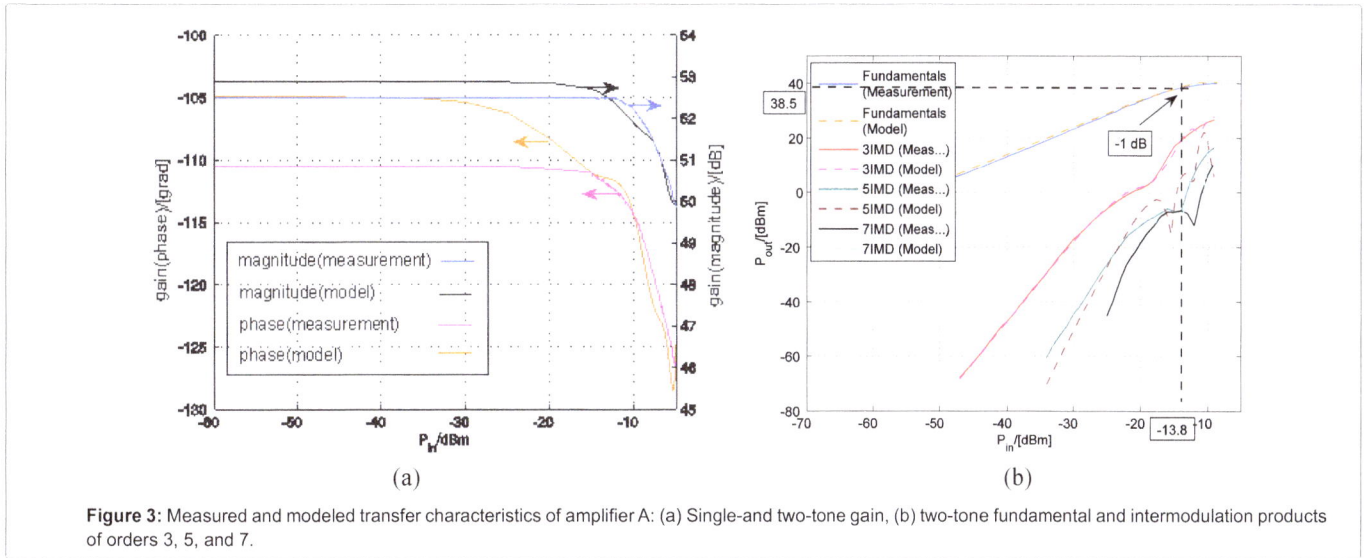

Figure 3: Measured and modeled transfer characteristics of amplifier A: (a) Single-and two-tone gain, (b) two-tone fundamental and intermodulation products of orders 3, 5, and 7.

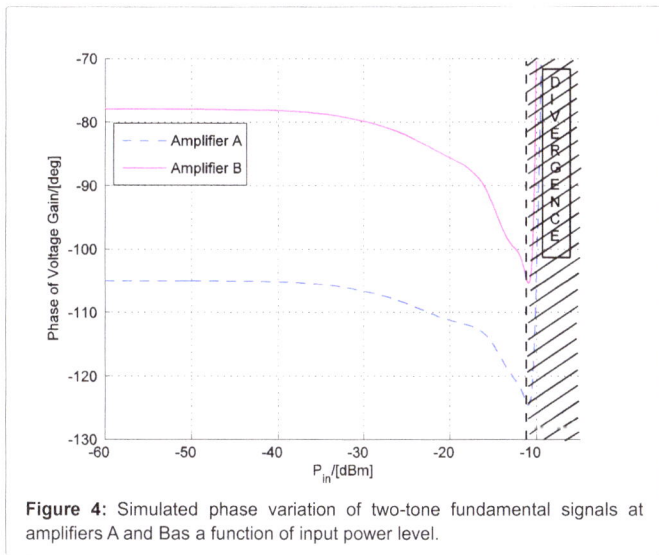

Figure 4: Simulated phase variation of two-tone fundamental signals at amplifiers A and B as a function of input power level.

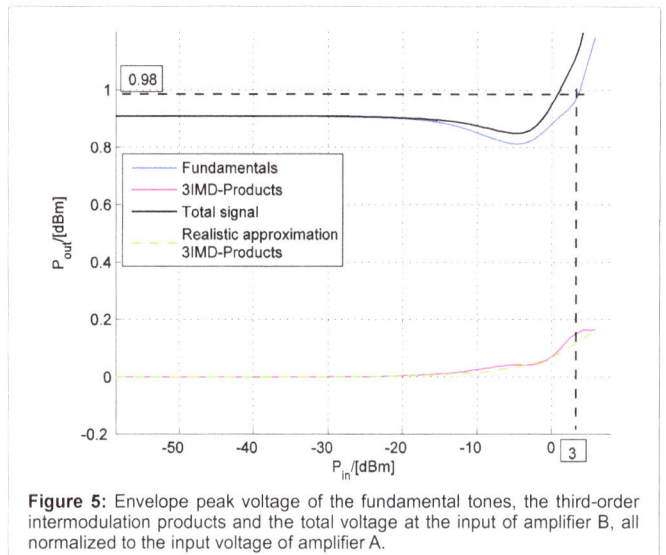

Figure 5: Envelope peak voltage of the fundamental tones, the third-order intermodulation products and the total voltage at the input of amplifier B, all normalized to the input voltage of amplifier A.

extent of the modeled amplifier characteristics, will be found as major sources of error in the simulation of the new circuit, section 4.

Simulation of the New Concept

In order to understand the interdependencies in the intermodulation cancellation of the new circuit, the experimental circuit was simulated using the Taylor-series models for the two individual amplifiers and using scattering matrix models for the other components in the circuit (perfect match for all components assumed. Due to the high volume of the output signals of the amplifier A and B, discrete numerical methods (with help DFT - Discrete Fourier Transform) of the entire circuit in has been simulated in MATLAB which is more flexible than other software. Other simulation programs like ADS have been shown restrictions to model [17].

One obvious deviation of the real circuit from the simplified concept of section 2 is the considerable insertion loss (1.6 dB) of the two delay lines: With reference to the principle of operation and designations of signals and test points given in section II, in the first loop, this insertion loss attenuates the input signal before the sampled signal from amplifier A gets subtracted, which requires a reduction of

the amplitude of the sampled signal ($k \wedge 2 < 2/\upsilon$) in order to achieve equal fundamental signal amplitudes at the input of both amplifiers. Also this reduces the amplitude of the sampled distortion signal $s^{\wedge\prime}(t)$ from amplifier A that fed into amplifier B and the resulting distortion signal at the output of amplifier B will been reduced accordingly.

On the upper signal path, the upper delay line attenuates the fundamental and distortion signals in the same way before they appear at the combiner. After the above-described adjustment of the first loop coupling from amplifier A to amplifier B, the signals at the input of amplifier B can be compared to the input signals of amplifier A. Figure 5 presents the envelope peak voltages of the fundamental tones, of the third-order intermodulation products and of the total signal, normalized to the envelope peak voltage of the fundamental two-tone signal at the input of amplifier A. It is seen that the fundamental tones have been adjusted to be equal in magnitude at both amplifiers for the "null" input power level of +3 dBm. At this input power level, the intermodulation products contribute already more than 10% of envelope voltage to the total input voltage of amplifier B, which presents a serious deviation from the assumption made in section II that the drive conditions of both amplifiers are basically equal.

Looking at lower drive levels, we find for the intermodulation products that the variation of voltage with drive power level exhibits an oscillatory error component as described in section 3; a more realistic indication of this variation is also given in the figure. The second notable effect is that the normalized fundamental signal voltage at amplifier B decays with reduced drive level. The reason becomes clear by inspecting Figure 3(a) and Figure 4: With reducing input power level the voltage gain of amplifier A increases and the insertion phase changes; thus, at the input of amplifier B, an increased signal sample from amplifier A is subtracted from the original input signal and creates a smaller fundamental signal component; the minimum around the drive level of -5 dBm is found to be due to the particular constellation of the amplification amplitude and phase variation with drive level.

Looking at the output side of both amplifiers, Figure 6 shows the amplitudes and phases of the fundamental and distortion signals at the power combiner: At the +3 dBm"null" drive level; we find the fundamental signal from amplifier B larger than that of amplifier A, which is due to the attenuation of upper delay line. At the same time, the two signals exhibit a considerable phase difference of 36° which in combination with the amplitude imbalance affects a combiner loss of about 0.5 dB; note that the power-added efficiency of the combiner circuit is further reduced by about 0.8 dB due to the dissipation loss of 1.6 dB from upper delay line affecting half the generated power of the power combiner.

Instead of for best combining efficiency, the delay line phase and the first loop coupling in this simulation were optimized for intermodulation suppression, seen by approximately equal intermodulation product amplitudes at the "null" drive level and a close to 180° phase difference; note the oscillatory variation in the intermodulation magnitude plot of Figure 6(a), which again can be attributed to the approximation error in the single amplifiers' Taylor series model. The intermodulation signal amplitudes are found approximately equal because the different effects of the insertion loss of the two delay lines in both loops nearly cancel, as both the intermodulation signals from amplifier A and of amplifier B get reduced in amplitude.

Again, turning to lower drive power levels, we observe about equal fundamental signal amplitudes at approx. -5 dBm input power, where Figure 5 has indicated a dip in the input signal of amplifier B, thus

compensating the amplitude imbalance due to the upper delay line attenuation. At even lower drive level, the input signal amplitude of amplifier B recovers from the dip and the fundamental signal from amplifier B becomes larger again than that of amplifier A. At the same time, due to the reduced fundamental signal drive of amplifier B relative to amplifier A, this amplifier produces considerably less intermodulation ($s''(t)$) than amplifier A, approximately a 3 dB reduction for 1 dB reduction in drive power. The combination of the two intermodulation components $-2s'(t)+s''(t)$ at the output of amplifier B thus increases in magnitude as the drive power is reduced and the gain of amplifier A increases. With growing difference in the two intermodulation contributions at the combiner, the circuit loses its cancellation effect.

Measurement of the Proposed Amplifier Circuit

Before building an experimental new amplifier circuit, two power amplifiers were assembled and tested: The amplifiers used 900 MHz silicon FET power modules MHW916 in cascade with preamplifiers which gave a saturated output power of about 14 W at a gain of about 53 dB. Measurement of the forward transmission group delay was performed with single-tone at small-signal level using a vector network analyzer HP8722C and the result was used to specify the length of the two delay lines in the new amplifier circuit. Measurements of the complex valued single-tone voltage gain (scattering coefficient S_{21}) as a function of input power level were performed using the vector network analyzer and two tone measurements (910 MHz and 911 MHz) of the fundamental signals and up to the seventh-order intermodulation product were performed using a spectrum analyzer HP8565E (Figures 7 and 8). Both sets of measurements were used for the modeling of the amplifiers transfer characteristics based on a Taylor-series expansion with complex coefficients.

The measured fundamental signals and up to the seventh-order intermodulation product, (a) sketch of setup and (b) photograph of bench equipment and amplifier circuit (Figure 8).

The setup Sketch of the experimental proposed amplifier circuit (called the feed forward power combiner circuit) is presented in Figure 9. The fundamental two-tone input signal is produced by combining two signal generators and the input signal split of 1:3 voltage ratio is realized by a -3 dB divider with a 10 dB attenuator in the upper signal

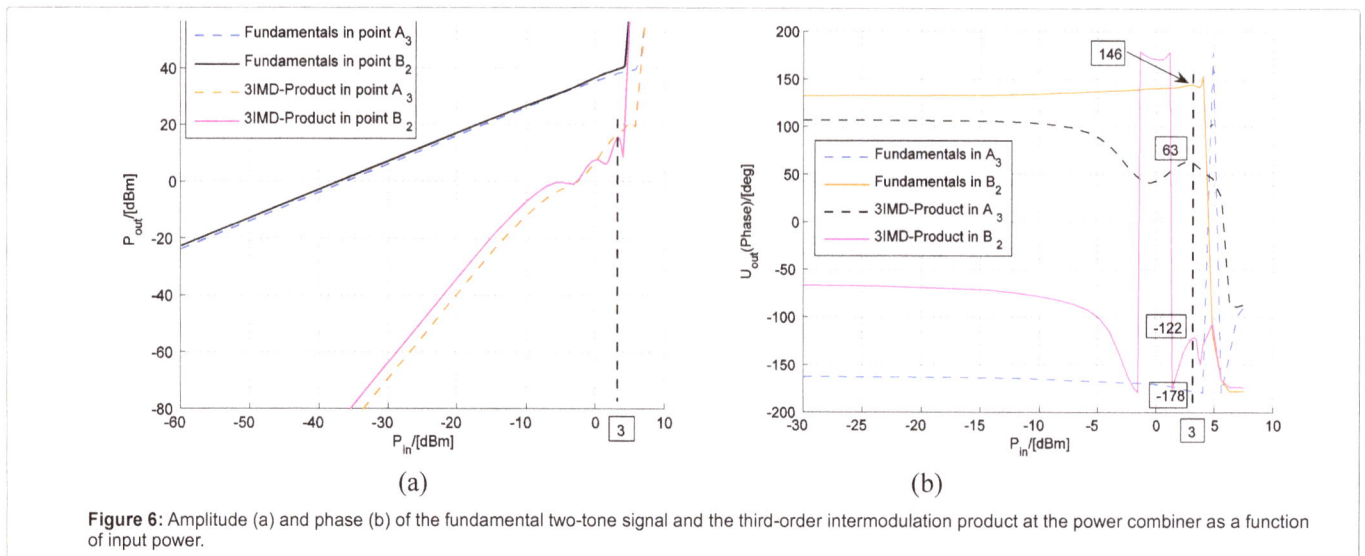

(a)

(b)

Figure 6: Amplitude (a) and phase (b) of the fundamental two-tone signal and the third-order intermodulation product at the power combiner as a function of input power.

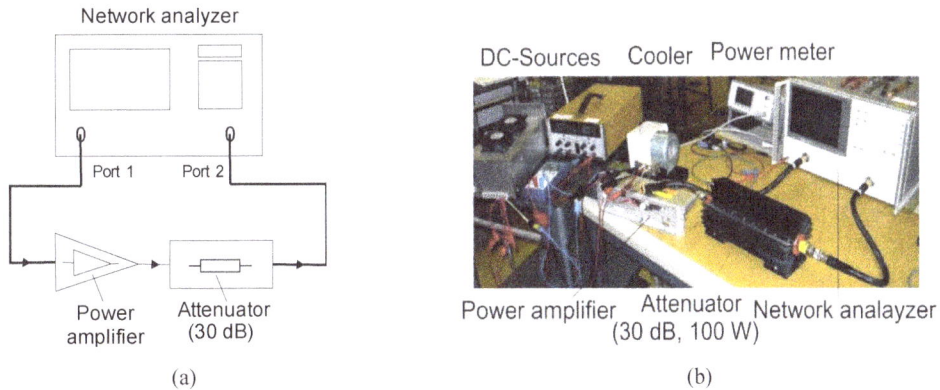

Figure 7: The experimental single-tone voltage gain, (a) sketch of setup and (b) photograph of bench equipment and amplifier circuit.

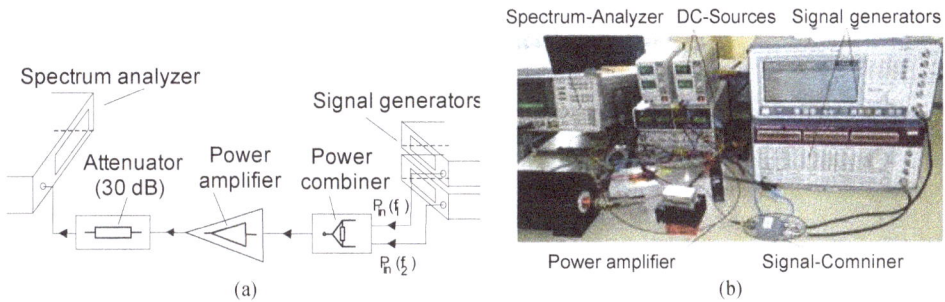

Figure 8: The measured fundamental signals and up to the seventh-order intermodulation product, (a) sketch of setup and (b) photograph of bench equipment and amplifier circuit.

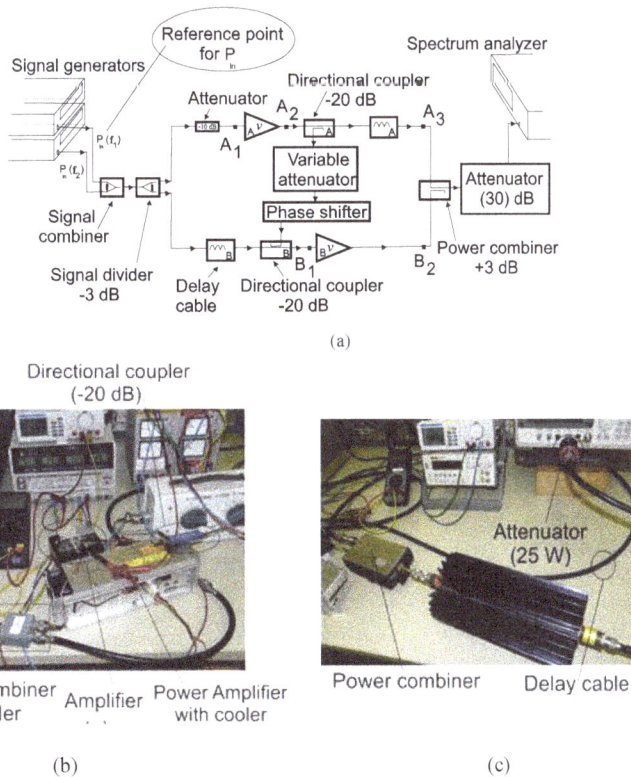

Figure 9: (a)The measured proposed amplifier circuit and (b) the sketch of setup and (c) photograph of bench equipment and amplifier circuit.

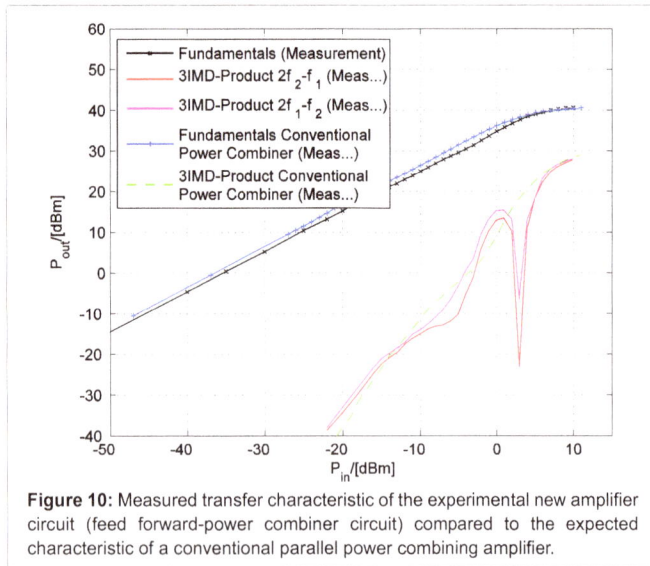

Figure 10: Measured transfer characteristic of the experimental new amplifier circuit (feed forward-power combiner circuit) compared to the expected characteristic of a conventional parallel power combining amplifier.

path to amplifier A. All components of the power combiner circuit are connected by coaxial cables which introduce some insertion loss and phase shift. In particular, the required time delays are realized by about 4 m long coaxial cables RG213 which introduce about 1.6 dB of extra loss. Between the two directional couplers a variable attenuator and a variable phase shifter are inserted in order to allow compensation of amplitude and phase offsets in the first loop. At the output side, a power attenuator is inserted between the power combiner and the spectrum analyzer in order to save the instrument from damage by high incident power.

Adjustment of the circuit turned out to be difficult: In particular, the variation of the voltage gain in magnitude and phase as a function of input power level, as seen in Figures 3 and 4, required selecting the drive power level for optimum linearity first of all. In our proof of concept experiment, we set the input power level to +3 dBm which corresponds to the 1 dB-saturation level of the individual power amplifiers.

With the drive signal level fixed, the signal at point B_1 at the input of amplifier B was observed using a spectrum analyzer and the upper delay line was adjusted and the variable attenuator and phase shifter were set such that the fundamental two-tone signals were approximately equal in level to the signals at point A_1 at the input to power amplifier A. This setting, at the same time gives approximately the correct pre-distortion level necessary for intermodulation cancellation at the second loop. Final adjustment of the variable attenuator and phase shifter was based on the measurement of the signals at the combiner output; either an optimum cancellation of the third-order intermodulation products could be achieved with the fundamental signals from amplifiers A and B slightly unequal in amplitude and phase or nearly equal fundamental signals could be achieved with considerable difference in phase and amplitude of the intermodulation products. When the amplifier circuit was adjusted for optimum intermodulation cancellation, phase- and amplitude deviations gave rise to a loss in fundamental signal output power of about 0.4 dB.

In Figure 10, the measurement of the proposed amplifier characteristic is presented and is compared to the expected characteristic of the conventional power combiner circuit using two amplifiers type "A" in parallel if excited at the same input power level as the amplifiers in the proposed amplifier circuit.

It is seen that the proposed amplifier achieves a notable extension of the linear range for the fundamental signals around the 1 dB compression level of the individual power amplifiers while the third-order intermodulation products are reduced by 25 and 45 dB at the +3 dBm input level (the two 3IMD-products are unequal, as explained in reference [19]). However, the cancellation is confined to a limited range of power levels around the "null"-input level and the intermodulation products level is not improved relative to the level of the conventional parallel power combiner outside this narrow region and is even worse in some parts of the input level range. This is a fundamental limitation of the feed forward power combiner compared to the conventional feed forward amplifier which was investigated by simulation in section 4. For a practical application of the feed forward power combiner concept, this means that in an operation mode with dynamically changing drive conditions, e.g., changing numbers and levels of modulated carriers as in a mobile communications base station amplifier, the loop adjustment has to be adapted dynamically also. Otherwise, under more static drive conditions, as, e.g., in TV-satellite power amplifiers, the feed forward cancellation concept could improve the linearity of present parallel power combining amplifiers with only moderate adaptivity requirements.

Conclusion

The limitation of the feed forward power combiner circuit was shown to be due to the saturation effect of the upper power amplifier, with its gain variation offsetting the balancing of the circuit loops. However, driving the power amplifiers into the saturation range is a necessary condition for high power efficiency. Power-added efficiency of our experimental amplifier was about 36% which is in contrast to around 10% efficiency of conventional FF-amplifiers. As a price, the critical drive level dependence of the combiner circuit may require higher adaptivity and control of the loop adjustments than a conventional feed forward amplifier, depending on the dynamics of the signals to amplify. By simulation, it can be shown that at lower drive powers the limitations get weaker as the intermodulation cancellation exhibits broader "null" and cancellation is improved even far outside the "null", similar to the characteristics of the conventional feed forward amplifier, yet losing the advantage in power efficiency.

References

1. Asif SZ (2007) Wireless Communications: evolution to 3G and Beyond. Artech House: 18-38.

2. Solbach K (2002) Feed Forward amplifier for GSM. University Duisburg-Essen, Germany, September.

3. Cavers J (1990) Amplifier linearization using a digital predistorter with fast adaptation and low memory requirements. IEEE Transactions on Vehicular Technology 39: 374-382.

4. Ogawa T, Iwasaki T, Maruyama H, Horiguchi K, Nakayama M, et al. (2004) High efficiency feed-forward amplifier using RF predistortion linearizer and the modified Doherty Amplifier. IEEE MTT-S Digest, SPC Electronics Corporation. Tokyo, Japan.

5. Cho KJ, Kim JH, Stapleton SP (2004) RF High Power Doherty amplifier for Improving the efficiency of a Feed Forward Linear amplifier. IEEE MTT-S International Microwave Symposium Digest3: 1-25.

6. Mucenieks L, Robertson C, Irvine B, Salvador N (2000) RF Power Amplifier Linearization using parallel Power Amplifier having intermod-complementing predistortion paths.

7. Black HS (1937) Wave Translation System.

8. Tegude FJ (2006) Automatic characteristic of the 3G-WCDMA mobile power amplifier, University Duisburg-Essen, Germany.

9. Pothecary N (1999) Feed forward linear power amplifiers, Artech House.

10. Kenington PB (2000) High-linearity RF amplifier design, Artech House, Inc. Norwood, MA, USA.

11. Eisenberg J, Altos L, Avis S (2002) Closed loop active cancellation technique(ACT)-based RF power linearization architecture.

12. Cho KJ, Kim JH, Stapleton SP (2004) RF High Power Doherty amplifier for Improving the efficiency of a Feed Forward Linear amplifier. IEEE MTT-S International Microwave Symposium Digest, Dept. of Radio Sci. & Eng., Kwangwoon Univ 2:

13. Ogawa T, Iwasaki T, Maruyama H, Horiguchi K, Nakayama M, et al. (2004) High efficiency feed-forward amplifier using RF predistortion linearizer and the modified Doherty Amplifie. IEEE MTT-S Digest, SPC Electronics Corporation 2: 537-540.

14. Chong E (2001) The Volterra Series and The Direct. Method of Distortion Analysis, University of Toronto, research work.

15. Vuolevi J, Rahkonen T (2003) Distortion in RF Power Amplifier, Artech House.

16. Aikio J (2007) Frequency Domain Model Fitting and Volterra Analysis Implemented on top of Harmonic Balance Simulation, Faculty of Technology, Department of Electrical and Information Engineering, University of Oulo, Oulo.

17. Motavalli MR (2010) Untersuchung einer Leistungs-kombinations-Schaltung mit Feed Forward-Linearisierung, Dissertation, University Duisburg-Essen, Germany.

18. Xiangqian G, Hongwen K, Hongxing C (2006) The least-square method in complex number domain, Beijing, China, March.

19. Carvalho N, Pedro JC (2002) A Comprehensive Explantion of Distortion Sideband Asymmetries. IEEE Transactions on Microwave Theory and Techniques 50: 2090-2101.

14

Alternate Relationship between Single Fiber Properties and Both of Fiber Microscopic and Physical Properties

Abeer S Arafa*

Cotton Research Institute, Agricultural Research Center, Egypt

Abstract

Current study was carried out in Cotton Research Institute, Agricultural Research Center, Egypt and Textechno company labs Monchengladbach, Germany during 2013 season. The materials used in this study comprised 16 different genotypes produced by cotton research institute. In addition to, two upland cotton varieties from Sudan. Aiming to study the effect of environmental conditions, genotypes and thier interaction on each characters under study beside studyig the relationship between single fiber prroperties and bundle properties. Giza 93 variety showed the lowest micronaire, fineness and area of cross section. On contrast, it showed the highest upper half mean. Giza 87 variety exhibited the highest bundle strength, the lowest reversals number per mm and the narrowest convolution angle. As for single fiber properties. Giza 45 variety followed by Giza 87 variety exhibited the lowest single fiber linear density readings. Single fiber strength is a little bit lower than bundle strength. Giza 87 variety surpassed the other genotypes on single fiber strength. Upland cotton varieties showed the worst bundle and single fiber properties. As maturity ratio increased bundle and single fiber properties improved and vice versa. According to, the relationship between single fiber prroperties and bundle properties, there was a nearly linear relation¬ship between fiber properties and single fiber properties. After excluding the weak correlated characters 3 linear multiple regression models for single fiber tenacity, single fiber linear density and single fiber elongation were produced. Fineness and degree of thickness were the predictor variables for single fiber strength parameters. However, fiber fineness per mtex and micronaire values were the most important factors for single fiber linear density, both of them proportionate directly with single fiber linear density. While, single fiber elongation contains 7 predictors i.e. Micronaire reading, UI, area of cross section, theta, fiber strength, fiber elongation and UHM characters. Nevertheless, it showed the weakest relation, UI, area of cross section, and fiber Strength proportionate inversely with single fiber elongation. The model is not reliable enough for single fiber elongation prediction.

Keywords: Cotton; Fiber; Single; Fineness; Strength; Elongation

Introduction

Quality is the ultimate goal of the cotton manufacturer because raw material costs are high up to 50% of the total manufacturing costs at the spinning mill. These costs decrease or increase depending on the fiber quality of the raw material.

The quality is a set of attributes, some of them are related to the bundle physical and mechanical characters each measured with fast and easy instruments like AFIS , HVI and Fibrotest while, the others are time consuming attributes such as single fiber characters which need some complicated instruments like Image analyzer, Favimat and Robot tester. In fact, the qualities of single or bundle fiber characters is a result of some genetic factors like fiber perimeter or diameter ,cellulose deposition order, the angle of deposition [1] some others are associated with the growing conditions like the amount of cellulose deposited inside the fiber which represents the fiber body.

On the other hand, the single fiber characters are an indicator for the bundle physical and mechanical characters is that the bundle breaking and elongation were shown to increase as the single fiber breaking elongation increased [2] something like a building consisting of bricks, walls, and then the rooms.

From a commercial and industrial point of view, cotton faces great competition with the other natural and synthetic textile fibers. It should be strong enough to compete with other natural and synthetic textile fibers [3].

Hence it is important:

1. Studing how much each characters affected by the environmental conditions, genotypes and the interaction between them.

2. Understanding how some microscopic characters are associated with bundle and single fiber characters like fiber perimeter or diameter which describes the fiber intrinsic or biological fineness that is controlled by genetics. Intrinsic fineness is completely different from the fineness in millitex or linear density as weight of unit length. When we deal with weight, we do weight of cellulose where the maturity and the growing conditions affect. So, if there are two fibers equally in the intrinsic fineness (diameter or the perimeter) the higher in maturity ratio will give higher millitex reading (Figure 1).

3. Studying the effect of wall thickening (Figure 2) and the structural properties like convolution angle which refers to spiral angle (the angle formed between the fiber long axis and the cellulose layer the more acute angle the higher, fiber strength and reversals per unit length (the point which the cellulose layer changed the deposition direction from clock wise direction to anti clock wise direction and vice versa. This forms weak points during the tenacity test (Figure 3) [4].

***Corresponding author:** Arafa SA, Cotton Research Institute, Agricultural Research Center, Egypt, E-mail: Sameh_owf@yahoo.com

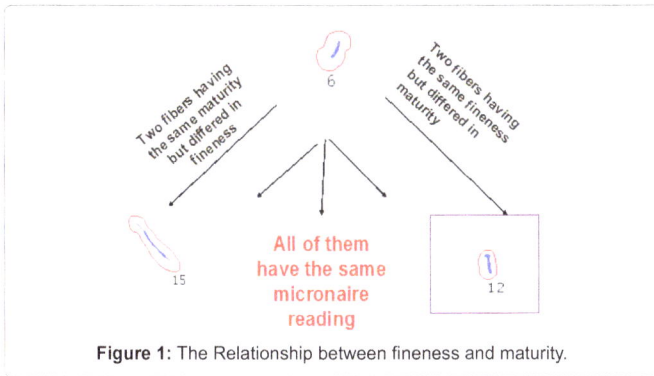

Figure 1: The Relationship between fineness and maturity.

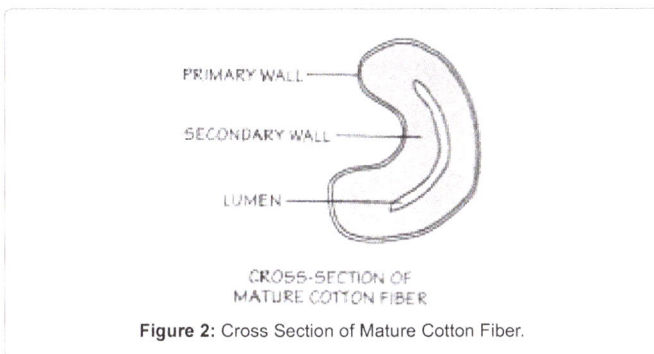

Figure 2: Cross Section of Mature Cotton Fiber.

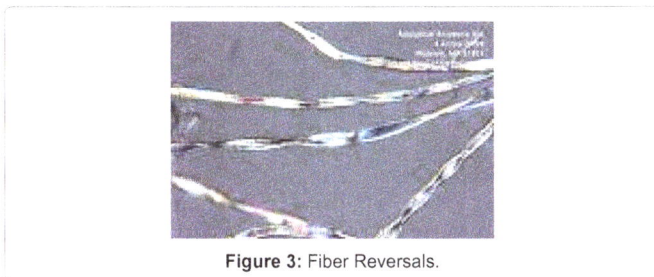

Figure 3: Fiber Reversals.

4. The relationship between single fiber properties and bundle properties.

The investigation was conducted during 2013 season. The materials used in this study comprised 16 different genotypes (G) named (Giza 88, Giza 92, Giza 93, [G.84 (G.70xG.51b)] defined as C1, Giza 45, Giza 87, Giza 80, Giza 90, G90xAus. - defined as C2, [G.83(G75x5844)] G.80 defined as C3, 10229xG86 defined as C4, Giza 86, green cotton and brown cotton) produced by cotton research institute. In addition to, (upland Sudan fine, upland Sudan coarse). Aiming to study the effect of inherent fiber characters on the single fiber properties. Under all the genotypes we used two maturity ratio levels (L1, L2) within each genotype were used to study the effect of the fiber maturity on the single fiber properties. Beside, study the effect of some structural properties on the behavior of the single fiber during the mechanical tests.

The investigation was conducted during 2013 season. The materials used in this study comprised 16 different genotypes (G) named (Giza 88, Giza 92, Giza 93, [G.84 (G.70xG.51b)] defined as C1, Giza 45, Giza 87, Giza 80, Giza 90, G90xAus. - defined as C2, [G.83(G75x5844)] G.80 defined as C3, 10229xG86 defined as C4, Giza 86, green cotton and brown cotton) produced by cotton research institute. In addition to, (upland Sudan fine, upland Sudan coarse). Aiming to study the effect of inherent fiber characters on the single fiber properties. Under all

the genotypes we used two maturity ratio levels (L1, L2) within each genotype were used to study the effect of the fiber maturity on the single fiber properties. Beside, study the effect of some structural properties on the behavior of the single fiber during the mechanical tests.

Studied characters

Microscopic characters

The cross sections and the Images were processed at the Textile Consolidation Fund, Alexandria, Egypt. While, the Image Analyzer located in the Fiber Structural and Microscopic lab, Cotton Research Institute, Giza. Was used to analyze the fiber cross section images to calculate fiber perimeter with [μ], fiber area of cross section (ASCW) in [μ]^2 and degree of thickening (θ). Number of reversals per mm, number of Convolutions per mm, and ribbon width in micron were tested using (G208 projection microscope according to ASTM D: 2130-1986). Convolution angle was calculated according to [5], convolution angle=(π/2 X Average ribbon width /C) where, C=Convolutions pitch length divided by the number of convolutions.

Fiber physical characters

The Micromat instrument was used to determine micronaire reading, maturity ratio (MR), hair weight (fiber linear density (millitex)) (ASTM-D;2818-1986). Fiber upper half mean length UHM (mm), length uniformity index UI, short fiber content SFC, fiber strength (g/tex) and fiber elongation (%) were measured by Fibrotest instrument in Textechno company labs Monchengladbach, Germany.

Single fiber characters

Single fiber measurements and force/elongation curves were performed using Favimate + and Favigraph instruments in Textechno company labs Monchengladbach Germany.

Complete randomized design (two ways ANOVA) was used to analyze the data statistically. The treatment means were compared using. L.S.D. test at 0.05% Level. Simple and multiple regression model were performed between fiber properties (X) variables, single fiber properties (Y) according to the procedures outlined by [6].

Results and Discussions

Data presented in Table 1, explained the effect of the genotypes, maturity level and their interaction on fineness and maturity parameters measurements. As to the micronaire reading, Giza 93 showed the lowest micronaire reading (2.65) followed by Giza 87 (2.70) these reading were fitted to their fineness readings (103.61 and 103.25 mtex) ,respectively. Micronaire reading looks similar to fineness in mtex both could not be good indicator for fineness because it expresses both of fineness and maturity. They are referring to the fineness when the comparison is done between the genotypes of the same maturity or they are referring to the maturity degree when the comparison is done between the two maturity level inside the same genotype, the relation between fineness and maturity is not that easy it is complicated. Thus, determining fineness and maturity parameters using image analyzer as a reference method was very important to explain the results under this study. Giza 93 and Giza 87 exhibited the lowest two readings for the area of cross section (71.07 and 79.70[μ]^2) respectively, which explained that the two pervious verities are the finest varieties comparing to the other genotypes. According to maturity ratio and theta values which, determine the amount of cellulose deposition or maturity degree. Giza 93 readings for maturity ratio and theta characters were (0.87 and 0.50), respectively.

Sample	Micromat measurements			Image analyzer measurements	
	mic	MR	Fineness	θ	ASCW[μ]²
G 88 L1	2.90	0.90	111.62	0.53	73.01
G 88 L2	3.65	1.03	137.59	0.62	98.56
G88mean	3.28	0.97	124.61	0.58	85.79
G 92 L1	3.00	0.85	105.90	0.52	73.00
G 92 L2	4.00	1.05	148.89	0.63	109.62
G92mean	3.50	0.95	127.39	0.58	91.31
G 93 L1	2.15	0.82	93.00	0.45	58.28
G 93 L2	3.15	0.91	114.22	0.55	83.86
G93mean	2.65	0.87	103.61	0.50	71.07
c1 L1	3.30	0.87	127.74	0.49	88.14
c1 L2	4.20	0.95	160.57	0.57	116.21
c1mean	3.75	0.91	144.15	0.53	102.18
G 45 L1	2.60	0.87	99.32	0.51	77.35
G 45 L2	3.20	1.00	120.26	0.60	85.27
G45mean	2.90	0.94	109.79	0.56	81.31
G 87 L1	2.40	0.83	94.69	0.48	72.19
G 87 L2	3.00	1.02	103.25	0.61	79.70
87mean	2.70	1.02	103.25	0.61	79.70
G 80 L1	3.15	0.81	146.63	0.44	83.86
G 80 L2	4.35	0.91	173.72	0.54	121.30
G80mean	3.75	0.86	160.18	0.49	102.58
G 90 L1	3.20	0.81	127.17	0.43	85.27
G 90 L2	3.75	0.93	144.02	0.54	101.66
G90mean	3.48	0.87	135.60	0.49	93.47
c2 L1	4.60	0.84	178.21	0.44	128.33
c2 L2	5.00	0.91	199.19	0.45	144.65
c2mean	4.80	0.88	188.70	0.45	136.49
c3 L1	3.90	0.89	160.15	0.55	101.63
c3 L2	4.40	0.98	176.46	0.59	123.02
c3mean	4.15	0.94	168.30	0.57	112.33
G 86 L1	3.90	0.81	153.26	0.44	106.39
G 86 L2	4.50	0.99	169.50	0.50	126.49
G86mean	4.20	0.90	161.38	0.47	116.44
c4 L1	3.50	0.85	145.70	0.45	92.19
c4 L2	3.85	0.91	152.32	0.50	104.80
c4mean	3.68	0.88	149.01	0.48	98.50
green L1	2.50	0.60	139.70	0.38	80.00
green L2	3.00	0.67	157.13	0.40	87.20
gre.mean	2.75	0.64	148.41	0.39	83.60
brown L1	2.90	0.91	117.78	0.48	101.69
brown L2	3.70	0.76	134.38	0.47	130.10
bro.mean	2.90	0.91	117.78	0.48	101.69
upland L1	3.30	0.64	197.78	0.35	131.70
upland L2	5.10	0.60	231.74	0.39	150.92
up.mean	4.20	0.62	214.76	0.37	141.31
L1 mean	*3.15*	*0.82*	*133.24*	*0.46*	*90.20*
L2 mean	3.92	0.90	154.88	0.53	110.89
LSD 0.05G	0.07	0.00	2.37	0.00	1.44
LSD 0.05 L	0.04	0.01	1.92	0.00	1.21
LSD 0.05 LxG	0.10	0.01	3.23	0.01	2.20

Table 1: Effect of cotton genotype, maturity level and their interaction on fiber fineness and maturity parameters.

However Giza 87 Varity showed (1.02 and 0.61) for these two characters, respectively. These two varieties exhibited high maturity ratio. Upland cottons showed the lowest maturity ratio reading (0.62) and the lowest theta reading (0.37) followed by the green colored cotton then Giza 80. On the other hand, Giza 90xAus. C2, showed the highest micronaire reading (4.8). While, the upland cottons exhibited the highest fineness in mtex reading (214.76). Indicating that micronaire

reading or fineness in mtex refers to the fineness, since it affected so much by maturity. So, their test should be accompanied by maturity test or image analyzer results.

Data of Fiber length and mechanical properties as affected by the genotype ,maturity level and their interaction are illustrated in Table 2, it is obvious from Table 2, that Giza 45 then Giza 93 varieties gave the highest upper half mean readings (35.28 and 35.248 mm), respectively.

Sample	Length parameters			Fiber mechanical characters	
	UHM(mm)	SFC	UI	strength(g/tex)	elongation%
G 88 L1	34.50	15.27	85.81	53.11	12.00
G 88 L2	35.76	10.77	85.95	58.54	13.82
G88mean	35.13	13.02	85.88	55.83	12.91
G 92 L1	30.32	16.46	82.01	50.19	11.31
G 92 L2	33.95	5.28	85.91	61.99	13.29
G92mean	32.14	10.87	83.96	56.09	12.30
G 93 L1	34.10	13.76	82.63	51.91	11.68
G 93 L2	36.38	11.26	86.06	55.81	12.68
G93mean	35.24	12.51	84.35	53.86	12.18
c1 L1	33.03	12.96	84.56	49.87	11.22
c1 L2	35.36	9.28	87.47	57.36	12.86
c1mean	34.20	11.12	86.02	53.62	12.04
G 45 L1	34.50	15.27	85.81	53.11	12.00
G 45 L2	36.05	20.39	87.51	55.84	12.93
G45mean	35.28	17.83	86.66	55.84	12.93
G 87 L1	33.59	9.36	81.84	54.85	11.11
G 87 L2	35.51	7.36	85.83	57.85	12.19
87mean	34.55	8.36	83.84	56.35	11.65
G 80 L1	29.02	18.33	81.68	35.96	11.74
G 80 L2	30.02	13.66	82.71	41.72	12.14
G80mean	29.52	16.00	82.20	38.84	11.94
G 90 L1	27.29	20.17	79.50	36.12	11.61
G 90 L2	30.46	15.63	80.73	35.9	11.25
G90mean	28.88	17.9	80.12	36.01	11.43
c2 L1	29.00	11.00	80.81	40.00	13.87
c2 L2	29.94	9.86	83.61	41.66	14.11
c2mean	29.47	10.43	82.21	40.83	13.99
c3 L1	30.00	12.14	81.84	33.00	12.17
c3 L2	30.31	11.81	82.25	36.24	12.56
c3mean	30.16	11.98	82.05	34.62	12.37
G 86 L1	30.42	6.51	84.96	50.1	13.77
G 86 L2	32.19	6.18	86.25	45.45	12.41
G86mean	31.31	6.35	85.61	47.78	13.09
c4 L1	31.31	11.21	83.65	46.00	12.31
c4 L2	32.26	10.13	85.62	47.18	12.48
c4 mean	31.79	10.67	84.64	46.59	12.4
green L1	29.14	13.12	81.32	33.14	10.00
green L2	29.37	10.71	83.17	35.20	10.30
gre.mean	29.26	11.92	82.25	34.17	10.15
brown L1	28.71	18.18	80.00	30.06	9.46
brown L2	29.97	18.96	81.45	35.01	9.83
bro. mean	29.34	18.57	80.73	32.54	9.65
upland L1	26.48	24.43	78.40	22.78	7.99
upland L2	27.69	11.21	81.83	25.22	8.52
up. mean	27.09	17.82	80.12	24.00	8.26
L1 mean	30.76	14.54	82.32	42.68	11.48
L2 mean	32.35	11.50	84.42	46.06	12.09
LSD 0.05 G	*0.24*	0.03	ns	0.44	0.02
LSD 0.05 L	*0.16*	0.01	ns	0.34	0.01
LSD 0.05 G.L	*0.26*	0.10	ns	n.s	0.04

Table 2: Effect of the genotype, maturity level and their interaction on fiber length and mechanical parameters as measured by Fibrotest instrument using HVI mode.

As for short fiber content, its worthy to mention that the SFC% measured by fibrotest instrument seems to be higher in content than those measured by the other instruments like fibrograph and sorters instruments. Data in Table 2, showed that the higher the maturity level the lower the short fiber content regardless the cotton genotype. (SFC %) being 11.50% for the higher level of maturity vs. 14.54% for the low maturity level. Immature fiber is easily to be broken during ginning resulting high SFC%. Regarding cotton genotype, Giza 86 variety showed the lower SFC% it is averaged 6.35%. While, the extra long cottons exhibited the higher SFC% it ranged from 8.36% in Giza 87 variety to 17.83% in Giza 45 variety and it ranged in upper Egyptian cottons from 10.43 % in c2 to 17.17.90% in Giza 90 variety. Colored cotton showed moderate SFC% but slightly higher than upper Egyptian varieties. Upland cotton showed the highest SFC% averaged 27.09%.

Sample	Convolutions no./mm	Reversals no./mm	Convolution angle
G 88 L1	3.61	1.41	13.00
G 88 L2	4.00	1.32	12.56
88mean	3.81	1.37	12.78
G 92 L1	3.52	1.41	15.38
G 92 L2	3.80	1.37	14.00
G92mean	3.66	1.39	14.69
G 93 L1	3.31	1.33	13.90
G 93 L2	3.42	1.27	13.80
G93mean	3.37	1.30	13.85
c1 L1	2.96	1.33	11.95
c1 L2	3.11	1.29	11.39
c1mean	3.04	1.31	11.67
G 45 L1	3.71	1.35	11.10
G 45 L2	3.98	1.32	10.58
G45mean	3. 85	1.32	10.58
G 87 L1	3.10	1.30	11.06
G 87 L2	3.79	1.30	9.32
87mean	3.45	1.30	10.22
G 80 L1	2.78	1.70	16.05
G 80 L2	3.32	1.50	15.03
G80mean	3.05	1.60	15.54
G 90 L1	3.45	2.10	15.57
G 90 L2	4.20	1.47	15.90
G90mean	3.75	1.79	15.73
c2 L1	3.21	1.71	16.30
c2 L2	3.55	1.66	15.70
c2mean	3.38	1.69	16.00
c3 L1	2.80	1.60	15.77
c3 L2	3.10	1.44	15.04
c3mean	2.95	1.52	15.41
G 86 L1	2.70	1.55	14.45
G 86 L2	3.60	1.53	13.68
G86mean	3.15	1.54	14.06
c4 L1	3.00	1.48	14.01
c4 L2	3.11	1.48	13.67
c4mean	3.06	1.48	13.84
green L1	2.70	1.90	17.01
green L2	2.81	1.70	16.66
gre.mean	2.76	1.80	16.84
brown L1	2.79	1.80	16.58
brown L2	3.00	1.70	16.36
bro.mean	2.90	1.75	16.47
UplandL1	2.54	2.11	17.06
upland L2	2.74	2.10	17.15
up.mean	2.64	2.10	17.11
L1 mean	3.08	1.61	14.61
L2 mean	3.44	1.50	14.06
LSD 0.05 G	0.13	0.02	0.03
LSD 0.05 L	0.15	0.01	0.01
LSD 0.05 GxL	ns	0.05	0.06

Table 3: Effect of the genotype, maturity level and their interaction on fiber microscopic characters.

Also, fiber strength could change the result in case of Giza 87 which gave the highest fiber strength and the lowest SFC (8.36) under the study. So, it is complex and its result affected by the inherent genes and the environmental conditions.

According to Uniformity index character, regardless it did show any significance due to the effect of the main factors or their interactions,

but the extra long genotype surpassed the long ones. Therefore, upland and color cotton gave the lowest reading of fiber UI. Regarding to the fiber mechanical properties, 87 gave the highest fiber strength (56.35 g/tex) followed by Giza 92 which gave approximately the same reading (56.09 g/tex), while Giza 45 achieved the third level of the fiber strength (55.84 g/tex). Giza 92 is known as the strongest Egyptian variety, but this may be ascribed to that the result of each genotype under study didn't express the standard reading because it's a mean of low and high maturity levels. In addition to, the number of weak points and the convolution angle degree as going to discus later. Consequently the environmental condition expressed by maturity level could affect the strength readings. It could be recognized from the low and high maturity levels strength reading at the end of the table (42.68 vs. 46.06 g/tex), respectively. It's worthy to mention that the effect of the environmental condition on the fiber length has a limited range. Thus, usually the extra long cultivars are the strongest genotypes. On contrast, upland cotton gave the lowest strength value (24.00 g/tex).

According to the percentage of fiber elongation, genotype, maturity level and their interaction significantly affected the percentage of fiber elongation Giza90xAus. Gave the highest value of fiber elongation (13.99%). In contrast, the lowest value was (8.26%) for upland cotton. The percentage of fiber elongation value of the high maturity level (12.09%) exhibited the low maturity level (11.48%). The interaction between Giza90xAus and the high maturity level regard as the highest elongation value (11.44%).

Table 3 presented the effect of genotype, maturity level and their interaction on fiber microscopic characters. As regard to the number of convolutions per mm. it is well known that the Egyptian cotton have the higher convolutions number comparing to the upland cotton. That's clear from Table 3, Giza 45 verity exhibited the highest value of convolutions number per mm (3.85). On the other hand, the upland cotton gave the minimum reading of the number of convolutions per mm (2.64). The high maturity level exhibited the higher convolutions number per mm than low maturity level (3.44 vs. 3.08). As for the reversals number per mm, reversal considered as weak points along the fiber where the fiber exposed to breakage when force applied along. Because it the point where the cellulose deposition layer reversed the direction from clock wise to anti-clock wise Figure 3. It affected by both genetic and environmental conditions. Egyptian cotton contains the lowest reversals number comparing to the upland cotton. Giza 87 showed the lowest reversals number per mm (1.30). In contrast the upland cotton gave the maximum value of the reversals per number (2.10). Maturity level also, affected the weak points the highest in maturity ratio the lowest in reversals number per mm and vice versa as shown down in Table 3, accordingly, the interaction between the upland cotton and the low maturity ratio gave the highest value of reversals number per mm (2.11). According to the convolution angle, convolution angle refers to the spiral angle. The Egyptian cotton is characterized by narrow spiral and convolution angles, furthermore the extra-long genotypes has narrower angle of cellulose deposition along the fiber axis comparing to the long genotypes. Giza 87 cultivar gave the lowest value of convolution angle (10.22°) that reflected positively on its strength as mentioned before in Table 2. On the contrary, the upland cotton gave the widest convolution angle (17.11°).

Table 4, showed the effect of genotype, maturity level and their interaction on single fiber properties, it's clear from Table 4, that the effect of the main factors and their interaction were significant on all single fiber characters under the table. As to the single fiber linear density per mtex, its trend was as similar as bundle linear density. It

Sample	Linear density(mtex)	Single fiber tenacity(g/tex)	Single fiber elongation%
G 88 L1	115.00	49.20	11.44
G 88 L2	141.00	51.61	12.04
88mean	128.00	50.41	11.74
G 92 L1	127.00	46.73	10.55
G 92 L2	139.00	55.98	11.18
G92mean	133.00	51.36	10.87
G 93 L1	105.00	45.66	11.89
G 93 L2	119.00	48.35	10.61
G93mean	112.00	47.01	11.25
c1 L1	148.00	49.22	10.98
c1 L2	159.00	50.10	11.38
c1mean	154.00	49.66	11.18
G 45 L1	100.00	51.00	10.00
G 45 L2	105.00	51.28	10.49
G45mean	103.00	51.14	10.25
G 87 L1	103.00	50.09	10.10
G 87 L2	113.00	54.48	10.66
G87mean	108.00	52.29	10.38
G 80 L1	161.00	36.25	13.12
G 80 L2	165.00	41.93	13.77
G80mean	163.00	39.09	13.45
G 90 L1	122.00	42.60	13.12
G 90 L2	151.00	40.37	14.49
G90mean	136.50	41.49	13.81
c2 L1	187.00	33.12	12.27
c2 L2	204.00	35.60	13.45
c2mean	195.50	34.36	12.86
c3 L1	151.00	48.75	15.35
c3 L2	160.00	50.51	16.20
c3mean	155.50	49.63	15.78
G 86 L1	147.00	33.78	13.03
G 86 L2	186.00	35.76	9.76
G86mean	166.50	34.77	11.40
c4 L1	143 .00	35.11	11.01
c4 L2	154.00	36.78	11.76
c4mean	149.00	35.95	11.39
green L1	115.00	24.44	10.00
green L2	135.00	26.64	10.15
gre.mean	125.00	25.54	10.08
brown L1	129.00	40.76	10.92
brown L2	111.00	45.99	11.98
bro.mean	120.00	43.38	11.45
upland L1	191.00	29.62	9.60
upland L2	231.00	22.38	13.54
up. mean	211.00	26.00	11.57
L1 mean	135.80	41.52	11.60
L2 mean	151.60	43.18	12.10
LSD 0.05 G	3.10	0.28	0..13
LSD 0.05 L	2.40	0.19	0.11
LSD 0.05 GxL	3.90	0.33	0.16

Table 4: Effect of the genotype, maturity level and their interaction on single fiber properties.

could be arranged in ascending order according to the fineness mtex as follows: 1- the extra-long genotypes, i.e. Giza 45 (103 mtex) followed by Giza 87 (108 mtex) then Giza 93 (112 mtex), 2- The brown colored cotton (120 mtex), 3- the long Egyptian genotypes, i.e. Giza 90 (136.50

mtex) and 10229xG86 or C4 (149 mtex), 4- the upland cotton (211 mtex). This may be ascribed to that Giza 45, Giza93 and Giza87 which belong to extra-long extra fine fiber gave the lowest area of cross section whether in the lower or in the higher maturity level, while Giza 90 and C4 gave the lowest long fiber areas. On the contrary, the upland cotton exhibited the highest value of the area of cross section as explained before . The interaction between Giza 45 and the lowest maturity level gave the lowest fiber linear density (100 mtex). On the other hand, the highest fiber linear density was obtained from the interaction between the upland cotton and the high maturity level (231.00 mtex).

As regards to the single fiber strength it could be recognized from Table 4, that single fiber strength is a little bit lower than bundle strength. Giza 87 recorded (52.29 g/tex) and surpassed the other genotypes on this trait. This may due to that Giza gave highest theta, maturity readings and the lowest structure properties, i.e. convolution angle and number of reversals per mm. on the other hand the upland cotton gave the lowest single fiber strength (26.00 g/tex) according to the reasons discussed above. The higher maturity level surpassed the lower one in single fiber tenacity property, respectively (43.18 vs. 41.52 g/tex). The interaction between Giza 87 and the highest maturity level gave the highest single fiber tenacity. On the other hand, the lowest single fiber tenacity was obtained from the interaction between the upland cotton and the low maturity level.

Concerning the percentage of single fiber elongation, it's noticeable that single fiber elongation did not behave as bundle elongation, some readings were higher than those of the bundle elongation like, Giza 80, Giza 90, C 3, brown cotton and upland cotton. In contrast, the rest genotypes get lower readings than bundle elongation. In general the low maturity ratio gave low elongation % comparing to the high maturity level within the same genotype. This could be detected from the interaction means as well as the overall maturity level means.

The relationship between single fiber properties and bundle properties

Researchers usually use Regression analysis as a common statistical method for estimation of the relation between Y variable and the x variables. At first, the types of relationship between fiber properties (x variables) and single fiber properties (Y variable) were checked individually by using curve estimation and correlation analysis. Statistical analysis indicated that there was a nearly linear relationship between fiber properties and single fiber properties. After excluding the weak correlated characters 3 linear multiple regression equations were:

a. The relationship between single fiber tenacity and fiber properties.

b. The relationship between single fiber linear density and fiber properties.

c. The relationship between single fiber elongation and fiber properties.

The relationship between single fiber tenacity and fiber properties: It's clear from Table 5, and Figures 4 to 16 that there were excellent relationship between single fiber tenacity and bundle tenacity ($R^2=0.8107$, $r=0.90$). While, both of Theta and the reversals number per mm gave nearly the same relation level ($r=0.87$). Also, convolution angle character and maturity ratio gave good relationship with single fiber tenacity. On the other hand, the weakest relationship was between single fiber tenacity and short fiber content ($R^2=0.0383$, $r=0.20$). In addition, there were direct relationship between single fiber tenacity and all the studied characters except for, short fiber content, fiber finesses,

character	equation	R^2	r
UHM	Y=2.0813×-22.559	0.4916	0.70
SFC	Y= - 0.3465×+47.869	0.0383	-0.20
UI	Y=1.7412×-102.26	0.2562	0.51
Fiber strength(g)tex	Y=0.645×+14.404	0.8107	0.90
Fiber elongation	Y=3.2271×+5.4688	0.4221	0.65
Fiber finesses	Y= - 0.1538×+65.525	0.3409	-0.60
MIC	Y= - 1.9691×+51.604	0.0559	-0.24
Theta	Y=103.59×-9.495	0.7561	0.87
Area of cross section	Y=-0.1908×+62.872	0.3048	-0.55
Convolutions	Y=8.2849×+16.588	0.4093	0.64
Reversals	Y= - 26.573×+83.814	0.7443	-0.87
Convolution angle	Y= - 0.9718×+54.751	0.5855	-0.77
Maturity	Y=62.934×-11.932	0.7420	0.86

r= Correlation
R^2=Determining factor

Table 5: Simple linear regression between each fiber property and Single fiber tenacity.

Measuring the relations using the simple regression and correlation is not satisfactory. Also, partial correlation between more than character is very important incase if they used as indicator for building multiple regression models. According to the previous reasons stepwise analysis was used to form the best model for single fiber

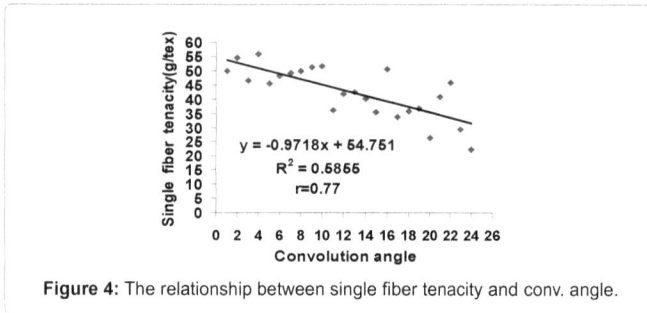

Figure 7: The relationship between single fiber tenacity and UHM.

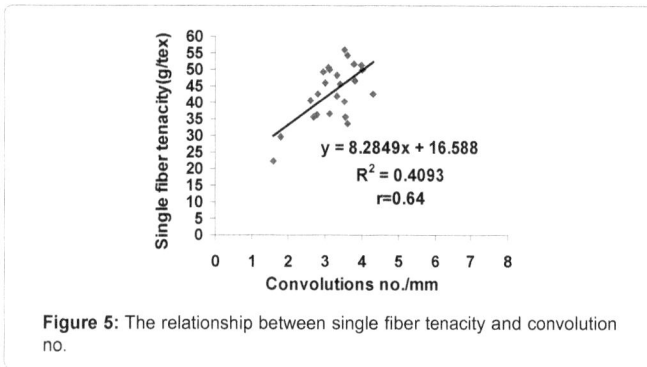

Figure 4: The relationship between single fiber tenacity and conv. angle.

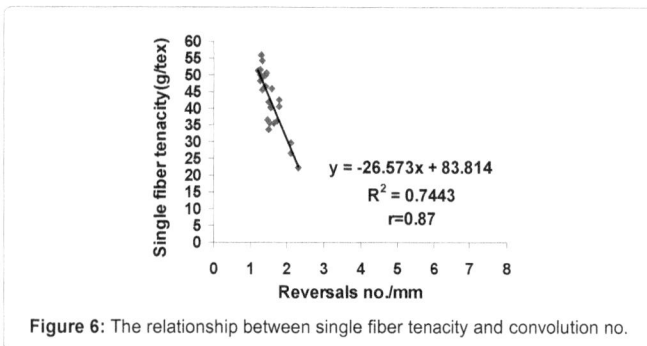

Figure 8: The relationship between single fiber tenacity and SFC.

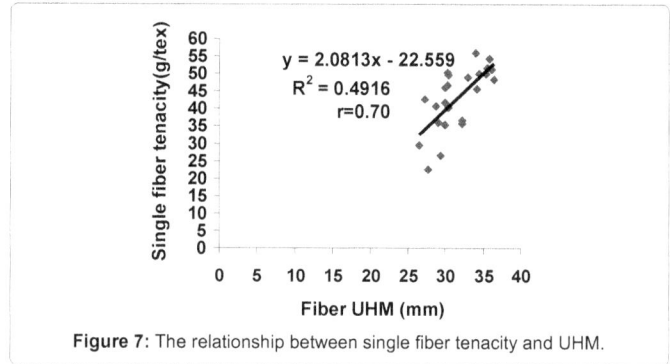

Figure 5: The relationship between single fiber tenacity and convolution no.

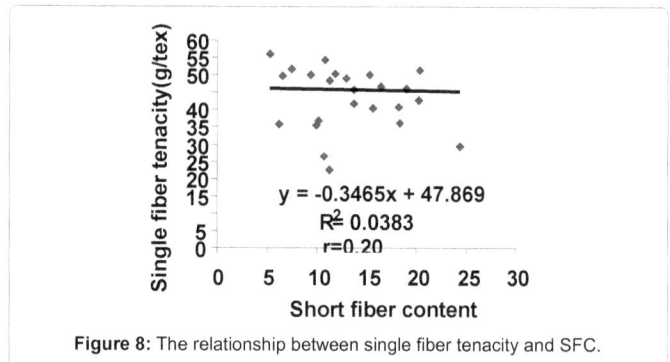

Figure 9: The relationship between single fiber tenacity and UI.

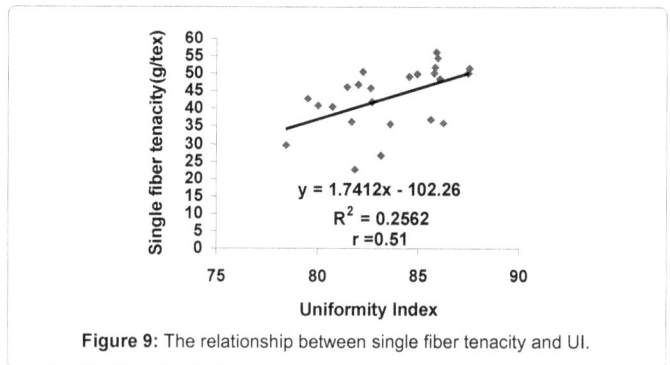

Figure 6: The relationship between single fiber tenacity and convolution no.

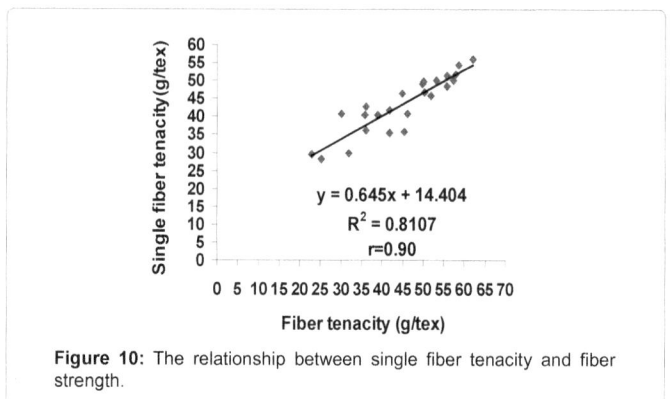

Figure 10: The relationship between single fiber tenacity and fiber strength.

micronaire reading, area of cross section, reversals and convolution angle; where, there was a kind of negative relationship between each of them and single fiber tenacity.

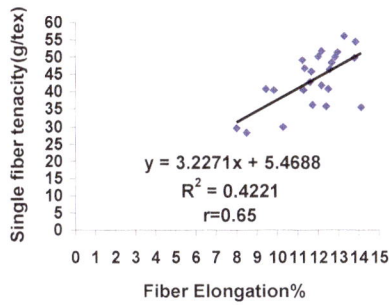

Figure 11: The relationship between single fiber tenacity and fiber elongation.

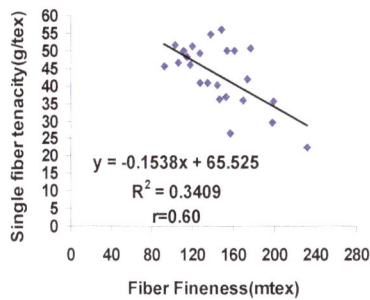

Figure 12: The relationship between single fiber tenacity and fiber fineness.

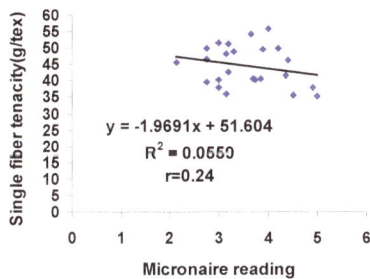

Figure 13: The relationship between single fiber tenacity and micronaire reading.

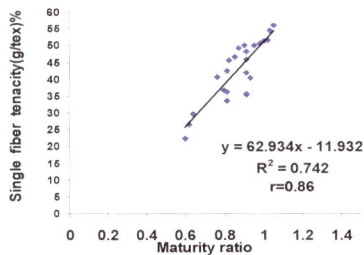

Figure 14: The relationship between single fiber tenacity and MR.

tenacity and the independent fiber characters. According to the result of the analysis, data in Table 6, indicated that microscopic characters, length and mechanical parameters were excluded from the model. In contrast, fineness and maturity parameters represented by theta and fineness with millitex were the predictors of the model. Table 7 and Figure 17, indicated the strong relationship between the single fiber

and the independent variables in the model with correlation= 0.936 and determining factor= 0.874. Table 8, shows regression coefficients of variables, t-values and significance level of theta and fineness variables. Arrangement of variables in the table indicates their relative importance for the model. Signs (+ or -) of regression coefficients of variables indicate the direction of influence.

This means fine and mature fiber increased the single fiber strength character.

The relationship between single fiber linear density and fiber properties: It's clear from Table 9, and Figures 18 to 30 that there were excellent relationship between single fiber linear density and fiber fineness per mtex (R^2=0.8847, r=0.94). While, both of micronaire reading and area of cross section gave nearly the same relation level as follows, respectively (R^2=0.6622, r=0.81 and R^2=0.6779, r=0.82.). On the contrary, short fiber content gave the weakest relationship to single fiber linear density (R^2=0.0352, r=0.19), all the characters under study proportionate adversely with single fiber linear density except, fiber finesses, micronaire reading, area of cross section, reversals and convolution angle.

After Appling the stepwise analyses regression coefficients, t-values and the partial correlation determined the excluded character to build up the best model describes the relationship between single fiber linear density and all the studied fiber properties. Its clear from Table 10, that all the character were excluded from the model except the micronaire reading and the fiber fineness per mtex characters. Thus, the model showed strong correlation r=0.946 and high determining factor=0.894 Table 11 and Figure 31

Obviously, fiber fineness per mtex was the most important factor for single fiber linear density. Micronaire value as an indicator for the specific surface area of the fiber was another important fiber parameter

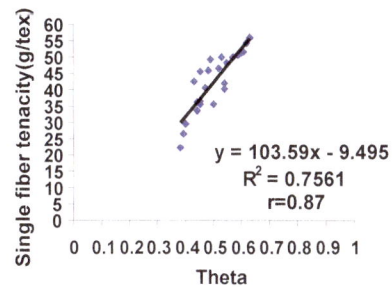

Figure 15: The relationship between single fiber tenacity and theta.

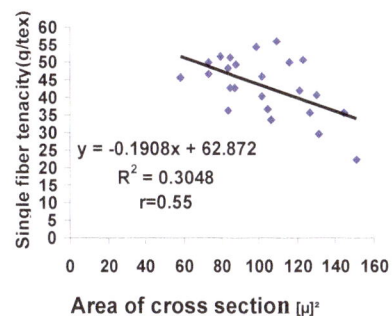

Figure 16: The relationship between single fiber tenacity and area of cross section.

	Model	Beta In	t	Sig.	Partial Correlation	Collinearity Statistics
						Tolerance
1	MIC	-.266a	-4.517	.000	-.563	1.000
	MR	.042a	.235	.815	.035	.161
	FI NN	-.335a	-5.802	.000	-.658	.859
	ASCW	-.251a	-4.050	.000	-.521	.961
	UHM	.213a	2.269	.028	.324	.512
	SF C	.159a	2.304	.026	.328	.945
	UI	-.038a	-.443	.660	-.067	.682
	STRENGTH	.201a	2.220	.032	.317	.554
	ELONGATI	-.037a	-.443	.660	-.067	.720
	CONV	.019a	.199	.843	.030	.567
	REVER	-.212a	-2.040	.047	-.294	.430
	CONVANGL	-.230a	-2.981	.005	-.410	.709
2	MIC	.104b	.795	.431	.120	.169
	MR	-.053b	-.388	.700	-.059	.159
	ASCW	.186b	1.497	.142	.223	.180
	UHM	-.017b	-.193	.848	-.029	.373
	SF C UI	.099b	1.804	.078	.265	.907
	STRENGTH	-.071b	-1.090	.282	-.164	.677
	ELONGATI	-.011b	-.126	.900	-.019	.415
	CONV	-.005b	-.078	.938	-.012	.715
	REVER	-.080b	-1.092	.281	-.164	.537
	CONVANGL	.023b	.240	.812	.037	.330
		-.082b	-1.160	.253	-.174	.569

Table 6: Excluded varibles and predictors for single fiber tenacity.

Model	r	R square	Std. Error	Significant
1	0.936	0.874	0.868	0.000

r = Correlation
R²=Determining factor

Table 7: Single fiber tenacity model summary.

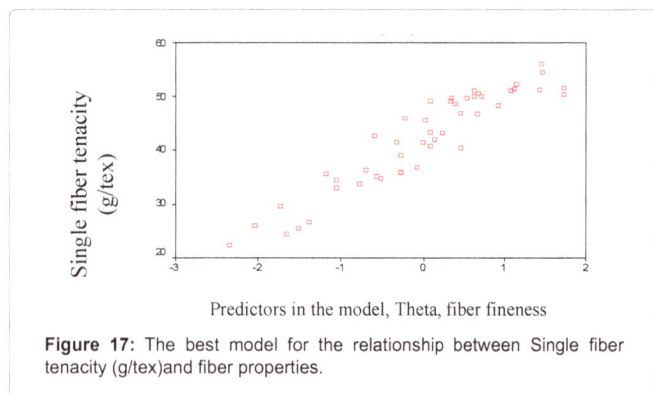

Figure 17: The best model for the relationship between Single fiber tenacity (g/tex)and fiber properties.

Statistical Parameter	Constant	theta	Fineness millitex
B*	8.165	95.22	-9.38E-02
Std. Error	5.016	7.277	0.016
T	1.628	13.085	-5.802
Significant	0.111	0.000	0.000

Table 8: Regression coefficients, t-values and significance level of the variables of the linear regression model for single fiber tenacity (g/tex).

for single fiber linear density, both of them proportionate directly with single fiber linear density, (Table 12).

character	equation	R^2	r
UHM	Y= - 5.2545x+312.78	0.2370	0.49-
SFC	Y= - 1.2075x+162.61	0.0352	0.19-
UI	Y= - 2.4339x+350.01	0.0379	0.19-
Fiber strength(g)tex	Y= - 1.4257x+210.21	0.2420	0.50-
Fiber elongation	Y= - 3.8406x +191.73	0.0357	0.19-
Fiber finesses	Y=0.8985x +15.396	0.8847	0.94
MIC	Y=36.6x +14.29	0.6622	0.81
Theta	Y= - 165.58x +229.86	0.1491	0.39-
Area of cross section	Y=1.1067x +32.929	0.6779	0.82
Convolutions	Y= - 38.582x +273.55	0.3563	0.60-
Reversals	Y=61.742x +50.833	0.3101	0.56
Convolution angle	Y=2.2722x +118.18	0.2470	0.50
Maturity	Y= - 97.747x +231.79	0.1323	0.36-

r= Correlation
R²=Determining factor

Table 9: Simple linear regression between each fiber property and Single linear density.

The relationship between single fiber elongation and fiber properties: It's clear from Table 13, and Figures 32 to 44 that the relationship between single fiber elongation and fiber properties weren't strong enough the strongest relation was for UHM (R^2=0.3298, r=0.57). This may attributes to that the linear regression could not fit the relation it could be quadratic or any type other than linear type.

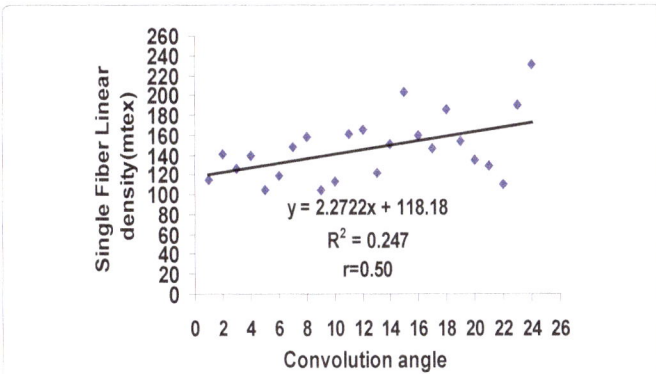

Figure 18: The relationship between single fiber linear density and conv. Angle.

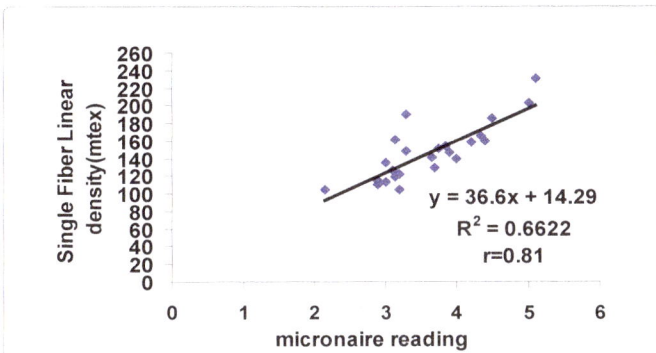

Figure 22: The relationship between single fiber linear density and theta.

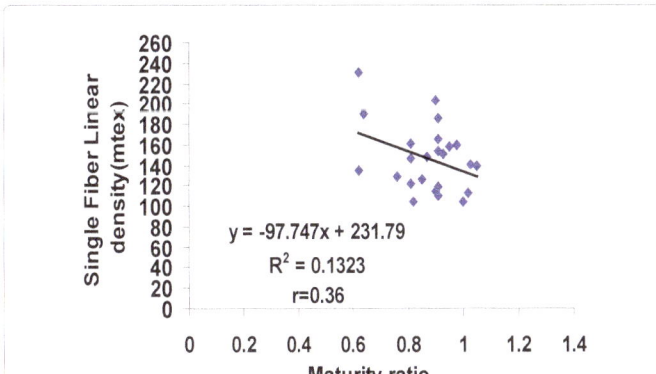

Figure 19: The relationship between single fiber linear density and micronaire reading.

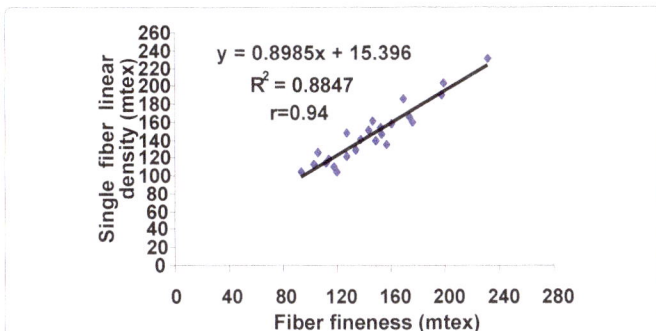

Figure 23: The relationship between single fiber linear density and area of cross section.

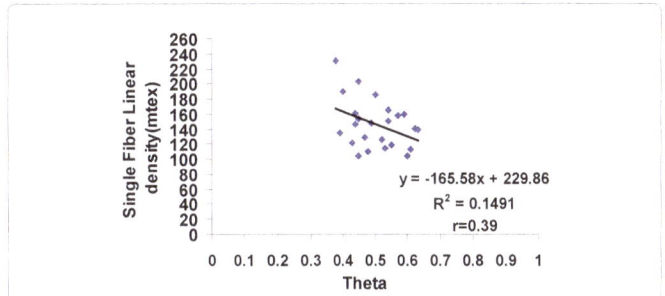

Figure 20: The relationship between single fiber linear density and MR.

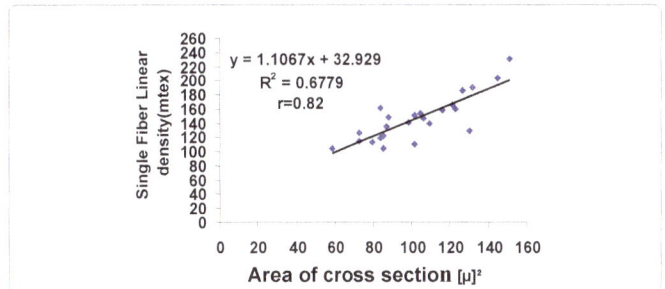

Figure 24: The relationship between single fiber linear density and UHM.

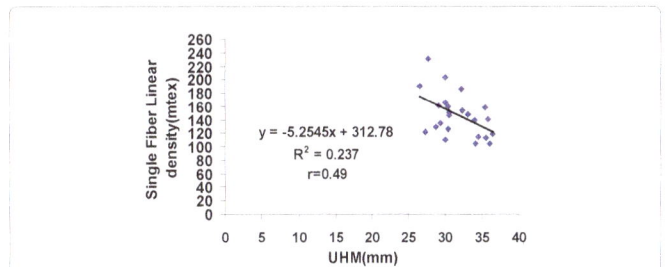

Figure 25: The relationship between single fiber linear density and SFC.

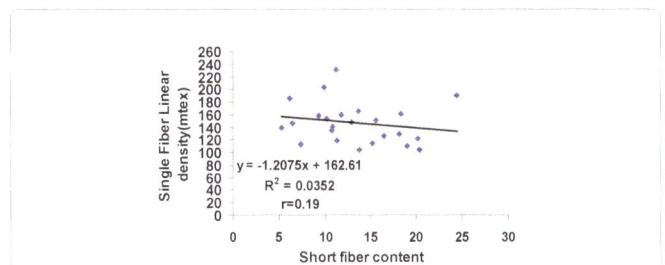

Figure 21: The relationship between single fiber linear density and fiber fineness.

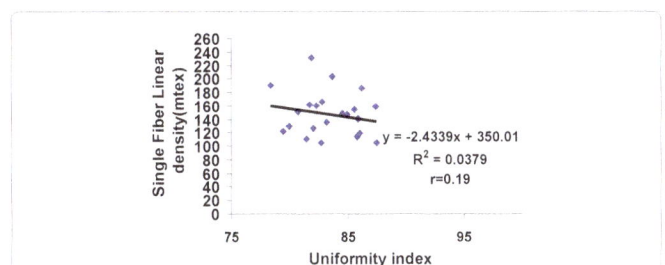

Figure 26: The relationship between single fiber linear density and UI.

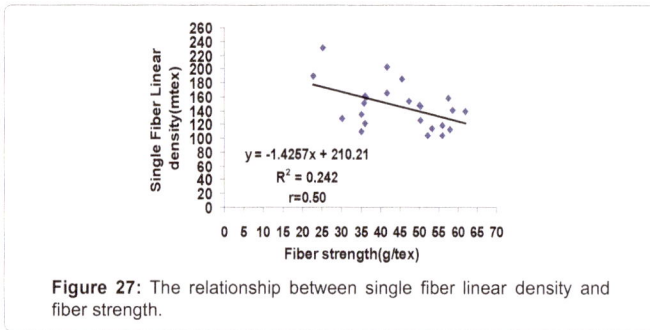

Figure 27: The relationship between single fiber linear density and fiber strength.

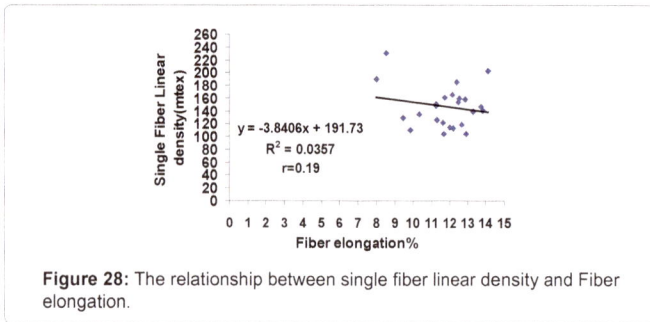

Figure 28: The relationship between single fiber linear density and Fiber elongation.

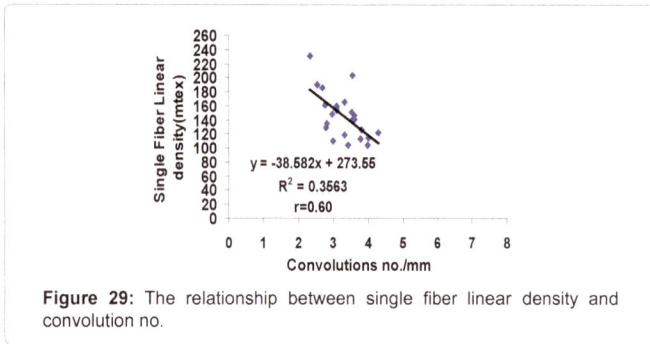

Figure 29: The relationship between single fiber linear density and convolution no.

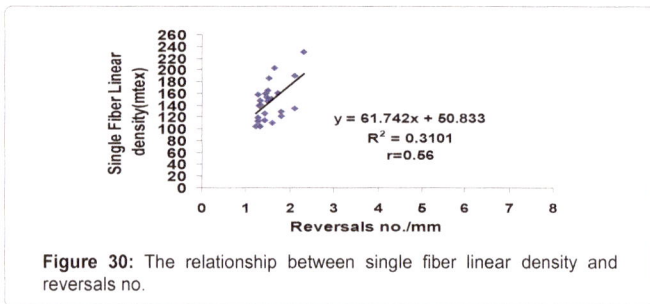

Figure 30: The relationship between single fiber linear density and reversals no.

All the characters under study proportionate directly with single fiber elongation except, short fiber content, UI, and fiber strength.

Table 14, represents the excluded varibles and predictors for single fiber elongation. This model surpassed the previous models on the number of the predictors variables, it contains 7 predictors. They are: Miconaire reading, UI, ASCW, theta, fiber strength, fiber elongation and UHM characters. Nevertheless, the model results and graph distribution illustrated in Table 15 and Figure 45, showed the lowest correlation value comparing to the others single characters (r=0.888 and determining factor=0.788) Mic, UI, ASCW were the most important factors for single fiber elongation. Theta values was other important fiber parameter for single fiber elongation, then fiber strength followed

by fiber elongation finally UHM ranked the last important character in the model. Table 16, Indicated that UI, ASCW, and fiber Strength proportionate inversely with single fiber elongation. However, the rest fiber properties proportionate directly with single fiber elongation. In

Model	Beta In	t	Sig.	Partial Correlation	Collinearity Statistics Tolerance
1 MIC	.213a	2.358	.023	.335	.294
MR	.095a	1.749	.087	.255	.849
THETA	.030a	.533	.597	.080	.859
ASCW	.106a	.922	.362	.138	.202
UHM	.048a	.740	.463	.111	.631
SF C	-.052a	-1.014	.316	-.151	.991
UI	.018a	.326	.746	.049	.923
STRENGTH	.097a	1.540	.131	.226	.644
ELONGATI	.060a	1.165	.250	.173	.983
CONV	.062a	1.109	.274	.165	.835
REVER	-.087a	-1.397	.169	-.206	.667
CONVANGL	-.074a	-1.214	.231	-.180	.698
2 MR	-.006b	-.065	.948	-.010	.334
THETA	-.089b	-1.290	.204	-.193	.495
ASCW	-.066b	-.491	.626	-.075	.134
UHM	-.021b	-.301	.765	-.046	.505
SF C	-.001b	-.019	.985	-.003	.796
UI	-.049b	-.851	.400	-.129	.722
STRENGTH	.022b	.286	.776	.044	.431
ELONGATI	-.047b	-.661	.512	-.100	.482
CONV	-.015b	-.232	.818	-.035	.560
REVER	-.003b	-.045	.965	-.007	.427
CONVANGL	-.028b	-.438	.663	-.067	.606

a. predictors in the model(constant), Fineness
b. predictors in the model(constant), Fineness, mic
c. Dependant variable: Single linear density

Table 10: Excluded varibles and predictors for single fiber linear density.

Model	r	R square	Std. Error	Significant
1	0.946	0.894	10.311	0.023

r= Correlation
R²=Determining factor

Table 11: Single fiber linear density model summary.

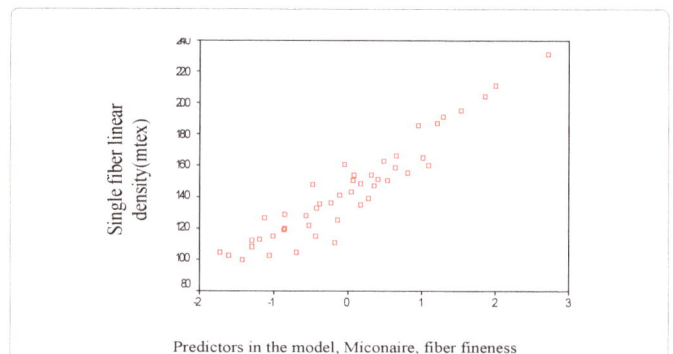

Predictors in the model, Miconaire, fiber fineness

Figure 31: The best model for the relationship between single fiber linear density (mtex) and fiber properties.

mic	Fineness millitex	Constant	Statistical Parameter
9.382	.7400	4.198	B*
3.979	.0880	7.807	Std. Error
2.358	8.402	.5380	T
.0230	0.000	.5930	Significant

Table 12: Regression coefficients, t-values and significance level of the variables of the linear regression model for single fiber linear density.

character	equation	R^2	r
UHM	Y=0.1669x+3.374	0.3298	0.57
SFC	Y= - 0.0149x+12.119	0.0021	-0.05
UI	Y= - 0.1823x+27.159	0.0840	-0.29
Fiber strength(g)tex	Y= - 0.041x+13.75	0.0790	-0.28
Fiber elongation	Y=0.1942x+9.6383	0.0361	0.19
Fiber finesses	Y=0.0157x+9.6326	0.1064	0.33
MIC	Y=0.9563x+8.4645	0.1787	0.24
Theta	Y=0.2625x+11.789	0.0001	0.01
Area of cross section	Y=0.0179x+10.083	0.0701	0.27
Convolutions	Y=0.2747x+11.017	0.0071	0.08
Reversals	Y=0.4292x+11.256	0.0059	0.07
Convolution angle	Y=0.0941x+10.528	0.0393	0.20
Maturity	Y=1.5633x+10.559	0.0138	0.12

r= Correlation
R^2=Determining factor

Table 13: Simple linear regression between each fiber property and the percentage of single fiber elongation.

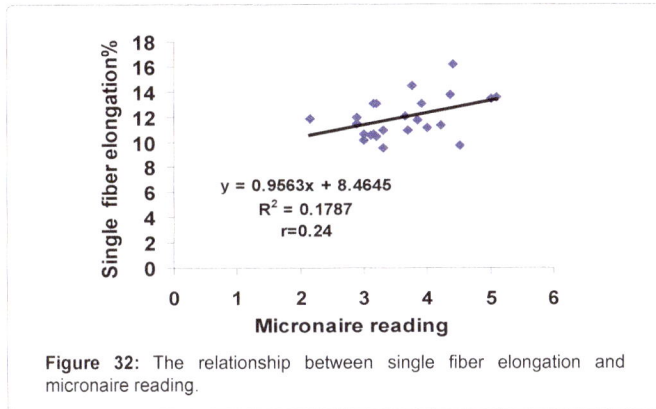

Figure 32: The relationship between single fiber elongation and micronaire reading.

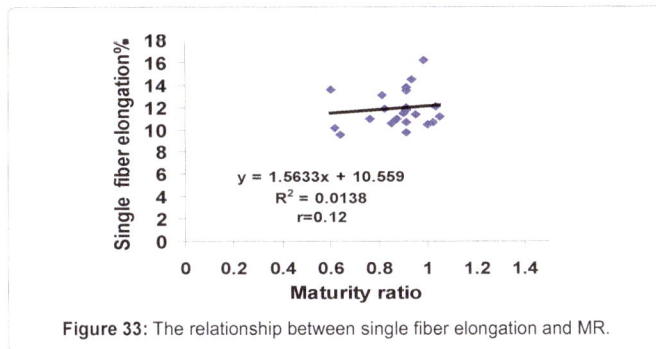

Figure 33: The relationship between single fiber elongation and MR.

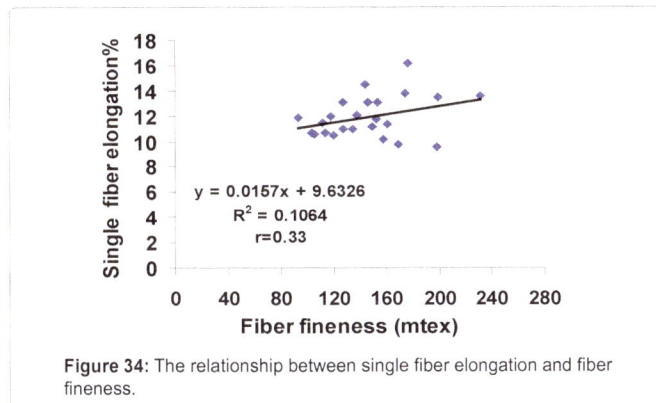

Figure 34: The relationship between single fiber elongation and fiber fineness.

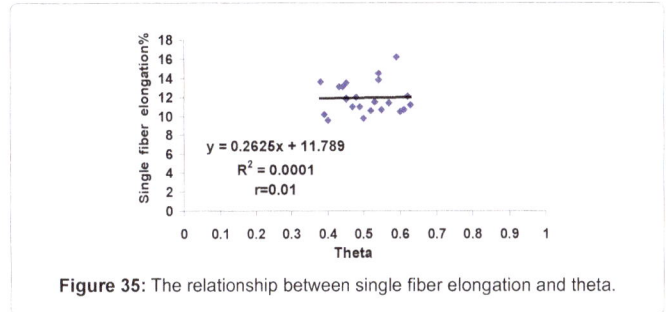

Figure 35: The relationship between single fiber elongation and theta.

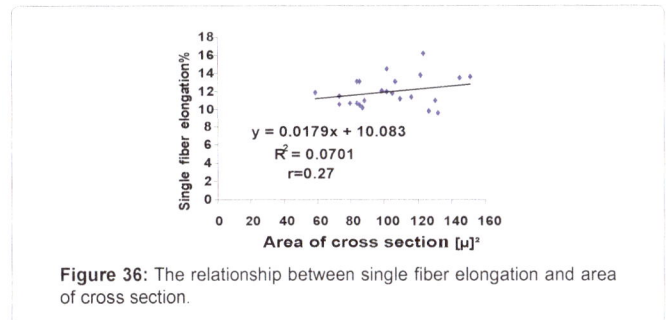

Figure 36: The relationship between single fiber elongation and area of cross section.

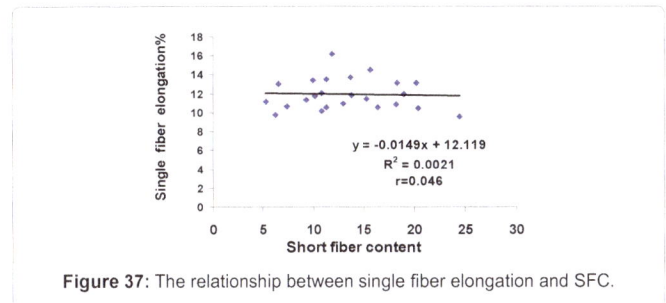

Figure 37: The relationship between single fiber elongation and SFC.

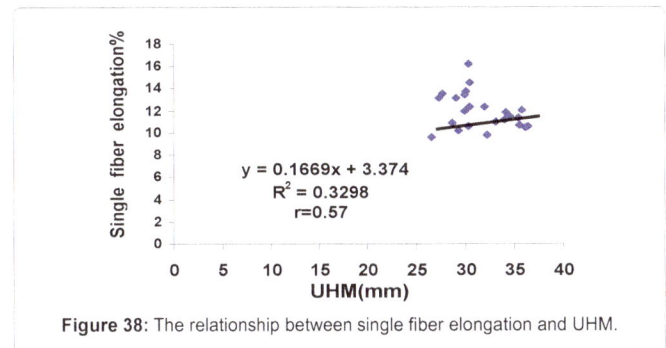

Figure 38: The relationship between single fiber elongation and UHM.

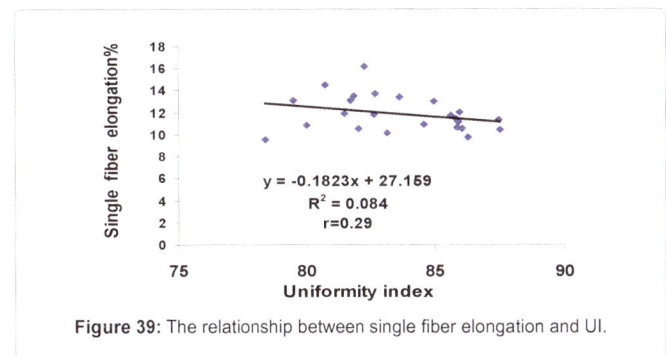

Figure 39: The relationship between single fiber elongation and UI.

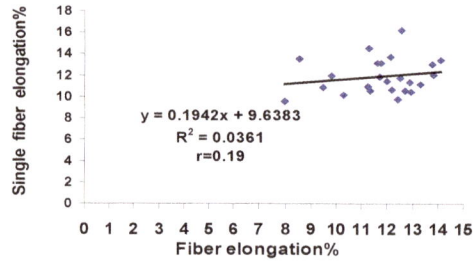

Figure 40: The relationship between single fiber elongation and fiber elongation.

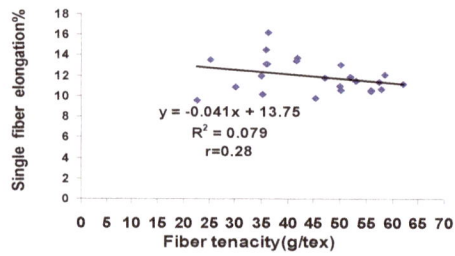

Figure 41: The relationship between single fiber elongation and strength.

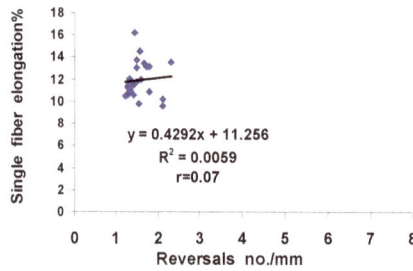

Figure 42: The relationship between single fiber elongation and reversals no.

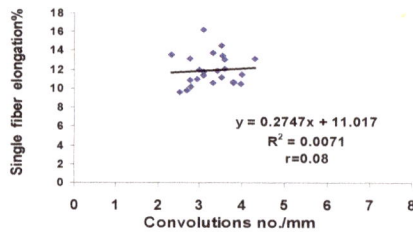

Figure 43: The relationship between single fiber elongation and convolution no.

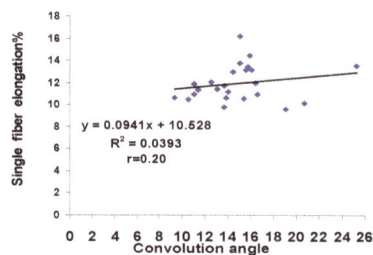

Figure 44: The relationship between single fiber elongation and conv. Angle.

Excluded Variables

Model	Beta In	t	Sig.	Partial Correlation	Collinearity Statistics Tolerance
1 MR	.119 a	.943	.351	.141	.996
FIN N	-.128 a	-.547	.587	-.082	.294
THETA	.118 a	.934	.356	.139	1.000
ASCW	-.504 a	-1.854	.071	-.269	.203
UHM	-.200 a	-1.531	.133	-.225	.899
SFC	.203 a	1.550	.128	.228	.899
UI	-.320 a	-2.716	.009	-.379	1.000
STRENGTH	-.241 a	-1.910	.063	-.277	.937
ELONGATI	.070 a	.527	.601	.079	.925
CONV	-.005 a	-.043	.966	-.006	.997
REVER	.031 a	.238	.813	.036	.952
CONVANGL	.157 a	1.200	.237	.178	.911
2 MR	.423 b	3.321	.002	.452	.695
FIN N	-.608 b	-2.517	.016	-.358	.212
THETA	.436 b	3.412	.001	.462	.682
ASCW UHM	-1.179 b	-4.719	.000	-.584	.150
SFC	.553 b	1.992	.053	.291	.169
STRENGTH	.026 b	.173	.863	.026	.630
ELONGATI	.153 b	.648	.521	.098	.252
CONV	.469 b	3.274	.002	.447	.553
REVER	.175 b	1.335	.189	.199	.793
CONVANGL	-.602 b	-3.416	.001	-.462	.360
	-.376 b	-1.757	.086	-.259	.290
3 MR	.274 c	2.341	.024	.340	.620
FIN N	-.222 c	-.953	.346	-.146	.173
THETA	.323 c	2.900	.006	.408	.643
UHM	.417 c	1.803	.079	.268	.166
SFC	.175 c	1.409	.166	.212	.594
STRENGTH	-.256 c	-1.225	.227	-.186	.211
ELONGATI	.144 c	.869	.390	.133	.340
CONV	.006 c	.053	.958	.008	.707
REVER	-.337 c	-1.948	.058	-.288	.293
CONVANGL	-.261 c	-1.454	.153	-.219	.284
4 MR	-.058 d	-.247	.806	-.039	.147
FIN N	.179 d	.689	.495	.107	.119
UHM	.013 d	.046	.964	.007	9.712E-02
SFC	.124 d	1.056	.297	.163	.578
STRENGT	-.633 d	-3.227	.002	-.450	.169
HELONGATI	.060 d	.381	.705	.059	.327
CONV	-.207 d	-1.708	.095	-.258	.520
REVER	-.060 d	-.280	.781	-.044	.177
CONVANGL	-.172 d	-1.007	.320	-.155	.273
5 MR	.129 e	.584	.563	.092	.137
FIN N	-.181 e	-.694	.492	-.109	9.704E-02
UHM	.547 e	1.906	.064	.289	7.428E-02
SFC	.007 e	.057	.955	.009	.510
ELONGATI	.345 e	2.276	.028	.339	.257
CONV	-.124 e	-1.077	.288	-.168	.487
REVER	-.352 e	-1.733	.091	-.264	.151
CONVANGL	-.314 e	-2.046	.047	-.308	.256
6 MR	-.237 f	-.913	.367	-.145	8.832E-02
FIN N	-.080 f	-.317	.753	-.051	9.376E-02
UHM	.578 f	2.136	.039	.324	7.412E-02
SFC	-.005 f	-.049	.961	-.008	.508
CONV	-.154 f	-1.408	.167	-.220	.481
REVER	-.305 f	-1.557	.128	-.242	.149
CONVANGL	-.295 f	-2.009	.051	-.306	.255
7 MR	-.199 g	-.796	.431	-.128	8.784E-02
FIN N	-.062 g	-.253	.801	-.041	9.363E-02
SFC	-.004 g	-.041	.967	-.007	.508
CONV	-.184 g	-1.766	.085	-.275	.474
REVER	-.173 g	-.831	.411	-.134	.127
CONVANGL	-.165 g	-.879	.385	-.141	.156

a. predictors in the model (constant), Mic, b. predictors in the model (constant), Mic, UI , c. predictors in the model (constant), Mic, UI, area of cross section,
d. predictors in the model (constant), Mic, UI, area of cross section, Theta
e. predictors in the model (constant), Mic, UI, area of cross section, Theta, Strength
f. predictors in the model (constant), Mic, UI, area of cross section, Theta, Strength, Elongation
g. predictors in the model (constant), Mic, UI, area of cross section, Theta, Strength, Elongation, UHM
h. dependant variable: single fiber elongation

Table 14: Excluded varibles and predictors for single fiber elongation.

Model	r	R square	Std. Error	Significant
1	0.888	0.788	0.7979	0.039

r= Correlation
R²=Determining factor

Table 15: single fiber elongation model summary.

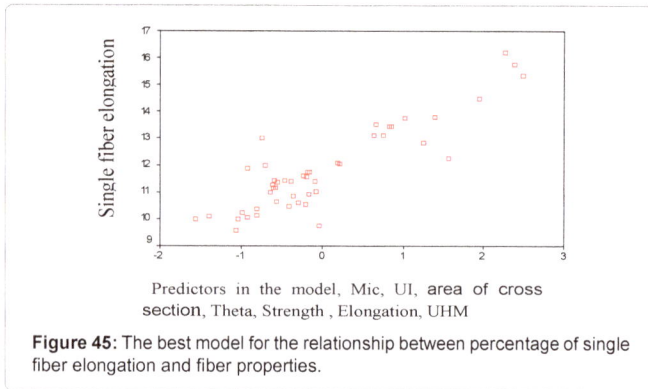

Predictors in the model, Mic, UI, area of cross
section, Theta, Strength , Elongation, UHM

Figure 45: The best model for the relationship between percentage of single fiber elongation and fiber properties.

Statistical Parameter	Constant	Mic	UI	Area of cross section	Theta	Strength	Elongation	UHM
B*	31.127	2.930	-0.407	-7.72E-02	8.076	-0.166	0.3950	0.3400
Std. Error	8.000	0.6280	0.1310	0.018	2.719	.0350	0.160	0.159
T	3.891	4.667	-3.103	-4.274	2.970	-4.741	2.471	2.136
Significant	0.000	0.000	0.004	0.000	0.005	0.018	0.000	0.039

Table 16: Regression coefficients, t-values and significance level of the variables of the linear regression model for percentage of single fiber elongation.

general the results of the single fiber elongation are absurd and couldn't be helpful as prediction.

Acknowledgement

I would like to extend my appreciation to Dr. Guntram Kugler Ph.D., General Manager Laboratory Projects, and all the member of Textechno Herbert Stein GmbH & Co. KG, Monchengladbach, Germany for testing the samples in their labs.

References

1. Harzallah O, Benzina H, Drean JY (2010) Physical and Mechanical Properties of Cotton Fibers: Single-fiber Failure. Text Res J 80: 1093-1102

2. Koo HJ, Jeong SH, Moon WS (2001) Study on the effects of single fiber tensile properties on bundle tensile properties through estimation of HVI bundle and toughness modulus. Fibers and Polymers 2: 22-25.

3. Thibodeaux DP, Herbert JJ, Deluca LB, Moraitis JS (1997) Relating single fiber measurements to cotton strength, structure and morphology. Beltwide cotton conference 1: 545-545.

4. Abdel- Salam ME (1999) The Egyptian cotton ''production, quality and marketing '' .El-Kalema press, Cairo.

5. Meredith R (1951) Cotton fiber tensile strength and X-ray orientation. J Text Inst Trans 42: 291-299.

6. Little TM, Hills FJ (1978) Agricultural Experimentation Design and Analysis. John Wiley and Sons. New York.

15

Studies on the Effect of Nano Zinc Treatment on Jute Fabric

Chattopadhyay DP[1]* and Patel BH[2]

Department of Textile Chemistry, Faculty of Technology and Engineering, The M S University of Baroda, Vadodara, India

Abstract

In the present investigation in-house synthesized colloidal nano-zinc was applied on jute fabric by pad-dry-cure technique. The fabric samples treated with zinc nano-particles were tested for changes in absorbency, air permeability, water permeability, electrical resistivity, antibacterial activity and dye ability. These properties were compared to their untreated counterpart. The treatment with zinc nano sol improves the physical properties along with the dye ability of jute fiber with direct dye. The treatment also enhances the antimicrobial efficacy of jute fiber.

Keywords: Antimicrobial; Absorbency; Air and Water permeability; Dyeing, Jute; Zinc Nano-sol

Introduction

Incorporation of nano-sized particles in textile fiber can generate special properties or cause enhancement of existing properties. Nano-size particles of metal and metal oxides can be applied on textiles to impart antibacterial property as well as to improve various functional properties like water repellence, soil-resistance, anti-static, anti-infrared, flame-retardancy etc [1-6].

Nano surface coating is one of the approach to the production of highly active surfaces to have UV-blocking, antimicrobial and self-cleaning properties. Works reported that the self-cleaning property can be imparted by nano-TiO_2/nano-ZnO coating [7], while nano-Ag imparts antimicrobial property. Nano-particle coating may affect the other fabric properties like dyeing, absorbency, air permeability and strength. The ZnO nano-particles were also reported to improve the antistatic, friction resistance properties of textiles [8,9]. Earlier, our research group synthesized and characterized the zinc nano-particles using chemical reduction reaction and reported their effect on physical and antimicrobial properties of cotton [10] and effect of Cu and Ag Nano sol on natural textiles [11-13]. In this work, jute fabric was coated with the in-house synthesized zinc Nano-particles which was synthesized as reported elsewhere [10]. The mechanical property of these fabrics was evaluated in addition to the effect on absorbency, air permeability, water permeability, electrical resistivity, dye ability and antibacterial properties.

Materials

In house synthesized Zinc nano-sol was applied to jute fabric. Change in functional properties due to the treatment was evaluated by standard methods.

Fabric

Plan weaves jute fabric ends/inch-13, picks/inch-10 and weight-465.11 gm/m^2 was supplied by Institute of jute technology, Kolkata, India for the study.

Dyes and chemicals

Three commercial direct dyes, namely C.I. Direct Red 9 (D1), C.I. Direct Blue 67 (D2) and C.I. Direct Green 6 (D3), In-house synthesized and TSC capped metallic nano zinc-colloid were used in this study without further purification. Gram positive *Staphylococcus aureus* (Lab collection) was used for evaluation of antimicrobial property test as per ASTM E-2149 method.

Equipments

• Scanning electron microscope (SEM, Model JSM-5610 LV, Japan)

• X-ray fluorescence spectrometer (EDX 800 Simadzu, Japan)

• Laboratory Constant temperature shaking water bath (Alliance enterprise, India).

• Laboratory two bowl automatic padding mangles (EEC, India).

• Launder-ometer (Digi. wash, Paramount Scientific Instruments., India).

• Xenon arc Fadometer, (FDA-R, Atlas, U.S.A.).

• Spectrophotometer interphased with computer colour matching system; Spectra scan 5100 (RT) (Premier colour scan instrument), India.

• Air permeability test: Metefem, type-FF12/A, no: 8838_002, made in Hungary.

• Antibacterial test equipments: Incubator cum oven, Laminar air flow (Hexatec Instruments Pvt. Ltd. Mumbai: Model: HIPL-042), Autoclave Equitron (Sr.No. NC11GC-2824), Shaker Incubator and UV spectrophotometer.

• Electrical surface resistivity test: KEITHLEY-614 Electrometer MET/K/16B, USA.

Experimental methods

Preparation of textile fabrics for nano treatment

The jute fabric was first treated in a bath containing 5gpl non-ionic detergent (Lissapol N) and 2 gpl sodium carbonate for 30 minutes at 80°C temperature. Then the fabrics were washed thoroughly in running water, neutralized and again washed thoroughly in running water.

*Corresponding author: Chattopadhyay DP, Department of Textile Chemistry, Faculty of Technology and Engineering, The M S University of Baroda, and Vadodara, India, E-mail: dpchat6@gmail.com

Finally, the pH of fabrics was checked to neutral.

Coating of samples with zinc nano colloids by pad–dry–cure method

Nano-colloidal solutions were applied to jute fabric samples by soaking them in the dispersion for 10 min and then squeezed on an automatic two bowl padding mangle operated at 15 rpm with a pressure of 1.75 kg/cm^2 using 2-dip-2-nip padding sequence at 70% expression. The padded substrates were air dried and finally cured at 120 ˚C for 20 min in a preheated curing oven.

Testing and Analysis

Physical testing

Before physical testing the samples were dried and conditioned at 65± 2% RH and 27 ± 2°C temperature.

Determination of tensile properties :

2 cm×8 cm fabric samples were tested at 100 mm/min traversing speed for the determination of breaking load, breaking elongation, stress and strain. The test was performed as per B.S. 2576:1959

Determination of crease recovery angle: The test was performed as per AATCC test method 66-2003.

Determination of bending length: The stiffness in terms of bending length of nano treated and untreated samples were measured as per AATCC Test Method 115-2005 using prolific stiffness tester (India).

Determination of absorbency by Drop test method: Absorbency of nano treated and untreated fabric sample were evaluated using AATCC Test Method 79-2000. The test was conducted in a standard atmospheric condition. Before the test the samples were kept in standard atmosphere having a relative humidity of 65 ± 2% RH and 27 ± 2°C temperature. Five tests were conducted for each sample. An average of five readings was reported.

Evaluation of water permittivity: This test is conducted for determining the hydraulic conductivity (water permeability) of textiles materials in terms of permittivity under standard testing conditions in uncompressed state. It shows flow of water perpendicular to plane of fabric which is very important to assess for filtration and drainage purpose. Permittivity is an indicator of the quantity of water that can pass through a textile material in an isolated condition, by using ASTM D 4491 water permeability test method. Constant head method has been selected for this test.

Permittivity Ψ was calculated as follows

$\Psi = QR_T/hAt$

Where

Ψ = Permittivity (S^{-1})

Q = Quantity of flow (mm^3)

h = Head of water on the specimen (mm)

A = Cross sectional area of test area specimen (mm^2)

t = Time for flow Q (S)

R_T = Temperature correlation factor determined using Equation (c)

$R_{T=} U_t/U_{20t}$

Where: U_t = Water viscosity at test temperature millipoises.

U_{20t} = Water viscosity at 20°C millipoises.

Evaluation of air permeability: The air permeability of treated and untreated fabric samples were measured on Metefem air permeability tester. The testing was carried out as per ASTM D 737 test method. The result of the test measured in m^3/m^2/h to three significant digits.

Evaluation of antibacterial activity

Soil burial test (AATCC Test Method 30-2004): The samples exposed to the action of micro-organism in compost soil for 7 days, at the end of the period, the specimens were tested for the change in tensile strength measured as per B.S. 2576:1959 test method on Instrument Model-Lloyd, LRX. Samples of 2 cm×8 cm were tested at 100 mm/min traversing speed.

ASTM E-2149 test method: Antimicrobial activity was measured as per ASTM E-2149 test methods. Antimicrobial test for jute fabrics with and without nano zinc was carried out with gram-positive *Staphylococcus aureus* and gram-negative *Escherichia coli* bacterium. The percentage reduction in bacterial growth was calculated using following equation

$$\text{Percent Reduction in bacterial growth} = \frac{(B-A)}{B} \times 100$$

Where,

A is the optical density of the inoculated test culture containing the treated sample.

B is the optical density of the inoculated test culture containing the untreated sample.

Determination of electrical surface resistivity: The purpose of this test method was to determine the electrical surface resistivity of fabric. The test was carried out as per AATCC test method 76-2005. The specimen size of the treated and untreated fabric samples were used as electrodes and the electrical surface resistivity of fabric samples were measured on KEITHLEY-614 electrometer. The instrument directly showed the electrical resistivity (R) value of the fabric samples.

Dyeing of untreated and treated jute with direct dyes: All dyeing were carried out in laboratory dyeing machine at a liquor to material ratio of 20: 1, in the presence of 50 gpl of sodium chloride and 20 gpl of sodium bicarbonate using 2.0% dye on the weight of the sample. The sample was initially treated in the dye bath at 40°C for 10 min. The temperature was slowly raised to boil over 30 min (2°C/min) and the dyeing was continued at boil for further 45 min. Finally the samples were washed and dried in air. For comparison purpose, untreated jute was also dyed under the same condition.

Evaluation of dyed samples: The colour strength values (*K/S* value) of the samples were measured using Spectra Scan 5100 (RT) spectrophotometer (Premium Colourscan Instruments, India). All the dyed samples were also evaluated for fastness to light by ISO 105-BO2 (1990) and washing by ISO 105-CO6 (C2S) methods.

Results and Discussion

The treatment with nano-particles was performed using pad-dry-cure technique. The distributions of nano-particles on the surface of the fabric were observed using SEM and elementally analyzed using XRF technique.

Characterization of zinc nano treated fabric

Figure 1 represents SEM photographs of jute fabrics, treated with zinc nano-particles. Deposition of zinc nano was done successfully using pad-dry-cure technique. Table 1 shows the amount of zinc on the fabric measured by XRF technique.

Effect of zinc nano treatment on physical properties

The effect of nano zinc treatment on the physical properties of jute is presented in Table 2. It is seen from the results that introduction of nano zinc particles into the structure of the fiber caused an improvement in the load bearing capacity of the fiber. Improvements in the bending length and crease recovery angle of treated fabrics were also observed. These may be due to the interference of zinc nano-particles with the polymer chain mobility. From the SEM microphotograph of nano zinc treated samples shown in Figure 1, it can be clearly seen that the zinc nano-particles are distributed on the surface of the treated samples, being distributed in polymer matrices nano-particles can carry load and increase the toughness and abrasion resistance of the polymer matrices and enhance tensile strength of treated fiber.

Effect of zinc nano treatment on fabric absorbency

The water absorbency of zinc nano treated and untreated fabric samples were presented in Table 3. From the results it is found that nano- zinc treatment caused a drop in water absorbency of the samples. The drop in absorbency was found to be 4.62% in case of treated jute compared to the untreated jute fabric. The nano-metal particles act as whiskers and resists immediate penetration of water droplets, which increased the time to get water molecules absorbed, inside the fabric structure.

Effect of zinc nano on water permeability of fabric

The results presented in Table 4 show that the water permeability of the treated samples was dropped after zinc-nano treatment since the nano-particles present in polymer matrix resists the flow of water through the fabric.

Effect of zinc nano on air permeability of fabric

The results presented in Table 5 were measured on Metefem for air permeability of zinc nano treated and untreated fabric samples. It shows that the air permeability through treated samples was reduced compared to untreated sample.

Effect of zinc nano on resistance against microbes

Antimicrobial activity on the fabric was measured by soil burial test and the results are shown in Table 6. It is clear that the zinc nano treatment was found to enhance the resistance of jute towards microbial attack when measured in terms of loss in breaking load due to soil burial

Sample	Amount of zinc, µg
Jute fabric	Nil
Jute fabric treated with zinc nano sol	0.15 µg

Table 1: Amount of zinc detected using XRF.

Sample	Breaking load (kgf)	Crease recovery angle °(deg)	Bending length (cm)
Jute (Control)	10.44	90	3.00
Jute treated with zinc nano sol	10.78 (+3.26)	101 (+12.22)	3.32 (+10.67)

Note: Values in the parenthesis indicate percentage change in physical properties.

Table 2: Effect of zinc nano treatment on physical properties of jute.

Fabric sample	Time (sec)	
	Untreated	Treated with Ag nano
Jute	13	13.6

Table 3: Absorbency of zinc nano treated and untreated fabrics

Sample	Ψ-Water Permeability(S^{-1})
Untreated Jute sample	0.3726
Treated with zinc nano sol	0.3659

Table 4: Water permeability of fabric before and after the zinc-nano treatment.

Sample	Air permeability $(m^3/h/m^2)$
Untreated Jute sample	884.28
Treated with zinc nano sol	859.24

Table 5: Air permeability of jute fabric before and after zinc nano sol treatment.

Sample	Breaking load, kgf	
	Before soil burial	After soil burial
Untreated jute sample	10.44	5.02
Jute pretreated with zinc nano particles	10.92	9.86

Table 6: Effect of zinc nanoparticles on resistance against microbial attack.

Fabric sample	Reduction in bacterial growth (%)
Jute (Control)	Nil
Zinc nano treated Jute	73.12

Table 7: Anti-bacterial effect of zinc nano treated fabric

Fabric sample	Electrical surface resistivity (ohm Ω)
Jute (Control)	>200 x 10^9
Zinc nano treated Jute	>200 x 10^9

Table 8: Electrical surface resistivity of jute fabrics.

test. The breaking load of untreated control samples were reduced due to bacterial damage during soil burial test whereas zinc nano-particle treated sample resist the bacterial attack. The treated samples were also tested against gram-positive *Staphylococcus aureus* bacteria. Anti bacterial efficiency of zinc nano treated samples against the original fabric sample is reported in Table 7. It is found from the results that the zinc nano treated jute have more resistance against bacterial growth compared to its untreated counterpart. The enhancement in the resistance towards bacterial attack may be attributed to the certain degree of sterilizing effect of the metal atoms. It is possible that zinc nano particles may get attached to the surface of the microbe's cell membrane enter inside the cell and destroy their metabolic function.

Effect of zinc nano treatment on electrical surface resistivity

The Electrical surface resistivity of treated and untreated fabric

Figure 1: Scanning electron microphotographs of untreated jute (UJ) and jute treated with zinc nano sol (ZJ) fabric.

Sample	Colour strength (K/S)					
	C.I. Direct Red 9		C.I. Direct Blue 67		C.I. Direct Green 6	
	Control	Zinc nano treated	Control	Zinc nano treated	Control	Zinc nano treated
Jute	14.15	14.36	16.43	16.57	17.26	17.88

Table 9: Effect of zinc nano-particles on colour strength of fabrics dyed with direct dye.

Sample	Fastness to											
	Light						Washing					
	C.I. Direct Red 9		C.I. Direct Blue 67		C.I. Direct Green 6		C.I. Direct Red 9		C.I. Direct Blue 67		C.I. Direct Green 6	
	C	T	C	T	C	T	C	T	C	T	C	T
Jute	2-3	3	3-4	3-4	2-3	3	3	3-4	3	3-4	3	4

Note: C – Control sample, T – Sample treated with zinc nano sol.

Table 10: Effect of zinc nanoparticles on fastness properties of samples dyed with direct dyes.

Pairs	Description	Reference Table	Mean	SD	t-value	p-value
1	Breaking load of jute (Control)	2	10.4400	.61500	-1.843	0.098
	Breaking load of jute treated with zinc nano sol	2	10.7800	.15492		
2	Crease recovery of jute (Control)	2	90.0000	2.19241	-18.517	0.000
	Crease recovery of jute treated with zinc nano sol	2	101.0000	1.50923		
3	Bending length of jute (Control)	2	3.0000	.40000	-1.829	0.101
	Bending length of jute treated with zinc nano sol	2	3.3200	.44920		
4	Absorbency of jute (Control)	3	13.0000	.50553	-2.491	0.034
	Absorbency of jute treated with zinc nano sol	3	13.6000	.55976		
5	Water permeability of jute (control)	4	.3720	.02658	0.611	0.556
	Water permeability of jute treated with zinc nano sol	4	.3650	.04767		
6	Air permeability of jute (Control)	5	884.2800	9.08195	8.492	0.000
	Air permeability of jute treated with zinc nano sol	5	859.2400	3.59605		
7	Breaking load of untreated jute (before soil burial)	6	10.4400	.61500	-2.478	0.035
	Breaking load of jute treated with zinc nano sol (before soil burial)	6	10.9200	.12293		
8	Breaking load of untreated jute (after soil burial)	6	5.0200	.35528	-43.700	0.000
	Breaking load of jute treated with zinc nano sol (after soil burial)	6	9.8600	.15055		
9	K/S of control jute dyed with C.I. Direct Red 9	9	14.1500	.17159	-2.973	0.016
	K/S of treated jute dyed with C.I. Direct Red 9	9	14.3600	.14298		
10	K/S of control jute dyed with C.I. Direct Blue 67	9	16.4300	.42960	-1.367	0.205
	K/S of treated jute dyed with C.I. Direct Blue 67	9	16.5700	.24967		
11	K/S of control jute dyed with C.I. Direct Green 6	9	17.2600	.45509	-4.439	0.002
	K/S of treated jute dyed with C.I. Direct Green 6	9	17.8800	.11353		

Note: p-value<=0.005, significant at 5% level of significance.

Table 11: Mean comparison between different control groups with respect to their treatment group using paired t-test statistics.

samples was measured on KEITHLEY614-Electrometer and the results are given in Table 8. There was no significant change in surface resistivity of jute was observed due to the incorporation of zinc nano sol.

Effect of zinc nano treatment on dyeing behavior

The zinc nano sol treated jute fabric was dyed with three direct dyes and compared with the untreated samples. The K/S values of the nano zinc pretreated samples were found to be higher (Table 9) than the corresponding untreated samples. The maximum improvement in color strength was observed with direct green 6 dyes. The higher K/S values of nano-treated samples indicate that the presence of nano metal particles increased the dye affinity towards the material. The zinc nanoparticles in the fabric thus acted as mordant. The better coupling of the dye and fiber is also reflected in the improvement of the color fastness properties (Tables 10 and 11), which is a major drawback of most direct dyes. Above table indicates the pair wise mean comparison of data sets using paired t-test statistics, look in to the results, it can

be concluded that pair numbers 2, 4, 6, 7, 8, 9, and 11 have significant difference because the calculated t-value is significant as p-value found to be less than 0.05 which means that there exist real differences between the mean values of the control and the mean values of the zinc nano sol treated samples for the corresponding properties which also include reduction in water absorbency.

Conclusions

Zinc nano-sol was successfully applied on jute fabric using pad-dry-cure technique. The treatment reduced the water permittivity and air permeability of jute fabric samples. After the treatment the fabric manifested better dye ability, higher fastness and enhanced bacterial resistance.

Acknowledgement

Authors are thankful to AICTE (RPS) for financial assistance to carry out this research work.

References

1. Qian L (2004) Applicaton of nanotechnology for high performance textiles. Journal of Textile and Apparel Technology and Management 4: 1-7.

2. Qian L (2004) Nanotechnology in Textiles: Recent Developments and Future Prospects. AATCC Review 4: 14.

3. Chattopadhyay DP (2006) Nanotechnology-the emerging trends. Textiles 33: 21-24.

4. Patel BH, Chattopadhyay DP (2007) Nano-particles and their uses in textiles. The Indian Textile Journal 118: 23-31.

5. Wong YW, Yuen CMW, Leung MYS, Kuand SKA, Lam HLI, et al. (2004) Nanotechnologies in textiles. Textile Asia 35: 5-27.

6. Lee HJ, Jeong SH (2004) Bacteriostasis of nanosized silver on polyester nonwovens. Textile Research Journal 74: 442.

7. Lamb R, Zhang H, Jones A, Postle R (2004) Proc. 83rd TIWC, Shanghai, China.

8. Jin H, Liu L, Gu L (2004) Proc. 83rd TIWC, Shanghai, China.

9. Das A, Kothari KV, Vandana N (2005) AUTEX Res. J 5: 133.

10. Chattopadhyay DP, Patel BH (2011) Modification of Cotton Textiles with Nanostructural Zinc Particles. Journal of Natural Fibers 8: 39-47.

11. Chattopadhyay DP, Patel BH (2010) Effect of nanosized colloidal copper on cotton fabric. Journal of Engineered Fiber Fabrics 5: 1-6.

12. Chattopadhyay DP, Patel BH (2009) Improvement in physical and dyeing properties of natural fibres through pre-treatment with silver nanoparticles. Indian Journal of Fibre and Textile Research. 34: 368-373.

13. Chattopadhyay DP, Patel BH (2012) Influence of copper nanocolloids on cotton fibre. Textile Asia 43: 19-21.

16

Effect of Stitch Length on the Physical Properties of Both Plain and 1 X 1 Rib Knitted Fabrics

Ichetaonye SI[1]*, Ichetaonye DN[1], Adeakin OAS[1], Tenebe OG[2], Yibowei ME[1] and Dawodu OH[1]

[1]*Department of Polymer and Textile Technology, Yaba College of Technology, Yaba, Lagos, Nigeria*
[2]*Department of Home Economics, Federal College of Education, Kontagora, Nigeria*

Abstract

The Effect of the Stitch length on the properties of plain and 1x1 rib knitted fabric was carried out. Fabrics with various stitch length determined by the stitch dial at 4, 6, 8 and 10 were knitted using the flat bed machine for plain knitted fabric and the v-bed machine for 1x1 rib knitted fabric separately. The knitted fabrics with stitch length (Plain) 0.67 cm, 0.84 cm, 0.86 cm, 0.94 cm and also (1×1 Rib) 0.8 cm, 1.04 cm, 1.06 cm and 1.08 cm were tested separately for linear density, weight, tightness factor, thickness factor, elastica factor and stitch density of the fabric and the results show that significantly they differ. The results show that the higher the stitch length the more loose the loops hence the structure of the fabric and the shorter the stitch length the tighter or compact the fabric is.

Keywords: Stitch length; Knitted fabric; Stitch dial; Flat bed; V-bed

Introduction

Knitted fabrics are textile structure assembled from basic construction unit called loops and there exist two basic technologies for manufacturing knitting structures namely weft and warp (Figure 1). The weft knitted fabric can be manufactured using circular or flat knitting machine and every form can have various configurations depending on the machine performance [1]. Thus, all weft knitted fabric are based on one of the three basic structures either plain, rib or purl and their stitches can also be used for decorative pattern designs [2]. The simplest weft structure produced by the needles of a single flat bed machine is called Plain knit or Jersey knit while the structures obtained from the production of a double flat bed machine inclined in an inverted v-shape at 90[0] is called Rib structure [2,3].

However, jersey is used predominantly for clothing manufacture and has different appearance on both side of the fabric. The fabric posses some high stretchy features, light weighted and most often used for T-shirts, dresses, woman's tops, ladies suiting as well as sucks [4]. In knitting, ribbing is a pattern in which vertical stripes of stockinette stitch alternates with vertical stripes of reverse stockinette. This is noted by (number of knit stitches) Figure 2 by (number of purl stitches) having the same appearance on both side of the fabric [5]. Rib structure causes fabric to pull – in due to its narrow width which can extend to the same width as the plain fabric used for producing welts (Figure 3), close fitting garments and can be composed of any structural combination of 1x1, 2x2, 3x3, etc [3].

It is interesting that the process of yarn conversion into a weft knitted fabric can be performed from only one yarn package, which is a substantial advantage regarding the process preparation. Hence, due to the curved structure of the loop (Figure 4), the weft knitted structure are mostly elastic, stretchable and easily deformed [6,7].

Objectives

1. To determine what effect does stitch length have on plain and 1x1 rib knitted fabric.

2. To compare their various physical properties.

3. To determine what environmental conditions are both fabrics preferably be used.

Experimental

Equipment and materials

1. Flat bed knitting machine

2. V-bed knitting machine

3. Pilling Box

5. Weighing Machine

6. Comb

7. Dead weight

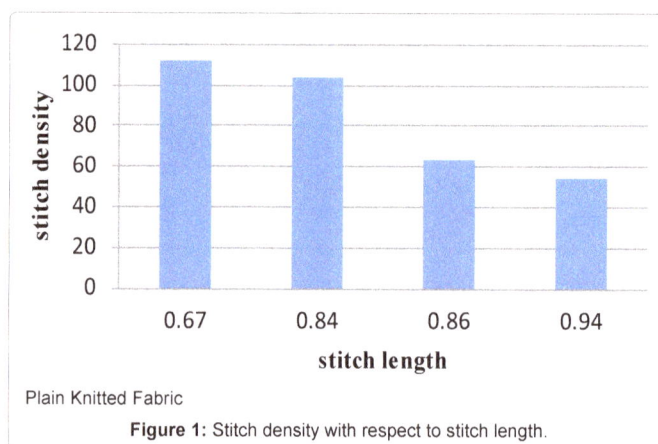

Plain Knitted Fabric

Figure 1: Stitch density with respect to stitch length.

*Corresponding author: Ichetaonye SI, Department of Polymer and Textile Technology, Yaba College of Technology, Yaba, Lagos, Nigeria
E-mail: ik4simon@yahoo.com

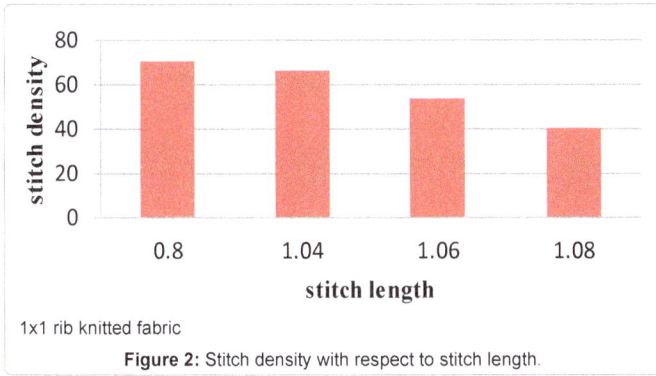

1x1 rib knitted fabric

Figure 2: Stitch density with respect to stitch length.

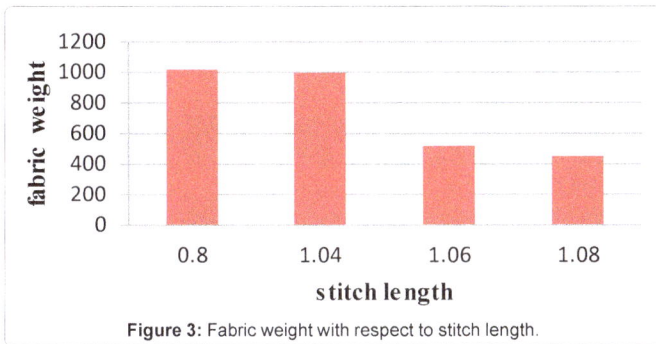

Figure 3: Fabric weight with respect to stitch length.

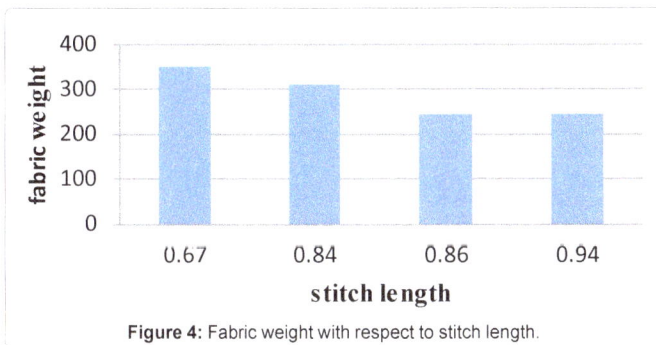

Figure 4: Fabric weight with respect to stitch length.

5. Cotton yarn (various colours and count)

6. Engine oil for lubrication

7. Alternating comb

Procedure for knitting

Plain knitted fabric on a single flat bed machine: A group of 50 adjacent needles about the middle of the machine were raised to the knitting zone. A knitting yarn was passed through various guides, tension devices and to the feeding point of the cam-box (Figure 5). With the yarn held at its tip underneath the cam-box, the needles were alternated and the cam-box was traversed across the needle bed from right to left. The knitting comb was hung on the course to facilitate the processes of knock-over, the alternated needle was brought back to the knitting zone and then 150 courses was knitted for stitch dial 4, 6, 8 and 10 respectively. Then the fabric was pressed-off.

1x1 Rib fabric on a v-bed machine: A group of 25 adjacent needles alternatively located about the middle of the machine on both the front and back needle bed were raised to the knitting zone (Figure 6). A knitting yarn was passed through various guides and tension devices

to the feeding point of the cam-box and with the yarn held underneath the cam-box at its tip, the cam-box was traversed once across the needle beds (Figure 7). The knitting comb was hung on the course, the diagonal raising cams was set to all knit, 150 courses was knitted for stitch dial 4, 6, 8 and 10 respectively and the fabric was pressed-off.

Measuring stitch length

After knitting 150 courses with 50 needles on the flat bed and the v-bed machine for each stitch dial and press-off, three (3) courses were unraveled from each separate sample until regular knitting is obtained. With the aid of a sharp scissors the last loop of the fabric selvedges are cut off, two (2) spots were made using a marker on both edges of the fabric, the yarn of the next course was removed and measured on a mater-rule. This was repeated three (3) times and the average was recorded which is the "COURSE LENGTH".

Mathematically

$$S.L = \frac{C.L}{N}$$

Where S.L = Stitch Length

C.L = Course Length

N = No. of Needles

This was repeated for stitch dial 4, 6, 8 and 10 respectively for both plain and rib fabric.

Tightness factor

$$T.F = \frac{\sqrt{T}}{L}$$

Where T.F = Tightness Factor

T = The linear density of yarn in Tex

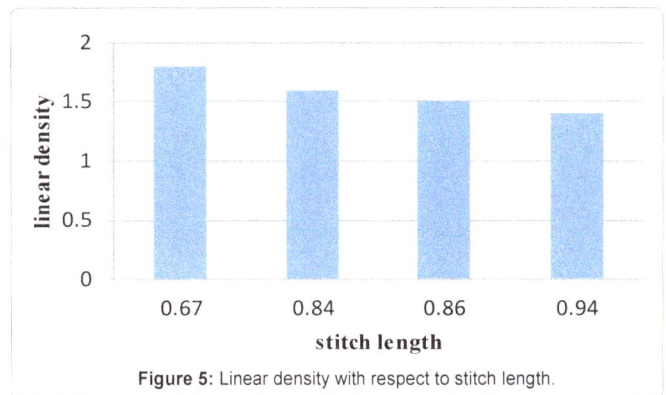

Figure 5: Linear density with respect to stitch length.

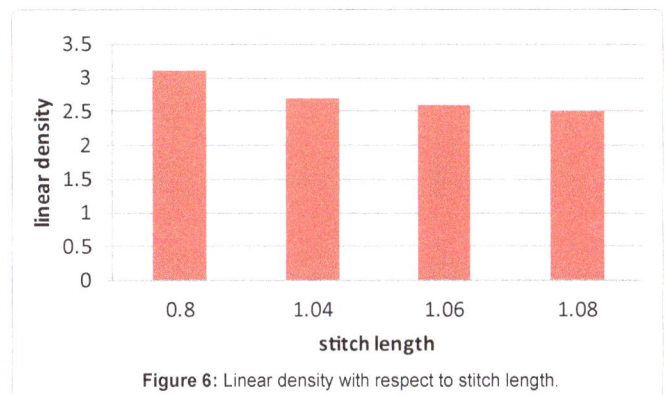

Figure 6: Linear density with respect to stitch length.

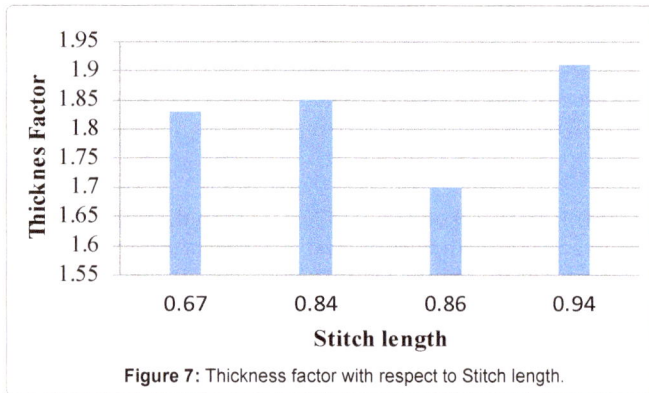

Figure 7: Thickness factor with respect to Stitch length.

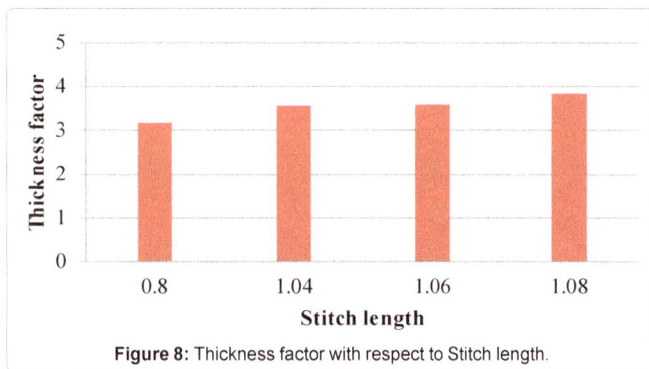

Figure 8: Thickness factor with respect to Stitch length.

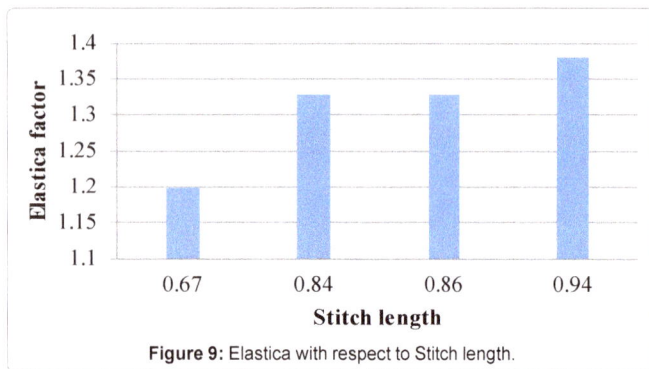

Figure 9: Elastica with respect to Stitch length.

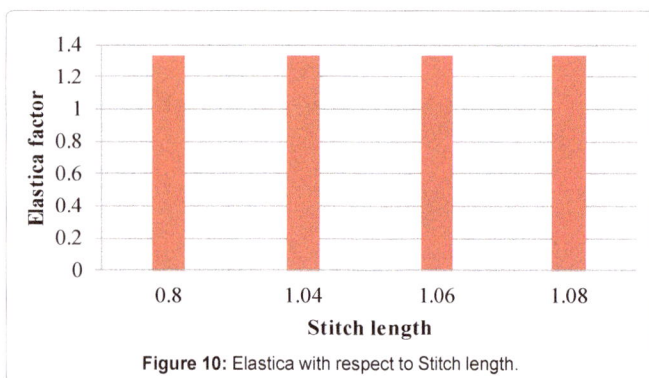

Figure 10: Elastica with respect to Stitch length.

L = Stitch length or loop length in cm

Elastica factor

Mathematically

$$E = \frac{b}{a}$$

Where b = the height of the loop

a = the width of the loop

Pilling

The pilling test for each fabric was carried out in ICI Pill Box. Four 5"x 5" samples of each loop length (2 in course direction, 2 in wale direction) were sewn and fitted round four rubber tubes (Figure 8). The cut ends were sealed with sellotape and brushed on the machine for 4 minutes ± 10 seconds, and then two are mounted face up on the second rotating platform while the other two are face down in contact with the first two. After 2 minutes ± 10 seconds of rubbing, the two upper specimens are evaluated for pilling.

Washing shrinkage

Eight (8) specimens, four (4) for each type (plain and rib) were spread on the table and a glass template of square size was placed each on them. There were six marks on the glass template with distance between two marks to be 50 mm (Figure 9). Each specimen were marked with a marker and then sewn by hand sewing machine. Each specimen was washed at 60°C for 90 minutes with liquid soap and then taken out for drying using the flat drying method. After drying, each specimen was re-weighed and the shrinkage percentage (%) was determined using

$$Shrinkage\ \% = \frac{(Lb - La)}{La} \times 100$$

Where L_b = Length of fabric before wash

L_a = Length of fabric after wash

Wbrasion resistance

Eight (8) circular specimens of 38 mm in diameter were cut using a cutting die i.e. four (4) each for plain and rib. The specimen was weighed to determine the pre-test mass for each dial (Figure 10). Four (4) specimen holders from the Martindala tester were removed and each cut specimen was placed with the technical face down into the gold ring. After which the handle was screwed back on and each assemble holders were placed into the machine, while the silver caps and black knobs are replaced. The required weight was added by resting the weights on the end of the handles (Kpa = 1 kilo pascal) then the counter system to record the desired movements was set to run a batch of 500. After the batch was completed, the specimen holders were taken off weighed and the percentage differences recorded.

Discussion

Judging from the structures and the various analyses, the following observations were made,

As the fabric got tighter, two situations were observed;

A. The stitch length size reduces because of the compactness of the fabric and the reduced stitch dial used.

B. The stitch density increases because of the small size of the stitch length of the knitted fabric based on its compactness.

As seen from Tables 1 and 2 the plain and the 1x1 rib knitted fabric for stitch density, linear density, fabric weight and tightness factor decrease as the stitch length increase in size due to the type of design in

Fabric Design	Stitch Dial	Stitch length (cm)	Stitch density	Fabric weigt (g/m²)	Linear density (tex)	Tightness factor (√T/L)	Thicknes factor (mm)	Elastica factor	Pilling Grade	Washing Shrinkage (%) Waleswise	Coursewise	Abrasion (g)
A	4	0.67	112	351.67	1.8	1.64	1.83	1.2	2.0	0	0	0.02
B	6	0.84	104	310.99	1.6	1.38	1.85	1.33	3.0	10	14	0.02
C	8	0.86	63	245.93	1.5	1.32	1.70	1.33	3.0	12	14	0.03
D	10	0.94	54	243.21	1.4	1.22	1.91	1.38	3.0	16	16	0.04

Table 1: Plain Knitted Fabric

Fabric Design	Stitch Dial	Stitch ength (cm)	Stitch density	Fabric weight (g/m²)	Linear density (tex)	Tightness factor (√T/L)	Thickness Factor(mm)	Elastica Factor	Pilling Grade	Washing Shrinkage (%) Waleswise	Coursewise	Abrasion (g)
A	4	0.8	70	1012.32	3.1	1.97	3.18	1.33	2.0	0	8	0.02
B	6	1.04	66	998.48	2.7	1.61	3.55	1.33	3.0	4	8	0.03
C	8	1.06	54	511.52	2.6	1.57	3.61	1.33	3.0	8	12	0.04
D	10	1.08	40	448.17	2.5	1.52	3.83	1.33	3.0	8	12	0.05

Table 2: 1x1 Rib Knitted Fabric

the order of tight, medium, slack and very slack.

Thus, the thickness factor, elastica, pilling, washing shrinkage as well as abrasion increased for plain and 1x1 rib knitted fabric which differs in chart when compared with others in the sense that the fabric gets bulkier even as stitch length increases.

However, rib fabric has a very high degree of elasticity in the course-wise direction and cannot be unraveled from the end knitted first because the sinker loops are securely anchored by the cross-meshing between face and reverse loop (Figure 11). This makes it suitable for making collars, necklines, cuffs, bottom edges of sweaters, knit hats, men's hosiery etc due to its firm and more relaxed state properties. While plain fabric can be used for making sheets, towels, sweaters, T-shirts, Men's underwear, dresses etc. Generally, there were slight differences between pilling along wales and courses, but it is too insignificant to warrant different grading. Fabric tends to pill a little more along the wales than along courses. More-over, the plain knitted fabric generally produced flatter pills lying closer to the fabric surface as well as having less abrasion effect compared to the 1x1 rib fabric whose pills were also fluffy and having a higher abrasion effect as the stitch length increases (Figure 12).

Furthermore, as the stitch length increase so as the elastic also having 1x1 rib exhibiting higher elastica compared to plain knitted fabric because relaxed 1x1 rib is theoretically twice as thick and half the width of an equivalent plain fabric, but it has twice as much width wise recoverable stretch. In practice 1x1 rib normally relaxes by approximately 30 percent compared with its knitting width [7-12].

Conclusion

The results analyses confirmed that, of all the parameters being studied, stitch length is the most important or decisive factors affecting the course per unit length (CPU) and wales per unit width (WPU) and also other variable such as tightness factor, gauge factor, linear density, stitch density, thickness factor, fabric weight etc.

However, the experimental result also shows that it was practically difficult to obtain the perfectly stable state for plain knitted fabric. This was not only because the knitted structure could not fully relaxed due to high internal restrictive force but also due to the fact that different drying processes and condition would result in different fabric dimension. Thus, stitch length has greater effect on both 1x1 rib and plain knitted fabric but more on 1x1 rib in the sense that the lesser the stitch length, the more compact the fabric structures is, the higher

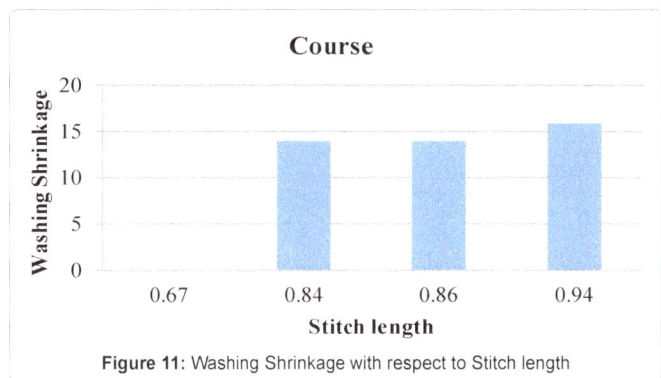

Figure 11: Washing Shrinkage with respect to Stitch length

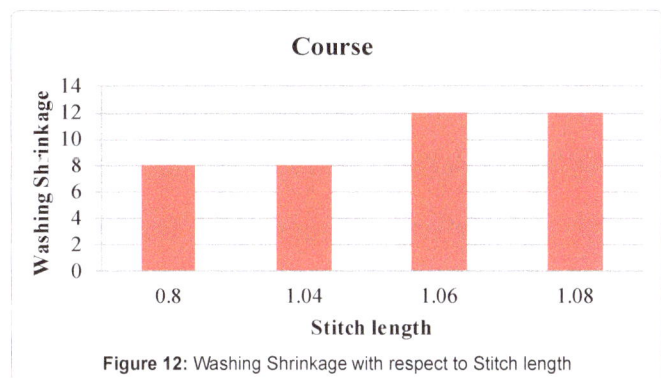

Figure 12: Washing Shrinkage with respect to Stitch length

the stitch density and fabric weight but the higher the stitch length, the more loose the fabric becomes and the lesser the stitch density and fabric weight.

References

1. David JS (1982) Knitting Tech.

2. Nutting TS, Leaf GA (1964) A Generalised Geometry of Weft-Knitted Fabrics. Journal of the Textile Institute 55: T45.

3. Poole HB, Brown P (1978) The Dimensions of 1×1 Rib Fabrics, Part I: Literature Survey. Textile Research J 48: 339-343.

4. Fletcher HM (1954) The Relationship of Geometry of Plain Knit cotton Fabric to Its Dimensional Change and Elastic Properties Textile Research J 24: 729.

5. Batchelor CW (1972) Double jersey Knitting and Patterning Hos Trade J 234.

6. Portrait of the Channel Islands.

7. Fletcher HM, Roberts SH (1965) Elastic Properties of Plain and Double Knit Cotton Fabrics Textile Research J 35: 497.

8. Smirfitt JA (1965) Worsted 1×1 Rib Fabrics, Part II Some Physical Properties. J Textile Institute 56: T298-T313.

9. Mundun DL (2002) The Geometry and Dimensional Properties of Plain Knit Fabrics - J Textile Institute 50: T448.

10. Dart T (1960) An Entropy Stress Study of various Textile Fibers. Textile Resesrch J 30: 372.

11. Abbott (1964) The Relationship between Fabric Structure and Ease – of – Care Performance of Cotton Fabrics. Textile research J 34 : 1049.

12. Hearle F (1969) Structural Mechanics of Fibres and Yarns and Fabrics 1: 403.

17

Mechanical Properties of Unidirectional Jute-Polyester Composite

Das S* and Bhowmick M

Central Institute for Research on Cotton Technology (ICAR), Adenwala Road, Matunga, Mumbai

Abstract

Unidirectional compressed jute fibre sheets are prepared by using raw jute and jute sliver at 120°C for 20 minute, at a pressure of 50 kg/cm². Compressed raw jute fibre sheet produced from raw jute reed and jute sliver were used for preparation of the composites in unsaturated polyester resin matrix. Three different fibre content 25 (w/w)%, 35 (w/w)%, 44 (w/w)% composites were fabricated by using raw jute and jute sliver. Tensile, flexural, ILSS and SEM properties of composites made from raw jute and jute sliver were studied. The Tensile, flexural and ILSS properties of composites made from raw jute was found to be higher than composites made from jute sliver.

Keywords: Composite; Jute; Resin; Flexural; Tensile; ILSS

Introduction

In a struggle to save our environment, researchers all over the world have been provoked to develop environment friendly composites of the natural fibre. In recent years, use of environment friendly composites is increasing which leads to replacing of natural fibre with inorganic fibres. Natural fibres have many advantages over synthetic fibres as reinforcement material with good specific strength, low cost, lower pollutant emissions; good energy recovery, biodegradable, eco-friendly renewable natural resource [1]. Jute is one of the most important lignocellulosic reinforcement material which is abundantly available in India. Cellulose is the basic component of jute fibre, long cellulosic fibrils embedded in a matrix of hemicelluloses and lignin and form ultimate jute cell. Jute cell is cemented together by lignin so as to form a long filament [2]. Jute plant grows 2.5-4 m in height. Jute fibres are 1 to 3.5 m long and are separated from the stalk by retting process [3]. The variation of ultimate jute cell length/breadth ratio is observed lengthwise along jute reed from bottom portion to tip portion [4]. Jute is a hydrophilic fibre and most of the common synthetic resins are hydrophobic in nature so fibre-matrix adhesion is poor. Several other disadvantages are low moisture resistance, poor wettability, poor dimensional stability, etc. Some physical, chemical and surface treatments may also be able to eliminate these problems which are described in the literature.

Several researchers have studied the physical and mechanical properties and improvement of mechanical and interfacial properties of jute fibre composites. Roe and Ansel [5] prepared the unidirectional jute composite in polyester resin matrix by the press-molding technique. The calculated fibre strength and modulus were 442 MPa and 55.5 GPa, respectively. They reported that polyester resin formed an intimate bond with the jute fibres up to a volume fraction of 0.6, above which the quantity of resin was insufficient to wet out the fibres completely. Then tensile strength and modulus of jute composite at volume fraction of 0.6 were 250 MPa and 35 GPa, respectively. Dash et al. [6] fabricated a low cost jute-polyester composite by using solution impregnation and hot curing methods. They fabricated composites by using both untreated and bleached jute slivers in polyester resin matrix with various percentages. It was observed that tensile and flexural properties of 60 wt% composites yielded the best results. Saha et al. [7] fabricated jute nonwoven fabric-polyester resin composites. They studied the dynamic mechanical and thermal properties of untreated and chemically modified jute-polyester composites. They reported that much stiffer and stronger composites can be prepared by using partial cyanoethylation of jute fibre. Cyanoethylation jute composites show better creep resistance at comparatively lower temperatures(up to 80°C)

whereas a reversed phenomenon is observed at higher temperatures (120°C and above). Ray et al. [8] treated jute fibres with 5% NaOH solution for four different time periods (2, 4, 6 and 8 h) and prepared jute composites in vinyl ester resin matrix. They reported that alkali treatment leads to improvement in the flexural strength and modulus of composites. They also reported that maximum improvements in the composite properties were observed with 4 h alkali-treated fibres. Gassan and Bledzki [9,10] reported that the treatment of the isometric jute yarns with 25% NaOH for 20 min resulted in an improvement of 120% and 150% in the tensile strength and modulus of the jute yarn. They reported a 60% improvement in the jute-epoxy composite's tensile properties, reinforced with these treated yarns. Tripathi et al. [11] studied the comparison of mechanical properties of jute-epoxy composites prepared by using untreated jute filament, sliver jute filament, bleached jute filament and mercerized jute filament as reinforcement.

Gowdaa and Naidu [12] studied the mechanical properties of woven jute fabric-reinforced polyester resin composites. Semsarzadeh et al. [13] studied the tensile strengths and impact energies of jute fibre-reinforced polyester composites. Doan et al. [14] studied the physical properties of jute - polypropylene composites and effect of maleic anhydride coupling agent on the properties of composite. Carvalho [15] investigated tensile and impact properties of different fabric style jute-polyester composites. He reported that plain weave jute composite have higher tensile strength than knitted jute composite and tensile strength of knit weave composite is independent of direction of test. Khan et al. [16] investigated effect of gamma and UV radiation on mechanical properties of 38 wt% fibre content jute-polyester composite. They enhanced tensile strength to 108% and bending strength to 58% of jute-polyester composites. They produced 2 mm thick polymer film by using mixture of 2-hydroxy ethyl methacrylate and aliphatic urethane diacrylate oligomer polymer on jute fabric. They used Co-60 gamma radiation for producing the polymer film on jute fabric as well as curing the composite. They modified surface of the bleached jute fabric by

---

*Corresponding author: Das S, Scientist, Central Institute for Research on Cotton Technology (ICAR), Adenwala Road, Matunga, Mumbai 400019
E-mail: sekhar.tex@gmail.com

using 254–133 nm UV radiation. They increased tensile strength to 150% and bending strength to 90% by using UV radiation. Khan and his team [17] studied mechanical, thermal, water uptake properties of jute fabric-gelatin bio composites. They reported that 50 wt% fibre content yielded the best results. They also reported that the tensile strength, tensile modulus, bending strength, bending modulus and impact strength of the 50 wt% jute fabric-gelatin composites were 85 MPa, 1.25 GPa, 140MPa, 140MPa and 9GPa and 9.5 kJ/m2, respectively. Sultana et al. [18] fabricated 35 wt% jute- polypropylene composites by using sodium Periodate oxidised raw Jute fibres. They reported an improvement of interfacial adhesion and bonding between the fibre and matrix and reduction of water absorption properties. Reddy [19] fabricated fully biodegradable jute composite in soy protein matrix and compared the mechanical properties with jute polypropylene composites. He reported that tensile and flexural properties of jute-soy protein are much higher than jute - polypropylene composites. Behera et al. [20] studied the physical properties of jute-nanoclay modified soy resin composite. Saw et al. [21] fabricated the jute-coir hybrid composites by using epoxy novolac resin and tested different physical properties of this composite.

Raw jute and jute sliver are voluminous materials so it is very difficult to fabricate composite by hand lay-up technique. In the excess resin extraction process for composite preparation, the jute fibres also come out with the extracted resin. In this study an innovative method was applied to avoid this difficulty. Jute fibre sheet was produced which is a compact form of the fibre. Mechanical properties such as tensile, flexural, ILSS and SEM of the composites made from raw jute and jute sliver were compared.

Experimental

Materials

The raw jute fibre reeds (TD-3 grade) from same bale were collected from local market in Kolkata, India. Unsaturated polyester resin manufactured by Swastik Interchem Pvt. Ltd. was purchased from Mumbai's local market. Cobalt octoate (accelerator) and methyl-ethyl-ketone peroxide (initiator) manufactured by Triveni Interchem Pvt. Ltd. also used and these were purchased from Mumbai.

Composite fabrication

Raw jute fibre reed of TD-3 grade (*C. olitorius*) was selected for producing jute sliver and jute fibre sheet. One portion of raw jute is used for producing raw jute fibre sheet and other portion is used to produce jute sliver. For producing jute sliver, raw Jute fibre reed was sprayed with 30% jute batching oil in water emulsion (2% oil on the weight of fibre) and the material was kept in a closed chamber for 24 hours. Then it was passed through a softener machine for softening the jute fibre. The soften fibres were then successfully passed through breaker card followed by a finisher card.

Raw jute fibre reed and jute sliver are voluminous so it is difficult to handle the reinforcing material during composite fabrication by hand layup technique. For producing compressed jute sheet, unidirectional raw jute fibre reeds and sliver were laid out by hand on a smooth steel plate. Then little amount of water was sprinkled on jute fibre and it was covered by a smooth steel plate and kept it in hot press at 120°C for 20 minute, at a pressure of 50 kg/cm² for getting uniformly compressed unidirectional jute sheet. In this process the fibre volume is reduced as the trapped air inside the fibre is forced out to get a compressed jute fibre sheet. Unidirectional jute fibre sheet-polyester resin composites were prepared by hand lay-up technique. The first coat of the resin

formulation [resin + 2% accelerator (o.w.r)* + 2% catalyst (o.w.r)*] was smeared on the polyester sheet which was kept on a smooth steel plate. Polyester sheet was used for quick and easy removal of the composite from the mould. A single fibre sheet was placed on top of the resin. Again resin was applied on the fibre sheet by using a brush. The next fibre sheet layer was placed on top of the first layer and the above process was repeated. After the final layer of fibre sheet was laid out, it was covered with a polyester sheet. The steel roller was used to even out the resin and to remove any entrapped air in the resin. A flat smooth steel plate was placed on top of polyester sheet to ensure a smooth surface. A compressive pressure of 5 kg/cm², 25 kg/cm² and 35 kg/cm² were applied on the mould to produce 25 (w/w)%, 35 (w/w)% and 44 (w/w)% composite respectively, at the time of curing. The specimens were also post cured at 70°C for 2 h after removal from the mould.

The composite samples having fibre content 25 (w/w)%, 35 (w/w)% and 44 (w/w)% obtained from the raw jute were labelled as RJ25C, RJ35C and RJ44C. Accordingly, sliver jute composites having the fibre content 25 (w/w)%, 35 (w/w)% and 44 (w/w)% were labelled as SJ25C, SJ35C and SJ44C, respectively.

Fibre fineness and bundle strength measurements

Jute fibre bundle strength and fineness were measured using "JTRL bundle strength tester" and "JTRL fibre fineness tester". Fibre fineness was measured by airflow method.

Mechanical testing

Tensile and flexural test: Tensile tests along with the fibre direction were carried out with ASTM D3039 for polyester and composite samples of 500 mm long, 15 mm wide and 4 mm thick in a universal testing machine (Instron) in order to determine the tensile properties. Test speed was 5 mm/min. Three point bend tests along with fibre direction were also performed in a universal testing machine (Instron) in accordance with ASTM D790 to measure the flexural strength of the polyester resin and composite samples. Test specimens having 150 mm length, width of 25 mm, thickness of 4 mm and a loading span of 64 mm was employed. The support span/specimen thickness ratio was 16:1 and the crosshead speed of 2 mm/min. Five samples were tested for both tensile and flexural strength.

Interlaminar shear strength: The short beam shear tests were carried out in a universal testing machine (Instron) in accordance with ASTM D2344-84 to evaluate inter-laminar shear strength (ILSS). The support span/specimen thickness ratio was 5:1 and cross head speed was 2 mm/min.

Five samples from each type of composites were tested for their mechanical properties and the data regarding their mechanical tests were expressed as mean ± standard deviation.

Electron microscopy

Fractured surfaces of the composites and the morphology of different fibres were studied with a Philips XL 30 scanning electron microscope (SEM) with an acceleration voltage of 10 kV. The sample surfaces were sputter coated with gold to avoid charging.

Results and Discussion

Raw jute fibre used for composite preparation was tested for fineness and bundle strength and are given in Table 1. Single jute fibre fineness was found to be 2.7 tex and the bundle strength of the jute fibres was 21.04 g/tex.

Fineness (tex)	2.70
Bundle Strength (gm/tex)	21.04

Table 1: Bundle strength and fibre fineness of raw jute fibre.

Material	Tensile strength (MPa)	Tensile modulus (GPa)	Tensile Strain (%)	Flexural strength (MPa)	Flexural modulus (GPa)	Flexural Strain (%)	ILSS (N/mm²)
Polyester resin	25 ± 7.31	0.95 ± 0.21	4.6 ± 0.18	31.5 ± 10.59	3.88 ± 1.05	2.01 ± 0.59	0.94 ± 0.32
RJ25C	80 ± 13.39	3.68 ± 0.48	4.5 ± 0.55	102 ± 16.23	9.42 ± 1.31	2.27 ± 0.15	12.79 ± 3.01
RJ35C	106 ± 16.30	4.83 ± 0.63	5.2 ± 0.83	124 ± 17.97	11.6 ± 1.65	3.49 ± 0.28	10.87 ± 1.51
RJ44C	122 ± 31.11	5.56 ± 0.67	4.8 ± 0.54	145 ± 21.94	15.41 ± 2.22	3.18 ± 0.31	10.18 ± 0.62
SJ25C	71 ± 11.93	3.24 ± 0.65	4.8 ± 0.59	85 ± 20.16	7.56 ± 1.36	2.61 ± 0.60	9.48 ± 1.29
SJ35C	89 ± 9.74	4.46 ± 0.45	5.4 ± 0.48	103 ± 14.64	10.64 ± 1.41	2.66 ± 0.57	8.45 ± 1.71
SJ44C	109 ± 16	4.89 ± 0.55	4.7 ± 0.54	112 ± 17.30	13.24 ± 2.12	2.57 ± 0.48	8.95 ± 0.93

Table 2: Comparison of mechanical properties of the composites made from raw jute and jute sliver.

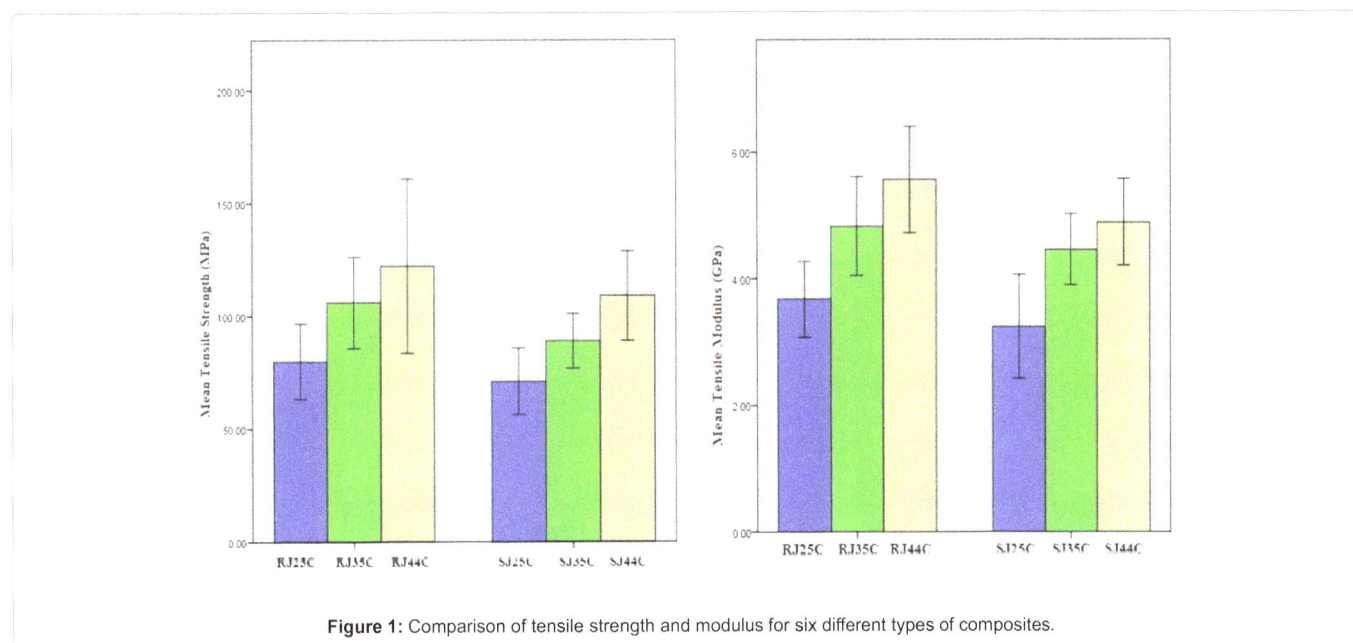

Figure 1: Comparison of tensile strength and modulus for six different types of composites.

Mechanical properties of polyester resin

Samples were prepared for tensile and flexural property determination of polyester resin for comparison purpose. It was found that the resin had a tensile strength of 25 MPa and tensile modulus of 0.95 GPa. The flexural strength of the cast resin was 31.5 MPa with flexural modulus of 3.88 GPa.

The tensile properties of the composite samples

Tensile, flexural and ILSS properties of the resin and composite samples are reported in the Table 2.

Samples were prepared for tensile property determination of polyester resin for comparison purpose. It was found that the resin had a tensile strength of 25 MPa and tensile modulus of 0.95 GPa.

It is well known that composite strength properties are mainly dependent on fibre loading (%), fibre strength and interfacial strength between fibre and matrix. Mechanical properties of composite give some indirect information about interfacial bonding between fibre and matrix. Figure 1 indicates the effect of increasing of fibre content to tensile properties of jute-polyester composites. For both raw and sliver jute composites show tensile strength and modulus increase with fibre loading percentage. It has been observed from the Table 2 that, RJ44C had tensile strength 15% higher than RJ35C and 52.5% higher than RJ25C composites and RJ35C had 32.5% higher tensile strength than RJ25C. Tensile modulus (5.56 GPa) of RJ44C was higher than RJ35C (4.83 GPa) and RJ25C (3.68 GPa).

It has been also observed from the Table 2 that, SJ44C had tensile strength 22.4% higher than SJ35C and 53.5% higher than SJ25C composites and SJ35C had 25.3% higher tensile strength than SJ25C. Tensile modulus (4.89 GPa) of SJ44C was higher than SJ35C (4.46 GPa) and SJ25C (3.24 GPa) Figure 2.

The tensile strength and modulus of raw jute composites had higher than sliver jute composites. R25C, R35C and R44C had 11.2%, 19.1% and 11.9% higher tensile strength than S25C, S35C and S44C and tensile modulus. Figure 3 of raw jute composites were also higher than sliver jute composites.

Flexural properties of jute composites

Flexural strength and modulus of a composite is dependent on the fibre strength and extreme layer of reinforcement plays a vital role. The crack always initiates on the tension side of the composite

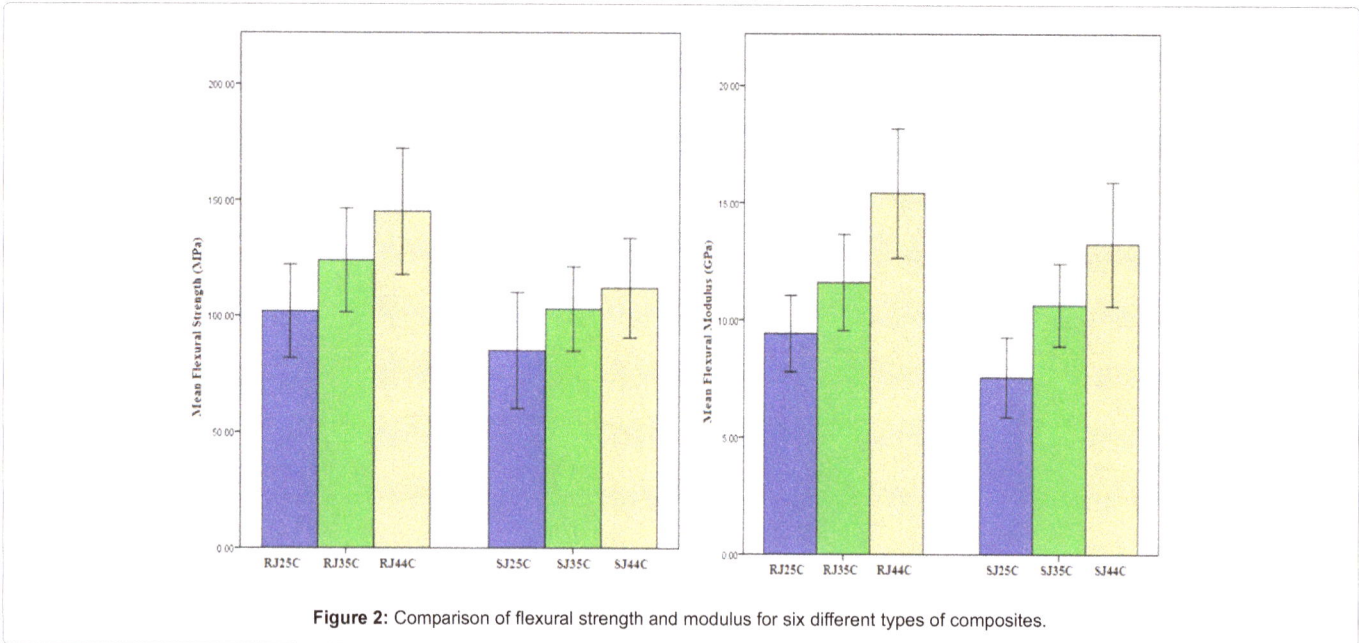

Figure 2: Comparison of flexural strength and modulus for six different types of composites.

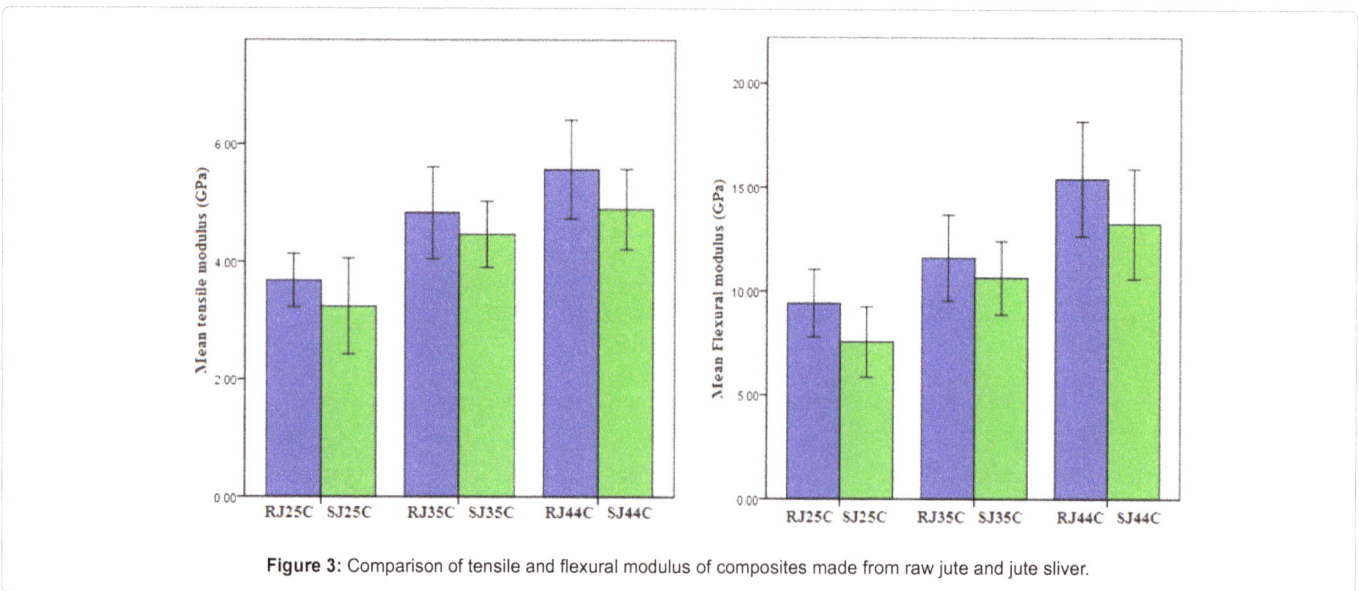

Figure 3: Comparison of tensile and flexural modulus of composites made from raw jute and jute sliver.

sample and slowly propagates in an upward direction. Normally, the flexural modulus is very sensitive to the matrix properties and fibre-matrix interfacial bonding [21]. It has been observed from the Table 2 that, polyester resin had flexural strength and modulus of 31.5MPa and 3.88GPa. Figure 2 indicates that flexural strength and modulus is directly proportional to fibre content of jute composites. From Table 2, it has been observed that RJ44C composite samples had 16.9% and 42.1% higher flexural strength than RJ35C and RJ25C composite samples and RJ35C had 21.55% higher flexural strength than RJ25C. Flexural modulus is also increased with fibre loading percentage, RJ44C (15.41 GPa) has higher modulus than RJ35C (11.6 GPa) and RJ25C (9.42 GPa). The same trend is also observed for jute sliver composites, SJ44C had 8.7% and 31.7% higher flexural strength than SJ35C and SJ25C, respectively and SJ35C had 21.1% higher flexural strength than SJ25C. Flexural modulus (13.24 GPa) of SJ44C was higher than SJ35C

(10.64 GPa) and SJ25C (7.56 GPa). The flexural strength and modulus of raw jute composites were higher than sliver jute composites. R25C, R35C and R44C had 20%, 20.3% and 16.9% higher tensile strength than S25C, S35C and S44C and flexural modulus. Figure 3 of raw jute composites were also higher than sliver jute composites.

Interlaminar shear strength

Short beam shear test is an effective method for explaining interfacial adhesion property of a composite. It is a 3-point short span length bend test, which generally create failure by inter-laminar shear. A large span length in bending test increases the maximum normal stress without affecting the inter-laminar shear stress and thereby increases the tendency for longitudinal failure. Inter-laminar shear failure initiated and propagated only when the span length is short. Maximum shear stress is observed in a beam at the mid-plane. So in

used to observe the failure surface after flexural failure of the samples. The SEM micrographs of the samples from raw jute are given in Figure 5.

It can be observed that the SEM micrographs clearly shows that the resin has penetrated the jute mats used as reinforcement material and it has totally surrounded each and every fibre. So, this signifies that the above mentioned process of composite preparation can be used for increasing fibre reinforcement percentage without compromising composite quality.

Conclusion

In this research a new method for production of compressed raw jute and jute sliver sheets preparation is described, which can be used as the reinforcement material to improve fibre percentage in composites produced. Among the two types of starting material, i.e., raw jute and jute sliver, it can be concluded that the composites made from raw jute have higher tensile and flexural properties compared to the composites made from jute sliver. ILSS properties of jute composite made from raw jute are also better than composite made from jute sliver. Jute fibre sheet would be economical than commonly used reinforcement material such as jute fabric and non woven.

References

1. Mukheijee A, Ganguly PK, Sur D (1993) Structural Mechanics of Jute: The Effects of Hemicellulose or Lignin Removal. J Textile Institute 84: 348-353.

2. Joshia SV, Drzalb LT, Mohanty AK (2004) Arorac S Are natural fiber composites environmentally superior to glass fiber reinforced composites? Composites Part A 35: 371-376.

3. Ghosh RK, Sreewongchai T, Nakasathien S, Phumichai C (2013) Phenotypic variation and the relationships among jute genotypes using morpho-agronomic traits and multivariate analysis. Australian Journal of Crop Science: 830-842.

4. Ali A (1989) Ultimate fibres in tossa jute (Corchures olitorius L.) and their utilization in selection Indian Acad Sci 99: 37-41.

5. Roe PJ, Ansell MP (1985) Jute-reinforced polyester composites.

6. Dash BN, Rana AK, Mishra HK, Nayak SK, Mishra SC, et al. (1999) Novel Low cost jute-polyester composites Part-I: processing, mechanical properties and SEM analysis .

7. Saha AK, Das S, Bhatta D, Mitra BC (1999) Study of Jute Fiber Reinforced Polyester Composites by Dynamic Mechanical Analysis. Journal of Applied Polymer Science 71: 1505-1513.

8. Ray D, Sarkar BK, Rana AK, Bos NR (2001) The mechanical properties of vinyl ester resin matrix composites reinforced with alkali treated jute fibres Composites Part A 32: 119.

9. Gassan J, Bledzki AK (1999) Alkali treatment of jute fibres Relationship between structure and mechanical properties. J Appl Polym Sci 71: 623.

10. Gassan J, Bledzki AK (1999) Possibilities for improving the mechanical properties of jute/epoxy composites by alkali treatment of fibers Comp Sci Tech 59 1303.

11. Tripathy SS, Dilandro LD, Fontanelli D, Marchetti A, Levita G (2000) Mechanical Properties of Jute Fibers and Interface Strength with an Epoxy Resin. Journal of Applied Polymer Science 75: 1585-1596.

12. Gowdaa TM, Naidu ACB, Chhayab R (1999) Some mechanical properties of untreated jute fabric-reinforced polyester Composites. Elsevier Science Ltd: 277-284.

13. Semsarzadeh MA, Lotfali AR, Mirzadeh H (1984) Jute-reinforced polyester structures. Polym Compos 5: 141.

14. Doan TTL, Gao SL, Ma der E (2006) Jute/polypropylene composites I Effect of matrix modification Composites Science and Technology 66: 952-963.

15. Carvalho LHD, Cavalcante JMF, Almeida JRM (2006) Comparison of the Mechanical Behavior of Plain Weave and Plain Weft Knit Jute Fabric-Polyester-Reinforced Composites. Polymer-Plastics Technology and Engineering 45: 791-797.

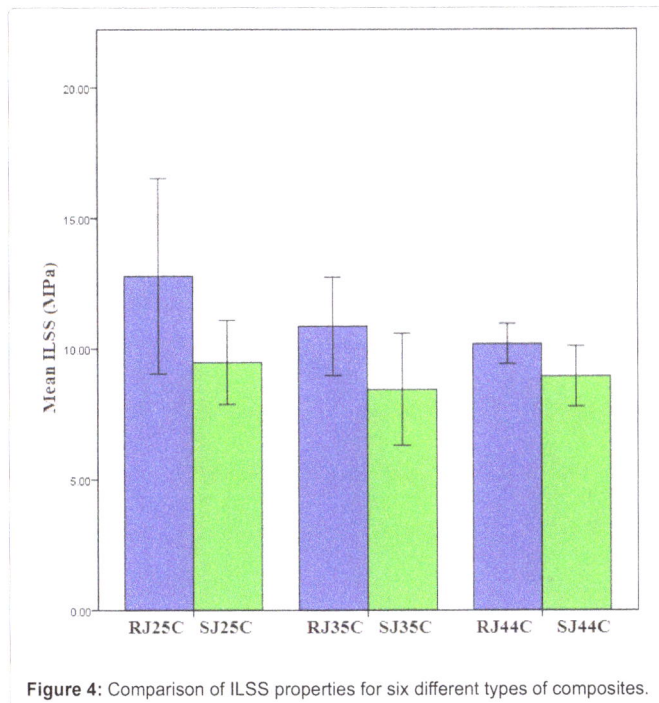

Figure 4: Comparison of ILSS properties for six different types of composites.

Figure 5: SEM micrograph of failure surface of the composites made from raw jute (A) and jute sliver (B).

short the shear test of failure determination consists of a crack running along the mid-plane of the beam so that crack plane is parallel to the longitudinal plane.

Interlaminar shear strength (ILSS) properties of the resin and composite samples are reported in the Table 2 which indicates that composites made from raw jute had higher ILSS strength than sliver jute composites for the three percentages of composites.

Figure 4 indicates comparison of ILLS properties of same fibre content composites with two types of composites. It was observed that composite made from raw jute have higher ILLS than sliver jute composites, for 25 (w/w)%, 35 (w/w)% and 44 (w/w)% fibre content composites.

Tensile strength, tensile modulus, flexural strength, flexural modulus and ILLS properties of jute composite made from raw jute was higher than composite made from jute sliver. The probable reason behind this is that raw jute fibre sheet has more oriented fibres than jute sliver. Jute sliver contain jute batching oil (2% oil on the weight of fibre) which affect bonding between fibre and polyester matrix.

Structural analysis of the composite samples

To study the structural morphology and to observe the failure structure of the composite samples scanning electron-microscopy was

16. Khan MA, Haque N, Al-Kafi A, Alam MN, Abedin MZ (2006) Jute Reinforced Polymer Composite by Gamma Radiation Effect of Surface Treatment with UV Radiation Polymer-Plastics Technology and Engineering 45: 607-613.

17. Khan MA, Islam T, Rahman MA, Islam JM, Ruhul A, et al. (2010) Thermal Mechanical and Morphological Characterization of Jute/Gelatin Composites Polymer-Plastics Technology and Engineering 49: 742-747.

18. Sultana S, Huque MM, Helali MM (2007) Studies on the Physicomechanical Properties of Sodium Periodate Oxidized Jute Reinforced Polypropylene (PP) Composites Polymer-Plastics Technology and Engineering 46: 385-391.

19. Reddy N, Yang Y (2011) Completely biodegradable soy protein–jute bio composites developed using water without any chemicals as plasticizer. Industrial Crops and Products 33: 35-41.

20. Behera AK, Avancha S, Sen R, Adhikari B (2013) Development and Characterization of Nanoclay-Modified Soy Resin-Based Jute Composite as an Eco-friendly/Green Product Polymer-Plastics Technology and Engineering 52: 833-840.

21. Saw SK, Akhtar K, Narendra Y, Singh AK (2014) Hybrid Composites Made from Jute/Coir Fibers Water Absorption Thickness Swelling Density Morphology and Mechanical Properties. J Natural Fibers 11: 39-53.

18

Brand Selection on Different Items - A Study for Investigation of Bangladeshi Young Customer Priority Level

Prasad RK*, Jannat F and Ali A

Department of AMT, BGMEA University of Fashion and Technology, Dhaka, Bangladesh

Abstract

Peoples mind, behavior and attitude are changed very quickly for the globalization and technology. Now a day's Bangladeshi young customers are well informed about the clothing items and they spent more time for this. This paper explores Bangladeshi young customers buying behavior when they go to purchase different fashion items like-t-shirt, polo shirt, shirt and pant. Data is collected from 350 people from different area of Dhaka, Savar and Gazipur. Data is analyzed as first priority and second priority label for each fashion item. This research helps to know which brand product are got most priority for a particular item, which brand is doing well business in Bangladeshi fashion market etc. Many fashion maker or brand developer is helped by that to improve their fashion marketing, increasing customer demands and build a strong relationship with the customer.

Keywords: Customer buying behavior; Clothing brand; Different item; First priority; Second priority

Introduction

Today, we live in consumer based society where consumer is known as the king of the market. The behavior of the customer affects the market size, brand name, fashion etc. [1]. Consumer buying behavior has dramatically changed in the last few decades. Consumers are better informed and have more choices in purchasing. Consumer's purchase decision is dependent on many different factors for example, brand equity brand loyalty [2-4]. Time tested concepts on brand loyalty and mass marketing, are being turned on their heads as they fail to gauge the purchase decision of new generation customers [5]. Variations in customers like and dislike, the cognitive growth and motivation force them to act in various ways in purchasing clothing's products [6,7]. Clothing is considered to be second skin of the body and interest in clothing is higher during the whole life. Bangladeshi people are very traditional in clothing choice [6]. Domestic clothing market of Bangladesh plays a very important role for the Bangladeshi young customer as they produce choice able clothing. Bangladeshi domestic clothing market is customer orienting market. Customer orientation marketing is a contemporary marketing technique based on customer needs and wants. This technique aims at producing items and goods that people willing to buy [8]. Bangladeshi domestic clothing market mainly produces t-shirt, polo shirt, woven shirt and denim pant for the Bangladeshi young customer. The customer everyone has a separate and elegant fashion sense which is mainly related to the clothing throughout the world. Clothing defines the personality, education, behavior and the way of thinking of the people [9]. Bangladeshi young customers are mainly dependent on different brand for different types of clothing as different brand produces quality full clothing product. Brand management holds the key in the modern apparel market [10]. Many young customer of Bangladesh like local product other than brand product as local product offers low cost to the customer. Although different brand produces almost same type of quality full clothing for the young customer but the customer has different brand choice for each type of clothing product. This paper focuses on the priority level of Bangladeshi young customer for different clothing product from different brand. The paper explores the first and second priority level of basic apparel product from 20 different brands of Bangladesh.

Brand and Brand Value

A brand, as defined by Keller, is "a product, but one that adds other dimensions that differentiate it in some way from other products designed to satisfy the same need [6,11]. A brand is a distinguishing name and/or symbol (such as logo, trademark, or package design) intended to identify the goods or services of either one seller or a group of sellers, and to differentiate those goods or services from those of competitors [12]. Brand values in industrial marketing are, therefore, the things which cause people to buy one company's product or service rather than another. They can be big issues - the core brand values or they can be small issues - noncore values. Brand values can relate to the products, pricing, the delivery or other aspect of service [12,13].

Methodology

Research approach: The aim of this study was to investigate the Bangladeshi young customer priority level of different product from different brand. The questionnaire has been made on the basis on some different clothing item and some popular brand of Bangladesh. The questionnaire was handed out to customer in the stores of different market.

Problem analysis: This study has been carried out to identify the brand priority for t-shirt, polo shirt, woven shirt and denim pant. It has been carried out by providing questionnaire to the target customers who are interested about clothing brand. The result of this study can be used to the brand developer of clothing item. This study also helps to the brand manufacturer to improve the quality of the existing product.

Develop questions: The questionnaire has been made by

*Corresponding author: Ripon Kumar Prasad, Department of AMT, BGMEA University of Fashion and Technology, Dhaka, Bangladesh
E-mail: srkprasadte10@gmail.com

considering the target customer gender, age, location, purchase capability etc. Closed ended questions has been developed with "tick mark" option for customer, so it is not time consuming and difficult for the customer. On the questionnaire there are 20 brand names which brands are very popular in Bangladesh. There is also an option "without brand" in the questionnaire as many customers may not be interested to brand for high price.

Data collection and sampling: The data was collected from 350 young customers to know their brand priority about t-shirt, polo-shirt, woven shirt and pant. Respondents were aged from 18-24 and located at dwelling places such as Dhaka, Gazipur, Narayangong as brand clothing item are available at central area and near about central area. The data has been collected also from the different University Campus such as DU, JU, BUFT, AIUB, NSU, EWU etc.

Data analysis: Data analysis is conducted through the process of quantitative content analysis. The data processed through Microsoft word and excel program.

Measure: A sample size of 350 young customers taken from different market places of Dhaka city to ensure adequate items to response ratio. Among the total respondents maximum are students also including employee which contains male and female both.

Results: From the findings a fashion designing and marketing can be estimated for any t-shirt, polo shirt, woven shirt and denim pant suppliers and apparel retail brands. And from the detailed discussion of clothing materials, the outcomes can be used as a base of "product development" for t-shirt, polo shirt, woven shirt and pant. This study can also used in the new brand development process.

Results and Discussions

First and second priority of brand for T-shirt purchase

The questionnaire have 20 top brand and without brand option. Among 350 people 73 people choose easy brand as first priority for purchasing t-shirt which covered 20.68% of total respondents. On the other hand, 71 people among 350 people choose easy brand t-shirt as second priority. Easy brand offers comparatively low price product to the customer. Peoples are also interested in quality which can be seen in purchasing yellow brand t-shirt. On the other hand availability of product is one major factor for purchasing any item. Easy brand product and without brand highly available in the market so that these brand product has more priority than other brand (Figure 1).

First and second priority of brand for polo shirt purchase

15.30% of total respondents choose easy brand polo shirt as first priorities where as 14.73% of total respondents choose easy brand polo shirt as second priority for purchase. Easy also offers comparatively low price than other as like t shirt. Most of the people choose easy brand for the low price and great availability in the market. 13.60% and 11.05% young customers were choosing without brand polo shirt for the first and second priority respectively for the low price offering. Some of the young customers choose high price polo shirt with good quality like yellow brand (Figure 2).

First and second priority of brand for woven shirt purchase

56 respondents out of 350 respondents choose without brand woven shirt due to low price with medium to high quality fabric. 15.6% choose without brand woven shirt as first priority and 10.76% choose it as second priority. Lower amount of respondents choose Richman brand t-shirt and polo shirt as first and second priority but in case of woven shirt no. of respondents priority level increase to Richman brand. Some of the respondent choose Easy brand woven shirt due lower price than other brand as like t-shirt and polo shirt. Respondents of yellow brand were increase for woven shirt than knit item due to high quality fabric with finer count yarn offered by yellow brand (Figure 3).

First and second priority of brand for denim pant purchase

In case of pant, no top class brand has the maximum percentage as first and second priority. 22.10% and 15.01% respondents choose without brand pant as first and second priority respectively. Without brand pant offers very lower price to customer for this reason respondents has more interest on without brand pant. On the other some of the respondents choose yellow, easy and Westecs brand pant although they have very high price compared to without brand (Figure 4).

From the overall discussion, it has been observed that for knit item garments respondents have more priority on easy brand. But in case of woven items respondents have more priority on without brand garments. Taste is one's own so that each respondent has separate brand choice for each item although some of customer choose same brand for all of items.

Conclusion

From the study that the authors performed and draw the conclusion that the young Bangladeshi fashion customer have different brand priority for different items which depends on many factors

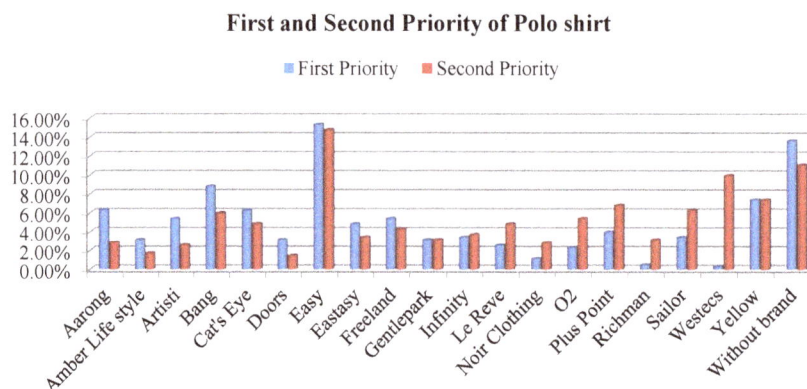

Figure 1: First and second priority of brand for T-shirt purchase.

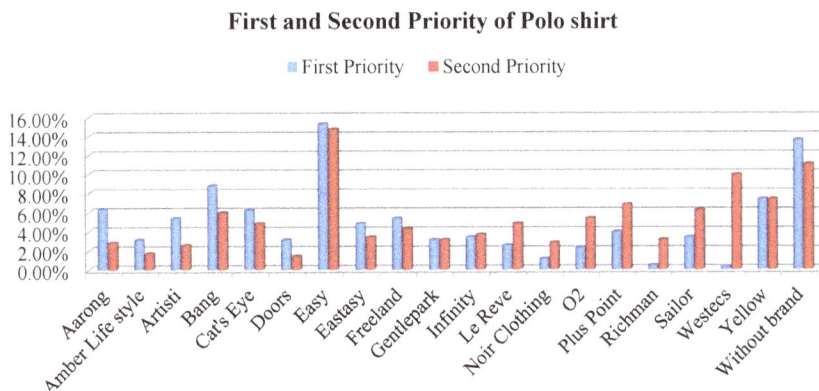

Figure 2: First and second priority of brand for polo shirt purchase.

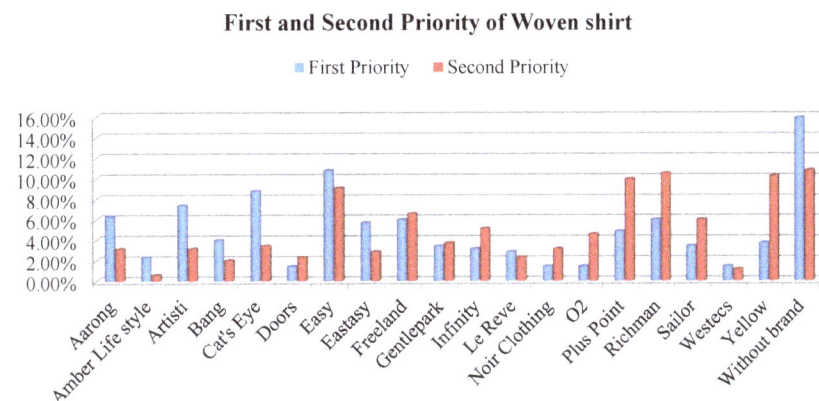

Figure 3: First and second priority of brand for woven shirt purchase.

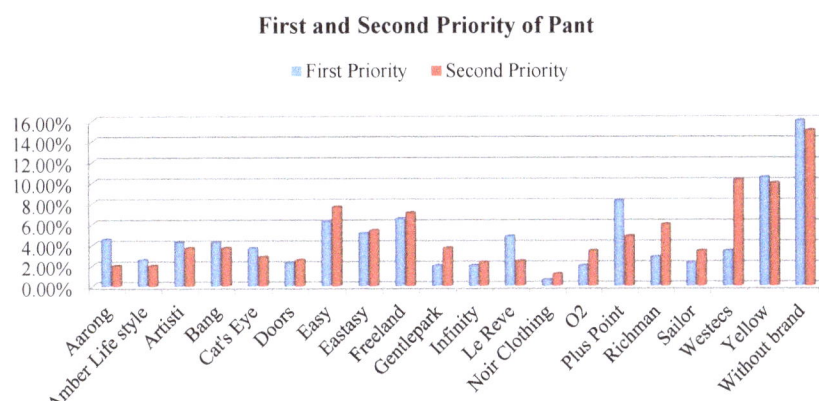

Figure 4: First and second priority of brand for pant purchase.

such as price, quality, availability, brand value etc. The results of this study also accelerate the fashion trend of the young customer where the study only shows the brand priority about the young customer perception for t-shirt, polo shirt, shirt and pant. The results show that 20.68% of customers choose the Easy brand as their first choice in purchasing t-shirt. 12.75% of customers choose Without Brand (Local) and 10.48% of customer choose Bang brand as their first choice of purchasing t-shirt at second and third stage. On the other hand 20.11% of customer choose Easy brand as their second choice in purchasing

t-shirt. In that case Yellow and Plus Point got 9.35% and 8.78% of total respondents of their second choice of purchasing t-shirt at second and third position. 15.30% of customers choose Easy brand as their first choice in purchasing polo shirt. 13.60% of customers choose Without Brand (Local) and 8.78% of customer choose Bang brand as their first choice of purchasing polo shirt at second and third position. On the other hand 14.73% of customer choose Easy brand as their second choice in purchasing polo shirt. In that case Without Brand (Local) and Yellow got 11.05% and 9.92% of total respondents of their second choice of purchasing polo shirt at second and third position. 15.86% of customers choose Without Brand (Local) as their first choice in purchasing shirt. 10.76% of customer choose Easy brand and 8.78% of customer choose Cat's Eye brand as their first choice of purchasing shirt at second and third position. On the other hand 10.76% of customers choose Without Brand (Local) as their second choice in purchasing shirt. In that case 10.48% and 10.20% of total respondents choose Richman and Yellow brand respectively of their second choice of purchasing shirt at second and third position. 22.10% of customers choose Without Brand (Local) as their first choice in purchasing denim pant. 10.48% of customer choose Yellow brand and 8.22% of customer choose Plus Point brand as their first choice for purchasing denim pant at second and third position. On the other hand 15.01% of customers choose Without Brand (Local) as their second choice in purchasing pant. In that case 10.20% and 9.92% of total respondents choose Westecs and Yellow of their second choice of purchasing pant at second and third position.

Acknowledgement

The authors are thankful to young customers of different places who shared their buying behavior and priority level in time of purchasing different types of fashion items.

References

1. Thakur A, Lamba, B (2013) Factors Influencing Readymade Apparel Purchase in Jalandhar City. Researchers World 4: 155.

2. Knox S, Walker D (2001) Measuring and Managing Brand Loyalty. Journal of Strategic Marketing 9: 111-129.

3. Myers C (2003) Managing Brand Equity: A look at the Impact of Attributes. Journal of Product and Brand Management 12: 39-52.

4. Rahman MS (2012) Young consumer's perception on foreign made fast moving consumer goods: the role of religiosity, spirituality and animosity. International Journal of Business and Management Science 5: 103.

5. Babu MG, Vani G, Panchanatham N (2010) Consumer Buying Behaviour.

6. Islam MM, Islam MM, Azim AYMA, Anwar MR, Uddin MM (2014) Customer perceptions in buying decision towards branded Bangladeshi local apparel products. European Scientific Journal.

7. Lalitha A, Ravikumar J, Padmavali K (2008) Brand preference of Men Wear. Indian Journal of Marketing 38: 33-36.

8. Sanad RA (2016) Consumer Attitude and Purchase Decision towards Textiles and Apparel Products. World Journal of Textile Engineering and Technology 2: 16-30.

9. Rajput N, Kesharwani S, Khanna A (2012) Consumers' attitude towards branded apparels: gender perspective. International Journal of Marketing Studie 4: 111.

10. Pandian MSK, Varathani SN, Keerthivasan V (2012) An Empirical study on consumer perception towards branded shirts in Trichy City. International Journal of Marketing, Financial Services and Management Research 20.

11. Biplab SB (1998) Hand Book of Marketing Management, Himalaya Publishing House.

12. Aaker DA (1991) Managing Brand Equity. New York: The Free Press.

13. Prasad RK, Arifuzzaman M (2016) Buying Behavior of Young Customers in Bangladesh-A Movement towards Investigation of Their Fashion Attributes. International Journal of Textile Science 5: 19-24.

19

Theoretical Prediction of Overall Porosity of Terry Woven Fabrics

Hemdan Abo-Taleb*, Heba El-Fowaty and Aly Sakr

Textile Engineering Department, Mansoura University, Mansoura, Egypt

Abstract

A new geometrical model of terry woven fabrics made with cotton yarn has been analyzed to understand the pile yarn path. Theoretical model has been created to predict the porosity of terry woven fabrics depending on their geometrical such as warp and weft spacing pile length pile height pile ratio (terry ratio) type of terry yarn count yarn crimp fabric thickness yarn density and fiber density. Cotton terry woven fabrics were produced with different parameters namely pile height picks per cm and weft yarn count their porosity was determined by measuring fabric density and fiber density and compared with the theoretical porosity. The validity of the theoretical model was confirmed by experimental results. Experiments show that this model gives good results.

Keywords: Porosity; Terry woven fabric; Theoretical model; Terry ratio; Pile height; Fabric thickness; Fibre density; Prediction

Nomenclature

Mf	Mass of fibres in the fabric g
MF	Mass of fabric g
H	Pile height cm
L	The length of loop of pile yarn cm^3
Vc	The volume of the cuboid of the fabric sample cm^3
VP	The volume of pile warp yarns in the cuboid cm^3
Vg	The volume of ground warp yarns in the cuboid cm^3
Vw	The volume of weft yarns in the cuboid cm^3
mp	The weight of pile yarns in the cuboid g
Nmp	The metric count of pile warp yarn Nm
Nmg	The metric count of ground warp yarn Nm
Nmf	The metric count of weft yarns Nm
ρ_{yx}	The density of ground warp yarn g/cm^3
ρ_{yf}	The density of filling yarn g/cm^3
C1	The crimp of ground warp yarn %
C2	The crimp of filling yarn %
Sw	Static water absorption %
Mw	The weight of wet samples g
Md	The weight of dry samples g
M	Fabric surface density g/m^2

Introduction

Porosity is a vital quality in such end – use applications as sport garments, underwear products T- shirts socks and others. Porosity also significantly influences the thermal comfort of the human body for the proper body temperature [1].

Porosity is the most important property that affects the absorption capacity of the material. Porosity is used to describe the porous structure of textile materials. It is defined as the ratio of the void space in a porous medium over the total bulk volume of the medium. It is a dimensionless quantity and can range between 0 and 1 [2]. The amount of porosity i.e., the volume fraction of voids within the fabric determines the capacity of a fabric to hold water the greater the porosity the more water the fabric can hold [3]. The porosity and size of capillary of terry woven fabric will influence its physical properties such as bulk density liquid uptake the mass transfer and the thermal conductivity.

It is logical to expect that fabric structure has an impact on porosity [4]. The structure of a textile contains pores between the fibres and the yarns. It is quite clear that pore dimension and distribution is a function of the fabric geometry. The yarn diameter surface formation techniques number of yarn threads per unit length (yarn density) pile height and number of pile loops for terry woven fabrics per unit area are the main factors affecting the porosity of textiles. The porosity of a fabric is connected with certain important features of it such as air permeability water permeability and dyeing properties etc. [5-7].

There are several different methods available for the assessment of the parameters of porosity, such as: geometrical methods [8-10] liquid intrusion methods [11-13] liquid extrusion methods [14-16] liquid through methods [17] etc. but they can only be used after a fabric is produced. An exception to this is the method based on an ideal geometrical model of an individual textile product as a porous material along with input data. Such a method does not need expensive laboratory equipment or sample weaving. Currently the results of this geometrical method based an ideal model of a porous structure do not compare well with real values determined by other methods. Some of the other methods used for estimating the porosity can only give truly very approximate values which may not by accurate enough. On the other hand some of them are not capable of estimating all the relevant porosity parameters.

*Corresponding author: Hemdan Abou-Taleb, Faculty of Engineering, Textile Engineering Department, Mansoura University, Mansoura, Egypt
E-mail: haboutaleb_mm@yahoo.com

All models that lead to the determination of the porosity of woven fabric include some simplifying assumptions which introduce some inaccuracies into the result. Therefore it is very difficult to find the optimal method that predicates the permeability or porosity of the woven fabric.

A lot of work has been done over the years to overcome the mentioned shortcomings. Some researchers used a more theoretical approach [18-22], while some used a more experimental approach [23-25].

Some previous studies investigated the relationship between porosity and structural characteristics of woven fabrics such as fabric weight thickness density or finishing process [26-35].

In previous studies to predict the porosity of woven fabrics only the porosity of interyarn interstices was considered and interstices of interfibre were omitted [28,36-40]. However Xu and Wang [41] used the stepwise regression method to explore the permeability of interfibre interstices. In the present study the overall porosity through interstices of fibers and yarns together was calculated. So more ealistic results can be achieved than previous studies.

Although the prediction of porosity of woven fabrics has been described previously in the literature [26,28,40,42], no work has yet been carried out to predict the porosity of terry woven fabrics.

In spite of the availability of other porosity testers very little work has been done to correlate the porosity of a woven fabric with its structural properties.

The aim of this paper is to demonstrate theoretical model created to predict the porosity of terry woven fabric. The model will enable the engineering of terry woven fabrics to a required porosity thus determining other properties such as absorbency rate and liquid retention.

The object of this study was selected because of a lack of research dealing with investigations into the behavior of terry structure when in contact with liquid in general.

In this study a new geometrical model has been analyzed to understand the pile yarn path of terry woven fabrics. This method is suitable for all types of terry woven fabrics.

Theoretical Model

Assumptions

• The pile yarns in the terry woven fabric lies in a zigzag path with triangle shapes for both front and back surfaces as shown in Figure 1.

• The air space in the pile yarn consists of very fine circular capillaries which are parallel to the pile yarn and have the same length as the pile yarn.

• Fibres are equally distributed within the yarns as there are no external compression forces.

Ideal geometrical model of a terry woven fabric

Our experiments were carried out with cotton terry woven fabrics that have pile loops on both sides. The reason for the selection was the popularity of terry fabrics in home, sauna and leisure textiles such as towels, dressing – gowns, slippers head gear etc. The structure and weave repeat can be seen in Figure 1a.

As seen in Figure 1a ground warp (G_1) which is up at the beginning

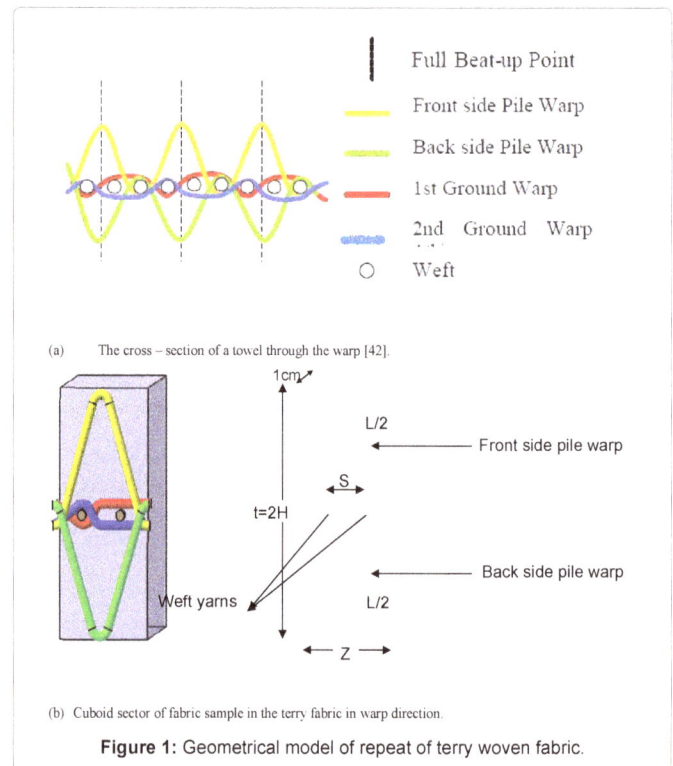

(a)   The cross – section of a towel through the warp [42].

(b)   Cuboid sector of fabric sample in the terry fabric in warp direction.

Figure 1: Geometrical model of repeat of terry woven fabric.

goes down and (G_2), which is first down and then goes upwards through the two yarns. Back side pile (Bp) warp is always opposite to the pile (Fp) warp of the front side. When the (Bp) warp makes the first loop on one side of the fabric a second loop will be formed on the other side. The (Fp) warp operates in a similar fashion.

In the present work the terry woven fabrics are assumed to be of different pile heights, different yarn diameters different warp spacing's and of different weft spacing's as a general case. The geometrical model of terry woven fabrics is shown in Figure 1.

A cuboid sector of fabric with thickness (t) and width (z) with 1 cm in length can be taken into consideration for calculating the overall porosity.

Each repeat of this geometrical model is consisted of front and back pile yarns, two ground warps and two weft yarns.

For the calculation of overall porosity, it was assumed that the pile warp yarns in the front and back sides of terry woven fabrics lies in a zigzag path as shown in Figure 1b. If the number of spaces (weft spacing) displaced from the beginning of pile warp yarn to the next pile on both front and back surfaces (floats) is S, then the yarn makes a rhombus layout as shown in Figure 1b.

The width of the cuboid Z can be given as follows:

$$Z = ns \times \text{weft spacing (P1)} = n_s/n_2 \text{ cm} \tag{1}$$

Where n_2 is picks per cm. n_s is number of spaces S

Theoretical model derivation

When a terry woven fabric is treated as a three – dimensional formation void spaces (pores) can be situated between fibres in the yarns and between warp and weft threads in the fabric.

Figure 1b shows an ideal geometrical model of the porous structure of a terry woven Fabric. To compare terry woven fabrics with porosity the following parameters are commonly used : area of the pore cross section pore area distribution pore density equivalent pore diameter maximum and minimum pore diameters pore length pore volume and the portion of the open area the portion of pore volume etc.

The overall porosity of a terry woven fabric (ε) can be theoretically calculated on the basis of packing factor (ϕ) as follows:

$$\varepsilon = 1 - \varphi \tag{2}$$

The fabric packing factor expresses the ratio of fibre volume (V_f) with regard to the fabric volume (V_F):

$$\varphi = \frac{\rho_F}{\rho_f} = \frac{m_F}{V_F} \times \frac{V_f}{m_f} \tag{3}$$

Where the symbols V(cm³) m (g) and ρ (g/cm³) stand for volume mass and specific density respectively. Subscripts f and F denote fibre and woven fabric.

The fabric mass is actually the mass of fibres used ($m_f = m_F$) so the expression given by Equation (3) can by simplified:

$$\varphi = V_f / V_F \tag{4}$$

By considering the geometry of a woven fabric shown in Figure 1 the volume of the cuboid (V_c) of the fabric sample can by calculated by using the following Equation (5).

V_c = length of fabric × width × thickness.

$$Vc = 1\ cm \times (n_s/n_2)\ cm \times t\ cm = (n_s/n_2)^* t\ cm^3 \tag{5}$$

Where fabric thickness (t) is measured under a pressure of 10 g/cm².

Thus the length of loop pile yarn (L) through the width (Z) can be calculated as follows:

$$L = 2\sqrt{H^2 + \left(\frac{z}{2}\right)^2} = 2\sqrt{\left(\frac{t}{2}\right)^2 + \left(\frac{Z}{2}\right)^2},cm$$

$$L = 2\sqrt{\frac{t^2}{4} + \left(\frac{n_S}{2n_2}\right)^2} = 2\sqrt{\frac{t^2}{4} + \frac{n_S^2}{4n_2^2}},cm \tag{6}$$

Where H is pile height in cm.

If we consider the fibre volume (V_p) is the volume of pile warp yarns in the cuboid and the weight (m_p) is the weight of pile yarns in the cuboid then.

Volume of material (Vp) in cm³ =

= weight of pile yarn (fibre)/fibre density

Weight of material (m_p) in grams =

= total pile yarn length in both sides in metre/pile yarn count (N_{mp})

$$\text{Thus, } V_p = \frac{2L(metre)}{N_{mp} \times \rho_f} \times \frac{ends/cm}{2},\quad cm^3 \tag{7}$$

Where N_{mp} is the metric count of pile warp yarn. As a result volume of ground warp yarns (V_g) in the cuboid can be calculated as follows:

$$V_g = \frac{\text{No.of ground warp yarns per cm} \times 1cm \times Z\,cm \times (1+\frac{C_1}{100})}{N_{mg} \times p_{yg}},cm^3 \tag{8}$$

Where

N_{mg} is   the metric count of ground warp yarn

ρ_{yg} is   the density of ground warp yarn g/cm³

C_1 is   the crimp of ground warp yarn %

Also volume of weft (filling) yarns (V_w) in the cuboid can be calculated as follows :

$$V_w = \frac{\text{Picks/cm} \times z \times 1\,cm \times (1+\frac{C_2}{100})}{N_{mf} \times 100 \times \rho_{yf}},cm^3 \tag{9}$$

Where

N_{mf} is   the metric count of weft (filling) yarns

ρ_{yf} is   the density of weft (filling) yarns, g/cm3

C_2 is   the crimp of weft (filling) yarns %

* **Yarn bulk density calculation:** Peirce [43] suggested that yarn bulk density (ρ_y) can be calculated as follows:

$$\rho_y = \rho_f (1 - \varepsilon_y)\ g/cm^3 \tag{10}$$

Where ρ_f = fibre density for cotton (1.54), g/cm³

ε_y = yarn porosity.

Theoretical approaches were used in order to estimate the porosity of the yarn (ε_y) within the fabric using certain assumptions based on yarn and fibre parameters.

The yarn porosity (ε_y) defined as the ratio of the pore area to the yarn cross – section area can be calculated by Equation (11) [44].

$$\varepsilon_y = 1 - \varphi y \tag{11}$$

Where φy = packing density of yarn i.e. the ratio of the total fibres area to the yarn area.

The circular cross – sectional theoretical diameter of the fibre (d_f) was calculated using Equation (12)

$$\text{Fibre tex} = \frac{\pi}{4}d_f^2 \times 10^5 \times \rho_f \tag{12}$$

Where d_f = fibre diameter cm

ρ_f = fibre density g/cm³

In Equation (11) to calculate the inter – fibre yarn porosity theoretically the packing density of the yarn was calculated by using the ratio of total fibres area to the total yarn area. The theoretical circular yarn diameter (d_y) was used to calculate the total yarn area (A_y). The total fibres area was calculated using the number of fibres in the cross – section (n) and the area of one fibre (A_f) assuming that the fibre diameters within the yarn cross – section were equal to an another.

The average number of fibes (n) in the cross – section of the yarn was found theoretically by using (equation 13).

$$n = \frac{Nmf}{Nmy} = \frac{\text{yarn tex}}{\text{fibre tex}} = \frac{\text{yarn tex}}{\frac{\pi}{4}d_f^2 \times 10^5 \times \rho_f} \tag{13}$$

Also by using Equations (13, 14, 15) the packing density of the yarn (φy) can be calculated based on the theoretical fibre (n) the area of the one fibre (A_f) whose cross – section was assumed to be circular and yarn area (A_y).

$$\varphi y = \frac{n \times Af}{Ay} = \frac{\text{yarn tex} \times \frac{\pi}{4} d_f^2}{\frac{\pi}{4} d_f^2 \times 10^5 \times \rho_f \times \frac{\pi}{4} d_y^2} \qquad (14)$$

$$\varphi y = \frac{\text{yarn tex}}{10^5 \times \rho_f \times \pi d_y^2 / 4} \qquad (15)$$

Where d_y = yarn diameter cm.

By assuming that the cross – section of the yarns are circular and the volume percentage of fibres is 60% [45] it is known that the level of twist the slope of the fibre and so on can influence this percentage so 60% is just a reasonable approximation. Thus yarn diameter can be obtained by the following equation:

$$dy = 0.01189 \sqrt{\frac{Den}{\rho_f \times 0.6}} \ \text{Mm} \qquad (16)$$

In which dy = yarn diameter in mm

Den = yarn denier and

ρ_f = fibre bensity g/cm³

By this way the theoretical yarn bulk density (ρ_y) was calculated by Equation (17) using Equations [1,2,6 and 7] based on the theoretical fibre number in the cross – section of the yarn (n) the theoretical fibre diameter (d_f) and the theoretical yarn diameter (d_y).

$$\rho_y = \frac{\text{yarn tex}}{10^5 \times \pi d_y^2 / 4} \ \text{g/cm}^3 \qquad (17)$$

Where d_y is yarn diameter in cm.

By substituting in Equations (8, 9) with the values of yarn bulk density (ρ_y) the values of V_g and V_w can be calculated easily.

As a result total volume of yarns (pile ground and weft) in the cuboid can be calculated as followes:

$$V_{total} = V_p + V_g + V_w \ \text{cm}^3 \qquad (18)$$

Then from Equation (2.4 and 18) total overall porosity can by calculated as follows:

$$\text{Porosity } (\varepsilon) = 1 - \frac{V\text{fibres (Vtotal)}}{\text{Volume of cuboid(Vc)}}$$

$$\text{Hence Porosity } (\varepsilon) = 1 - \frac{(V_p + V_g + V_w)}{(n_s / n^2) \times t \times 1 \text{cm}} \qquad (19)$$

In the design of a porous fabric for a particular application e.g terry woven fabric it is most likely that porosity (ε) and pile height (H) will be selected first. As a result ε (fabric porosity) H (pile height) d_y (effective yarn diameter parameter) and n (number of yarns per fabric opening in warp and filling). As a resultε, H dy and n have been selected as design constants. From these it is desired to predict the yarn size (Tex), and Fabric setting (ends and picks/cm) required to provide a fabric having the given pile height and porosity.

Fabric Manufacture and Testing

In the experimental part of this study in order to compare the values of overall porosity calculated from theoretical modeling and that using experimental overall porosity six woven terry samples at different pile heights weft densities and weft linear densities were carried out.

The pile of the terry fabrics used in this research was constructed on both sides of the fabric as can be seen in Figure 1a.

Generally this structure consists of three components namely pile warp yarn ground warp yarn and weft yarn which from the terry woven fabric. The specifications of each fabric are summarized in Table 1. The pile and ground warp density was 24 ends/cm and the values of weft density range from 10 to 20 picks/cm.

The pile and ground warps were plied cotton yarns of 24/2 Ne whereas the values of weft counts range from 12/1 to 20/1 Ne and were made from cotton yarns. Pile loops were embedded using basic 3-pick terry toweling and manufactured with different pile heights.

Warps are ordered throughout the fabric width 2:2 piles and ground warps. In 2:2 warp order each two ground warp ends are followed by two pile warp ends. In Figure 2, the weave notation of 3 weft pile is given in 2:2 warp orders.

The cotton terry fabrics used in the experimental work were woven by Eng. Mahmoud Mohamed El-Fowaty, Chairman of the Board of Directors of (El-Fowaty Tex) Company in Aga Egypt.

For the experimental investigation of porosity and static water absorption different types of terry woven fabrics with constant set of warp and varying set of weft and varying yarn fineness and varying warp pile height were produced on PROMATECH Vamatex Leonardo Dyna Terry model Double Flexible Rapier 230 cm Terry Weaving m/c. with a Jacquard shedding mechanism using the Staubley – CX870.

Produced fabrics are intended to be used for face towel where water absorbency is a necessary product feature. The incorporation of these hydrophobic samples aims to verify the versatility of the test. These woven fabric samples (10 cm × 10 cm) were conditioned in the test environment (20+2°C and 65+2% relative humidity) for at least 24 hours before testing.

The experimental evaluation of the porosity was carried out by measuring the fabric weight, fabric thickness and by calculating the fabric bulk density from the length the width and the thickness of the terry woven fabrics.

The porosity of the fabric was determined according to Equation (20) with reference to Hsieh's work [45].

$$\text{Porosity}(\varepsilon) = 1 - \frac{\text{Fabric weight}(g/cm^2)/\text{Fabric thickhess}(cm)}{\text{Bulk density of fiber}(g/cm^3)} \qquad (20)$$

The static water absorption was measured according to method [46]. The samples were conditioned in laboratory conditions cut into pieces (10 cm × 10 cm) and then weighed (Md). After that the samples were kept for one minute in distilled water. After being removed from

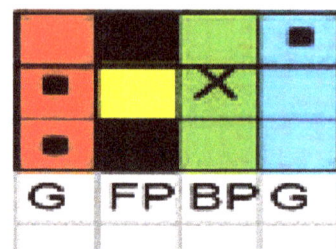

Figure 2: The weave repeat of the terry fabric (Basic 3-pick terry weave in 2:2 warp order [42].

the water they were hung for three minutes to remove excess water and the weight of the wet samples (M_w) was measured. An electronic balance was used in the weight measurements. The static water absorption (S_w) was calculated using the following formula:

$$S_w = (M_w - M_d) / M_d \times 100 \%$$ (21)

Also, the static water absorption or (liquid retention capacity) as a function of experimental overall porosity (ε) distilled water density (0.997 g/cm³), ρ_w and fibre density (ρ_f) can be given as [47].

$$S_l = \frac{\rho_w}{\rho_f}\left(\frac{\varepsilon}{1-\varepsilon}\right)$$ (22)

The proposed mechanism by Hsieh of water absorption depends on pore size pore size distribution pore connectivity and total pore volume.

The fabric's thickness (t) and surface density (M) was measured in accordance with ISO 5084: 1996 [48] and LST ISO 3801:1998 [49] respectively. Fabric properties were given in Table 2.

Results and Discussion

The results of overall porosity and static water absorption at different pile heights picks per cm and weft yarn counts are presented in tabulated form in Table 2. It could be seen from this table that the value of both overall porosity and static water absorption varies very rapidly from one variant to the other. In order to predict the overall porosity values for terry woven fabrics the values of pile loop length (L) and yarn density (ρ_y) are calculated And from a knowledge of yarn weft count (N_{mf}) picks per cm(n_2) yarns crimp (c_1, c_2) and fabric thickness (t) the volume of pile yarns (V_p) and volume of ground warp yarns (V_g) and volume of weft yarns (V_f) per unit volume (cuboid) as

a new developed equations were calculated and expressed in part of the theoretical model. From Equation (8) it could also be seen that the average overall porosity is affected by the pile height weft yarn diameter and picks spacing. Table 2 gives the calculated and experimental values of overall porosity.

When Table 1 was examined it was seen that fabric sample No. (3) has the smallest fabric weight weft density terry (pile) ratio weft diameter and number of piles per unit area. So it is possible that fabric No. (3) has the highest overall porosity through the interfibre volume of pile yarns (Vp) ground warp yarns (Vg) and weft yarns (V$_w$). On the contrary, Sample No. (2) has the smallest overall porosity and static water absorption percentage. However in this study since the volume of air voids between the yarns and through the interstices in the fibres was calculated theoretical values of overall porosity were obtained very close to the experimental results as shown in Figure 3.

The match is close which is also indicated by the high values of correlation coefficients and R^2 was obtained from the statistical analysis. Figure 4 shows the correlation plots between the predicted and measured values (X- axis – measured – values y- axis – predicted values). In Figure 4 the fabric overall porosity is compared with the corresponding fabric overall porosity calculated from the fitted Equation (19) to the different tested fabrics. The values of overall porosity were correlated well (R=0.9915) with a slope of 1.325 and an intercept of 0.275.

Static water absorption in terry woven fabrics is related to overall porosity. As discussed earlier variability in overall porosity affected the static water absorption values. Also as demonstrated in Figure 5 there is a positive linear relationship between the predicted overall porosity and static water absorption. Thus the higher overall porosity the higher

Sample no	Fabric Weight g/m²	Fabric thickness at 10 g/cm², mm	No of picks/cm	Weft yarn count				Pile * (terry) ratio	Pile ** height, mm	No of piles /cm² for each face	Yarn crimp	
				Ne	Nm	Tex	Den				Wrap c₁	Wrap c₂
1	337	12300	10	12	20.28	46.2	442.9	3.545	5.318	20	5.74	6.53
2	366	0.7629	20	12	20.28	46.2	442.9	2.645	1.984	42	8.03	7.92
3	266	11371	10	20	33.902	29.525	265.75	2.604	3.906	20	4.72	8.49
4	366	11897	15	12	20.28	49.2	442.9	4.253	4.253	30	5.14	6.48
5	299	12394	10	16	27.04	36.906	332.19	3.113	4.670	20	5.93	5.39
6	361	12434	15	16	27.04	36.906	332.19	3.545	3.545	20	5.88	5.71

*Pile ratio= length of pile yarn/metre of towel [47].

$$**\text{Pile height} = \left[\frac{pile\ ratio}{pick/cm}\right] \times \frac{1}{2} \times type\ of\ terry, cm$$

Table 1: Fabric Specification of terry woven fabrics.

Sample No	Fabric thickness at 10 g/ cm², cm	No of picks/ cm	volume of fabric cuboid cm³	Loop length of pile yarn metre ×10⁵	Weft yarn diametre, mm	Weft yarn bulk density, g/cm³	calculated, cm⁵×10⁵				Overall porosity ( )		Static water absorption %	Liquid retention capacity SI
							V$_p$	V$_g$	V$_w$	Total	Theoritical	Experimental		
			Eq.(5)	Eq.(6)	Eq.(16)	Eq.(17)	Eq.(7)	Eq.(8)	Eq.(9)	Eq.(18)	Eq.(19)	Eq.(20)	Eq.(21)	Eq.(22)
1	0.13200	10	0.024600	2.34796	0.2603	0.924	1.80432	1.35429	1.13701	4.29562	0.82538	0.822	374.48	2.9897
2	0.07629	20	0.007629	1.25778	0.2603	0.924	0.96656	0.69181	1.15181	2.8107	0.63165	0.688	255.74	1.4276
3	0.11371	10	0.022742	2.30065	0.2016	0.9245	1.76796	1.34122	0.69404	3.80322	0.83277	0.848	374.06	3.6118
4	0.11897	15	0.015863	1.78694	0.2603	0.924	1.37320	0.89777	1.12574	3.39671	0.78587	0.800	345.58	2.5896
5	0.12394	10	0.024788	2.35289	0.2254	0.9249	1.80811	1.35672	0.84281	4.00764	0.83832	0.843	370.23	3.4762
6	0.12434	15	0.016579	1.82314	0.2254	0.9249	1.40101	0.90405	0.84537	3.15043	0.80997	0.811	379.78	2.7780

Table 2: Comparison of experimental result with the calculated values of overall porosity.

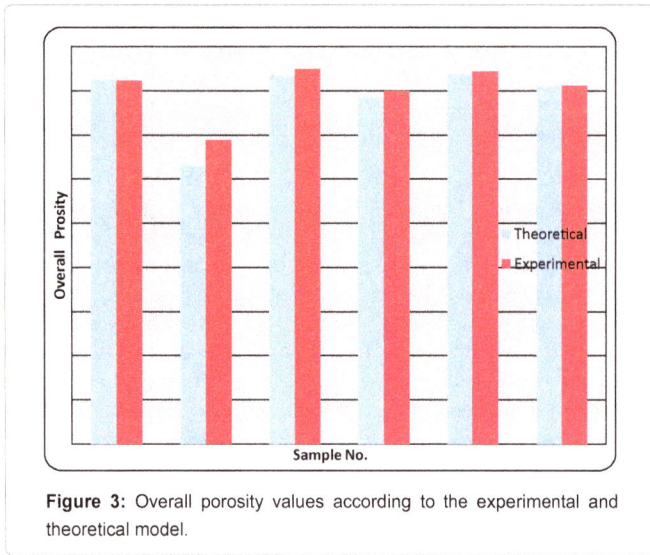

Figure 3: Overall porosity values according to the experimental and theoretical model.

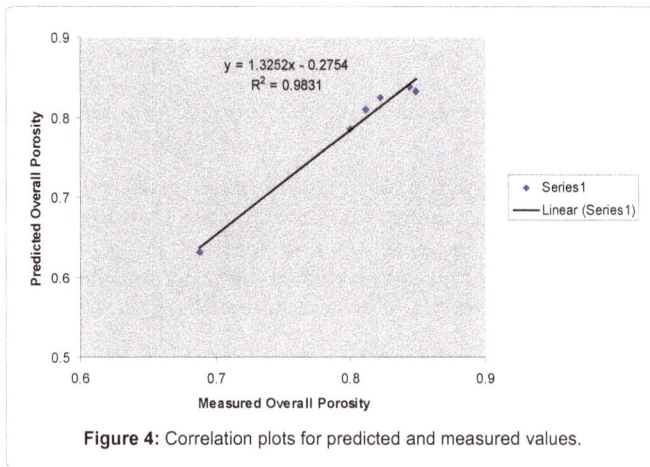

Figure 4: Correlation plots for predicted and measured values.

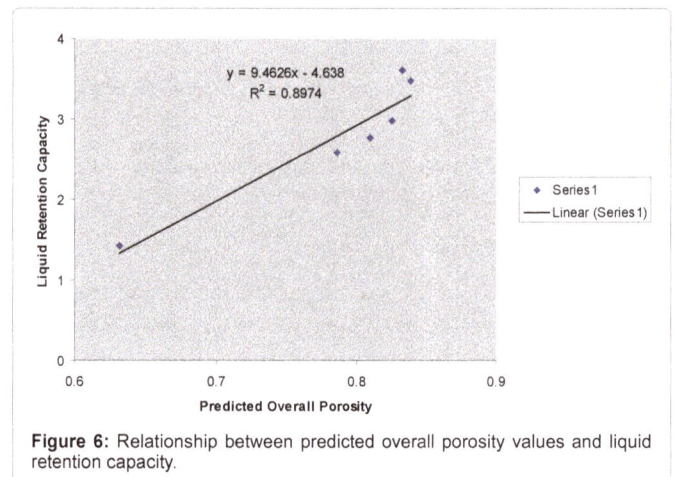

Figure 5: Relationship between predicted overall porosity values and static water absorption.

static water absorption.

Also the correlation between the calculated values of porosity obtained through the geometrical model and the experimental results of liquid retention capacity obtained through testing show high (R) value as shown in Figure 6. This indicates the functionality of the model for predicting the capillary flow of liquid through terry fabrics

in terms of fibre and yarn structure parameters. The results obtained through the theoretical model are somewhat overestimated compared the experimentally obtained results, which can be related to the approximation made to the yarn thickness (yarn diameter) based on an average diameter value. But in the theoretical approach the circular yarn diameter (dy) must be calculated using the major and minor diameter of the yarn at constancy yarn perimeter. Nevertheless the experimental and theoretical values are of the same ranking order.

Conclusion

In the case of calculating the overall porosity of terry woven fabrics air voids volume between not only yarns but also fibres must be considered. Thus in this study air voids volume or volume of yarns and fibres within the fabric and yarns respectively constituting the yarn structure were calculated theoretically and it was tried to determine the overall porosity of terry woven fabrics.

Due to differences between ideal and real geometry and random variation of fabric structure there are not exactly dependences between experimental overall porosity and predicted overall porosity values. However closeness of the results of the predictions based on the calculated values from the theoretical model and experimental values show that our model can be used successfully for the prediction of overall porosity for different terry woven fabrics. This model is practically simple and efficient.

The overall porosity and static water absorption are strongly related to each other. If a fabric has very high porosity it can by assumed that it can absorb large amount of water and wet easily. Because it was found that nearly positive linear relationship exists between overall porosity and static water absorption values (R=0.9824).

It could be assumed that this developed model (overall porosity) is applicable for predicting static water absorption of other woven types produced with different fibre types.

Before manufacturing theoretical model is aimed to establish and predict the value of overall porosity via static water absorption and some fabric properties.

The overall porosity was calculated for a geometrical model of terry woven fabric by means of knowing its geometrical parameters. The comparison between both calculated and measured values was then possible. The good agreement shown by Table 2 and Figure 4 supports the validity of the derived Equation (11) at least within the range of

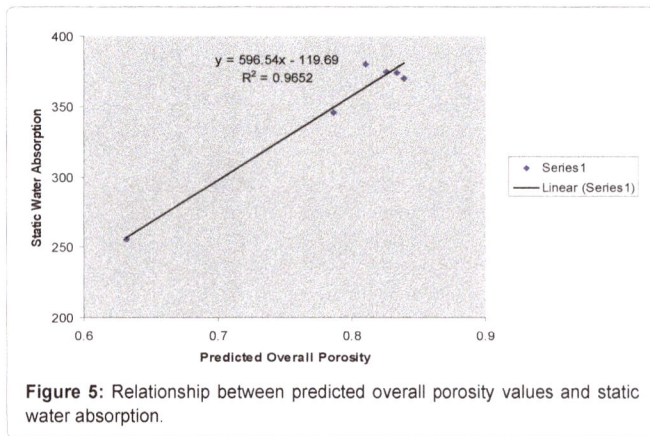

Figure 6: Relationship between predicted overall porosity values and liquid retention capacity.

fabric specification used. From such study a good prediction of fabric overall porosity could be calculated with Knowledge of fabric thickness yarn count warp and weft spacing and pile height.

Acknowledgement

We would like to acknowledge Eng. Mahmoud Mohamed El-Fowaty Chairman of the Board of Directors of (El-Fowaty Tex) Company in Aga Egypt for his assistance in manufacturing and supplying the terry woven fabrics used in the experiments.

References

1. Mavruz S, Ogulata RT (2009) Investigation and statistical prediction of air permeability of cotton knitted fabrics. Textil ve Konfeksiyon 19: 29-38

2. Chatterjee PK (2011) Absorbency Textile Science and Technology : Absorbency 7 New York : Elsevier

3. Haghi AK. (Heat and Mass Transfer in Textiles), WSEAS Press.

4. Ciukas R, Abramaviciute J (2010) Investigation of the air permeability of socks knitted from yarns with peculiar properties. Fibres and Textiles in Eastern Europe 78: 84-88.

5. Cay A , Vassiliadis S, Rangoussi M, Tarakcioglu I (2004) On the use of image processing techniques for the estimation of the porosity of textile fabrics. International Journal of Signal Processing 1: 51-54.

6. Cay A, Tarakcioglu I (2008) Relation between fabric porosity and vacuum extraction efficiency: Energy issues. Journal of the Textile Institute 99: 499-504.

7. Ogulata RT, Mavruz S (2010) Investigation of porosity and air permeability values of plain knitted fabrics. Fibres and Textiles in Eastern Europe 18: 71-75.

8. Matteson MJ, Orr C (1987) Filtration – Principles and Practices, Marcel Dekker New York.

9. Piekaar HW, Clarenburg LA (1967) Aerosol Filters – Pore size distribution in fibrous filters .Chemical Engineering Science 22: 1399-1408.

10. Dubrovski DP, Brezocnik M (2002) Using Genetic Programming to Predict the Macroporosity of Woven Cotton Fabrics. Text Res J 72: 187-194.

11. Dosmar M, Wolber P, Bracht K, Weibel P (1993) A new in – place integrity test for hydrophobic membrane filters. Filtration and Separation 30: 305-308.

12. Rucinski K, Caronia A, McNeil R (1986) Filter media characterization by mercury intrusion. Tappi Journal 23: 121-123.

13. Rebenfeld L , Miller B (1995) Research Approach to Quantification of Flow Behaviour in Non – wovens, J Text Inst 86: 241-251.

14. Miller B, Tyomkin I (1986) An extended Range liquid extrusion Method of Determining Pore Size Distributions. Text Res J 56: 35-40.

15. Miller B, Tyomkin I (1986) Methods for Determining Pore Size Distributions In Non – woven Materials. Proceedings of Tappi Nonwoven Conference.

16. Rushton A, Green DJ (1968) The Analysis of of Textile Filter Media . Filtration and Separation 6: 516-522.

17. Hssenboehler CB (1984) A New Method for Pore Structure Analysis Using Air Flow. Text Res J 54: 252-261.

18. Peirce FT, Womerskey JR (1978) Textile Institute, Manchester, England.

19. Love L (1954) Graphical Relationships in Cloth Geometry for plain Twill and Sateen Weaves. Tex Res J 24: 1073-1083.

20. Kemp A (1958) An Extension of Peirce's loth Geometry to the Treatment of Non – circular Threads. J Textile Inst 49: 44.

21. Hamilton JB (1964) A General System of Woven – Fabric Geometry. J Textile Inst 55: 66-82.

22. Weiner LI (1971) Textile Fabric Design Tables. Technomatic Publishing Co Inc Stamford, CN.

23. Seyam A, El-Shiekh A (1993) Mechanics of Woven Fabrics Part III: Critical Review of Weavability Limit studies. Text Res J 63: 371-378.

24. Gee NC (1953) Cloth Setting and Setting Theories. Textile Manu 80: 399-401.

25. Brierley S (1931) Theory and Practise of Cloth Setting. Textile Manu 58: 47-49.

26. Dubrovski PD (2000) Volume porosity of woven fabrics. Text Res J 70: 915-919.

27. Dubrovski PD (2000) A geometrical model to predict the macroporosity of woven fabrics. J Text Inst 92: 288-298.

28. Militky J, Vik M, Vikova M, Kremenakova D (2010) Influence of fabric construction on their porosity and permeability.

29. Dubrovski PD (2002) Using gentic programming to predict the macroporosity of cotton woven fabrics. Text Res J 72: 187-194

30. Turam RB, Okur A, Deveci R, Acikel M (2012) Predicting the inter – yarn porosity by image analysis method. Text Res J 82: 1720-1728.

31. Robertson AF (1950) Air Porosity of open weave fabric. Text Res J.

32. Dimitrovski K (1996) new method for determination of porosity in textiles.

33. Gooijer H, Warmoeskerken MMCG, Wassink JG (2003) Flow Resistance of Textile Material, Part I: Monofilament Fabrics. Text Res J 73: 437-443.

34. Szosland J (1999) Identification of structure of Inter – thread channels in model of woven Fabrics. Fibres and Textiles in Eastern Europe 7: 41-445.

35. Szosland J (2003) Modelling the structure barrier abitity of woven fabrics. Autex Res J 2: 102-110.

36. Cay A, Atav R, Duran K (2007) Effects of warp – weft density variation and fabric porosity of the cotton fabric on their colour in reactive dyeing. Fibres & Textiles in Eastern Europe 15: 91-94.

37. Kulichenko AV (2005) Theoretical analysis calculation and prediction of the air permeability of textiles. Fibre Chemistry 37: 371-379.

38. Ogulate RT (2006) Air permeability of woven fabrics. Journal of Textile and Apparel Technology and Management 5: 1-10

39. Ogulate RT, Koc E (1999) Determination of air permeability of woven fabrics. Tekstil and Teknik 175: 50-52.

40. Elnashar EA (2005) Volume Porosity and Permeabilty in Double – Layer Woven Fabrics. AUTEX Research Journal 5: 207-2018.

41. Xu G, Wang F (2005) Prediction of the permeability of woven fabrics. Journal of Industrial Textiles 34 : 243-254.

42. Terry weaving, terry–towel.pdf.

43. Blinov I, Belay S (1988) Design of Woven Fabrics. Mir Publishers Moscow

44. (1982) Textile Materials: Textile Industry Publishing House, Beijing China.

45. Hsieh YL (1995) Liquid Transport in fabric structures. Text Res J 65: 299-307.

46. Karahan M, Eren R (2006) Experimental Investigation of the Effect of Fabric Parameters on Static Water Absorption in Terry Fabrics. Fibres and Textiles in Eastern Europe 14: 159-163.

47. Textiles-Determination of thickness of Textiles and Textile Products.

48. Textiles-Woven Fabrics-Determination of Mass per unit Length and Mass per unit Area.

49. Hsiesh YL (1995) Text Res J 65: 299.

20

Effect of Crêpe Texture on Tensile Properties of Cotton Fabric under Varied Relative Humidity

Htike HH, Kang J, Yokura H and Sukigara S*

Kyoto Institute of Technology, Kyoto, Japan

Abstract

The physical properties of cotton crêpe fabrics with highly twisted weft yarn (2200 twists per meter) were evaluated after special embossed finishing to generate piqué. The effect of piqué was investigated as the main factor affecting the tensile properties of the sample fabrics under varied relative humidity. First, the mechanical and surface properties were measured using the Kawabata Evaluation System for Fabrics under standard room conditions, and then, the other physical properties were examined. The existence of piqué in the crêpe texture significantly changed the samples' tensile properties, such as tensile energy and extensibility at maximum applied load (EM) in the weft direction. Additional tensile measurements were carried out under varied relative humidity (10-90% RH) at 25°C. The crêpe fabrics with piqué had higher EM and residual strain values than the non-piqué samples under all RH conditions. In particular, the effect of piqué on EM became obvious at 90% RH.

Keywords: Crêpe texture; Tensile properties; Relative humidity; Embossed finishing

Introduction

An advantage of textiles is that their mechanical properties can be modified by varying the constituent yarns and weave structure. Crêpe fabrics made from highly twisted yarns exhibit a variety of wrinkly structures. Such fabrics are produced by using (1) hard-twist filling yarns, (2) chemical treatment, (3) crêpe weaves, and (4) embossing [1]. Japanese cotton crêpe, called *Takashima chijimi*, is made by employing twisted yarns in both the warp and the weft; however, the weft yarn twist is more than 2000 twists per meter. Such cotton fabrics, which combine hard-twist filling yarns and weave density, are popular materials for men's summer wear. Consumer demand for more attractive crêpe fabrics and new products signifies a shift in preference from casual cotton wear to more formal outerwear, women's dresses, and summer jackets. Accordingly, uniformly ribbed crêpe texture has been designed via additional embossed finishing of basic crêpe weave fabrics with piqué. The number of uniform ribs in piqué crêpe fabrics may govern not only the design but also the physical properties of the fabrics.

From the perspective of fabric weave design, the fundamental relationships among crêpe appearance, yarn twist level, yarn count, and finishing have been reported [2-6]. These findings are useful for achieving the required crinkly design with respect to the effect of twisted yarn shrinkage on fabric crinkling after finishing [7,8]. Yokura et al. tried to establish the silhouette and handle design of silk chirimen fabrics for women's thin dresses on the basis of the mechanical properties of these fabrics [9]. In previous reports, the mechanical properties of crêpe fabrics have been measured under standard temperature and humidity conditions (e.g., around 20 ± 3°C and 65% RH). However, cotton fabrics are moisture-sensitive materials; thus, the effect of environmental humidity cannot be ignored. One advantage of crêpe fabrics is their high extensibility along the weft direction. In addition, it is necessary to investigate the effect of relative humidity on the tensile deformation of crêpe fabrics used in attire that is comfortable for hot and humid weather. In this study, the surface and mechanical properties of cotton crêpe fabrics were measured using the Kawabata Evaluation System for Fabrics (KES-FB) in order to investigate the most distinctive characteristics of these fabrics. Then, the tensile extensibility and recovery of the fabrics were measured under varied relative humidity conditions. The objective of this study was to determine the influence of relative humidity on the tensile properties of piqué crêpe fabrics.

Materials and Methods

Samples

Crêpe appearance is governed by the yarn twist level and finishing. A plain weave gray fabric (S1,40Ne:147.64dtex) was finished by four different processes in order to change its appearance, as shown in Figure 1. Three samples (S1-2, S1-3, and S1-4) were processed with an embossing roller, and two of them (S1-2 and S1-3) were additionally accompanied with piqué. S1-5 was treated via normal finishing without embossing. Plain weave fabric (S0) using the same warp yarn as the crêpe fabric was also produced for comparison. The specifications of the samples are listed in Table 1.

Figure 2 shows microscopic images of the fabrics captured using a 3D microscope (VR-3000 series, Keyence). Traces along the warp and weft yarn directions as well as those between the yarns are shown in this figure. The images of samples S1-4 and S1-5 show a randomly bumpy appearance, whereas uniformly ribbed structures are seen for samples S1-2 and S1-3.

Measurement of physical properties of samples

The physical properties of the samples were examined at 25 ± 2°C and 50–60% RH by using KES-FB testers (Kato Tech Co., Ltd.). The specific characteristic values, measurement conditions, and KES-FB testers are listed in Table 2. The properties of standard-size samples (20 cm × 20 cm) were measured three times in the warp direction and three times in the weft directions with a different position used for each measurement.

*Corresponding author: Sachiko Sukigara, Kyoto Institute of Technology, Kyoto, Japan, E-mail: sukigara@kit.ac.jp

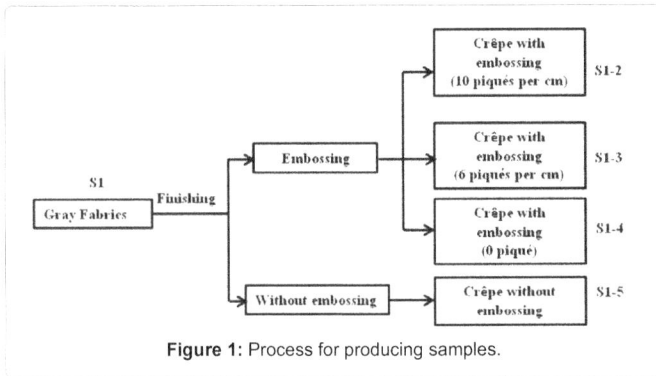

Figure 1: Process for producing samples.

Sample no.	Structure	Weave density/cm		Weight (mg/cm²)	Thickness (mm)
		Ends	Picks		
S0	Plain	38	29	9.91	0.46
S1	Gray	27	21	8.09	0.49
No. of piqué per cm					
S1-2	Crêpe with embossing	10	40	22	10.64	0.93
S1-3	Crêpe with embossing	6	36	22	9.85	0.91
S1-4	Crêpe with embossing	0	34	22	9.71	0.76
S1-5	Crêpe without embossing		32	22	9.67	0.97

Yarn twist - S0: 1000 twists per meter (warp and weft)
S1~S1-5: warp, 1000 twists per meter; weft, 2200 twists per meter
Yarn linear density - warp, weft: 147.64 dtex

Table 1: Specifications of fabric samples.

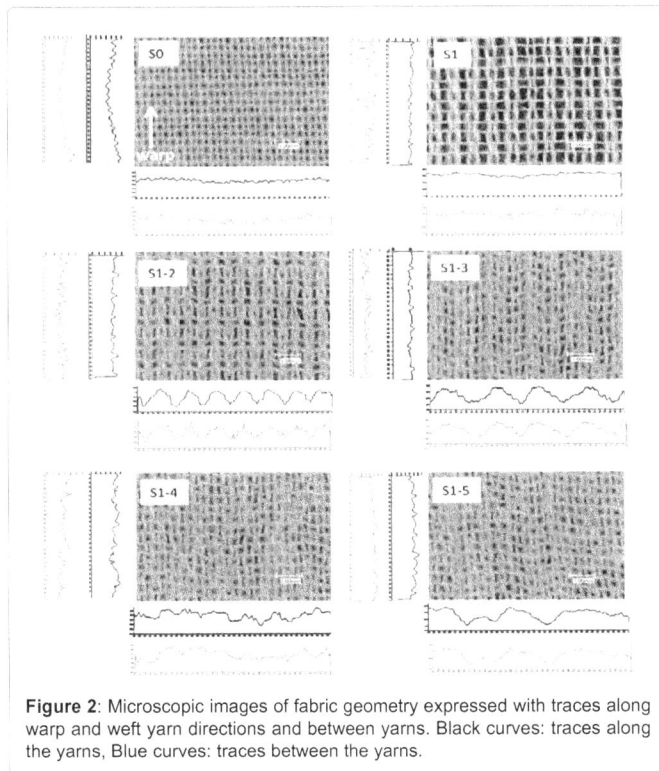

Figure 2: Microscopic images of fabric geometry expressed with traces along warp and weft yarn directions and between yarns. Black curves: traces along the yarns, Blue curves: traces between the yarns.

Fabric tensile test

Figure 3 shows a schematic of the sample preparation and conditioning for the tensile test. First, the prepared samples (1 cm × 6 cm) were dried in an infrared moisture determination balance (FD-720, Kett Electric Laboratory) at 100°C for 3 min. Then, the samples were transferred to a chamber with a precisely controlled humidity supply (SRG-10R-AS, Daiichikagaku, Japan) and conditioned for 10 min under 10% RH. The humidity was increased to preset values of 40%, 60%, 70%, 80%, and 90% RH at 25°C, under which the samples were conditioned for 1 h. For the wet condition, the samples were immersed in water at 25°C for 10 min and measurements were then carried out in the water. The tensile properties were measured using the KES-G1S (Kato Tech Co., Ltd.), which equipped with a humidity control chamber. Loading-unloading cyclic tests were carried out three times at a maximum tensile load of 49 N/m and tensile speed of 0.1 mm/s. The gauge length was 5 cm and three specimens per sample were measured under each RH condition.

Characteristic values such as extensibility at 49 N (EM, Percentage), tensile energy up to EM (WT, J/m²), resilience (RT, Percentage) and initial linearity (LT) were calculated following the KES system characterization [10].

Results and Discussion

Effect of crêpe appearance on mechanical and surface properties

Figure 4 compares the physical properties of the plain and crêpe fabrics. The horizontal axis represents the difference between the plain fabric (S0) and the crêpe fabrics as a percentage.

- **Tensile properties:** Extensibility (EM2) and tensile energy (WT2) values of all the crêpe fabrics in the weft direction were significantly larger than those in the warp direction. Furthermore, EM2 and WT2 of the crêpe fabrics were larger than those of the plain fabric.

Under stretching due to tensile force, the wavy parts of the crêpe fabrics were flattened. This phenomenon was more apparent in uniformly ribbed crêpe (i.e., crêpe with piqué structure). This made the piqué fabrics more extensible: the higher the number of piqués, the more extensible the fabric. S1-2 (10 piqués per centimeter) had larger EM2 and WT2 values than S1-3 (6 piqués per centimeter). The equidistant ribbed piqué structure of S1-2 and S1-3 made these fabrics extensible by over 400% as compared to S0. In addition, S1-2 and S1-3 showed larger EM2 and WT2 values than S1-4 and S1-5 (crêpe without piqué). Tensile resilience (RT) reflects the ability of a fabric to recover from tensile deformation. The RT2 values of S1-2, S1-3, S1-4, and S1-5 were 20.00%, 26.98%, 27.05%, and 23.76% lower than the RT2 value of S0, respectively. Thus, higher extensibility was accompanied by lower recovery.

- **Bending property:** The extensibility and bending rigidity of crêpe are influenced by the curvature of interlacing of the weft yarns on the warp yarns as well as by small crinkles on the base fabric. In the case of bending, differences between the piqué and non-piqué fabrics were observed in the warp direction. The warp bending rigidity (B1) and bending hysteresis (2HB1) values of S1-2 and S1-3 were larger than those of S1-4 and S1-5. When the piqué cords along the warp direction were bent, an additional bending moment was created. On the other hand, the waved crêpe fabric was easily bent in the weft direction as compared to the plain fabric. Bending hysteresis represents both the friction between yarns and viscoelastic behavior. The 2HB values of the

Properties	Symbol	Characteristic value	Unit	Measuring conditions	KES machines
Tensile	EM	Strain at maximum load	%	Strip biaxial deformation Maximum load: 49 N/m	KES-FB1
	LT	Linearity	none		
	WT	Tensile energy	J/m²	Speed: 0.1 mm/s	
	RT	Resilience	%		
Bending	B	Bending rigidity	µNm	Pure bending Maximum curvature K: ± 250 m⁻¹	KES-FB2-S
	2HB	Hysteresis of bending moment	mN		
Shearing	G	Shear stiffness	N/m	Shear deformation under constant tension of 9.8 N/m	KES-FB1
	2HG	Hysteresis of shear force at 8.7 mrad	N/m		
	2HG5	Hysteresis of shear force at 87 mrad	N/m		
Compression	LC	Linearity	none	Maximum pressure: 0.98 kPa Rate of compression: 20 µm/s	KES-G5
	WC	Compression energy	J/m²		
	RC	Resilience	%		
Surface	MIU	Coefficient of friction	none	Ten steel piano wires (diameter of a wire: 0.5 mm, length: 1 cm) Contact force: 0.49 N	KES-SE-STP
	MMD	Mean deviation of MIU	J/m²		
	SMD	Geometrical roughness	µm	A steel piano wire (diameter: 0.5 mm, length: 5 mm) Contact force: 0.1 N	KES-SE
Air Resistance	R	Air resistance	kPa·s/m		KES-F8-AP1
Thickness	T0	Thickness at 49 Pa	mm		
Weight	W	Weight per unit area	g/m²		

Table 2: Physical properties of fabrics measured with the KES-FB system.

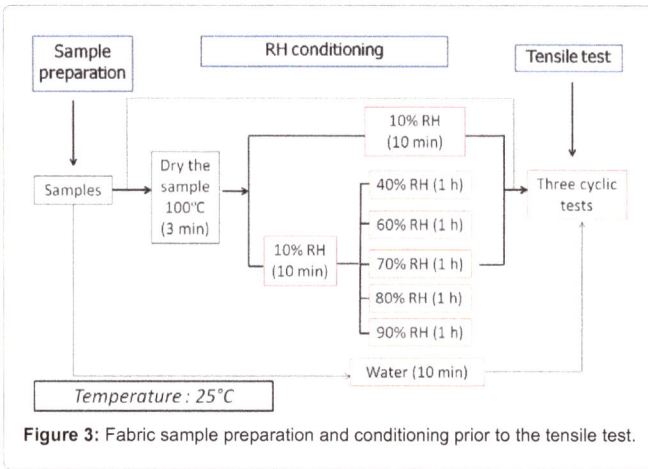

Figure 3: Fabric sample preparation and conditioning prior to the tensile test.

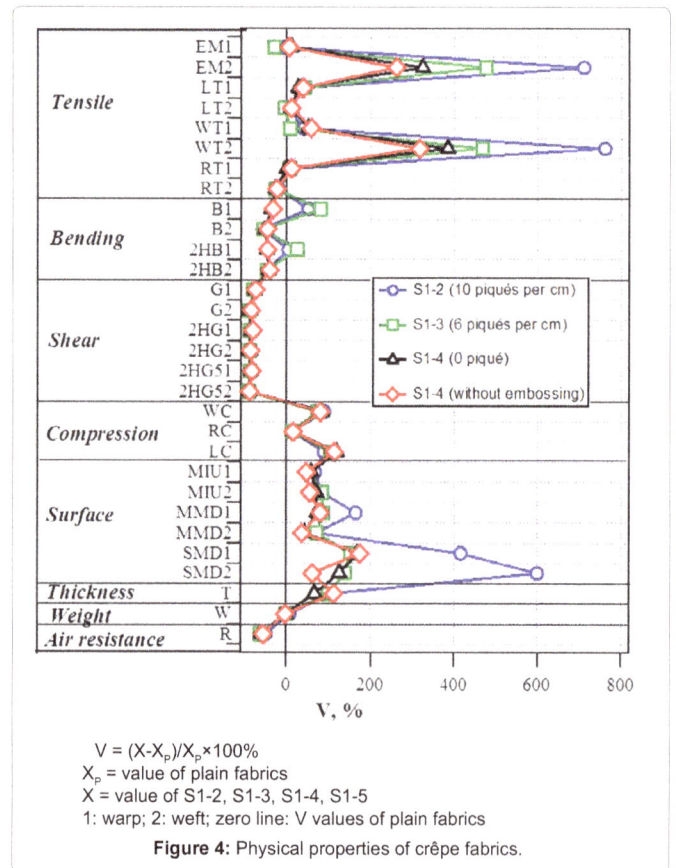

$V = (X-X_p)/X_p \times 100\%$
X_p = value of plain fabrics
X = value of S1-2, S1-3, S1-4, S1-5
1: warp; 2: weft; zero line: V values of plain fabrics

Figure 4: Physical properties of crêpe fabrics.

crêpe and piqué crêpe structures were the same as or smaller than those of the plain fabric.

- **Shear and compression properties:** The shear stiffness and hysteresis of all the crêpe fabrics were smaller than those of the plain fabric. Furthermore, the thickness of the crêpe fabrics was greater than that of the plain fabric because of the bumpy appearance of the crêpe fabrics, as shown in Figure 2. This geometry appeared because of the larger compression energy (WC) of the crêpe fabrics.

- **Surface property:** The surface property values of all the crêpe samples were larger than those of the plain samples in both the warp and the weft directions. In particular, the surface roughness values of S1-2 (SMD1 and SMD2) were notably larger than those of the other three fabric samples. The surface roughness was measured under a contact force of 0.1 N; thus fine piqué shape was maintained, and the piqué fabrics showed larger SMD values than the other fabrics.

- **Air resistance:** The air resistance of all the crêpe fabrics was smaller than that of the plain fabrics. Greater spacing was observed between the yarns of the crêpe fabrics, as shown in Figure 2. The average air resistance values of S1-2, S1-3, and S1-4 were 0.083, 0.063, and 0.066 kPa.s/m, respectively. Air resistance increased as the piqué became finer.

Effect of relative humidity on fabric tensile properties

As shown in Figure 4, among all the mechanical properties, tensile properties of the samples showed the largest differences. Therefore, the effect of relative humidity on tensile properties was investigated further. The tensile extensibility (EM) in the weft direction of the crêpe fabrics was larger than that in the warp direction. The effect of humidity on EM was measured in both the warp and the weft directions under 10% RH and 90% RH. The difference between the EM values under 10% RH and 90% RH in the warp direction was in the range of 0.2–0.8%; thus, the effect of humidity was small. The subsequent discussion focuses on the fabric tensile properties in the weft direction.

Figure 5 shows the force-extension curves in the weft direction of S1-5 under 90% RH. The total energy loss in the first cycle curve is comparatively larger than that in the other curves. The difference between the second and third curves is small. This tendency was observed for all the samples under all the humidity conditions. After the first tensile curve, the residual strain was defined as $\Delta\varepsilon$ in order to evaluate the irreversible strain after the extension, because once the crêpe fabrics were extended, the subsequent deformation would be small. This tendency was observed for all the samples.

In this study, all the samples were made from the same yarns of the same cotton fibers. Thus, the main source of the differences in tensile properties was the surface geometry of the fabrics.

Figure 6 shows the average EM-1 values of three specimens of each sample fabric plotted against relative humidity (10–90% RH). The extensibility of the samples increased with relative humidity because of the nature of cotton fiber [11]. The EM-1 values of S1-2 (10 piqués per centimeter) and S1-3 (6 piqués per centimeter) were larger than those of S1-4 and S1-5 (without piqué) under all the RH conditions, mainly because the piqué shape increased the extensibility. According to the surface trace in Figure 2, the piqué height of S1-2 was the same as that of S1-3, but the trace length of S1-2 was greater than that of S1-3. In Figure 6, the difference between these two samples (S1-2 and S1-3) is clearer over 60% RH. In our previous study, the EM values of the same yarn used for weft markedly increased over 70% RH. A similar tendency was observed for sample S1-2.

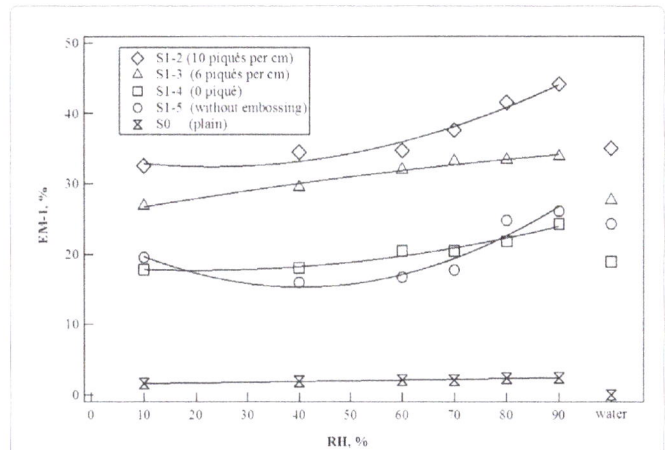

Figure 6: Effect of RH on first-cycle tensile extensibility (EM-1) of the fabric samples.

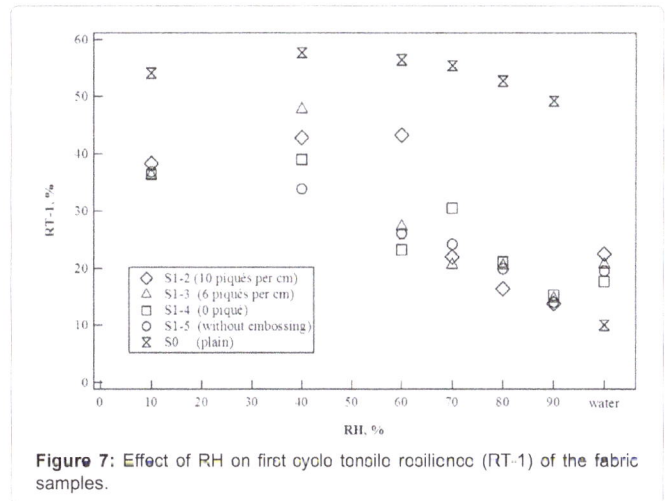

Figure 7: Effect of RH on first cycle tensile resilience (RT-1) of the fabric samples.

In water, the EM-1 values of all the samples were smaller than those under 10% RH. Water penetrating into the fabric might increase the friction between the yarns and make the fabrics less extensible.

The relationship between RH and RT-1 is shown in Figure 7. The tensile recovery shows an inverse trend with relative humidity. Significant changes in RT-1 were observed in the range of 40% -70% RH at 25 °C. The differences among the crêpe samples were greater in this humidity range than outside it. In water, the RT-1 values of S1-2 and S1-3 were the same as those of S1-4 and S1-5, whereas their EM values were larger. Although the large extensibility of piqué crêpe fabrics (S1-2 and S1-3) is beneficial, their poor recovery must be considered when using these fabrics for clothing. In light of the irreversible recovery, the residual strain was investigated.

Figure 8 shows that the $\Delta\varepsilon$ values of the piqué fabrics (S1-2 and S1-3) were larger than those of the non-piqué fabrics (S1-4 and S1-5). This difference became considerably larger when the RH was increased from 70% to 90%. Additional piqué seemed to increase the extensibility and maintain the recovery below 70% RH. Table 3 lists the values of residual strain against tensile extension under different RH conditions for all the fabrics. The values of $\Delta\varepsilon/(EM\text{-}1)$ for the crêpe fabrics at 90% RH were in the range of 0.5–0.6, indicating that the recovery of piqué decreases as humidity increases. From the viewpoint of clothing that allows free

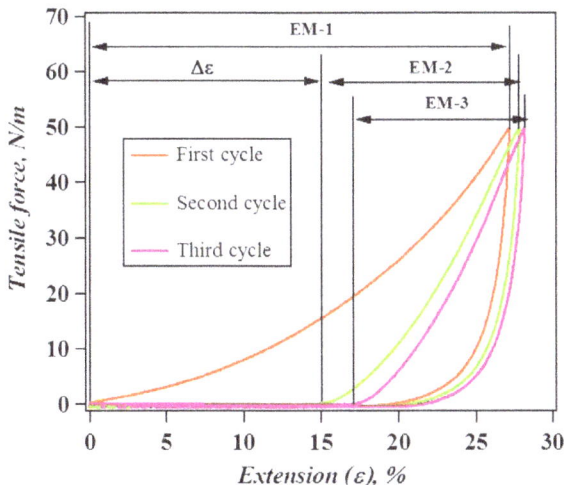

Figure 5: Tensile load vs. extension curves of S1-5 at 90% RH, 25°C (EM-1: First-cycle extensibility; EM-2: Second-cycle extensibility; EM-3: Third-cycle extensibility; $\Delta\varepsilon$: residual strain).

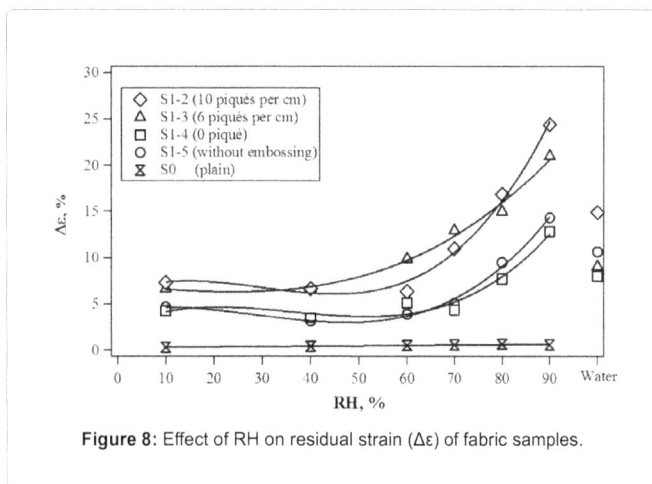

Figure 8: Effect of RH on residual strain ($\Delta\varepsilon$) of fabric samples.

Humidity, RH, %	$\Delta\varepsilon/$(EM-1)				
	S0	S1-2	S1-3	S1-4	S1-5
10	0.18	0.22	0.25	0.24	0.24
40	0.21	0.19	0.23	0.19	0.20
60	0.24	0.18	0.31	0.25	0.23
70	0.27	0.29	0.39	0.22	0.29
80	0.27	0.41	0.45	0.35	0.38
90	0.24	0.55	0.62	0.53	0.55

Table 3: Residual strain $\Delta\varepsilon$ against maximum tensile extension EM-1.

body movement, high fabric extensibility is attractive. The extensibility of S1-2 was relatively stable in the humidity range below 70% RH. Thus, except under very high humidity, the extensibility of crêpe fabric with piqué can be used in designing new fabrics for women's clothing.

Conclusion

In this study, four crêpe fabrics (three embossed and one non-embossed) were used to investigate the effect of piqué and crinkle on tensile extensibility, resilience, and residual strain under varied relative humidity and in water at 25°C.

The results of KES-FB measurements in a standard environment showed that the piqué appearance governs some physical properties, namely, EM2, WT2, B1, and 2HB1. The shear stiffness and hysteresis values of the crêpe fabrics were smaller than those of the plain fabric, but no difference was observed between the piqué and non-piqué fabrics. The most notable difference between the piqué crêpe and the non-

piqué fabrics was their tensile extensibility. The effect of environmental humidity on the tensile properties of the fabrics was investigated, especially in the weft direction.

Piqué crêpe fabrics (S1-2:10 piqués per centimeter, S1-3: 6 piqués per centimeter) showed larger tensile extensibility (EM-1) and residual strain ($\Delta\varepsilon$) than crinkled fabrics without piqué (S1-4) under all RH conditions. Even at 90% RH, the piqué structure was maintained, which indicates the high extensibility of piqué fabrics as compared to crinkled fabrics without piqué. However, the tensile resilience of the piqué fabrics decreased and their residual strain increased over 80% RH. In conclusion, piqué crêpe is applicable to new fabrics for women's clothing, except under very high humidity.

Acknowledgement

We would like to express our gratitude to Mr. Shiro Takahashi, Takahashi Textile Co. Ltd., Shiga, Japan for providing the fabric samples used in this study. This work was supported by JSPS KAKENHI Grant Number 24220012 and 15H01764.

References

1. Celanese A (2001) Complete Textile Glossary. Celanese Acetate LLC, New York.

2. Yamashita S, Takatera M, Shinohara A (1997) Influence of fabric structure and twisted yarn on cotton crepe design part 2: Effect of weaving condition on shrinkage, rib shapes and number of ribs. Journal of the Textile Machinery Society of Japan 50: T155-163.

3. Ishikura H, Kase S, Nakajima M (1991) Study on crinkle design of crepe part 1: Crinkling mechanism of crepe. Textile Machinery Society of Japan Transactions 44: 62-69.

4. Ishikura H, Kase S, Nakajima M (1992) Study on crinkle design of crepe part 2: Analysis of the phase difference between contiguous hard twisted yarn in crepe. Journal of the Textile Machinery Society of Japan 45: T49-T58.

5. Ishikura H, Kase S, Nakajima M (1992) Study on Crinkle Design of Crepe Part 3: Experiment on the helix numbers in crepe fabrics. Journal of the Textile Machinery Society of Japan 45: T65-T70.

6. Ishikura H, Yang L, Kase S, Nakajima M (1992) Study on crinkle design of crepe part 4: Experiment on the helix numbers in crepe fabrics. Journal of the Textile Machinery Society of Japan 45: T154-T164.

7. Yang XH, Li DG (2007) Evaluation and control principle of the crepe effect on fabrics. Textile Research Journal 77: 779-784.

8. Ishikura H (1988) Study on Qualitative Improvement on Crepe summer Wear. Journal of the Textile Machinery Society of Japan 41: T169-T176.

9. Yokura H, Endo S, Sukigara S (2013) Silhouette and handle design of silk chirimen fabrics for women's thin dress based on the mechanical properties. Journal of Textile Engineering 59: 133-139.

10. Raheel M (1995) Modern textile characterization methods: Objective Measurement of Fabric Hand. (1st edn), Marcel Dekker Inc.

11. Morton WE, Hearle JWS (2008) Physical properties of textile fibers (4th edn). The Textile Institute, Wood head Publishing Limited, Cambridge, England.

21

Generation Mean Analysis of Fibre Quality Characters in Upland Cotton (*Gossypium hirsutum* L.)

Gawande NG*, Deosarkar DV and Kalyanker SV

Department of Agricultural Botany, Vasantrao Naik Marathwada Krishi Vidyapeeth, Parbhani, India

Abstract

Generation mean analysis was carried out to estimate the nature and magnitude of gene effects for fibre quality traits in three crosses of upland cotton (*Gossypium hirsutum* L). The presence of epistasis was reflected by scaling tests and inadequacy of simple additive-dominance model for most of the characters studied. The results obtained revealed that the nature and magnitude of gene effects differed in different crosses and showed importance of additive as well as non-additive gene effects in the inheritance of different characters. In view of the parallel role of additive and non-additive gene effects in the inheritance of different characters, selection in the segregating generations should be delayed when dominance gene effects would have diminished or sophisticated selection procedures as recurrent selection and population improvement programmes may be followed. However, additive gene effects may be fixed with respect to some specific traits such as micronair value and fibre strength.

Keywords: Generation mean; Scaling tests; Recurrent selection

Introduction

Cotton, which has been reputed as "Queen of the fibre plants" is an important fibre crop in India Cotton is an important fibre crop being used in the textile industry. Over 90% of cotton grown in the world is *Gossypium hirsutum* L. or upland cotton. It plays a key role in the national economy by way of its contribution in trade, industry, employment and foreign exchange earnings. The average productivity of cotton in India is the lowest among cotton growing nations of the world. In order to increase the yield potential, it is desirable to efficiently utilize the available genetic variability. Genetic analysis of quantitative traits further helps to elucidate the nature and magnitude of genetic variation present in the population. The estimates of gene effects in a plant improvement programme have a direct bearing upon the choice of breeding procedure to be followed. Additive gene effects are useful in the development of pure lines whereas dominance and epistatic effects can be used to exploit hybrid vigour. In upland cotton, various studies have been conducted to study the nature and magnitude of gene effects in the inheritance of different quantitative characters and involvement of both additive and non-additive gene effects have been reported by many workers [1,2]. In the present study, additive, dominance and epistatic gene effects were estimated by Generation Mean analysis for fibre quality traits in three crosses of upland cotton (*Gossypium hirsutum* L).

Materials and Methods

The experimental material comprised of three hybrids namely BN-1 × AC-738 (NHH-44), NH-615 × NH-625 (NHH-206), AK-32 × DHY 286-1 (AKH HY2). The three hybrids viz., NHH-44, NHH-206 and AKH HY-2 were selfed and backcrossed to their parents to get seeds of F2 and backcross generations viz., BC1 and BC2. Thus, six generations (P1, P2, F1, F2, BC1 and BC2) were developed for each of these three cross combination during *kharif* 2013-14. Each cross was grown in a separate experiment in a randomized block design with three replications during *kharif*, 2014-15 at experimental farm of Department of Agriculture Botany, VNMKV, Parbhani. The plants were spaced 90 cm between the rows and 30 cm within rows. All the recommended cultural practices were adopted to raise a healthy crop. Ten plants from each of parents and F1s, 20 plants from BC1 and BC2 generations and 50 plants from F2 populations in each replication were randomly selected for recording data for upper half mean length (mm), ginning (%),

micronaire value, fibre strength, uniformity index and fibre maturity coefficient. The scaling test (A, B and C) were calculated for each trait to detect adequacy of additive-dominance model or presence of non-allelic interaction according to Hayman and Mather [3]. The adequacy of additive-dominance model was tested by joint scaling test of Cavalli [4]. The six parameters (m, d, h, i, j and l) were computed according to Hayman [5].

Result and Discussion

The analysis of variance for all characters studied in three crosses of cotton is presented in Table 1. The analysis of variance revealed that the mean square due to crosses were significant for all the traits except for fibre maturity coefficient in all three crosses. The character which failed to show significant variation among the generation in respective crosses was not subjected to further genetic analysis of generation means. Mean data (Table 2) on various characters recorded on 6 generations viz., P1, P2, F1, F2, BC1 and BC2 for 3 cross combinations were subjected to scaling test and joint scaling test. Significance of at least one of the scaling tests (Table 3) revealed the presence of non-allelic interactions for all the traits in 3 crosses.

For upper half mean length (mm), the A, B, C, scaling tests were significant for all the crosses except C scale in C1 cross. The chi square values were significant for the all the crosses. The additive (d) gene effects were negatively significant in all crosses except C1. The dominance (h) gene effects were negatively significant in crosses C1 and C3. As regards, digenic interaction all the components recorded significant effect (except (j) component in C1 cross) with higher magnitude of values for (l) components. The dominance (h) and dominance × dominance (l)

*Corresponding author: Nitin G Gawande, Department of Agricultural Botany, Vasantrao Naik Marathwada Krishi Vidyapeeth, Parbhani, India
E-mail: nitingawande570@gmail.com

Character	Source of variation	d.f.	NHH-44	NHHH-206	AKH HY-2
Upper half mean length (mm)	Replication	1	0.77	0.09	0.07
	generation	5	2.52*	4.18**	6.85**
	Error	5	0.44	0.09	0.31
Ginning (%)	Replication	1	0.538	0.080	0.0919
	generation	5	0.885*	1.209*	1.092*
	Error	5	0.163	0.117	0.206
Micronair value	Replication	1	0.006	0.007	0.007
	generation	5	0.059*	0.071*	0.093**
	Error	5	0.005	0.012	0.008
Fibre strength	Replication	1	0.48	1.76	0.21
	generation	5	5.84**	9.08**	6.28*
	Error	5	0.34	0.38	0.68
Uniformity index	Replication	1	0.27	1.92	0.05
	generation	5	3.72**	12.31**	9.58**
	Error	5	0.21	0.85	0.16
Fibre maturity coefficient	Replication	1	0.00013	0.00013	0.00021
	generation	5	0.00021^{NS}	0.00016^{NS}	0.00011^{NS}
	Error	5	0.00013	0.00021	0.00011

* and ** significant at 5% and 1% respectively.

Table 1: Analysis of variance for crosses and generations for different characters in cotton.

Character	Cross	P1	P2	F1	F2	BC1	BC2	SE±	CD (5%)
Upper half mean length (mm)	NHH-44	22.24	22.11	24.57	23.42	21.67	21.88	0.66	1.70
	NHH-206	23.88	20.51	23.46	21.44	20.78	21.05	0.31	0.79
	AKH HY-2	23.74	23.26	25.74	24.15	20.23	22.38	0.56	1.43
Ginning (%)	NHH-44	33.43	34.15	35.29	34.17	34.77	34.92	0.404	1.039
	NHH-206	38.43	39.34	40.08	38.09	39.21	39.84	0.342	0.879
	AKH HY-2	33.92	33.57	35.28	33.31	34.05	34.74	0.454	1.166
Micronair value	NHH-44	3.24	3.29	3.72	3.51	3.42	3.45	0.068	0.174
	NHH-206	3.54	3.62	3.89	3.44	3.34	3.61	0.108	0.278
	AKH HY-2	3.79	3.64	4.2	3.66	3.94	3.7	0.088	0.226
Fibre strength	NHH-44	21.20	20.10	23.60	23.20	20.50	19.40	0.59	1.51
	NHH-206	24.60	21.70	25.50	23.80	19.60	23.10	0.62	1.58
	AKH HY-2	23.40	21.10	26.20	24.10	22.40	22.30	0.82	2.12
Uniformity index	NHH-44	75.50	73.70	75.30	76.30	72.90	73.40	0.45	1.17
	NHH-206	71.20	71.90	77.20	73.60	70.70	71.00	0.92	2.37
	AKH HY-2	73.20	75.30	77.10	76.10	71.40	72.80	0.40	1.02

* and ** significant at 5% and 1% respectively.

Table 2: Mean performance of parents, F1, F2 and back cross generations of three cotton crosses for fiber quality characters.

Parameter		C1 (NHH44)	C2 (NHH206)	C3 (AKH Hy- 2)
Scaling test	A	-3.470** ± 0.351	-5.780** ± 0.150	-9.020** ± 0.082
	B	-2.920** ± 0.166	-1.870** ± 0.196	-4.240** ± 0.284
	C	0.190 ± 0.390	-5.550** ± 0.306	-1.880** ± 0.291
Joint scaling test	X^2	1171.53**	17178.60**	10775.80**
Digenic model	m	23.420** ± 0.207	21.440** ± 0.145	24.150** ± 0.185
	d	-0.210 ± 0.190	-0.270** ± 0.012	-2.150** ± 0.138
	h	-4.185** ± 0.610	-0.835 ± 0.433	-9.140** ± 0.530
	I	-6.580** ± 0.082	-2.100** ± 0.144	-11.380** ± 0.123
	J	-0.275 ± 0.385	-1.955** ± 0.129	-2.390** ± 0.293
	l	12.970** ± 0.408	9.750** ± 0.398	24.640** ± 0.351
Type of epitasis		Duplicate	Duplicate	Duplicate

* and ** significant at 5% and 1% respectively.

Table 3: Scaling test, joint scaling test and estimation of gene effects by digenic model in respect of Upper half mean length (UHML).

components were in the opposite direction in all the crosses in dicated duplicate type of epistasis (Table 3). The ginning % indicated that the A, B, C scaling tests and chi-square values were significant in all the crosses. As regards main gene components, additive (d) component recorded significantly negative effect in all the crosses while dominance (h) component recorded positive significant effect in all the crosses. As regards digenic gene interaction, all the components exhibited significant effects with additive x additive (i) component exhibiting the positive direction and in considerable magnitude in all the crosses. The dominance (h) and dominance × dominance (l) components were in the opposite direction in all the crosses indicated duplicate type of epistasis (Table 4).

The chi- square values were significant for all the crosses under study for micronair value. All crosses have significant individual scaling tests (except C scale in C1 cross). The main effect and digenic interaction effects were also significant for all the crosses under study except (i) in C2 cross, (j) component in C1 cross and (i) component in C3 cross. The additive (d) and dominance (h) effects were in positive direction with magnitude of later was higher than former component in cross C3. The additive (d) effects were in negative direction and significant in cross C1 and C2 while in dominance (h) effect had higher magnitude and positive significant in cross C1 and C2. As regards to digenic interaction, additive × additive (i) component had positive effect in cross C2 and C3 except C1 crosses. The C1 and C2 crosses exhibited negative additive × dominance (j) effect. The dominance × dominance (l) components had positive significant effect in crosses C1 and C2 showed complimentary epistasis except C3 (Table 5). The chi-square values were significant in all the three crosses with significant A, B, and C scaling tests for fibre strength. The additive effects (d) were significant in cross C1 and C2 while dominance effect (h) were negatively significant in crosses C1 and C2. Among the epistasis gene interaction, additive × additive (i) components were negatively significant in all the three crosses. The interaction additive × dominance (j) effects were negatively significant in cross C2 and C3. The dominance × dominance (l) components was positively significant in all the crosses under study. The dominance (h) and dominance x dominance (l) components were in the opposite direction in all the crosses indicated duplicate type of epistasis (Table 6). Uniformity index recoded the scaling tests A, B, C and chi-square values significant an all the three crosses. The additive (d) effect and dominance (h) effects were in negative direction and significant in all the crosses except cross C2. Among the epistasis interaction additive x additive (i) component negatively significant in all three crosses. The (j) component negatively significant in cross C1 and had little I magnitude than other components. The dominance x dominance (l) component had higher magnitude while found to be positively significant in all the crosses. The (h) and (l) components were in the opposite direction in all the crosses indicated duplicate type of epistasis (Table 7).

Dominant gene effect appeared to be more important for upper half mean length (mm) (cross 2 and 3), ginning (%) (cross 1, 2 and 3), micronair value (cross 3), fibre strength (cross 1 and2) and uniformity index (cross 1, 2 and 3). The contribution of the parent to dominance

Parameter		C1 (NHH44)	C2 (NHH206)	C3 (AKH Hy- 2)
Scaling test	A	-0.120** ± 0.036	-0.750** ± 0.072	-0.110** ± 0.038
	B	-0.110* ± 0.048	-0.290** ± 0.068	-0.440** ± 0.039
	C	0.070 ± 0.053	-1.180** ± 0.053	-1.190** ± 0.070
Joint scaling test	X^2	5265.09**	2583.88**	988.50**
Digenic model	m	3.510** ± 0.047	3.440** ± 0.091	3.660** ± 0.044
	d	-0.030** ± 0.008	-0.270** ± 0.014	0.240** ± 0.017
	h	0.155 ± 0.142	0.450 ± 0.273	1.125** ± 0.112
	I	-0.300** ± 0.046	0.140 ± 0.090	0.640** ± 0.041
	J	-0.005 ± 0.048	-0.230** ± 0.093	0.165** ± 0.041
	I	0.530** ± 0.106	0.900* ± 0.187	-0.090 ± 0.083
Type of epitasis		Complimentary	Complimentary	Duplicate

* and ** significant at 5% and 1% respectively.

Table 5: Scaling test, joint scaling test and estimation of gene effects by digenic model in respect of micronair value.

Parameter		C1 (NHH44)	C2 (NHH206)	C3 (AKH Hy- 2)
Scaling test	A	-3.800** ± 0.576	-10.900** ± 0.341	-4.800** ± 0.374
	B	-4.900** ± 0.190	-1.000** ± 0.311	-2.700** ± 0.336
	C	4.300** ± 0.198	-2.100** ± 0.666	-0.500** ± 0.566
Joint scaling test	X^2	817.63**	1329.26**	660.72**
Digenic model	m	23.200** ± 0.590	23.800** ± 0.608	24.100** ± 0.686
	d	1.100** ± 0.071	-3.500** ± 0.101	0.100 ± 0.106
	h	-10.050** ± 1.767	-7.450** ± 1.437	-3.050 ± 1.715
	I	-13.000** ± 0.586	-9.800** ± 0.599	-7.000** ± 0.678
	J	0.550 ± 0.600	-4.950** ± 0.375	-1.050* ± 0.495
	I	21.700** ± 1.183	21.700** ± 0.919	14.500** ± 1.058
Type of epitasis		Duplicate	Duplicate	Duplicate

* and** significant at 5% and 1% respectively.

Table 6: Scaling test, joint scaling test and estimation of gene effects by digenic model in respect of fiber strength.

Parameter		C1 (NHH44)	C2 (NHH206)	C3 (AKH Hy- 2)
Scaling test	A	-5.000** ± 0.077	-7.000** ± 0.405	-7.500** ± 0.212
	B	-2.200** ± 0.161	-7.100** ± 0.312	6.800** ± 0.205
	C	5.400** ± 0.413	-3.100** ± 0.870	1.700** ± 0.343
Joint scaling test	X^2	2053.30**	1093.78**	1051.79**
Digenic model	m	76.300** ± 0.423	73.600** ± 0.675	76.100** ± 0.215
	d	-0.500** ± 0.050	-0.300 ± 0.168	-1.400** ± 0.035
	h	-11.900** ± 0.911	-5.350** ± 1.492	-13.150** ± 0.613
	I	-12.600** ± 0.420	-11.000** ± 0.654	-16.000** ± 0.212
	J	-1.400** ± 0.173	0.050 ± 0.391	-0.350 ± 0.192
	I	19.800** ± 0.501	25.100** ± 0.957	30.300** ± 0.496
Type of epitasis		Duplicate	Duplicate	Duplicate

* and** significant at 5% and 1% respectively.

Table 7: Scaling test, joint scaling test and estimation of gene effects by digenic model in respect of uniformity index.

effects varies according to trait. The sign for dominance effects is a function of the F1 mean value in relation to the mid parental value and indicates which parent is contributing to the dominance effect [6]. Whereas, absence of significant values for [h]component signifies no dominance genetic differences or presence of bi-directional dominance between 2 parents and the dominant effects seemed to be not important in the genetic control of these crosses [7]. Whereas additive gene effects [d] were found to be important for the traits fibre strength (cross 1). Jagtap [8] stated that when additive effects are larger than non-additive ones, selection in early segregating generations would be effective. For exploiting these characters pedigree selection may also be suitable. Similar conclusions were drawn by Srinivasan et al. [9], Refaey and

Parameter		C1 (NHH44)	C2 (NHH206)	C3 (AKH Hy- 2)
Scaling test	A	0.320* ± 0.148	-0.090 ± 0.138	-1.100** ± 0.180
	B	0.400* ± 0.179	0.260* ± 0.140	0.630** ± 0.169
	C	-1.980** ± 0.345	-5.570** ± 0.278	-4.810** ± 0.516
Joint scaling test	X^2	141.28**	806.82**	1886.59**
Digenic model	m	34.170** ± 0.394	38.090** ± 0.267	33.310** ± 0.530
	d	-0.150** ± 0.030	-0.630** ± 0.044	-0.690** ± 0.042
	h	3.950** ± 0.926	6.935** ± 0.634	5.875** ± 1.163
	I	2.700** ± 0.392	5.740** ± 0.263	4.340** ± 0.528
	J	-0.040 ± 0.223	-0.175 ± 0.168	-0.865** ± 0.223
	I	-3.420** ± 0.552	-5.910** ± 0.399	-3.870** ± 0.661
Type of epitasis		Duplicate	Duplicate	Duplicate

* and ** significant at 5% and 1% respectively.

Table 4: Scaling test, joint Scaling test and estimation of gene effects by digenic model in respect of ginning percentage.

Razek [10] and Esmail [11] for one or other fiber quality traits. Even though, the traits upper half mean length (mm) and fibre strength (cross 2) were found to be influenced by additive and dominance gene actions, magnitude of [h] reveals the importance of dominance gene action in inheritance of these traits.

The traits, upper half mean length (mm) (cross 1, 2 and 3), ginning percentage (cross 1, 2 and 3), micronaire value (cross 3), fibre strength (cross 1, 2 and 3) and uniformity index (cross 1, 2 and 3), were possessing opposite sign of [h] and [l] indicating the role of duplicate gene action controlling these traits. Similar results were obtained by Rajendrakumar and Raveendran [12] for 2.5% span length and Bhatti et al. [13] and Refaey and Razek [10] and Srinivasan et al. [9] for 2.5% span length, bundle strength, uniformity raito and micronaire value. Such type of duplicate type of gene action would limit the range of variability and thus slow down the pace of progress. In such situations, heterosis breeding would be advantageous. Since only two crosses for micronaire value (cross 1 and 2), the signs of [h] were similar to the [l] type of epistasis, it was concluded that complementary type of interaction was present. These variations can be exploited by simple pedigree selection. Mass selection for several early generation aimed at the improvement of heterozygous population by modifying the frequencies of desirable gene followed by single plant selection in the resulting material would be cheapest and quickest procedure. However, the presence of non-fixable (h, j and l) component together with duplicate type of epistasis may cause delay in the improvement in this trait through selection in early generations.

It is clear from the result that epistasis cannot be ignored when establishing a new breeding programme to improve cotton population for economic traits. The inheritance of all the traits studied was controlled by additive and non-additive genetic effects, with greater value of dominance gene effect than the additive one in most cases. Among the non-additive effect, the other fixable component, i.e., additive × additive (i) type of interaction was also significant and constitutes a major portion of gene effects; therefore it may be possible to exploit it. The conclusion drawn in the present investigation can be compared with those reported in cotton by other workers. Jagtap [8] stated that when additive effects are larger than non-additive ones,

selection in early generation would be effective, while if the non-additive portions are larger than additive one, the improvement of the character need intensive selection through later generation. The evidence of non-allelic interaction was reported by Refaey and Razek [10] for fibre traits.

References

1. Phogat DS, Singh DP (2000) Genetics of gossypol and fibre characters in American cotton (G. hirsutum L.). J Cotton Resb Dev 14: 23-36.

2. Patel KG, Patel RB, Patel MI, Kumar V (2007) Genetics of yield, fibre quality and their implications in breeding of interspecific cross derivatives of cotton. J Cotton Res Dev 21: 153-157.

3. Hayman BI, Mather K (1955) The description of genetic interactions in continuous variation. Biometrics 11: 69-82.

4. Cavalli LL (1952) An analysis of linkage in quantitative inheritance. In: Reeve ECR, Waddington CH (eds.) Quantitative Inheritance, HMSO, London.

5. Hayman BI (1958) The separation of epistatic from additive and dominance variation in generation means. Heredity 12: 371-390.

6. Cukadar OB, Miller JF (1997) Inheritance of the stay green trait in sunflower. Crop Sci 37: 150-153.

7. Haleem SHMA, Metwali EMR, Felail AMM (2010) Genetic analysis of yield and its component in some Egyptian cotton (*Gossypium barbadense* L.) varieties. World J Agric Sci 6: 615-621.

8. Jagtap DR (1986) Combining ability in upland cotton. Indian J Agric Sci 56: 833-840.

9. Srinivasan K, Rajashekaran R, Mahalingam G (2013) Genetic analysis for quantitative and quality characters in three single crosses of upland cotton. Not Sci Biol 5: 450-453.

10. EL-Refaey RA, Razek UA (2013) Generation mean analysis for yield, its components and quality characteristics in four crosses of Egyptian cotton (*Gossypium barbadense* L.). Asian J Crop Sci 5: 153-156.

11. Esmail RM (2007) Genetic analysis of yield and its contributing traits in two intra specific cotton crosses. J Appl Sci Res 3: 2075-2080.

12. Rajendrakumar P, Raveendran TS (1999) Genetic analysis of yield and quality traits in upland cotton (*Gossypium hirsutum L.*) through generation mean analysis. Indian J Agric Sci 69: 538-540.

13. Bhatti MA, Azhar FM, Alvi AW (2006) Estimation of additive, dominance and epistatic components of genetic variation in fiber quality characters of upland cotton grown in salinized conditions. Int J Agric Biol 8: 824-827.

22

Effect of Winding Tension Parameters and Rewinding Passages at Winding on Cotton Yarn Properties

Bagwan AS[1]*, Rajput A[1], Dalal S[2] and Aakade A[2]

[1]Center for Textile Functions, Mukesh Patel School of Technology, Management, Engineering, Shirpur, India
[2]Maral Overseas Khalghat, Nimrani, India

Abstract

In present study, the trials were undertaken on winding machine to study the effect of winding tension process variables on yarn properties. The winding tension was varied at 3 levels and five yarn counts were selected: two medium counts (30s and 36s) and three coarse counts (20's, 24's, and 26's). In all, 15 trials were conducted. The yarn samples were assessed for important yarn properties, such as imperfections, classimat faults, RKm and hairiness, Elongation, IPI, Package density and optimizing tension parameters of individual count for obtaining quality yarn.

Keywords: Unevenness; RKm; Hairiness; Winding tensions; Elongation; Yarn imperfections

Introduction

Yarn tension control is a vital parameter for quality and efficiency in textile processes. It determines the build and structure of the package, and it has a significant influence on productivity of various processes such as winding, twisting and cabling, and in subsequent processes such as weaving and knitting [1]. However, a great majority of yarns are tensioned by primitive means (textile parameters), resulting in loss of production and inferior quality. The ideal solution for most processes is the ability to set the yarn tension to a desired level and to be assured that it does not change over time at any yarn position. However, this ideal is rarely achieved. In Most cases, it almost becomes a game of trying to find out how much tension variations a process can tolerate and then adjusting the process speed accordingly. With precise tension control, many processes can run at least 30% faster and have the added benefit of quality improvement at the same time [2-6].

Material and Method

In order to study effect of the winding tension variations on yarn count at winding machine and the effect of number of winding passages on yarn parameter, Five different yarn counts were selected on winding machine [5-9].

Auto coner – 338, two were medium counts (30s and 36s) and three were coarse counts (20's, 24's, and 26's). In all, 15 trials were conducted with varying tensions and increasing winding passages and analyze the properties of the yarn .The yarn samples were assessed for important yarn properties, such as imperfections, classimat faults, RKm and hairiness, Elongation, IPI, Package density and optimizing tension parameters of individual count for obtaining quality yarn [10-16]. Following cotton used to prepare various yarns counts (Table 1).

Results and Discussion

From Figure 1 it can be observed that the overall yarn quality in terms of imperfections, and hairiness were not improved significantly with lower breaks at winding. The yarns (20's) made from 100% cotton showed particular trend, improvement found in RKm, Elongation, total IPI value. It has been observed that at 650cn tension level RKm, elongation, winding cuts founds to be improved. This is due to the better control on winding tensions. From the results, it can be inferred that the improvement is more distinct at winding tension 650cn which gives improvement in yarn properties of 20's count.

From Figure 2 it is clear that at, rewinding the overall yarn quality in terms of imperfections, and hairiness not improved significantly with lower breaks at winding. The yarns (20's) made from 100% cotton shows particular trend, at rewinding and improvement found in, Elongation, total IPI value. It has been observed that tension level elongation was at 650cn, with an improvement in winding cuts significantly. This is due to the better yarn control on rewinding tensions. From the results, it can be inferred that, the improvement is more distinct in at rewinding tension 650cn, which gives improvement in yarn properties of 20's count. By this the total cuts /100 km found to be decreased at rewinding when compared to winding.

Mixing	Property range
Mean Length	27.3
Micronaire	4.28
Strength	32.4
Elongation	6.8
Trash	3.14

Table 1: Properties of cotton.

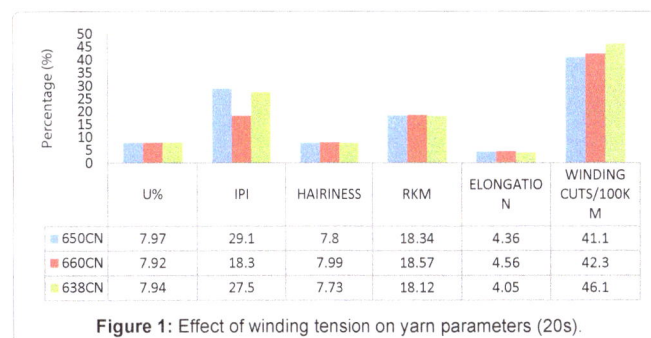

	U%	IPI	HAIRINESS	RKM	ELONGATION	WINDING CUTS/100K M
650CN	7.97	29.1	7.8	18.34	4.36	41.1
660CN	7.92	18.3	7.99	18.57	4.56	42.3
638CN	7.94	27.5	7.73	18.12	4.05	46.1

Figure 1: Effect of winding tension on yarn parameters (20s).

*Corresponding author: Bagwan AS, Center for Textile Functions, Mukesh Patel School of Technology, Management, Engineering, Shirpur, India
E-mail: abdulsalaambagwan@gmail.com

From Figure 3 it is clear, that the overall yarn quality in terms of imperfections, and hairiness not improved significantly with lower breaks at winding. The yarns (24's) made from 100% cotton shows particular trend, and improvement found in RKm, Elongation, total IPI value. It has been observed that at 610cn tension level RKm, elongation, winding cuts founds to be improved. This is due to optimization of tension level at winding. From the results, it can be inferred that the improvement were more distinct at winding tension 610cn which gives improvement in yarn properties of 24's count, as compare to 600cn and 590cn tension at winding stage. It was also observed that total cuts /100 km found to be decreased at winding.

From Figure 4 it was observed that at rewinding the overall yarn quality in terms of Hairiness and Elongation not improved significantly with lower breaks at winding. The yarns (24's) made from 100% cotton shows particular trend, improvement found in RKm, total IPI value. It has been observed that at 610cn tension level RKm, IPI level, winding cuts founds to be improved. This is due to the better control at rewinding tensions. From the results, it can be inferred that the improvement is more distinct in at winding tension 600 CN which gives improvement in yarn properties of 24's count.

From Figure 5 it was observed that, the overall yarn quality in terms of imperfections, and hairiness improved significantly with lower breaks at winding. The yarns (26's) made from 100% cotton shows particular

trend, the improvement not found in RKm, Elongation value. It has been observed that at 29cn tension level RKm, elongation, winding cuts founds to be improved. This is due to the better control on winding tensions. From the results, it can be inferred that the improvement is more distinct in at winding tension 29 CN which gives improvement in yarn properties of 26's count.

From Figure 6 it was observed that at rewinding the overall yarn quality in terms of imperfections, and hairiness improved significantly with lower breaks at winding. The yarns (26's) made from 100% cotton shows particular trend, the improvement is not found in RKm, Elongation value. It has been observed that at 29 CN tension level RKm, elongation, winding cuts founds to be improved. This is due to the better control on winding tensions. From the results, it can be inferred that the improvement is more distinct in at winding tension 29 CN which gives improvement in yarn properties of 26's count.

From Figure 7 it was observed that the overall yarn quality in terms of imperfections, RKm and improved significantly with lower breaks at winding. The yarns (30's) made from 100% cotton shows particular trend, the improvement is not found in Hairiness, Elongation value. It has been observed that at 25 CN tension level RKm, elongation, winding cuts founds to be improved. This is due to the better control on winding tensions. From the results, it can be inferred that the improvement is more distinct in at winding tension 25 CN which gives improvement in yarn properties of 30's count. As compare to lower tension level in winding.

From Figure 8 it was observed that at rewinding the overall yarn quality in terms of imperfections, RKm improved significantly with lower breaks at winding. The yarns (30's) made from 100% cotton shows particular trend, the improvement is not found in Hairiness, Elongation value. It has been observed that at 25 CN tension level RKm, elongation, winding cuts founds to be improved. This is due to the better control on winding tensions. From the results, it can be

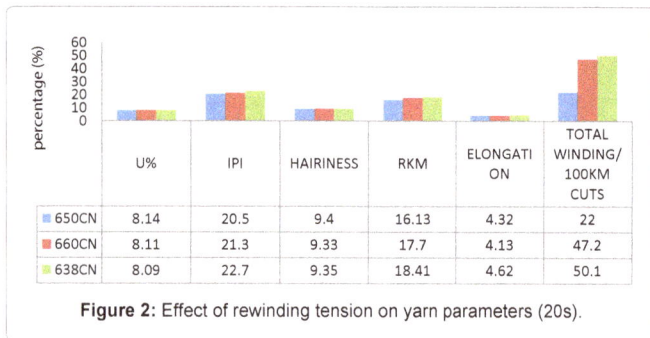

	U%	IPI	HAIRINESS	RKM	ELONGATION	TOTAL WINDING/100KM CUTS
650CN	8.14	20.5	9.4	16.13	4.32	22
660CN	8.11	21.3	9.33	17.7	4.13	47.2
638CN	8.09	22.7	9.35	18.41	4.62	50.1

Figure 2: Effect of rewinding tension on yarn parameters (20s).

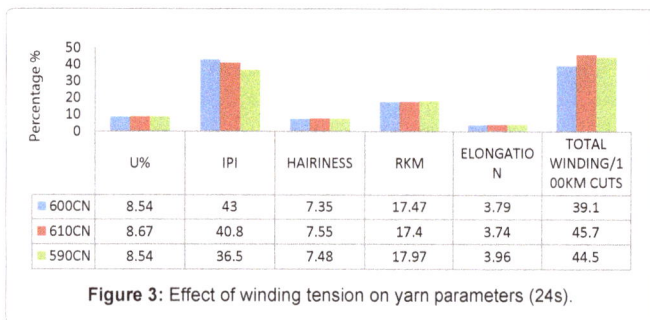

	U%	IPI	HAIRINESS	RKM	ELONGATION	TOTAL WINDING/100KM CUTS
600CN	8.54	43	7.35	17.47	3.79	39.1
610CN	8.67	40.8	7.55	17.4	3.74	45.7
590CN	8.54	36.5	7.48	17.97	3.96	44.5

Figure 3: Effect of winding tension on yarn parameters (24s).

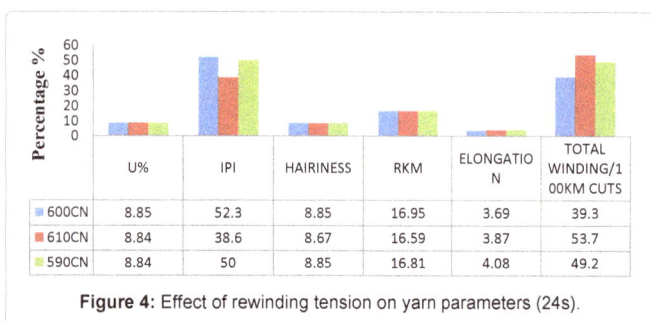

	U%	IPI	HAIRINESS	RKM	ELONGATION	TOTAL WINDING/100KM CUTS
600CN	8.85	52.3	8.85	16.95	3.69	39.3
610CN	8.84	38.6	8.67	16.59	3.87	53.7
590CN	8.84	50	8.85	16.81	4.08	49.2

Figure 4: Effect of rewinding tension on yarn parameters (24s).

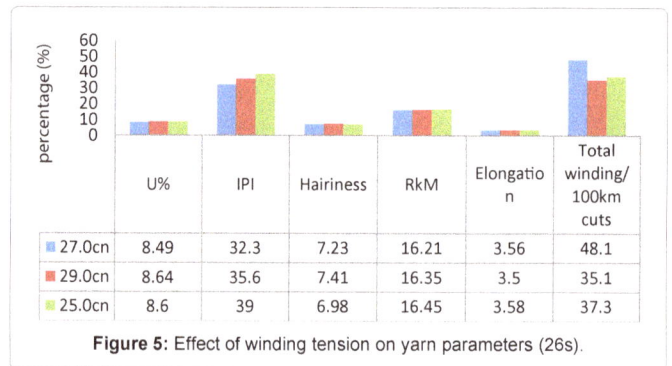

	U%	IPI	Hairiness	RkM	Elongation	Total winding/100km cuts
27.0cn	8.49	32.3	7.23	16.21	3.56	48.1
29.0cn	8.64	35.6	7.41	16.35	3.5	35.1
25.0cn	8.6	39	6.98	16.45	3.58	37.3

Figure 5: Effect of winding tension on yarn parameters (26s).

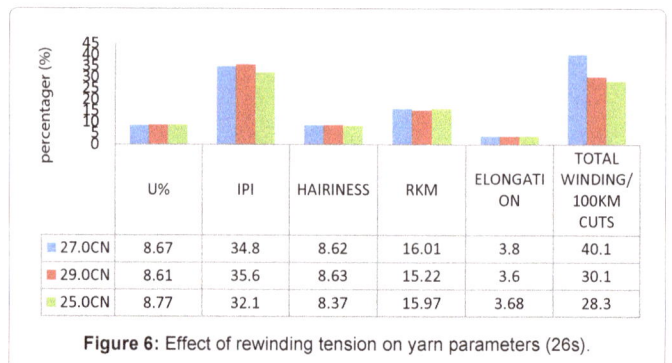

	U%	IPI	HAIRINESS	RKM	ELONGATION	TOTAL WINDING/100KM CUTS
27.0CN	8.67	34.8	8.62	16.01	3.8	40.1
29.0CN	8.61	35.6	8.63	15.22	3.6	30.1
25.0CN	8.77	32.1	8.37	15.97	3.68	28.3

Figure 6: Effect of rewinding tension on yarn parameters (26s).

inferred that the improvement is more distinct in at winding tension 25 CN which gives improvement in yarn properties of 30's count.

From Figure 9 it was observed that the overall yarn quality in terms of imperfections, and hairiness improved significantly with lower breaks at winding. The yarns (36's) made from 100% cotton shows particular trend, There is no improvement found in RKm, Elongation value. It has been observed that at 19 CN tension level RKm, elongation, winding cuts founds to be not improved. This is due to the better control on winding tensions. From the results, it can be inferred that the improvement is more distinct in at winding tension 19 CN which gives improvement in yarn properties of 36's count.

From Figure 10 it was observed that the overall yarn quality in terms of imperfections, and hairiness improved significantly with lower breaks at winding. The yarns (36's) made from 100% cotton shows particular trend, There is no improvement found in RKm, Elongation value. It has been observed that at 19 CN tension level RKm, elongation, winding cuts founds to be not improved. This is due to the better control on winding tensions. From the results, it can be inferred that the improvement is more distinct in at winding tension 19 CN which gives improvement in yarn properties of 36's count.

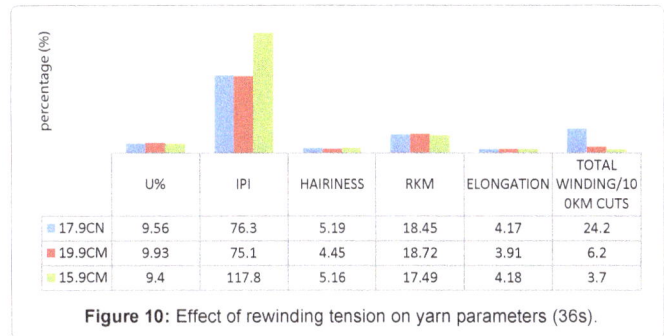

	U%	IPI	HAIRINESS	RKM	ELONGATION	TOTAL WINDING/100KM CUTS
23.0CN	8.99	54.3	6.52	16.95	3.65	40.1
25.0CN	8.74	41.5	6.7	18.3	4.19	35.1
20.0CN	8.95	43.3	6.63	17.33	3.91	36.3

Figure 7: Effect of winding tension on yarn parameters (30s).

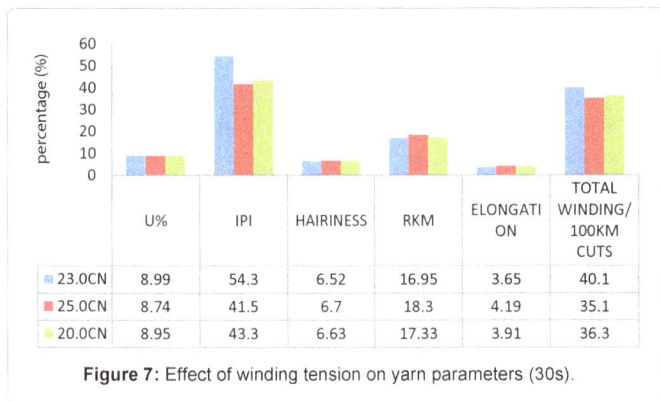

	U%	IPI	HAIRINESS	RKM	ELONGATION	TOTAL WINDING/100KM CUTS
23.0CN	9.35	69.5	7.76	16.27	3.7	30.1
25.0CN	9.41	61.8	7.71	17.08	3.7	25.1
20.0CN	9.47	68.6	7.55	16.87	3.99	20.3

Figure 8: Effect of rewinding tension on yarn parameters (30s).

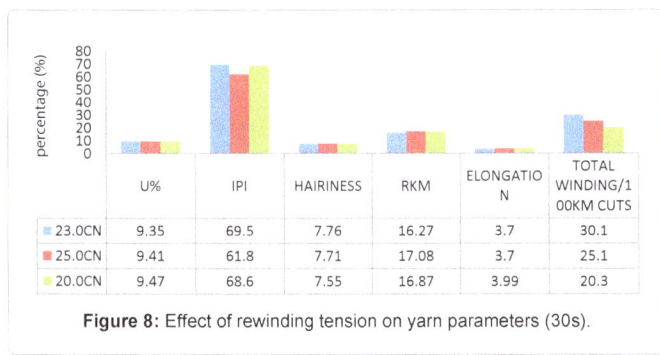

	U%	IPI	HAIRINESS	RKM	ELONGATION	TOTAL WINDING/100KM CUTS
17.9CN	9.36	89.1	4.58	18.69	4.14	24.2
19.9CN	9.63	68.5	4.47	18.49	3.7	18.9
15.9CN	9.65	78.9	6.63	17.33	3.91	30.6

Figure 9: Effect of winding tension on yarn parameters (36s).

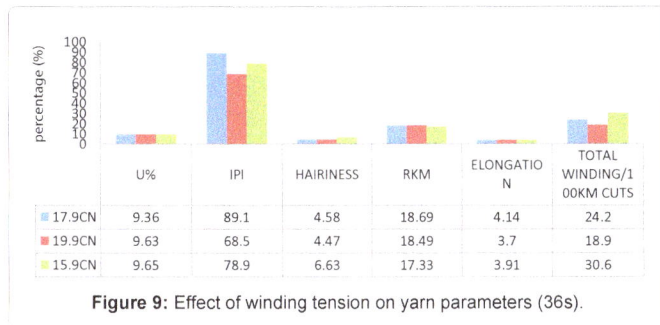

	U%	IPI	HAIRINESS	RKM	ELONGATION	TOTAL WINDING/100KM CUTS
17.9CN	9.56	76.3	5.19	18.45	4.17	24.2
19.9CM	9.93	75.1	4.45	18.72	3.91	6.2
15.9CM	9.4	117.8	5.16	17.49	4.18	3.7

Figure 10: Effect of rewinding tension on yarn parameters (36s).

Conclusion

Reference to Figures 1-6, Present investigations summarized that, As increase the tension level at winding machine on 20^s, 24^s, 26^s, 30^s, 36^s count RKm, Elongation not improved significantly but Total IPI value and winding cuts found improved significantly. For coarser and medium finer count, among the three tension levels, lower tension contribute towards optimum Rkm value and improvement in winding cuts.

At Rewinding for all the counts, number winding cuts were reduced, because in winding all the objectionable fault were removed in previous winding passage and As decreases the tension level at winding machine for all five counts the unevenness, RKm, elongation were not improved significantly but total IPI and winding cuts/100km reduced at lower tension levels at rewinding stage.

Acknowledgement

The authors acknowledged valuable support received from Director, Mukesh Patel School of Technology, Management, Engineering Shirpur, Associate Dean MPSTME and Principal, Center for Textile functions. MPSTME Shirpur and Technical Vice President of Maral overseas Khalghat, Indore, Madhya Pradesh India.

References

1. Pilley PR, Hariharan R (1984) Effect of Processing Factors on the Incidence of Yarn Faults in Spinning. Indian J Text Res 2: 100-105.

2. Desai AN, Patil VK, Balasubramanium N (1994) Effect of Ring Yarn Hairiness on Imperfections in Wound Yarns. BTRA SCAN 25: 5.

3. Gupte AA, Balasubramanium N (1991) Influence of Imperfections on U% and Cv% of Yarn. BTRA Scan: 4-10.

4. Desai AN, Jade BD, Paradkar TL, Balasubramanium N (1997) Wound Yarn Properties with Special Reference to Yarn Imperfections.

5. Chaudhari A (1994) Comparison of Clearing Efficiency and the Splicing Performance of Autoconer 138 and Schlafhorst 238. Textile Trends 35.

6. Rust JP, Peykamian S (1992) Critical Study on Yarn Hairiness. TRJ: 685.

7. Singh Arya PP, Gurbaxani CR, Moghe SS (1998) Yarn Hairiness and Its Measurement. ITJ: 125.

8. Basu A, Gotipamul R (2004) Effect of some ring spinning and winding parameters on extra sensitive yarn imperfections. Indian Journal of Fibre and Textile Research 30: 211-214.

9. Srivastava SK (1994) Spinners Hand Book of Quality Control.

10. Kimothi PD (1974) Process Control in Weaving. Ahmedabad Textile Industry's Research Association, Ahmedabad, India.

11. Paliwal MC. Winding. In: Ananthan TV, Paliwal MC (eds.) Weaving Tablet I.

12. Basu A (1998) Yarn Quality Changes. Indian Text J 108: 12.

13. Uster Statistics (1999) Zellweger Uster.

14. Furter R (1982) Evenness Testing in Yarn Production: Part-I. Textile Institute, Manchester, UK.

23

Aloe Vera Leaf Gel Extract for Antibacterial and Softness Properties of Cotton

Ibrahim W[1], Sarwar Z[2], Abid S[2], Munir U[2] and Azeem A[2]*

[1]*Department of Postgraduate Studies, National Textile University, Faisalabad, Pakistan*
[2]*Department of Textile Processing, National Textile University, Faisalabad, Pakistan*

Abstract

Natural plants extract for antimicrobial of textile finishing is a vital and potential area of current and future aspects therefore has greater market value. Cotton has long been recognized as media to support the growth of microorganisms such as bacteria and fungi. Among all the natural antimicrobial agents the plant products comprise the major segment. The objective of this work was to investigate the antibacterial properties of Aloe Vera leaf gel extract on cotton, also its effects on the performance parameters of fabric. Softness properties were also imparted on the fabric as it is the inherit property of Aloe Vera leaf gel. Best antimicrobial properties were achieved by pad dry method with high concentrations as compared to coating method. The softness of pad-dry samples was prominent against coated fabric. Whiteness Index decreased less in pad-dry as compared to coated samples.

Keywords: Microorganisms; Antimicrobial; Aloe vera; Inhibition zone; Whiteness index

Introduction

Textile finishing involves treating a textile material in such a way that the product has the desired functional properties required for its intended use and therefore has greater market value. The desired properties may include the fabric's dimensions and their stability, its weight, drape, appearance, softness and handle, as well as any required functional properties such as resistance to creasing, flames, water, oil, dirt or bacteria. For the treatment of diseases inhibitory chemicals employed to kill micro-organisms or prevent their growth. Today, however, with increased knowledge of the causative agents of various infectious diseases, antibiotic has come to denote a broader range of antimicrobial compounds, including anti-fungal and other compounds [1].

The microorganisms are found almost everywhere in the environment and can multiply quickly when basic requirements, such as moisture, nutrients and temperature are met. Most synthetic fibers, due to their high hydrophobicity, are more resistant to attacks by microorganisms than natural fibers. Proteins in keratinous fibers and carbohydrates in cotton can act as nutrients and energy sources under certain conditions. Soil, dust, solutes from sweat and some textile finishes can also be nutrient sources for microorganisms [2].

For these reasons, it is highly desirable that the growth of microbes on textiles be minimized during their use and storage. Consumers' demand for hygienic clothing and active wear has created a substantial market for antimicrobial textile products. Estimations have shown that the production of antimicrobial textiles was in the magnitude of 30,000 tones in Western Europe and 100,000 tones worldwide in 2000. Furthermore, it was estimated that the production increased by more than 15% of year in Western Europe between 2001 and 2005, making it one of the fastest growing sectors of the textile market. Sportswear, socks, shoe linings and lingerie accounted for 85% of the total production [2].

Obtaining the greatest benefit, an ideal antimicrobial treatment of textiles should satisfy several requirements. Firstly, it should be effective against a broad spectrum of bacterial and fungal species, but at the same time exhibit low toxicity to consumers. Secondly, the finishing should be durable to laundering, dry cleaning and hot pressing, a greatest challenge as textile products are subjected to repeated washing during their life. Thirdly, the finishing should not have negative effect on the

quality (e.g., physical strength and handle) or appearance of the textile. Finally, the finishing should preferably be compatible with chemical processes such as dyeing, be cost effective and not produce harmful substances to the environment. One further consideration is that the antimicrobial finishing of textiles should not kill the resident flora of nonpathogenic bacteria on the skin of the wearer [3].

Healing power of most of plants has been used since ancient times ranging 250000-500000 species on earth. Aloe Vera is a succulent plant species that is found only in cultivation, having no naturally occurring populations [4]. In the last few decades with the increase in new antimicrobial fiber technology, a range of synthetic antimicrobial products such as triclosan, metals and their salts organometallics and their quaternary ammonium compounds have been developed. Although the synthetic antimicrobial agents are very effective against the growth of many microbes and give a durable effect on textiles but they are cause of the concern due to the associated side effects, action on non-targeted areas and cause water pollution. Hence there is a great demand of antimicrobial agents based on natural ecofriendly agents which not only helps to improve antimicrobial effect but fulfill statutory requirements by regulating agencies [5].

Regarding the current research Wiyong et al. [6] investigated an antibacterial activity of apatite-coated titanium dioxide (TiO_2) against bacteria. They suggested that the presence of apatite-coated TiO_2 shows antibacterial activity for textile applications [6]. Mahesh et al. focused on plant based natural dyes and other bioactive natural extract in textile coating as antimicrobial textile finish has gained significant momentum [7]. Ramsamy et al. worked on natural extract finish *'punica guranatum'* for dyeability and antimicrobial properties on

*Corresponding author: Azeem A, Department of Postgraduate Studies, National Textile University, Faisalabad, Pakistan
E-mail: abdulazeem88@hotmail.com

cotton fabrics and reported natural extract dyed fabric has prominent antimicrobial activity with clear zero inhibition zone and wash durability sustain up to 10 washes [8]. Joshi et al. [1] reviewed bioactive agents based on natural products like chitosan, natural dyes, neem extracts and other herbal products for antimicrobial finished textiles [3].

Kedarnath et al. [9] tested the antimicrobial activity of Aloe Vera extract against pathogenic bacteria. Methanol extract has showed maximum inhibitory activity against *E. coli* and *Candida* [9]. A study was done to identify, quantify, and compare the phytochemical contents, antioxidant capacities, and antibacterial activities of Aloe vera lyophilized leaf gel (LGE) and 95% ethanol leaf gel extracts by Fatemeh [10].

The current research aspects of Aloe Vera leaf gel extracts and its application to textile material for both antibacterial cum softness properties has not been reported yet in the previous works so far. In present manuscript, natural plant Aloe Vera extract is applied on pure cotton fabric and its antimicrobial, mechanical and softness properties were analyzed.

Methods and Materials

Plain weave (1/1) 100% cotton fabric with 126.2 GSM, 40 Ne warp and weft count, 115 ends/inch and 90 picks/inch was used in this research work.

Chemicals

Aloe Vera (Aloe-emodin1,8-dihydroxy-3(hydroxyl-methyl)-9,10 anthracene dione) was taken from plants nursery at university. Methanol (CH_3OH) from Merk Millipore Chemicals was used for Aloe Vera extraction, lab scale wetting agent, Lutexal Hit Plus (acrylate based polymer, synthetic thickening agent) by BASF Chemicals for generating viscosity of gel paste and Printofix Binder 77 N (acrylate based binder) from Clariant Chemicals were used for 100% cotton fabric.

Equipment

a) Processing equipment: Lab scale Soxhlet Extractor by PCSIR, Stentor, padder and flat-bed printing machines by Tsujii Dyeing Machine Manufacturing Co. Japan were used for performing experiments.

b) Testing equipment: Spectrophotometer and bending testor (Gretag Macbeth Ltd., England), digital weight balance (Kyoto, Japan), lab scale coefficient of friction tester, magnifying glass and sample cutter testing equipment were utilized.

c) Testing equipment for antimicrobial finish: Lab scale petri dish, sterilized bottles, auto clave and incubator (Daihan Scientific, Korea), laminar flow (Biological Medical Services, USA) and pipettes (Socorex) equipment were specifically used for antimicrobial testing.

d) Design of experiment: Padding pressure was maintained at 1.5 bars and the padder moving at speed of 10 rpm. Table 1 shows design of experiment for padding and coating.

For maintaining acidic pH 5.5 drops of citric acid if necessary. Pick up was kept at 70-80%. Time used for drying of fabric was 60 sec. Time used for curing of fabric was 300 sec. Temperature used for Aloe Vera extract finish was 100°C for drying for coating of fabric.

e) Soxhlet Extraction Method: The solid cubes of Aloe Vera leaf (inside material) as shown in Figure 1 containing some of the desired

	Concentration	Temperature	No. of Samples
Padding	150 (g/l)	100°C	9
		120°C	
		140°C	
	200 (g/l)	100°C	
		120°C	
		140°C	
	250 (g/l)	100°C	
		120°C	
		140°C	
Coating	150 (g/l)	120°C	9
		140°C	
		160°C	
	200 (g/l)	120°C	
		140°C	
		160°C	
	250 (g/l)	120°C	
		140°C	
		160°C	

Table 1: Design of experiment.

Figure 1: Aloe Vera gel cubes.

compounds (mainly anthraquinones) were placed inside a thimble made from thick filter paper, which was loaded into the main chamber of the Soxhlet extractor. The extraction solvent i.e., Methanol was taken into a distillation flask and the Soxhlet extractor was placed onto this flask. The methanol in the distillation flask was heated to reflux and its vapors travel up a distillation arm, and floods into the chamber housing the thimble of solid Aloe Vera cubes. The condenser ensures that every methanol vapor cools, and drips back down into the chamber housing the solid Aloe Vera cubes (Figure 1). The chamber containing the Aloe Vera material got slowly filled with warm methanol. Some of the desired compounds get dissolved in the warm methanol. When the Soxhlet chamber was almost full, the chamber was automatically emptied by a siphon side arm, with the methanol running back down to the distillation flask. The thimble ensures that the rapid motion of the methanol does not transport any solid material to the still pot. This cycle was repeated for 4-5 times until the cubes gets small and the desired materials are extracted.

During each cycle, a portion of the non-volatile compound dissolves in the solvent. After many cycles the desired compound (anthraquinones and polysaccharides) was concentrated in the distillation flask. The advantage of this system was that instead of many portions of warm methanol being passed through the sample, just one batch of methanol is recycled.

After extraction, the methanol was removed, by exposing it to open air. As methanol is highly volatile liquid and evaporates fast at room temperature, leaving behind the desired materials. The non-soluble portion of the extracted solid was left in the thimble.

Application methods

Application of Aloe Vera extract finish was done by the following methods i.e., Pad → Dry and Coating → Dry → Curing. First, samples of 15 × 15 inches were cut according to the processing provisions of lab scale stenter available at wet processing lab, and dimensional requirements for tensile and tear test instruments (warp and weft wise). Second step was preparation of recipe according to design of experiment. Liquor amount taken for each sample was 400 mL. Then application of finish was done through padding followed by drying at 100°C, 120°C and 140°C for Aloe Vera extract finish for antimicrobial finish respectively. After application testing of samples was done and results were analyzed.

Testing

As Aloe Vera extract is applied on a cotton fabric for both antibacterial and softness properties, so some mechanical performance of textile testing methods involves.

a) Anti-microbial testing: The test method used was parallel streak method AATCC-147 was used to visualize the antimicrobial activity of Aloe Vera leaf gel extract applied on the cotton fabric by estimating the inhibition zone of developed culture.

b) Co-efficient of friction testing: Coefficient of friction testing measures the ease with which two surfaces in contact can slide past one another where there are two different values associated with the coefficient of friction static and dynamic ASTM D 1894.

c) Bending tester: The standard method used for bending test was ASTM D 1388. Testing was done using tensile strength tester.

d) Whiteness CIE testing: Whiteness of samples was determined as per AATCC 110-2005. Testing was done using Gretagmacbeth Color Eye 7000 A.

Results and Discussion

Antimicrobial results

Antimicrobial activity of pad-dry samples treated with different concentrations and dried at different temperatures is shown in Figure 2a-c. It can be seen from these Figures that Aloe Vera exhibits its activity against *Staphylococcus aureus* and *E. coli* to reasonable extent. The zone

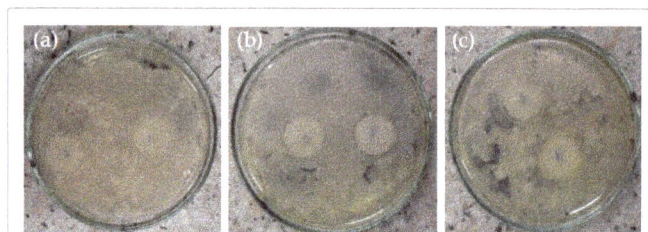

Figure 2: (a) Antibacterial activity of sample A and C (b) Antibacterial activity of sample D and F (c) Antibacterial activity of sample G and I.

of inhibition can be clearly seen from these figures; the bacteria grows on agar solution but when it approaches the fabric samples; the Aloe Vera bleeds from the fabric and shows resistance against the bacteria and there is almost no growth surrounding the bacteria.

The zone of inhibition for Aloe Vera is less, when compared to the zone of inhibition of other synthetic antibacterial agents. This can be understood from the point that Aloe Vera gel consists of almost 200 components or ingredients and its antibacterial agents may be present in less quantity as compared to other components.

It is also clear from this experiment that Aloe Vera, a natural product in its pure form works against bacteria and can be used for those purposes where extent of bacteria is not as strong as Malaria causing bacteria. In Figure 2c the Aloe Vera shows maximum resistance, its zone of inhibition is clearer as compared to Figure 2a and 2b. The reason is that in these samples the Aloe Vera concentration is higher as compared to the concentrations of Figure 2a and Figure 2b. Coating samples did not show any zone of inhibition, because the binder did not allow the Aloe Vera to leave the fabric so that it could work against the bacteria approaching them.

Coefficient of friction results

Coefficient of friction comprises two types of tests i.e., coefficient of static friction and coefficient of dynamic friction. Coefficient of static and dynamic friction for both pad-dry and coating samples padded with different concentrations and dried at different temperatures (Table 2).

Coefficient of static and dynamic friction (pad dry). It is shown by the Figure 3 that there is no effect of temperature on the standard/

	Standard Fabric	Drying Temp	150 g/l	200 g/l	250 g/l
Pad dry		100°C	0.565	0.566	0.576
	0.544	120°C	0.577	0.58	0.582
		140°C	0.574	0.588	0.599
		100°C	0.497	0.495	0.492
	0.514	120°C	0.487	0.48	0.491
		140°C	0.486	0.475	0.49
Coating	0.568	120°C	0.599	0.605	0.651
	0.601	140°C	0.601	0.602	0.61
	0.571	160°C	0.56	0.57	0.576
	0.509	120°C	0.516	0.513	0.599
	0.555	140°C	0.499	0.53	0.553
	0.519	160°C	0.494	0.541	0.576

Table 2: Coefficient of static and dynamic friction.

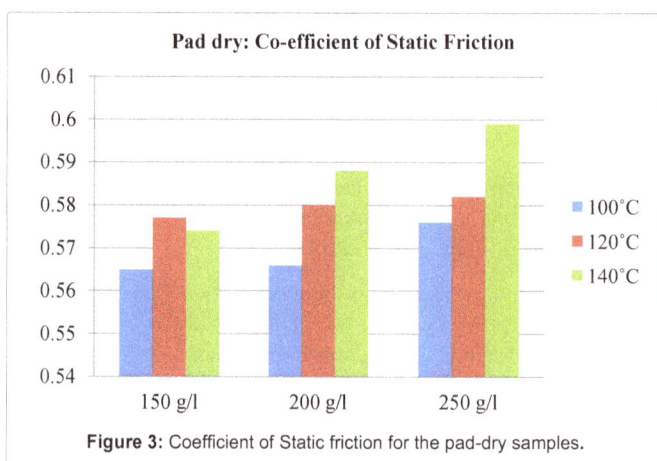

Figure 3: Coefficient of Static friction for the pad-dry samples.

unfinished fabric as they all have same co-efficient of static friction. There is an increase in trend as the concentration is increased also the trend increases with the raise in temperature.

It is shown by the Figure 4 that the coefficient of dynamic friction of standard/unfinished fabric is very high as compared to any of the finished fabric, treated with Aloe Vera. This shows that Aloe Vera exhibits softness and gives soft feel on cotton fabric results in lowering of co-efficient of dynamic friction. Also, increase in centration gives more softness and increase in temperature has reverse effect on it as Aloe Vera gets graded and cotton gets stiff at high temperature.

Place about here

Co-efficient of static and dynamic friction (coating). It can be seen from the Figure 5 that standard fabric has low coefficient of static friction at 120°C but high at 140°C and again low at 160°C. The treated samples showed an increase in trend at 120°C due to increase in concentration of Aloe Vera. The results of co-efficient of static friction at 140°C and 160°C are almost same for all concentrations (Figure 5).

It is shown by the Figure 6 that the standard/unfinished samples at all temperatures have greater coefficient of dynamic frictions as compared to 150 g/kg samples because the Aloe Vera is in low concentration and its softness property is diminished due to presence of binder and thickeners. At high concentrations, the Aloe Vera shows its softness properties as it over comes some of the hardness effects of the binder and thickeners present at the same concentrations in the recipe.

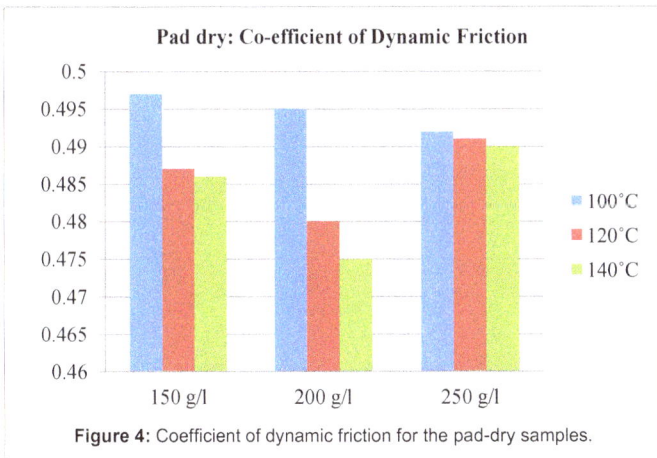
Figure 4: Coefficient of dynamic friction for the pad-dry samples.

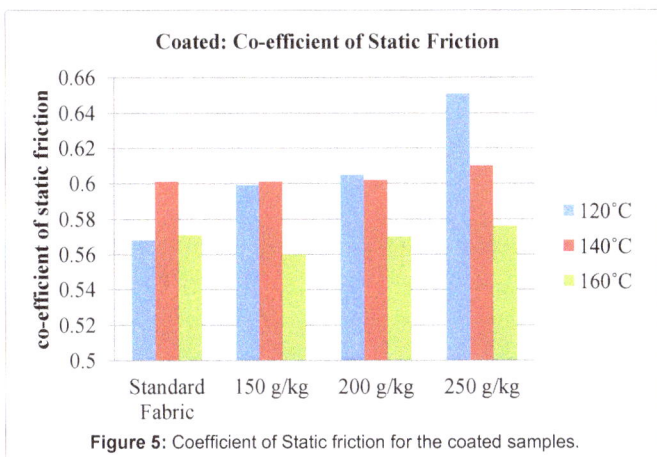
Figure 5: Coefficient of Static friction for the coated samples.

Whiteness index

Whiteness index for pad-dry samples padded with different concentrations and dried at different temperatures (Table 3).

Pad dry samples

It can be seen from the Figure 7 that an increase in concentration of Aloe Vera dims the whiteness of the finished samples as Aloe Vera is not transparent and has some light green color in gel form. Also within the same concentration the whiteness decreases with the increase in temperature because at high temperatures Aloe Vera gets yellow type color due to dehydration.

Coated samples

It is shown by the Figure 8 that standard fabric having binder and thickener and no Aloe Vera gives decrease in whiteness as temperature is increased. Curing is done for 5 minutes; cellulose turns yellow at high temperature if it is exposed for long time, also thickener and

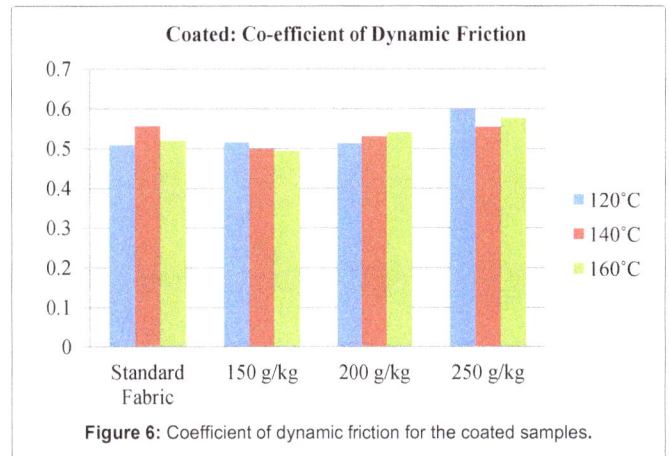
Figure 6: Coefficient of dynamic friction for the coated samples.

	Standard Fabric	Drying Temp	150 g/l	200 g/l	250 g/l
Pad-dry	0	100°C	67.59	67.49	67.08
	68.72	120°C	67.24	66.65	66.15
	0	140°C	66.82	66.01	65.49
Coating	68.34	120°C	66.13	65.99	63.34
	61.85	140°C	58.92	56.88	61.85
	60.86	160°C	48.9	47.23	60.86

Table 3: Whiteness index for samples.

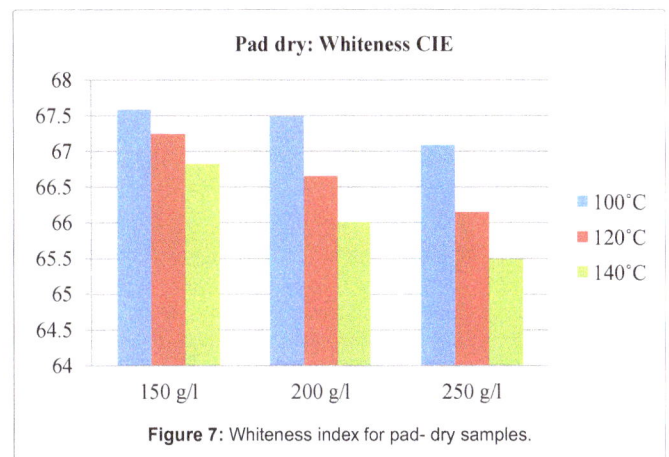
Figure 7: Whiteness index for pad- dry samples.

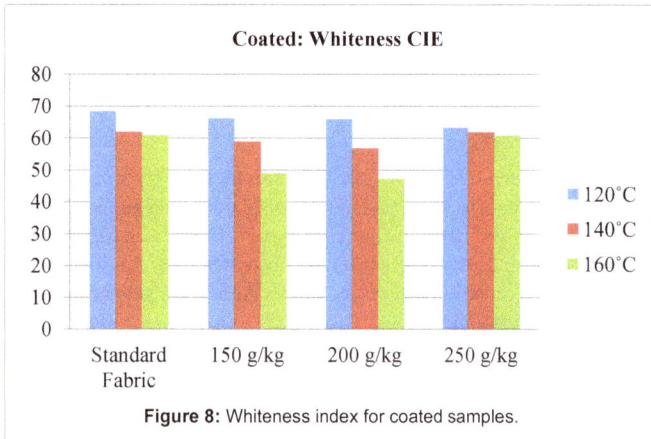

Figure 8: Whiteness index for coated samples.

		Standard Fabric	Drying Temp	150 g/l	200 g/l	250 g/l
Pad dry	Warp direction	0	100°C	4.32	4.49	4.36
		4.46	120°C	5.13	5.33	4.38
		0	140°C	4.11	4.47	4.08
	Weft direction	0	100°C	4.31	4.16	3.8
		4.33	120°C	4.71	4.53	4.7
		0	140°C	4.12	5.13	4.5
Coating	Warp direction	5.46	120°C	4.91	5.31	5.06
		5.47	140°C	5.28	5.32	5.48
		5.63	160°C	5.15	5.33	5.08
	Weft direction	5.05	120°C	4.8	4.8	4.45
		4.87	140°C	4.86	4.71	4.55
		5.13	160°C	4.55	4.61	4.65

Table 4: Bending length of samples.

binder assists in this result. And for other samples with different concentrations of Aloe Vera, whiteness is decreased with increase of temperature and concentration.

Bending length tests - stiffness test

Bending length of pad-dry and coating with respect to warp and weft wise samples, with different concentrations and dried at different temperatures (Table 4).

Pad dry

a) Warp wise bending length. It is clear from Figure 9 that bending length decreases as the concentration of Aloe Vera increases. Aloe Vera has softening effect and so its decreases the bending length of the finished fabric.

b) Weft wise bending length. From Figure 10 the temperature difference did not give any clear trend in the stiffness of the fabric. But the concentration increase gives a decrease in bending length of the fabric.

Coating

a) Warp wise bending length: The bending length of coated samples is greater as compared to the pad-dry samples as in Figure 11. This is due to the presence of binder and thickener in the recipe that stiffens the fabric. Also, the bending length of different concentrations is less compared to the standard fabric (containing only binder and thickener, no Aloe Vera) because the Aloe Vera present in the different concentrations lowers the stiffness of the fabric. Temperature has no obvious effect on the bending length of finished samples.

b) Weft wise bending length: In Figure 12 the bending length in the weft direction is less as compared to the warp direction because the weft yarns are more relaxed as compared to the warp yarns. Increase in concentration of Aloe Vera decreases the bending length also. The temperature tends no trends for bending length.

Place about here

Conclusion

Antibacterial and softness effect from Aloe Vera leaf gel extract on cotton fabric has been evaluated. From this work, we concluded that the best antibacterial properties could be achieved by pad dry method with high concentrations of Aloe Vera extract while coated samples with extract could not give such zone of inhibition. The softness of

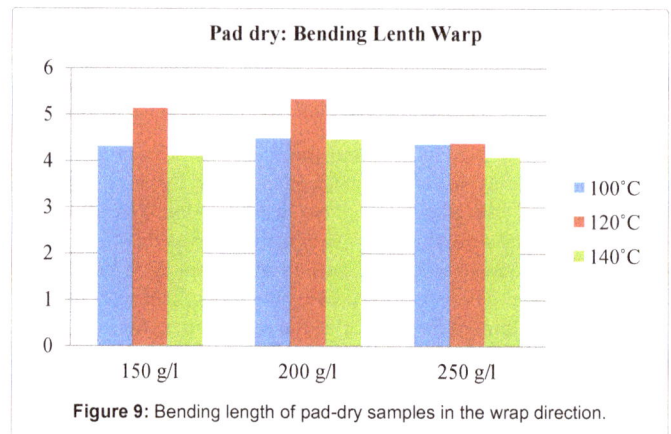

Figure 9: Bending length of pad-dry samples in the wrap direction.

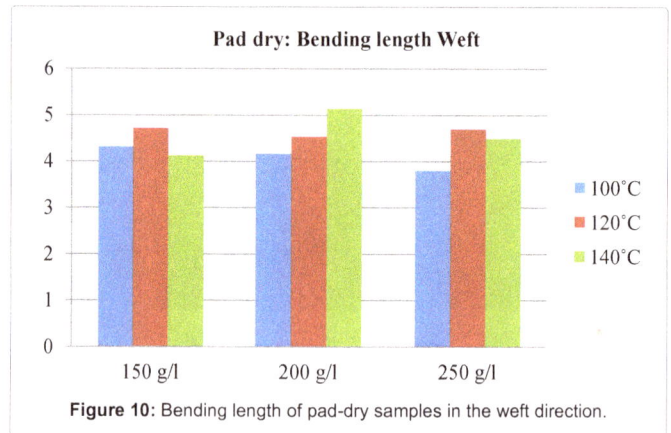

Figure 10: Bending length of pad-dry samples in the weft direction.

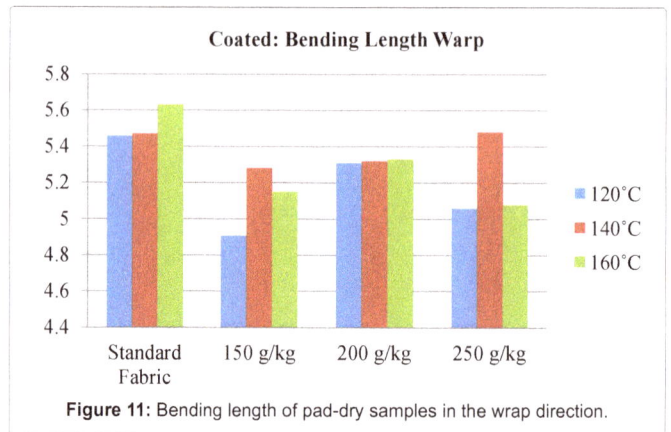

Figure 11: Bending length of pad-dry samples in the wrap direction.

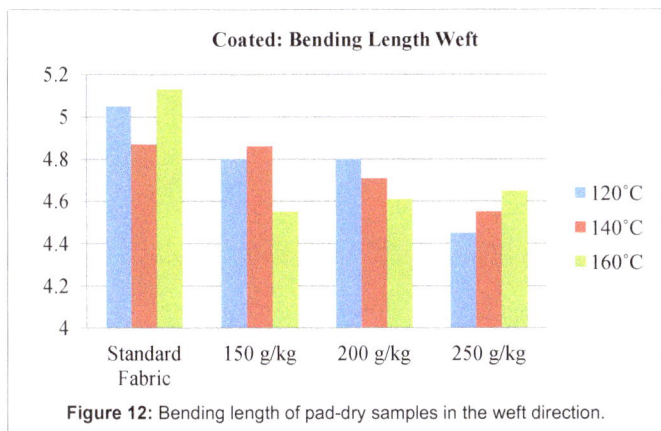

Figure 12: Bending length of pad-dry samples in the weft direction.

pad-dry samples is prominent against coated fabric as thickeners and binder adds in stiffness. Whiteness Index decreases less in pad-dry as compared to coated samples. Aloe Vera itself is not transparent and bares light green color, imparting some change in color of fabric, even when not cured at high temperature. The stiffness of coated fabric is greater as compared to padded samples, the main reason for that is binder and thickener.

Future Work

Aloe Vera can be mixed in the printing paste, but its compatibility with the auxiliaries must be checked before its application for textile printing. A careful selection should be made while choosing auxiliaries for coating so it could not affect the mechanical properties or whiteness of the fabric. Aloe Vera can be applied to the dyed fabric, for this shade change must be considered in different compatible natural plant extracts that are antimicrobial could be used for multi effects.

Conflict of Interest

We, the all authors verify and declared that there is no conflict of interest among any other organizations.

References

1. Joshi R, Purwar M (2004) Recent Developments in Antimicrobial Finishing of Textiles. AATCC Review 4: 22-26.

2. Hipler P, Elsner UC (2006) Antimicrobials and the Skin Physiological and Pathological Flora. Current Problems in Dermatology 33: 35-41.

3. Joshi M, Ali SW, Purwar R (2009) Ecofriendly antimicrobial finishing of textiles using bioactive agents based on natural products. Indian Journal of Fiber and Textile Research 34: 295-304.

4. Kumar DS, Arulselvan P, Sethilkumar GP, Subramanian S (2006) In vitro antibacterial and antifungal activities of ethanolic extract of aloe vera leaf gel. Journal of Plant Science 4: 348-354.

5. Ramachandra CT, Rao PS (2008) Processing of Aloe Vera Leaf Gel: A Review, American Journal of Agriculture and Biological Sciences 3: 502-510.

6. Wiyong K, Vichuta L, Suvimol S, Uracha R (2009) Antibacterial effect of apatite-coated titanium dioxide for textiles applications. Nanomedicine 5: 240-249.

7. Mahesh S, Manjunatha AH, Vijay GK (2011) Studies on Antimicrobial Textile Finish Using Certain Plant Natural Products. International Conference on Advances in Biotechnology and Pharmaceutical Sciences, Bangkok: 253-258.

8. Ramasamy R, Balakumar C, Kalaivani J, Sivakumar R (2011) Dyeability and antibacterial properties of cotton fabrics finished with punica granatum extracts. Journal of Textile Apparel and Technology Management 7: 1-12.

9. Kamble KKM, Vishwanath B, Patil CCS (2013) Antimicrobial activity of aloe Vera leaf extract. International Journal of Applied Biology and Pharmaceutical Technology 4: 286-290.

10. Fatemeh NB (2013) Antibacterial activities and antioxidant capacity of Aloe Vera. Organic and Medicinal Chemistry Letters 3: 1-8.

24

Development of High-performance Single Layer Weft Knitted Structures for Cut and Puncture Protection

Fangueiro R[1]*, Carvalho R[1], Silveira D[1], Ferreira N[2], Ferreira C[2], Monteiro F[2] and Sampaio S[1]

[1]Center for Textile Science and Technology, University of Minho, Guimarães, Portugal
[2]A Ferreira e Filhos S.A., Vizela, Portugal

Abstract

The number of violent situations against security agents using cut and puncture elements, like knives and needles, are increasing daily all over the world. There is a real need for the development of flexible materials able to protect these security professionals without compromising their comfort in different working conditions. The aim of this study is to develop single-layer weft knitted fabrics for cut and puncture protection to be used as protective clothing. Three different weft knitted structures were selected (single jersey, crepe, and moss tuck stitch) in order to study the effect of the yarn arrangement on cut and puncture performance. Different knitted fabrics have been produced with different materials including ultra-high molecular weight polyethylene, para-aramid, high tenacity polyester, high tenacity polypropylene and high tenacity polyamide. In order to study the performance of each structure in combination with the different materials, samples were tested according to EN 388. It was proved that due to the tuck stitches, crepe and moss structures improved cut and puncture performances in comparison with jersey structures. The conical puncture resistance was mainly attributed to the structure (high friction between the yarns and within the yarns due to the dense structure), the knife puncture resistance was mainly attributed to the strength of the yarns, and the cut resistance was mainly attributed to the structure (stretching and thickness) and to the mechanical properties of the yarns. Based on the experimental results and according to EN 388 test classification, moss tuck stitch structure produced with ultra-high molecular weight polyethylene presents the highest cut resistance, classified as level 5, and the highest conical puncture resistance, level 4. Crepe structure produced with para-aramid presents the highest knife puncture resistance. It is suggested the use of single layer moss tuck stitch fabric produced with ultra-high molecular weight polyethylene as protective clothing.

Keywords: Cut; Puncture; Protection; Tuck stitch; Ultra-high molecular weight polyethylene (UHMW-PE)

Introduction

The number of violent situations against security agents using cut and puncture elements, like knives and needles, are increasing daily all over the world. There is a real need for the development of flexible materials that are able to protect these professionals without compromising their comfort in different working conditions. Materials normally used in the equipment for personnel protection are usually based high performance fibers like glass, carbon or ultra-high molecular weight polyethylene (UHMW-PE) blended with conventional fibers like polyester, cotton or polyamide to provide the required comfort and flexibility [1]. Ideally, such garments should be flexible, pliable, soft and cut/abrasion resistant. Unfortunately, any improvement in the cut and/or abrasion resistance has usually been at the sacrifice of the other properties. In addition, if puncture resistance is needed, multiple layers of woven fabric are typically required, particularly made from high strength performance yarns, such as aramid [2]. Often, in knit fabrics, puncture resistance has been extremely difficult, if not impossible, to achieve due to knit stitches often being able to have mobility thus "robbing" yarn from adjacent stitches to open a hole in the fabric, without cutting or tearing the yarns [2]. Knitted fabric was seldom commended to be used as stab-resistant materials. But studies have also pointed out that the protective material based on knitted structure had the features of low weight, better designability, fulfilling wide-area protection, etc., [3]. Polo et al. reported that the multi-layer knitted fabric could absorb penetration energy, and possessed a fairly well shearing resistance, of which stitches locked the knife to stop penetrating before the fabric was destroyed completely [4]. Yao et al. [5] claimed that weft-knitted structure could resist stronger penetration force through the deformation of weft loops and self-locking, anyhow, it was self-evident that fabric had a larger deformation, and a deeper penetration. Lijuan et al. [6] investigated the structure and property of stab resistant warp-knitted single-face fabric. The study found that the underloop structure peculiar to warp-knitted fabric could stabilize the stitch, and added the yarn's agglomeration around knife edge, which had an obvious advantage in penetration force and yarn strength efficiency. The above studies revealed that the textile structure mainly suffered shearing and tensile action when the knife penetrated into the fabric. High-strength and good shearing-resistant fibers combining with tight textile structure contributed to a good stab resistance. Besides, the fabric distortion could absorb the penetration energy which could improve the stab.

The aim of this study is to develop single-layer weft knitted fabrics for cutting and puncture protection to be used as protective clothing for police agents, body-guards, etc. Different knitted fabrics have been produced in an electronic flat knitting machine: single jersey, crepe, and moss tuck stitch, using variations on normal and tuck loops in the coursewise and walewise directions. The selection of these structures was justified by their distinct deformation geometry. Previous studies showed that low deformation structures could absorb penetration energy easily, which could improve the stab resistance [4-6]. These two technological parameters are indicated in the literature as among the

*Corresponding author: Raul Fangueiro, Center for Textile Science and Technology, University of Minho, Guimarães, Portugal
E-mail: rfangueiro@civil.uminho.pt

most important factors of influence for the knitted fabrics.

Herein, it will be shown how the structure has a clear influence on elongation due to the specific geometry of each structure (jersey, crepe and moss) and how it improves stab resistance performance.

Fabrics have been produced with different materials including ultra-high molecular weight polyethylene (UHMW-PE), para-aramid (p-AR), high tenacity polyester (PES HT), high tenacity polypropylene (PP HT) and high tenacity polyamide (PA HT). In order to study the performance of each structure in combination with the different materials, samples were tested under shear and puncture, according to EN 388.

Materials and Methods

Yarns

High performance fibres based yarns, namely high tenacity polyester (PES HT), high tenacity polyamide (PA HT), high tenacity polypropylene (PP HT), ultra high molecular weight polyethylene (UHMW-PE), and para-aramid were used to compare the stab and puncture resistance performances of different weft knitted fabrics. Five specimens of high performance fibres based yarns were tested according to NP EN 2060, ASTM 3108, ASTM 3412 and NP EN 2062, to determine their yarn linear density, coefficient of friction and tensile properties. The yarn testing results are summarized in Table 1.

Knitted fabrics

Single jersey, crepe and moss tuck stitch weft knitted fabrics were produced for this research (Figure 1) on a E10 Stoll electronic flat knitting machine, with similar adjustment parameters namely cams settings, yarn feeding tension and fabric take down. In single jersey structure just normal loops on a single needle bed are used; in the crepe structure normal and tuck loops are combined in a single needle bed; in the moss tuck stitch structure the front and back needle beds are utilized to combine tuck and normal loops. Table 2 shows the physical characteristics of the weft knitted fabrics produced.

Puncture and cut

The puncture resistance using conical and knife probes was evaluated using a Hounsfield H 100 KS universal equipment (Figure 2a), with a cross-head speed of 100 mm/min and with 20 mm distance between the probe and the specimen, according to EN388 standard. The probe (Figures 2b and 2c) is pushed through the material at a fixed speed and the force required to penetrate through the material is measured. The maximum load, extension and energy of penetration were measured and averaged for five samples. Load-displacement curves were established and the deformation of the knitted fabrics in the penetration area analyzed.

Cut resistance performance was evaluated according to standard EN388 for two samples with five cuts each, using Coup test device (Figure 3). This device consists of a circular free-rotating blade, under pressure from a standard weight (5N), which is moved backwards and forwards over the surface of the specimen over a fixed stroke length. Number of cycles required for the blade to cut through the material is used to calculate the Cut Index, varying from 0 to 5 (5 represents the highest performance level).

Results and Discussion

Puncture with a conical probe

Conical probe puncture occurs in four different stages: (i) contact pressure of the tip of the conical probe against a fabric; (ii) slippage of the tip into both within yarns and between yarns, resulting in puncture; (iii) friction of the fabric against the conical section of the point; and (iv) slippage of the conical section through the fabric [7,8]. Resistance to conical probe puncture is however a much more complex problem as it also requires full consideration of the tightness of the fabric and of the profile of the puncture element. A dense structure is beneficial in resisting the conical puncture probe [9]. A high density fabric restricts the mobility of the yarns, which induces the increase of the inter-yarn friction and due to the friction within and between the yarns that increases the absorption energy. When the conical probe cannot perforate the fabric, the yarns are easily deformed through slipping over each other in the walewise and coursewise directions to allow absorbing more energy and increasing the resisting force for probe penetration, leading to a higher puncture resistance.

As can be seen in Table 3, structure 1 (single jersey) presents higher energy absorption than structure 2 (crepe) except for UHMW-PE. These results may be explained based on the lower elongation of the crepe structure (structure 2) compared to the jersey structure (structure 1). The single layer structure 2 during the resisting force for probe penetration absorbs less energy and it might be due to the tuck stitches within the structures that are at higher stress and that makes the yarns to reach faster the breaking point. As for structure 3 there is an increase in the absorption energy when compared to structure 2 and to structure 1. Structure 3 when stretched causes the face loop courses to cover the reverse loop courses, making the fabric twice as thick as single jersey [7] which facilitates energy dissipation from the conical probe impactor. The other variable influencing the puncture resistance with the conical probe but to a lesser degree is the mechanical properties of the yarn. The most prominent energy absorption changes were recorded for samples UHMW-PE structure 3, UHMW-PE structure 2 and UHMW-PE structure 1. Energy absorption for fabric UHMW-PE structure 3 increased in the order of 344% when compared to UHMW-PE structure 2, and structure 2 increased in order of 10% when compared to UHMW-PE structure 1, emphasising the influence

Yarn material	Linear Density (Tex)	Coefficient of Friction (μ)	Tenacity (N/Tex)	Breaking Extension (%)	Initial Modulus (GPa)	Displacement at maximum force (mm)	Breaking Load (N)
PA HT	47 (± 1.7%)	[0.43-0.70]	0.70 (± 4%)	16.15 (± 5%)	47.3 (± 5%)	80.72 (± 5%)	67.5 (± 4%)
PES HT	55 (± 1.6%)	[0.45-0.70]	0.53 (± 2%)	12.6 (± 6%)	50.2 (± 7%)	63.00 (± 6%)	29.54 (± 2%)
PP HT	110 (± 0.7%)	[0.35-0.55]	0.64 (± 5%)	19.56 (± 9%)	11.8 (± 17%)	97.81 (± 9%)	71.68 (± 5%)
UHMW-PE	176 (± 0.9%)	[0.31-0.37]	2.70 (± 7%)	8.56 (± 36%)	45.3 (± 38%)	43.01 (± 35%)	476.01 (± 7%)
p-AR	173 (± 0.4%)	[0.25-0.35]	1.65 (± 3%)	4.37 (± 4%)	47.7 (± 31%)	21.83 (± 4%)	285.64 (± 3%)

Table 1: Yarn testing results.

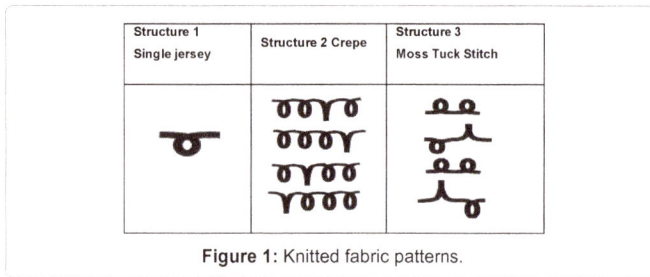

Figure 1: Knitted fabric patterns.

Structure	Yarns	Aerial mass (g/m²)	Loop Length (i)/100 wales (cm)		Density		Tightness factor
			Normal	Tuck	wales/ cm	courses/ cm	
Structure 1 Jersey	PES HT	185.44 (± 7%)	0.66 (± 2%)		5 (± 10%)	7 (± 12%)	11.24
	PA HT	160.19 (± 8%)	0.71 (± 0%)		6 (± 0%)	7 (± 6%)	9.72
	PP HT	429.42 (± 5%)	0.70 (± 0%)		5 (± 10%)	7 (± 8%)	14.16
	UHMW-PE	519.08 (± 7%)	0.88 (± 4%)		4 (± 0%)	5 (± 10%)	15.01
	p-AR	437.75 (± 4%)	0.93 (± 4%)		4 (± 11%)	6 (± 8%)	14.16
Structure 2 Crepe	PES HT	287.75 (± 8%)	0.44 (± 9%)		8 (± 7%)	16 (± 4%)	16.86
	PA HT	245.83 (± 24%)	0.46 (± 9%)		7 (± 6%)	15 (± 3%)	14.87
	PP HT	527.75 (± 8%)	0.56 (± 3%)		7 (± 0%)	14 (± 3%)	18.8
	UHMW-PE	664.33 (± 2%)	0.7 (± 1%)		6 (± 0%)	10 (± 4%)	18.87
	p-AR	729.92 (± 14%)	0.7 (± 1%)		7 (± 0%)	12 (± 7%)	18.71
Structure 3 Moss Tuck Stitch	PES HT	532.33 (± 4%)	0.47 (± 2)	0.47 (± 3%)	8 (± 5%)	8 (± 5%)	15.73
	PA HT	426.75 (± 8%)	0.50 (± 1%)	0.51 (± 1%)	9 (± 0%)	10 (± 5%)	13.61
	PP HT	852.75 (± 2%)	0.62 (± 1%)	0.60 (± 2%)	7 (± 0%)	8 (± 0%)	17.15
	UHMW-PE	1762.67 (± 5%)	0.68 (± 0%)	0.82 (± 1%)	6 (± 0%)	8 (± 0%)	17.65
	p-AR	1196.83 (± 4%)	0.62 (± 2%)	0.75 (± 3%)	7 (± 0%)	13 (± 4%)	19.18

Table 2: Characteristics of the weft knitted fabrics.

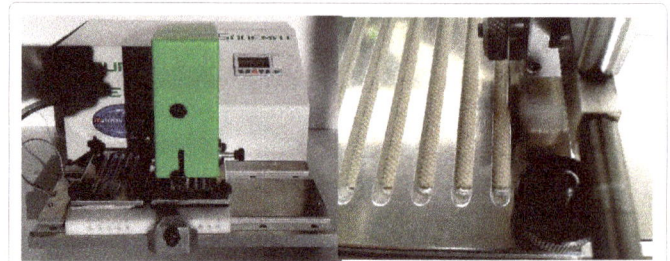

Figure 2: (a) Puncture testing set-up, (b) conical probe, (c) knife probe.

of the structure on the stab resistance. Of all the samples, UHMW-PE structure 3 presents the highest energy absorption capacity.

Figure 4 shows the load-displacement curves for the weft knitted samples tested. Highest energy absorption fabric, UHMW-PE structure 3, had maximum resistance of 579 N at 13 mm. Graphics show that the structure has a stronger influence on the conical probe puncture behaviour.

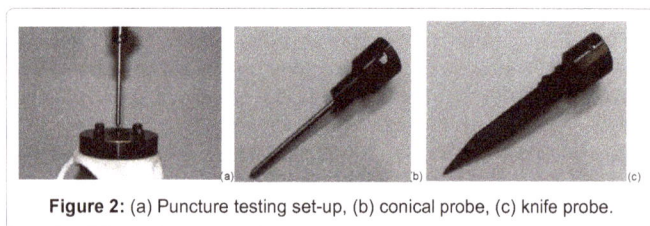

Figure 3: Coup testing device.

Puncture probes	Structure	Yarns	Displacement at max. load (mm)	Maximum Load (N)	Energy (J)
Conical probe	Structure 1 jersey	PES HT	10.180	132.82	0.3321
		PA HT	14.928	131.25	0.3957
		PP HT	13.932	398.46	0.8632
		UHMW-PE	13.956	314	0.3741
		p-AR	16.080	581.72	1.3400
	Structure 2 crepe	PES HT	11.500	172.4	0.2501
		PA HT	18.610	95.71	0.2990
		PP HT	13.800	364.8	0.3648
		UHMW-PE	14.082	124.802	0.4111
		p-AR	15.630	255.54	0.5740
	Structure 3 moss tuck stitch	PES HT	11.500	254.4	0.4281
		PA HT	15.280	211.36	0.7102
		PP HT	12.920	536.3	0.5363
		UHMW-PE	13.152	579.16	1.8201
		p-AR	11.862	338.64	0.7646

Table 3: Puncture testing results.

Images of structure 1, structure 2 and structure 3 produced with PP HT, p- AR and UHMW-PE yarns after conical probe puncture are shown in Figure 5. Structure 1 produced with PP HT shows a permanent hole in the fabric that persisted when the conical probe was removed, it seems that conical probe penetrates the fabrics after fracturing fibres and yarns, as seen in Figures 5a and 5b illustrates damage area of PP HT structure 2 fabric, showing a hole left on the fabric after puncture and some fibres and yarns fractured. The hole of PP HT structure 2 is smaller than the hole of PP HT structure 1 as structure 2 presents tighter structure. The tuck stitches acted holding the structure and limiting the area of damage. From Figure 5c it is seen that few fibres and yarns are fractured and the hole created by the test is very small. The tight and thick structure 3 in combination with the low initial modulus of PP HT yarns permit elastic deformations that are recoverable upon removal of conical probe. Figure 5d shows no fractured fibres and yarns as the conical probe penetrated the fabric primarily by displacing them without damaging. P-AR structure 1 is an open structure produced with a high tenacity yarn thus fabric damage is more likely to be from structure deformation rather than fracturing fibres and yarns. Figure 5e shows few fractured fibres and yarns with a hole in the fabric. In Figure 5f it is evident more fibres and yarn damage when comparing with Figure 5e and the hole from fabric in Figure 5f is smaller than from Figure 5e. It may be that p-AR structure 3 offered more resistance to puncture; fibres and yarns could not be pushed to side and the yarns were fractured in order to allow conical probe to pass through the fabric. Figures 5g-5i follow the same trend as Figures 5d-5f, respectively. Major differences are seen in Figures 5i and 5f; UHMW-PE structure 3 offered the highest puncture resistance of all samples but it is observed that the part of the fabric that was deflated

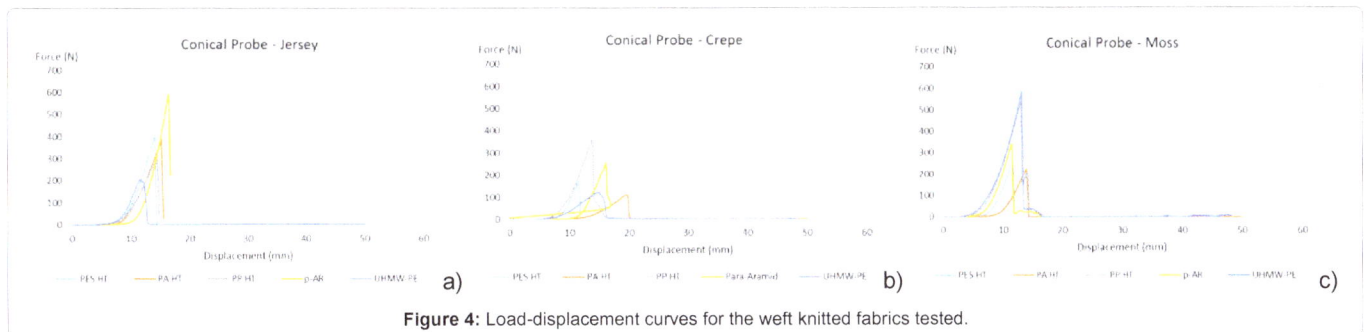

Figure 4: Load-displacement curves for the weft knitted fabrics tested.

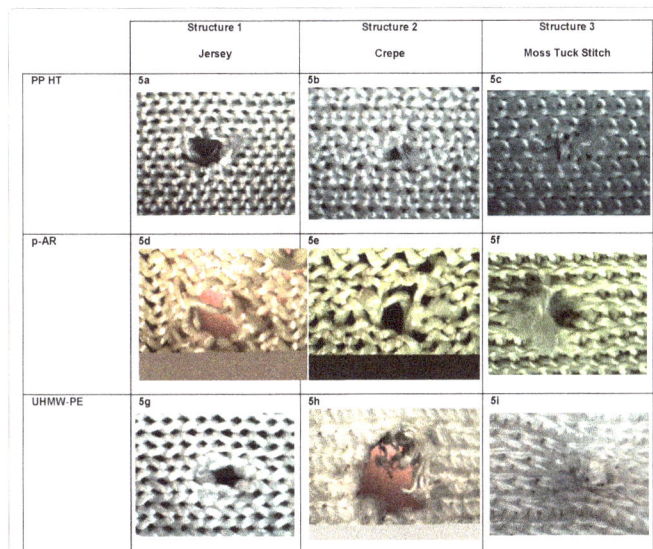

Figure 5: Images of structure 1, structure 2 and structure 3 produced with PP HT, p- AR and UHMW-PE yarns after conical probe resistance test.

Puncture probes	Structure	Yarns	Displacement at max. load (mm)	Maximum load (N)	Energy (J)
Knife probe	Structure 1 jersey	PES HT	8.908	11.47	0.0198
		PA HT	8.938	13.26	0.0238
		PP HT	8.308	36.462	0.0658
		UHMW-PE	17.202	56.806	0.3468
		Para-aramid	15.968	70.17	0.2642
	Structure 2 "crepe"	PES HT	12.330	17.675	0.0542
		PA HT	21.010	23.98	0.0902
		PP HT	26.130	58.5	0.4121
		UHMW-PE	14.398	2.25	0.0084
		Para-aramid	21.060	111.086	0.5035
	Structure 3 "Moss tuck stitch"	PES HT	7.980	42.65	0.1257
		PA HT	16.020	43.62	0.1480
		PP HT	21.480	85.1	0.4820
		UHMW-PE	6.882	48.146	0.1048
		Para-aramid	17.444	89.216	0.3862

Table 4: Puncture testing results.

after conical probe puncture was not lifted back up, and it might be due to the UHMW-PE yarn that it is a yarn with very high tenacity but very high initial modulus which does not permit elastic deformation recover. It is quite clear from Figure 5 that, in general, all fabrics from structure 1 have the most fabric damage and thus offering less puncture resistance and that all fabrics from structure 3 have the least fabric damage offering highest puncture resistance performance.

Puncture with a knife probe

During knife probe puncture through a fabric, fibre in contact with the knife edge has high stress and tends to be stretched, and the yarns will move in weft and warp directions, allowing the edge of the knife to pass through. The fabric will be deformed under the build-up force, and the edge of the knife will start to cut the yarn. The resistance to puncture of a knife depends on the material, the cross-section of the yarns, their strength and the support points in the base fabric [10]. Although a tight structure with higher density is helpful to resist puncture, the little slippage space may be a disadvantage for yarns, because they are easily cut directly by the blade of the knife [9].

In general, the energy absorption increases from structure 1 to structure 2 and from structure 2 to structure 3 samples, as can be seen in Table 4. The moss fabrics present the best knife puncture resistance, followed by the crepe fabrics. These might be explained by the tightness factor showed by crepe structures (Table 2), and the influence of the

tuck stitches. Anyhow p-AR structure 2 sample showed the highest knife puncture resistance. Structure 3, being a more compact structure, it was expected to present the highest knife puncture resistance, however its tight structure with higher density might explain these failure result.

Fabric damage under the knife is mainly dependent on the mechanical properties of the yarn and less influenced by the fabric structure itself, as can be seen in Figure 6. Highest energy absorption fabric, p-AR structure 2, presents maximum resistance of 111 N at 21 mm.

Analysing images of structure 1, structure 2 and structure 3 produced with PP HT, p- AR and UHMW-PE yarns after knife probe puncture (Figure 7), it is quite clear that in all the fabrics the area damaged after the knife puncture is higher than that after the conical probe puncture. From Figure 7a it is found that the dominant failure mechanism is fibre fracture due to cutting; cutting blade of the knife has caused severe and extensive fibre fracture. Significant damage is observed in PP HT structure 2 (Figure 7b). In case of PP HT structure 3, fibre and yarn cutting is evident, but the magnitude of damage is less compared to PP HT structure 2 and to PPHT structure 1. This might be due to the thickness (twice as jersey) of structure 3, with this additional "layer" structure deformation is less evident. Figures 7d-7f followed the same trend as Figures 7a-7c respectively, but overall the puncture damage is smaller and it might be due to the mechanical properties of p-AR yarn. P-AR yarn is a high tenacity yarn and has a moderate initial modulus which requires more energy to be cut and fractured and offers more elasticity to move the yarns from under the knife, minimizing

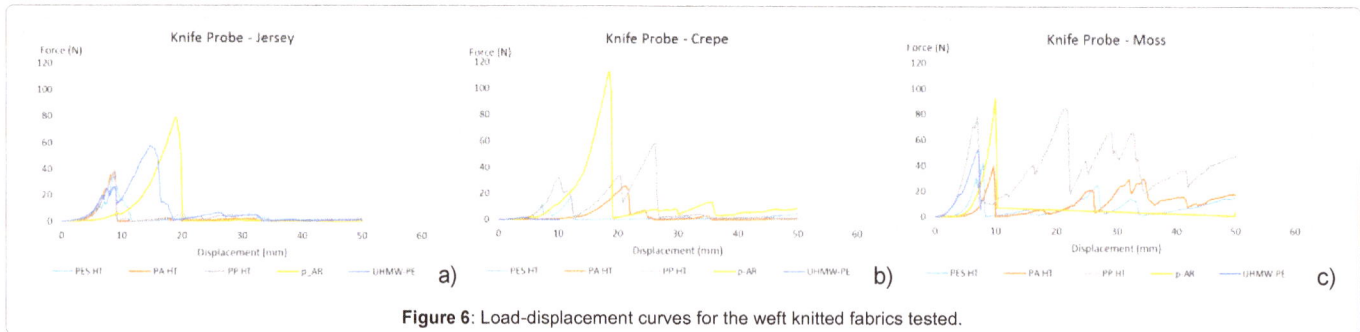

Figure 6: Load-displacement curves for the weft knitted fabrics tested.

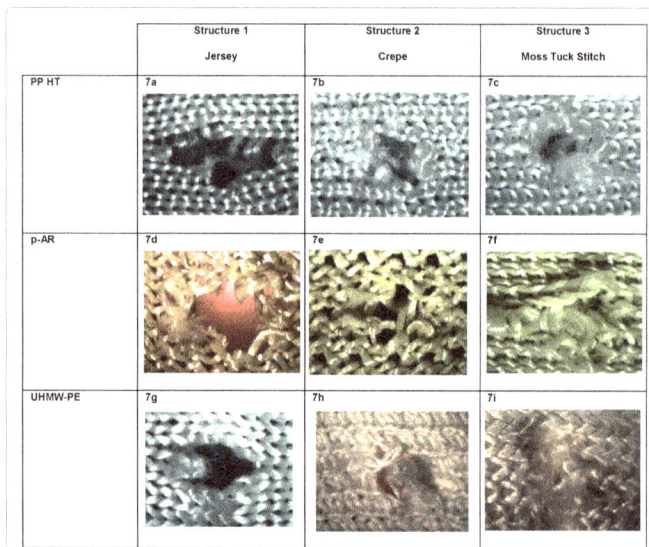

Figure 7: Jersey, crepe and moss knitted fabrics produced with PP HT, p- AR and UHMW-PE yarns after knife probe resistance test.

Yarns	Structure 1 (Jersey)		Structure 2 (Crepe)		Structure 3 (Moss tuck stitch)	
	Level	Cut index	Level	Cut index	Level	Cut index
PES HT	1	2.4	3	7.1	4	10.5
PA HT	1	2.1	3	5.4	2	3.4
PP HT	1	2.1	1	1.9	2	4.5
UHMW-PE	4	3.4	4	15.4	5	33.4
p-AR	2	10.9	5	21.1	5	48.4

Table 5: Structures cut index according to EN388.

the fabric damage. Fabric damage presented in Figures 7g-7i are comparable to fabric damage presented in Figures 7d-7f, respectively. The big difference being on the energy required to cut and fracture the yarns, fabrics produced with UHMW-PE yarn requires less energy than fabrics produced with p-AR yarns. UHMW-PE yarn is a high tenacity yarn and has a high initial modulus, which offers less elasticity to move the yarns from under the knife and the knife penetration occurs sooner.

Therefore the results confirm the literature review as the conical puncture resistance of a fabric is mainly attributed to the friction between the yarns and within the yarns, and the knife puncture resistance of a fabric is mainly attributed to the strength of the yarns [11].

Cut resistance

Cut resistance is defined as the ability of a material to resist damage when challenged with a moving sharp edge. The results presented in

Table 5, show the influence of the fabrics structure and of the yarn type. UHMW_PE moss tuck stitch fabrics present the highest cut resistance performance while the jersey fabrics the lowest. It is clear from Table 4 that the cut resistance index of UHMW_PE moss tuck stitch has increased considerably to the extent of 129%, when compared with the same fabric made of crepe structure, and 344% when compared with jersey structure. The increase in cut resistance is attributed to the good stretch (the yarn moves and the blade actually slides across the yarn without catching it to cut) and thickness of structure 3, in addition to the tensile strength of the yarn.

Conclusions

The effect of the weft knitted structure on the fabric puncture and cut resistance has not been investigated enough. Most work has focused particularly on multi-layered weft knitted fabrics [3,12]. The present study took into consideration the puncture and cut resistance behaviour of single-layer weft knitted fabrics (jersey, crepe and moss), produced with five types of high performance yarns (PES HT, PA HT, PP HT, UHMW-PE, and p-AR) in order to investigate the effect of the fabric structure on the mechanical protective performances. It was proved that due to the tuck stitches, crepe and moss structures improved cut and puncture performances in comparison with jersey structures. The conical puncture resistance of a single-layer knitted fabric was mainly attributed to the structure (high friction between the yarns and within the yarns due to the dense structure), the knife puncture resistance was mainly attributed to the strength of the yarns, and the cut resistance was mainly attributed to the structure (stretching and thickness) and to the yarns strength.

Based on the experimental results and on the EN 388 test classification, UHMW-PE moss tuck stitch structure present the highest cut resistance - level 5, and the highest conical puncture resistance - level 4. p-AR crepe structure presents the highest knife puncture resistance.

According to the results it is suggested the use of UHMW-PE single layer moss tuck stitch structure as protective clothing. Thermo-physiological comfort properties, such as air permeability, water vapour permeability, thermal resistance, wick ability, absorbency, drying rate are being assessed on the UHMW-PE single-layer moss tuck stitch in order to study the relationship between comfort and protection.

Acknowledgments

This project has been funded by COMPETE/QREN under the Project No. 2014/38320.

References

1. Dolez PI, Vu-Khanh T (2009) Recent developments and needs in materials used for personal protective equipment and their testing. International Journal of Occupational Safety and Ergonomics 15: 347-362.

2. Garcia V (2014) Sediment distribution patterns on the Galicia-Minho continental shelf.

3. Xuhong M, Xiangyong K, Gaoming J (2013) The experimental research on stab resistance of warp-knitted spacer fabric. Journal of Industrial Textiles 43: 281-301.

4. Flambard X, Polo J (2004) Stab resistance of multi-layers knitted structures. J Adv Mater 36: 30-35.

5. Xiaolin Y, Guanxiong O, Yaming J (2011) Research on the stab resistant mechanism of the weft knitted fabrics. TianJin University, China.

6. Lijuan L, Gaoming J, Xuhong M (2011) Structure and properties of stab resistant warp knitted fabric. J Text Res 32: 48-51.

7. Spencer DJ (2001) Knitting Technology: A Comprehensive Handbook and Practical Guide (3rd edn.). Woodhead Publishing Limited, Cambridge, England.

8. Triki E, Nguyen-Tri P, Gauvin C, Azaiez M, Vu-Khanh T, et al. (2015) Combined puncture/cutting of elastomer membranes by pointed blades: Charaterization of mechanisms. Journal of Applied Polymer Science.

9. Miao X, Jiang G, Kong X, Zhao S (2014) Experimental Investigation on the Stab Resistance of Warp Knitted Fabrics. Fibres & textiles in Eastern Europe 5: 65-70.

10. El Messiry M (2014) Investigation of Puncture Behaviour of Flexible Silk Fabric Composites for Soft Body Armour. Fibres & textiles in Eastern Europe 5: 71-76.

11. Gong X, Xu Y, Zhu W, Xuan S, Jiang W, et al. (2014) Study of the knife stab and puncture-resistant performance for shear thickening fluid enhanced fabric. Journal of Composite Materials 48: 641-657.

12. Alpyildiz T, Rochery M, Kurbak A, Flambard X (2011) Stab and cut resistance of knitted structures: a comparative study. Textile Research Journal 81: 205-214.

25

The Impact of Technical Textiles on Health and Wellbeing Current Developments and Future Possibilities

Diane E*

Faculty of Sports Science, Leisure and Nutrition, Liverpool John Moores University, UK

Abstract

he impacts of technical textiles upon society have been influential and have fundamentally impacted upon the structure of everyday human life. As populations continue to age, patient expectations for implant performance will continue to rise. The potential for high performing bio-medical materials is being realized and will continue to advance the evolution towards more fibre based approaches to device development that will perform more effectively and comfortably both outside and inside the human body. Medical textile breakthroughs have provided a broad array of implantable devices including vascular grafts, surgical mesh, heart valve components, orthopaedic sutures and fabric scaffolds designed to aid tissue repair. Technical textiles have the potential to perform better, last longer and increase comfort in the body as well as alleviating concerns around more traditional implanted materials such as metal. Although minimal invasive approaches to cardiovascular and orthopaedic procedures can be promoted with the use of technical textiles, they can be used in a more innovative way than simply repairing or replacing damaged tissue as surgeons can now use their ability to aid regrowth and natural functioning of damaged areas within the body. Global recessions and healthcare cost increases have had a marked global impact on every institution and organisation involved with the delivery of healthcare services and products. The NHS is constantly looking to make savings without compromising quality and this is placing the technical textile industry in a dynamic position to respond to these economic pressures. This paper examines key areas in which innovation in textile technology is promoting health and wellbeing. It considers the commercial and academic context in which these innovations are being developed, examines the main sectors in which these innovations are being directed, and concludes by suggesting which aspects of health and wellbeing are likely to gain the most from the application of technical textiles in the short and long term.

Keywords: Technical textiles; Healthcare; Wound treatment; Textile; Innovation; Biomimicry

Introduction

Textile use in health and wellbeing is nothing new. The Edwin Smith papyrus of 1550 BC provides the first recorded use of cloth bandages, woven metal known as chain mail provided some protection from the 3rd century BC in Europe whilst the Mongols used silk shirts as a barrier to infection in the Asian steppe during the 13th century AD. The acceleration in the possibilities offered by textiles occurred at the turn of the 20th century with the invention of the first synthetic fibres such as nylon, allowing man to select the qualities desired in a textile rather than adapt those offered by natural materials such as wool, cotton and silk.

By the latter half of the 20th century the science of synthetic fabric manufacture had reached such a level of sophistication that a new term was required to differentiate fabrics developed for utilitarian as opposed to aesthetic use: technical textiles. The Textile Institute of Manchester provides the clearest definition of what constitutes a 'technical textile': "Materials and products intended for end-uses other than non-protective clothing, household furnishing, and floor covering, where the fabric or fibrous component is selected principally but not exclusively for its performance and properties as opposed to its aesthetic or decorative characteristics."

Early technical textiles owed their qualities to innovative use of materials, such as polymers. An obvious example with applications for health and wellbeing is Kevlar, a polymer derived fibre invented by Stephanie Kwolek of the Du Pont Corporation in 1964. Five times the tensile strength of steel, Kevlar is used for preventative safety such as the creation of stronger brake pads and cables, and for bulletproof vests. Indeed it has proved so successful in saving lives when used for the latter function that its manufacturer Du Pont has created a Kevlar Survivors' Club.

In the 21st century innovation in technical textile materials continues, indeed it is accelerating. Even more exciting however, are the possibilities offered by new areas of development in synthetic fibres – nanotechnology, fibre coating, non-woven materials, wearable electronics and the increasing use of biomimicry to name the most notable. Reflecting the needs of a world where access to technology continues to be extremely unequal, encouraging research has also been undertaken into how technical textiles can supplement the qualities of relatively cheap and accessible natural fabrics such as wool.

Driving forces behind technical textile innovation

The main areas in which these developments are focussed will be considered in Section 2. Prior to this let us consider the institutions that are driving the rapid pace of technical textile development in the field of health and wellbeing. The remit of these institutions is clearly a major factor in directing research. '

a) Multinationals: American multinationals such as Du Pont can be seen to have led the development of technical textiles during the 20th century. Today they are still actively researching in this field, albeit with an emphasis on evolving existing materials rather than developing

---

*Corresponding author: Diane E, Faculty of Sports Science, Leisure and Nutrition, Liverpool John Moores University, UK
E-mail: D.Eldridge@ljmu.ac.uk

new ones given the commercial consequences of rendering one of their own products obsolete. A good example of this is the adaptation of Du Pont's Teflon for coating spectacle lenses.

b) Specialist companies: One of the more positive consequences of globalisation is that fabric manufacturers in first world countries have been forced to become technical specialists order to survive international competition. In so doing they have become driving forces of innovation in the technical textile sector, combining a commitment to new ideas with commercial realism. An example is Baltex, the descendent of a 19th century Derbyshire knitting concern. Unable to compete in the standard wool market, Baltex have brought their knitting expertise to development of advanced technical textiles such as their XD Spacer fabrics which provide protection against impact from ballistic threats.

c) Academic institutions: There is a considerable global body of research devoted to technical textiles. This includes first world countries with a traditional of textile manufacture and innovation such as Britain e.g., Manchester's Textile Institute and the University of Bradford and countries which are large scale growers and manufacturers of fabrics today such as India e.g. The Indian Institute of Technology [1]. In broad terms, the former tend towards research into leading edge innovation in technical textiles, whilst the latter can be seen to focus on adaptation of technical textile technology to suit manufacturing and less economically developed areas.

d) Medical bodies: State medical bodes such as Britain's National Health Service and charities such as *Medicin sans Frontiers* provide the bridge between technical textile innovation and the commercial issues associated with their manufacture. Medical Bodies' most significant contribution is their ability to provide projections of current and future health and well-being issues that might be addressed by technical textiles. For instance it was at the instigation of third world medical charities that 'nano meshes' have been developed to filter out bacteria such as gastroenteritis from water supplies.

e) Pressure groups: These organisations draw public, commercial and academic attention to issues that could be addressed through the use of technical textiles. The activities of Age Concern in Britain for example drew attention to the needs of the over-60 age group of the British population with regards to outdoor activities such as walking or jogging as a means of promoting health. The University of Wales now acts as a link between research and commerce in meeting this demand using technical textiles [2].

Main Areas of Technical Textile Innovation

Treatment of wounds

The use of textile bandages to protect wounds from infection is an ancient art. Technical textiles have transformed the bandage from a passive to an active component in healing. An established example is the introduction of alginate fibres derived from seaweed into the part of a dressing in contact with the wound encouraging a 'moist healing' process through the formation of a water soluble gel [3]. More advanced is the use of plasma coatings evolved by the Indian Institute of Technology which are used to affect the surface behaviour of textiles rendering them either hydrophilic (water absorbing) or hydrophobic (water repelling) as the particular nature of the wound requires [4].

Intracorporeal treatment

Non-degradable textiles have a wide variety of applications in internal treatment of the human body. Several current developments have tremendous potential to address specific health concerns. The need for hygienic delivery of non-refrigerated medicine and vaccines in the third world has been addressed both by new technical textiles and by innovative use of natural fabrics. An example of the former is German textile manufacturer Schoeller's iLoad reloadable drug delivery fabric which uses a carrier material to unload medication into the skin when stimulated by moisture, vibration, warmth and perspiration. An instance of the latter is the innovative use of silk by Vaxess Technologies of the USA for the hygienic transport of vaccines. Fibroin, a silk protein is used as a stabilizing agent in the preparation of dessicated vaccines which are rehydrated at point of use [5].

Implants into the body represent an established use of intracorporeal textiles with very broad applications. Age and disease related degenerative conditions as well as the consequences of trauma are all commonly treated by established textiles in the first world – artificial joints are commonly manufactured from polyethylene, artificial corneas from polymethyl methacrylate and artificial skin from a long chain polymer called chitin for example. Existing materials are biocompatible with the human body and thus the problem of the body rejecting the implant has largely been resolved. Deterioration in strength and functionality of the implant owing to the hostile environment of the human body remains an issue however; polyamide commonly used for artificial tendons can lose much of its strength within two years [6]. Clearly investigation into materials for longer lasting implants may provide a significant avenue for technical textile development.

Perhaps the most remarkable current development is the use of microscopic quantities of a nanofibre in the treatment of cancer *within* the bloodstream by researchers at the University of California, Los Angeles (UCLA). One of the main causes of cancer spread through the body are circulating tumour cells (CTCs) which use the bloodstream to establish new tumours. The 'NanoVelcro' chip acts like Velcro, trapping and isolating passing CTCs enabling them to be analysed and individual treatments tailored to the individual [5].

Extracorporeal

Extracorporeal (outside the body) treatments involve the use of mechanical organs, typically for blood dialysis. The innovative use of established textiles has been crucial to the successful deployment of these machines as they form the basis of filters that mimic the actions of membranes within human organs – cellophane for example replicates the function of the kidney in removing waste materials [6]. As it becomes possible to produce artificial versions of a greater variety of bodily organs, technical textiles will undoubtedly have a key role to play in future health.

Monitoring

Traditionally monitoring of the body has been undertaken by gel based wire sensors which are intrusive, unwieldy and difficult to use when a subject is moving. Electronic technical textiles make a fabric based approach to monitoring possible. British company, SmartLife's HealthVest system is an established form of this system currently employed by Manchester Royal Infirmary which uses dry sensors built into a garment to measure heart rate, ECG morphology, respiratory rate and bodily fluid flows. More advanced monitoring fabrics are currently either at the developmental or pre-production stage which essentially offers a wearable electronic hub which can be programmed via wireless technology to monitor a wide variety of bodily functions.

Filtering

Increasing urbanisation is making the global population evermore susceptible to airborne contamination including pollution and germs. Fabric face masks are a standard feature of life in many of the world's more polluted cities, particularly in the Far East. Technical textiles offer more sophisticated means to mitigate the effects of air pollution however, often at source. The use of textile membrane for the contentious issue of CO_2 capture for example, is being investigated by Australian research organisation CO2CRC. A porous membrane woven from hollow, solvent filled fibres of polymer diffuses and absorbs CO_2 into a treatable liquid Elsewhere an ingenious crowd based solution to air pollution in city centres has been suggested by Catalytic Clothing namely coating the fabric of standard high street clothing with nano-particles of photo catalyst titanium dioxide able to trap airborne particulate which can then be disposed of harmlessly through washing.

The use of nanotechnology in technical textiles can provide beneficial filtering in other areas in addition to air pollution. Carbon nanotubes are now routinely used in water filters in the third world removing viruses such as polio.

Protection

Protection from harm through protective clothing is perhaps the most obvious and widest field of application for technical textiles in the area of health and wellbeing. In general this clothing can be subdivided into the following areas

a) **Contamination:** Industrial and disaster workers have to operate in toxic environments in which bio-hazard suits are needed. Traditional suits made of impermeable materials have been known to cause problems of sweating, misting and heat exhaustion. However the use of laminate materials such as Cloropel, a chlorinated polyethylene allied to positive air pressure within the suits has made 'breathable' [7] bio-hazard protection materials possible. Given the growing demand for environmental clean up operations such as the treatment of radioactive materials in the former Soviet Union for instance, it is likely that increased research will be devoted to developing ever more lightweight and unobtrusive anti-contamination fabrics.

b) **Impact:** A modern version of traditional armour, impact protection fabrics counter the interrelated risks of explosions, stabbing, shooting and blunt impact. Armed forces and police are the traditional customers, and in view of the threat from Improvised Explosive Devices (IEDs) in modern warfare, there is a particularly pressing need for development in this area.

Body armour made of materials such as Kevlar have now been replaced by second generation ballistic fabrics made of High Performance Polyethylene (HPPE) such as Dyneema which give increased resistance to projectiles such as bullets and shrapnel[3]. Extra panels of ceramic can be added to increase ballistic protection to the point where the wearer is effectively bullet proof. The trade off is twofold – such armour is extremely heavy, a major problem is a fast moving, hot combat environment such as Afghanistan, secondly as a consequence of this weight issue only the head and torso can be protected, hence the high proportion of limb injuries encountered. NATO governments are very aware of this issue and their procurement agencies such as Britain's Integrated Soldier System Executive are driving innovation with substantial investment. Immediate developments include the enhancement of low-tech materials to protect a wider proportion of the body – the British Pelvic Protection System for instance is based on treated silk in the medieval manner which provides lightweight protection against shrapnel. The most promising long term

development is to combine high performance polyethylene with a layer of gel that hardens when struck, so-called liquid armour (www.military.com) currently being researched by BAE Systems Ltd and the US Army Research Laboratory which promises to be 50% or less in weight compared to existing systems.

Technical textiles have impact applications, indeed any of the wide variety of professions and activities requiring the use of sharp blades could be rendered safer by its development. In the case of stab protection Dyneema HPPE has just been approved for use in cut resistant gloves by the European Union for abattoir workers and butchers, whilst users of chainsaws utilise technical textiles such as Prolar for leg, hand and groin protection which snag the saw blade. However, these civilian applications tend to be evolutions of military practise rather than drivers of change in their own right.

c) **Temperature:** Like impact, the requirements of the principle end users are driving rapid innovation in the field of temperature resistant clothing, in this case fire services and the motor racing industry. The same requirements apply to the characteristics of these materials too, namely that they be as light and unrestrictive of movement as possible. Both heat resistance and flame retardance are required in this application. Traditionally man-made fibres have been unsuitable for this application on account of their relatively low melting points (polypropylene melts at 150°C, polyester at 255°C), instead treated natural fibres have found favour – the Royal Navy using hydroxide treated cotton known as Proban whilst foundry workers often use wool treated with hexafluoride known as Zirpro Wool [8]. Treated natural fibres have many advantages such as relative cheapness and breathability. However by their very nature they can only be developed to a certain point, and both the fire service and the motor racing industry have temperature requirements beyond their capabilities.

Du Pont's Nomex is the current material of choice – an Aromatic Polyamide or Aramid able to endure temperatures of up to 400°C without charring. Other technical textiles designed specifically for fire fighter units have even more impressive heat resistant properties such as Rhone-Poulenc's polyamide-imide fibre [8]. In all cases however, although these materials are comparatively lightweight, breathability remains an issue, and it is in this regard that further technical textile development would be most desirable. Laminates may well be the solution -a laminate of synthetic knitwear able to withstand temperatures of over 1000°C for up to 12 seconds was developed by Lamination Technologies Inc of Pennsylvania and is used in the clothing of Andy Green, the driver of the rocket propelled Bloodhound supersonic car. Significantly this laminate traps air pockets using an advanced knitting process which not only improves temperature resistance but also breathability.

d) **Ultraviolet:** Increased concerns regarding the dangers of ultraviolet light (UV) have led to research into UV filtering in fabrics, particularly beach and swim wear. Cost, comfort, environmental footprint and the demands of fashion are significant factors in these textiles. Given these factors, a naturally derived coating on conventional nylon and polyester is the preferred mode of development, exemplified by Cocona a fibre derived from coconut waste that combines UV and odour absorbent properties [7].

Fitness

The use of technical textiles in filtering harmful sunlight and for monitoring has already been discussed. Beyond these applications the most promising area of development at present is in the area of outdoor pursuits, and in particular making such pursuits more accessible to the West's ageing population. Comfort is the paramount factor here, and in the development of outdoor clothing *biomimicry* is the driving force in research – learning from the ways that natural materials are formed. As the founder of the American Biomimicry Guild Dr Janine

Benyus states, 'Nature has been innovating for 38 billion years, which has given enough time to establish what works and what doesn't [9] An early example of this process is Velcro, whose hooks were inspired by burdock burrs. More recently, Schoeller Textil AG of Switzerland have launched Nanosphere, an 'adsorbent' fabric featuring a rough micro-texture that repels water in the same manner as lotus and nasturtium leaves [9] Schoeller have also addressed the issue of perspiration via biomimicry with its C Change textile which features a membrane which reacts to changing temperature and moisture by opening to allow the escape of heat and water vapour in the manner of pine cones [10].

Investment in the outdoor clothing sector is rising as the market changes, particularly in Europe where the demand for outdoor clothing from the over 50s is rapidly outstripping those of the younger market. Aside from biomimicry, 'smart textiles' are also being developed by the University of Wales – seamless garments incorporating separate panels with different attributes including anti-chafing, insulation and even electronic heating elements. The use of textiles embedded electronics clearly has a great deal to contribute this sector, in addition to clothing and monitoring a wide variety of other applications are also in development such as the solar cell fabric, already available in the form of self-lighting tents, which may well be able to power electrical devfices fin ftThe medfium fterm ffufture [11-15].

Where next?

All the developments in technical textiles explored in this paper have substantial backing and either has or is likely to be manufactured. Which has the greatest potential to affect health and wellbeing depends on context more than technology. In terms of improving the largest number of lives innovative use of affordable natural fibres modified by the addition of technical textiles would seem to offer the greatest potential, as these are accessible to third world relief organisations whose impact is arguably greater than any other health and wellbeing institution –' Vaxess Technologies' desiccated silk protein based vaccine being an excellent example. In terms of the greatest impact in first world countries, nanotechnology fabrics such as Nanovelcro clearly have immense potential in the intracorproeal treatment of common conditions such as cancer and as such may be regarded as the leading aspect [16-20].

Conclusion

Overall however, the most enduring contribution of the global acceleration of technical textile development is not any one innovation but the growth of a pool of expertise. Corporations, charities and academic institutions in the USA, France, Britain, Switzerland, Australia, India and Bangladesh now maintain substantial workforces

dedicated to the future development of technical textiles. In so doing they have ensured that mankind's present and future health and wellbeing will be improved by this fascinating field of research and innovation.

References

1. Baltex (2013) Baltex Ltd Shaping the future of technical textiles, Ilveston.

2. Design ageing well (2011) Smarter outdoor clothing for active ageing: Newsletter for new markets in wearable technology. University of Wales.

3. Miraftab M (2011) Technical fibres. In: Handbook of Technical Textiles, Horrocks AR, Anand SC editors, Cambridge.

4. Gupta B, Saxena S, Grover N, Ray A (2010) Plasma-treated yarns for biomedical applications. In: Technical Yarns: Industrial and Medical Applications, Alagirusamy A, Das A editors, Cambridge.

5. Smart textiles and Nanotechnology (2016) The news service for textile futures.

6. Anand S, Rigby A (2011) 'Medical textiles'. In: Handbook of Technical Textiles, Horrocks AR and Anand SC editors. Cambridge .

7. O'Mahoney M (2011) Advanced Textiles for Health and Wellbeing, London.

8. Scott R (2011) Textiles in Defence. In: Handbook of Technical Textiles, Horrocks AR, Anand SC editors, Cambridge.

9. Colchester C (2007) Textiles Today: A global survey of trends and traditions, London.

10. Adam S, Sarchet P (2017) A User's guide to nanotechnology' The Guardian.

11. Bajaj P (2011) Heat and flame protection. In: Handbook of Technical Textiles, Horrocks AR and Anand SC editors, Cambridge.

12. McCann J, Bougourd J, Stevens K (2009) 'Design for Ageing Well: Newsletter for new markets in wearable technology' University of Wales.

13. Ozler L(2008) Techno textiles 2: Revolutionary fabrics for fashion and design.

14. Campbell M (2011)'Knitwear defies 1000 °C flames' in New Scientist.

15. Holmes D (2011) Textiles for survival'. In: Handbook of Technical Textiles, Horrocks AR, Anand SC editors, Cambridge.

16. Schoeller Textil (1988) Schoeller Textil 'iLoad -Reloadable drug delivery fabric from Schoeller'

17. Wilson A (2010) Sixth senses for safety' in Future Materials.

18. Wilson A (2010) Wearables help develop a sixth sense for safety and protection' in Technical Textiles International.

19. Vincenzine Pand Paradiso R (2010) Improving the Quality of life of the ageing population using a technology enabled garment system: Newsletter for new markets in wearable technology University of Wales.

20. News Report (2011) First steps in bringing outdoor product design to a new market: Newsletter for new markets in wearable technology', Newport University of Wales

26

Direct Carbonization of High-performance Aromatic Polymers and the Production of Activated Carbon Fibers

Yutaka Kawahara[1,2]*, Shunsuke Otoyama[1], Kazuyoshi Yamamoto[3], Hiroyuki Wakizaka[4], Yutaka Shinahara[5], Hideki Hoshiro[6], Noboru Ishibashi[3] and Norio Iwashita[7]

[1]Division of Environmental Engineering Science, Gunma University, Tenjin-cho, Kiryu 376-8515, Japan

[2]The Center for Fiber and Textile Science, Kyoto Institute of Technology, Matsugasaki, Sakyo-ku, Kyoto 606-8585, Japan

[3]Research Lab., Carbo-tec. Co. Ltd., 305, Creation Core Kyoto Mikuruma, Kajii, Kamigyo-ku, Kyoto 602-0841, Japan

[4]North Eastern Industrial Research Center of Shiga Prefecture, 27-39, Mitsuyamotomachi, Nagahama, Shiga 526-0024, Japan

[5]Nippon Felt Co. Ltd., Saitama Mill, 88, Haramamuro, Kounosu, Saitama 365-0043, Japan

[6]Kuraray Co. Ltd., North Umeda Hankyu Building Office Tower, 8-1, Kakudacho, Kita-ku, Osaka 530-8611, Japan

[7]National Institute of Advanced Industrial Science and Technology, 16-1, Onokawa, Tsukuba, Ibaragi 305-8569, Japan

Abstract

For expanding the utilization of several high-performance aromatic polymeric fibers, e.g., poly p-phenylene benzobisoxazole (PBO, Zylon®), poly p-phenylene terephthalamide (PTA, Kevlar 29®), and polyarylate (PA, Vectran®), direct carbonizing and graphitizing behaviors have been investigated. The PBO-based carbon fiber showed a typical radial texture on its fracture surface, and the graphitization degree (P1) reached 0.35 and crystallite sizes of Lc(002), La(110) after graphitization exceeded 30 nm. On the other hand, the P1 indices of the graphitized carbon fibers from PTA and PA were no more than 0.15. However, a low P1 value is preferable for the production of activated carbon fibers (ACF). In addition, on the surface of the PA-based carbon fibers produced at 900°C, some fine mesh-like morphologies were observed indicating the formation of a porous carbon structure. In contrast, for the PTA-based carbon fibers the development of radial texture could be seen only partially on the fracture surface, and porous morphologies were not recognizable. It was confirmed that the direct carbonization was enough to convert PA fibers into ACF. The BET surface area of the PA-based carbon fibers increased up to 900~1,000 m²/g after the direct carbonization at 900 °C, and exceeded 1,000 m²/g easily when activated.

Keywords: Activated carbon fibers; Aromatic polymers; Carbonization; Graphitization; Poly p-phenylene benzobisoxazole; Poly p-phenylene terephthalamide; Polyarylate

Introduction

Nowadays, polyacrylonitrile (PAN) and pitch, a heavy fraction of petroleum, are the two major starting materials for producing high-performance carbon fibers. For most high-volume applications, however, it is necessary to consider their production costs due to the lengthy, multistep process for the conversion of starting precursor fibers into high-performance carbon fibers. During the processing, the precursor fibers require oxidative stabilization prior to carbonization heat-treatment. This oxidative stabilization step is slow and affects the final mechanical properties of the carbon fibers. Therefore, direct carbonization has been tried using high-performance aromatic polymeric fibers, e.g., poly p-phenylene benzobisoxazole (PBO) [1-5], and Kevlar [6]. Especially, PBO-based carbon fibers obtained by the direct carbonization have once attracted remarkable attention. However, it was difficult to produce high-performance carbon fibers because the flaws present in the precursor fibers persisted throughout the carbonization and caused tensile failures of the carbonized fibers. Moreover, the release of nitrogen over 1,400°C affected the tensile strengths of the carbonized fibers [2,3]. Therefore, the graphitizing properties of these high-performance aromatic polymeric fibers have not been fully studied so far although they may be useful for the preparation of not only the carbon fibers but also activated carbon fibers (ACF), electrodes of secondary batteries, etc. For the production of these carbon goods from aromatic polymers, simple direct carbonization process is attractive to save the production cost. For this purpose, it is necessary to clarify their carbonizing and graphitizing behaviors.

At present, non-woven fabrics made of the high-performance aromatic polymeric fibers are often used as heat resistant or thermal insulating cushions, e.g., the roller covers in the rolling process of aluminum rod production. Therefore, for these purposes, also, it is important to understand the carbonization behavior of these super fibers. Moreover, in the production process of non-woven fabrics, the generation of raw edges that are severed and wasted is inevitable, which increases the production costs. However, these waste ends made of high-performance aromatic polymeric fibers may have a potential for the production of carbon fibers via the direct carbonization. Although it is not easy to produce carbon fibers with excellent tensile properties, carbon fibers with specific properties, e.g., heat resistance, absorption properties, electrical conductivity [2], chemical resistance, could be producible.

In this paper, direct carbonization experiments were attempted using high-performance aromatic polymeric fibers, i.e., PBO (Zylon®), poly p-phenylene terephthalamide (PTA, Kevlar 29®), and polyarylate

*Corresponding author: Yutaka Kawahara, Professor, Division of Environmental Engineering Science, Gunma University, 1-5-1, Tenjin-cho, Kiryu 376-8515, Japan
E-mail: kawahara@gunma-u.ac.jp

(PA, Vectran®), and then their graphitizing properties were compared firstly. The PBO-based carbon fiber showed a larger graphitization degree (P1) and crystallite sizes. In contrast, the P1 indices of the carbon fibers from PTA and PA were fairly small. However, it was confirmed that carbon fibers with BET surface areas of ~1,000 m²/g could be produced via the direct carbonization from PA fibers.

Materials and Methods

Materials

The PBO and PTA fibers were raw edges of non-woven fabrics supplied by Nippon Felt Co., Ltd. (Saitama, Japan), and the PA fibers were from Kuraray Co., Ltd. (Osaka, Japan). The PA fibers used were produced from the melt spinning of a co-polymer of 1,4 hydroxybenzoate and 2,6 hydroxynaphthanoate in the monomer ratio of 73/27, respectively, followed by the thermosetting treatment to elevate the melting temperature of the fibers.

Carbonization and activation at the laboratory-scale

The starting fibers of 4 to 5 g, released from contact with each other, were heated at a rate of 10°C/min, and then kept at 900°C for 60 min using an electric furnace having a uniform temperature zone of about 60 mm in length under nitrogen gas flowing at 1 L/min followed by cooling to room temperature.

To activate the carbon fibers, a rotary kiln (internal volume of ca. 0.75 L) was used. First, ca. 2 g of the fibers was heated to 900°C at a heating rate of 10°C/min under nitrogen gas flowing at 0.6 L/min, and then kept at 900°C for 10 min for activation while water vapor was added instead of nitrogen gas.

Heat-treatment at 3,000°C

Carbonized samples were heated almost linearly from room temperature to 3,000°C in 140 min, and were kept at 3,000°C for 30 min, and then left to cool to room temperature using an electric furnace (SCC-U-120/203/135, Kurata Giken Co., Konan, Shiga, Japan); argon gas was flowing throughout.

Thermogravimetric analysis/differential thermal analysis (TG/DTA) measurement

TG/DTA measurements were performed using an analyzer (WS-002, Bruker AXS K.K., Yokohama, Japan) on 10 mg of sample at a heating rate of 20°C/min up to 1,000°C under nitrogen gas flowing at 50 mL/min.

X-ray diffraction

Wide-angle X-ray diffraction (WAXD) was performed on isotropic samples which were obtained by grinding and mixing the carbonized fibers with high-purity silicon powder as the inner standard. The X-ray source was CuK$_\alpha$ radiation.

The degree of graphitization (P1), defined as the probability for adjacent hexagonal carbon layers to have the positional correlation in graphite, was determined from the Fourier coefficients of both the 101 and 112 WAXD peak profiles measured using step-scanning with a step of 0.01 °C and an accumulation time of 30 s [7,8]. The average interlayer spacing was calculated using the Bragg's equation from the peak diffraction angles of the carbons determined by referring to the diffraction of the silicon standard. The average interlayer spacings determined from the 002, 004, and 006 diffractions were denoted as d_{002}(002), d_{002}(004), and d_{002}(006), respectively. The crystallite sizes

parallel and perpendicular to the c-axis (Lc and La) were calculated from the full-width at half-maximum for the diffraction peaks of the carbons, which were corrected for the instrumental peak broadening by referring to the diffractions of the silicon standard. The size parameters determined from the 002, 004, 006, and 110 diffractions using the Scherrer's equation were denoted as Lc(002), Lc(004), Lc(006), and La(110), respectively. The standard deviation of the distribution in the interlayer spacing (σ_c) and the true crystallite size parallel to the c-axis (L_0) were determined using the Hosemann's equation [9].

$$[1/Lc(00l)]^2 = [1/L_0]^2 + \pi^4\sigma_c^4 l^4/[16d_{002}(004)^6] \tag{1}$$

where l is the order of the diffraction.

Adsorption characterization

The nitrogen gas adsorption capacity was determined using an adsorption measurement instrument (BELSORP-18plus, BEL Inc. Jpn, Toyonaka, Japan). The pore surface area (S) was determined using the Brunauer–Emmett–Teller (BET) plot [10]. The pore volume (V) was determined from the amount of nitrogen adsorbed at a relative pressure of 0.99. The mean pore diameter (D) was calculated as $D = 4V/S$ by assuming that the pores were uniform non-intersecting cylindrical capillaries.

Results and Discussion

Carbonization behavior

TG and DTA curves for the PBO fibers are shown in Figure 1. It is seen from the TG curve that the thermal decomposition of PBO fibers occurred in two steps, i.e., over 420°C and 660°C, similarly as in previous data [1,2]. The carbon yield at 1,000°C was ca. 60%. It has been reported that heat-treatment at 600°C is effective to improve the level of crystallinity in core regions of PBO fibers [1]. However, an exothermic peak corresponding to the crystallization of PBO molecules was not observed. On the other hand, it has been demonstrated that PBO fibers begin to pass from the ordered initial state to an amorphous state when heated over approximately 600°C [2], and the carbonization of PBO fibers can be modeled as a free-radical polymerization [4]. Thus, the sudden decrease in the TG curve over 660°C probably means that free-radical polymerization was occurring accompanying the thermal cracking of the PBO backbones. Thereby, the DTA exothermic peak at 678°C can be attributed to this thermal cracking and the peak at 750°C to the polycondensation to form carbon structure.

The surface of the PBO-based carbon fibers produced at 900C is shown in Figure 2a. The striations along the fiber axis are due to the development of a radial structure. No periodic banded structure along

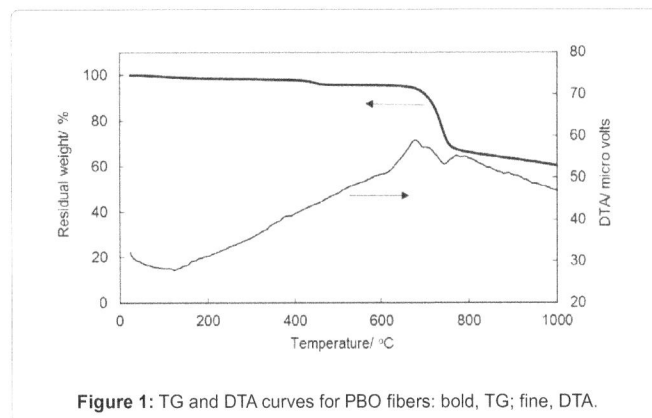

Figure 1: TG and DTA curves for PBO fibers: bold, TG; fine, DTA.

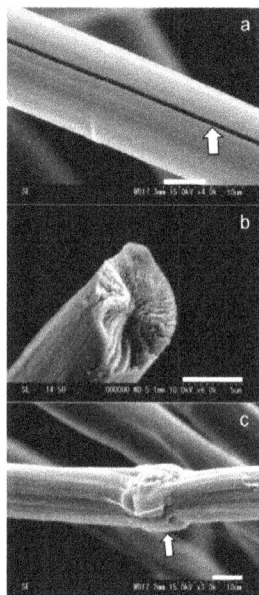

Figure 2: SEM images for PBO-based carbon fibers: (a) fissure running along the fiber axis (when carbonized at 900°C); (b) and (c), radial texture in the fracture surface and a swelling generated by the nitrogen puffing, respectively (when carbonized at 3,000°C). The scale bars are 5 μm.

the fiber axis could be recognized. These features coincide with the previous work on PBO-based carbon fibers [1]. It is well known that PBO-based carbon fibers will show a fracture surface having carbon layers stacked radially from the center [2]. In addition, most of the PBO-based carbon fibers exhibit so-called "open wedge texture" when heated over 900°C [5]. Therefore the fissure running along the fiber axis (refer to the arrow in Figure 2a) seems to be brought about by the development of the radial structure. The radial texture in the fracture surface became obvious with elevating the heat-treatment temperature. The fracture surface of PBO-based carbon fiber produced at 3,000°C showed a typical radial pattern (see Figure 2b). Moreover, another peculiar morphology, like a swelling, interfering with the running of striations (refer to the arrow in Figure 2c) could be observed at intervals on the surface of PBO-based carbon fibers produced at 3000°C. Such swelling is probably due to the nitrogen puffing or linear expansion of the carbon layers that occurs under released conditions over 1,400°C [2].

TG and DTA curves for the PTA fibers are shown in Figure 3. The TG curve obtained suggests that the thermal decomposition of PTA fibers proceeded in three steps, i.e., over 400, 540, and 600°C. The carbon yield at 1,000°C was 28.9% and still decreasing. In the case of Kevlar 49 fibers, however, the abrupt one-step weight decrease was observed between 450°C and 550°C [6]. On the other hand, the DTA curve corresponding to the temperature range of the 2nd step in the TG curve showed an exothermic shoulder and a peak at 570°C, and 610°C, respectively, while there was no such remarkable DTA shoulder or peak for the other steps. It is seen that the polycondensation reactions abruptly occurred and was concentrated in the 2nd step for the PTA fibers, i.e., 540～630°C, a little higher than the thermal decomposition temperature range of Kevlar 49 fibers.

An SEM image of the broken end of a PTA-based carbon fibers produced at 3,000°C is shown in Figure 4a. A clear banded structure on

the fracture surface with a periodicity of ~500 nm could be recognized along the fiber axis, similarly as in a previous work on Kevlar 49-based carbon fibers [6]. In the case of the pristine Kevlar 49 fibers, the higher-order structure was described as radially arranged pleated sheets alternating the crystallite orientation periodically in a similar length of ~500 nm along the fiber axis [11]. This inherent arrangement of crystallites will affect the stacking of carbon layers when carbonized. Therefore, the PTA-based carbon fibers showed periodic bands on the fracture surface. In the fracture surface of PTA-based carbon fibers heat-treated at 3,000°C (Figure 4b), the development of radial structure could be partially recognized. Overall, however, the texture was not so marked as expected from the higher-order structural model of Kevlar 49 fibers described above.

TG and DTA curves for the PA fibers are shown in Figure 5. From the TG curve it seems that the thermal decomposition of PA fibers proceeded in one step, abruptly, between 490 and 550°C, and almost ceased at 1,000°C. The carbon yield at 1,000°C was 33.4%. However, the DTA peaks and shoulders over 540°C indicate that the polycondensation reactions proceeded in several steps competitively and interactively, then only a monotonic decrease in the TG curve was seemingly observed.

The PA fibers were produced from melt spinning followed by a thermosetting treatment which could enhance the thermal stability, accompanying by crystal modification from psudo-hexagonal of as-spun fibers to orthorhombic [12], and increasing the average molecular

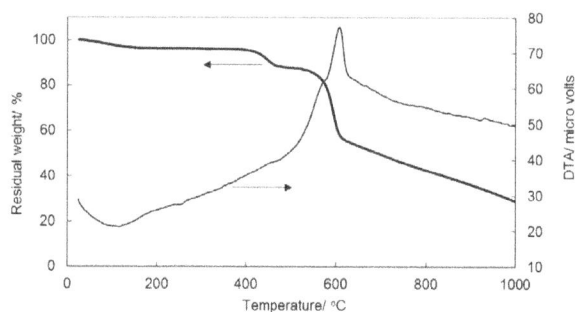

Figure 3: TG and DTA curves for PTA fibers: bold, TG; fine, DTA.

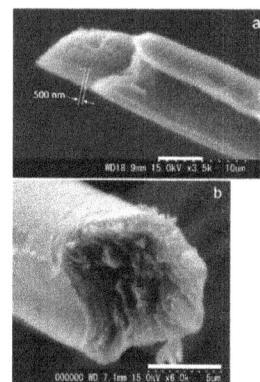

Figure 4: SEM images for PTA-based carbon fibers produced at 3,000°C: (a) broken end, (b) fracture surface. The scale bars are 5 μm.

weight through the solid-state polymerization (SSP) that could elevate the melting temperature of the heat-treated fibers [13]. The small endotherm at ~318°C (refer to an arrow in Figure 5) can therefore be attributed to the melting of crystallites in the heat-treated PA fibers. However, the increase of viscosity in the amorphous regions induced by SSP was large enough to prevent the fibers from fusing.

The surface of the PA-based carbon fibers produced at 900°C is shown in Figure 6a. The formation of bands on the exterior surface was much more obvious compared with the PTA-based carbon fibers. This is probably related to the fact that the pristine fibers contain the banded structure generated by the orientation of crystallites of liquid crystalline PA [14]. Moreover some fine mesh-like morphologies could also be observed (refer to an arrow in Figure 6a). Such morphologies will be advantageous to increase the BET surface area if the PA-based carbon fibers are applied to the production of ACF. The surface of PA-based carbon fibers produced at 3,000°C is shown in Figure 6b. The banded structure on the surface of fiber was not so marked and the fracture surface texture became featureless. This is probably related to the low graphitization degree, P1, of PA (Table 1). The particles deposited on the surface of the fiber are probably due to the chemical vapor deposition reactions with decomposition gases.

For the preparation of the roller cushions in the rolling process of aluminum rod production, PBO fiber is preferable due to its high heat resistance to 600°C. However, PBO fiber has been shown here to not be appropriate for the production of carbon fibers free from fissures or swellings. On the other hand, the porous morphologies observed

Sample	PBO	Kevlar29	Polyarylate	Cellulose [16]
$P1$	0.35	0.15	0.13	0.13
$d002(002)$, nm	0.3366	0.3398	0.3404	0.3393
$d002(004)$, nm	0.3360	0.3397	0.3390	0.3388
$d002(006)$, nm	0.3361	–	–	–
$Lc(002)$, nm	30	11	5	14
$Lc(004)$, nm	20	7	4	8
$Lc(006)$, nm	16	–	–	–
$La(110)$, nm	39	6	3	11
σ_c, nm	0.006	0.011	0.012	0.010
L_0, nm	31	7	5	15

Table 1: Crystallite structure of the carbon fibers produced at 3,000°C.

on PA-based carbon fibers suggest the potential for the development of ACF.

Graphitizing properties of aromatic polymers

In general, the graphitizing property of the starting material will determine its industrial usability as carbon products. That is, the composition of the starting material and the chemical structure of each component affect the carbon structure and its yield. Carbon is usually classified into graphitizing and non-graphitizing types [15]. The graphitizing property can be estimated by measuring the graphitization degree, P1. In fact, carbonaceous materials with large P1 tend to be transformed into a charcoal with graphite-like structure. In the case of pure graphite, P1 equals one. Such densely packed structure, however, is not suitable for the production of ACF because the diffusion of activation gases, such as water vapor or CO_2, into the carbonaceous materials would be inhibited and an ACF displaying only limited adsorption-capacity forms. Therefore, the graphitization degree should be controlled within a certain range, e.g., P1~0.1, for ACF applications. Another important property determining the applicability to ACF is a capability of organizing a porous carbon structure. Therefore, the PA fibers with these two important properties were more preferable for the starting material to produce ACF compared with the PTA yielding non-graphitizing solid carbon without a porous carbon structure.

The graphitizing heat-treatment on the carbon fibers produced at 900°C was conducted at 3,000°C, and the P1 indices were estimated using WAXD. The P1 index, 002 spacings based on the various orders and the corresponding crystallite sizes are listed in Table 1. The 006 diffraction peak could not be observed except for PBO due to disordering in the stacking of adjacent hexagonal carbon layers. Therefore, the standard deviation of the distribution in the interlayer spacing (σ_c) was enlarged for the PTA, PA, and cellulose [16]. The P1 value became the largest for the PBO-based carbon fibers that showed the typical radial texture on their fracture surface. On the other hand, the PTA and PA, with smaller P1 indices, should be categorized into non-graphitizing starting materials.

From the viewpoint of graphitizing properties, the PBO fibers, with higher P1 index, were better converted into short carbon fibers after graphitization and could be used as fillers to give functionalities to composite materials, such as electrical conductivity, rigidity, etc. As for the PTA fibers, the fibers were non-graphitizable and could be used as general-performance carbon fibers after appropriate carbonization by controlling the heating rate, the heat-treatment temperature etc.

Production of ACF by direct carbonization of PA fibers

Cellulosic materials that have been classified as non-graphitizing

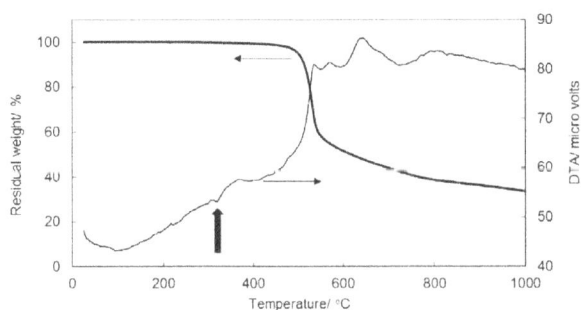

Figure 5: TG and DTA curves for PA fibers: bold, TG; fine, DTA.

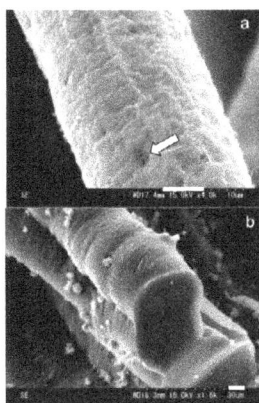

Figure 6: SEM images for PA-based carbon fibers produced at (a) 900°C, (b) 3,000°C. The scale bars are 5 μm.

Carbonization at 900°C (N_2, 1 h)	
Yield, %	39.3
BET area, m^2/g	927
BET volume, mL/g	0.37
Mean pore diameter, nm	1.61
Activation at 900°C (H_2O, 10 min)	
Yield, %	58.8
BET area, m^2/g	1,339
BET volume, mL/g	0.56
Mean pore diameter, nm	1.67

Table 2: Yields and surface characteristics of polyarylate-based carbon fibers.

carbon are often used as starting materials in the production of ACF destined for drinking water and/or wastewater control systems. Non-graphitizing carbons are highly attractive for ACF production to ensure an effective reaction with activation gases, leading to porous structures with large adsorption capacity. Activated carbon materials with BET surface area exceeding 1,000 m^2/g can easily be produced from woody biomass [17,18]. When the graphitizing property and crystal parameters of PA-based carbon fibers were compared with those of a cellulose-based one here, cellulosic materials extracted from reed grass (Table 1) [16], they ranged within similar values. In addition, the PA fibers showed porous morphologies when carbonized (Figure 6a), which would be advantageous for the production of ACF although to our knowledge, never tried so far.

Yields and surface characteristics of the PA-based carbon fibers are listed in Table 2. The yield of ACF was defined as the relative mass of ACF against the mass of carbon fibers before activation. The BET surface area reached 1,339 m^2/g when activated to the weight-loss of ca. 40%. However, it is seen that merely carbonization was enough to gain a BET surface area to 900~1,000 m^2/g. Moreover, the mean pore diameter was almost comparable to woody biomass-based ACs of 1.8~2.1 nm [16,18], implying that the direct carbonization is enough to convert PA fibers into ACF destined for drinking water and/or wastewater control systems.

Conclusion

Carbonizing and graphitizing behaviors for three high-performance aromatic polymers have been investigated. The PBO-based carbon fiber showed the typical radial texture on its fracture surface, and the P1 index reached 0.35 and crystallite sizes of Lc(002), La(110) after graphitization exceeded 30 nm. On the other hand, the P1 indices of the graphitized carbon fibers from PTA and PA were no more than 0.15. Therefore PTA and PA could be categorized as yielding non-graphitizing carbon. However, a low P1 value is preferable for the production of ACF. A starting material with a high P1 value tends to be transformed into a carbon with densely packed structure, which is not suitable for the production of ACF because the diffusion of activation gases, such as water vapor or CO_2, into the carbonaceous materials tends to be inhibited and an ACF displaying only limited adsorption-capacity forms. Moreover, a starting material should have a capability of organizing a porous carbon structure. On the surface of PA-based carbon fibers produced at 900°C, some fine mesh-like morphologies were observed, which suggests that carbon fibers with porous structure could be produced from the PA fibers. From this structural viewpoint, the PA fibers are more preferable to the PTA. The BET surface area was measured on the PA-based carbon fibers and it was found that the value increased up to 900~1,000 m^2/g after carbonization at 900°C, and exceeded 1,000 m^2/g easily when activated. It can be concluded, at least, that direct carbonization is enough to convert PA fibers into ACF.

References

1. Young RJ, Day RJ, Zakikhani M (1990) The structure and deformation behavior of poly(p-phenylene benzobisoxazole) fibers. J Mater Sci 25: 127-136.

2. Newell JA, Rogers DK, Edie DD, Fain CC (1994) Direct carbonization of PBO fiber. Carbon 32: 651-658.

3. Newell JA, Edie DD (1996) Factors limiting the tensile strength of PBO-based carbon fibers. Carbon 34: 551-560.

4. Newell JA, Edie DD, Fuller ELJr (1996) Kinetics of carbonization and graphitization of PBO fiiber. J Appl Polym Sci 60: 825-832.

5. Kaburagi Y, Yokoi K, Yoshida A, Hishiyama Y (2005) Highly graphitized carbon fiber prepared from poly(p-phenylene-benzo-bis-oxazole) fiber. Tanso 217: 111-114.

6. Tomizuka I, Isoda Y, Amamiya Y (1981) Carbon fibre from a high-modulus polyamide fibre (Kevlar). Tanso 106: 93-101.

7. Houska CR, Warren BE (1954) X-Ray study of the graphitization of carbon black. J Appl Phys 25: 1503-1509.

8. Noda T, Iwatsuki M, Inagaki M (1966) Changes of probabilities P1, $P_{ABA, PABC}$ with heat treatment of carbons. Tanso 47: 14-23.

9. Shioya M, Takaku A (1989) Characterization of the structure of carbon fibers by wide-angle and small-angle X-ray scatterings. Tanso 139: 189-198.

10. Adamson AW (1991) Physical chemistry of surfaces. John Wiley and Sons, New York, USA.

11. Dobb MG, Johnson DJ, Saville BP (1977) Supramolecular structure of a high-modulus polyaromatic fiber (Kevlar 49). J Polym Sci, Polym. Phys Ed 15: 2201-2211.

12. Cohen EK, Marom G, Weinberg A, Wachtel E, Migliaresi C, et al. (2007) Microstructure and nematic transition in thermotropic liquid crystalline fibers and their single polymer composites. Polym Adv Technol 18: 771-779.

13. Yamamoto Y, Nakagawa J (2009) The structure and properties of high-modulus, high-tenacity Vectran™ fibers.

14. Taylor JE, Romo-Uribe A, Libera MR (2003) Molecular orientation gradients in thermotropic liquid crystalline fiber. Polym Adv Technol 14: 595-600.

15. Oberlin A, Oberlin M (1983) Graphitizability of carbonaceous materials as studied by TEM and X-ray diffraction. J Microsc 132: 353-363.

16. Kawahara Y, Yamamoto K, Wakisaka H, Izutsu K, Shioya M, et al. (2009) Carbonaceous adsorbents produced from coffee lees. J Mater Sci 44: 1137-1139.

17. Ishibashi N, Yamamoto K, Wakisaka H, Kawahara Y (2014) Influence of the hydrothermal pre-treatments on the adsorption characteristics of activated carbons from woods. J Polym and the Environ 22: 267-271.

18. Kawahara Y, Izumoto M, Nishikawa G, Wakisaka H, Iwashita N, et al. (2006) Carbonization behavior of reed grass and production of activated carbon. Sen'i Gakkaishi 62: 242-244.

27

Study on Different Techniques of Fabricating Conductive Fabrics for Developing Wearable Electronics Garments

Ashok Kumar L*

Department of Electrical and Electronic Engineering, PSG College of Technology, Coimbatore, Tamilnadu, India

Abstract

The mainstream research discourses and practices have been transforming into interdisciplinary studies recently. The combination of electronics and textiles can be classified as: wearable electronics, textronics and fibertronics, based on degree of integration. This paper discussed about development of cotton wrapped nichrome yarn, copper core conductive yarn, optical core conductive yarn POF of different diameters to produce, nichrome fabric, copper core conductive fabric, optical core conductive fabric and teleintimation fabric. The yarn and the fabrics were tested using a dedicated test rig developed for this research work. This work provides the platform for the methodology of developing conductive yarns and fabric for the wearable electronics product development to progress on the issue of user acceptability.

Keywords: Conductive fabric; Copper core yarn; Optical core yarn; Wearable electronic

Introduction

Textiles are used in everyday life, for example as garment to protect ourselves from heat or cold, fabrics covering the surfaces of floors, or the upholstery of car seats. On the other hand electronic devices are spreading – but still some people do not have the knowledge to use them. So the next step to spread these electronics is to improve the user interfaces. Wearable Electronics could offer improved interfaces and make it easier for the user to accept electronic devices in everyday life.

Integrating electronic sensors and actuators into such textiles could be useful for a wide range of applications. Unfortunately, it is difficult to identify the needs without knowledge on the application in these fields. There is little interaction between the electronics-in-textiles community and respective industries. Today's textile conductors for electronics in textiles were not designed for the purpose of transmitting energy or data but rather for anti-static and anti-bacterial purposes. These threads need to be improved regarding conductance, process ability, reliability and signal transmission capability. Also isolation of textile integrated conductors has been neglected although it is required by most applications.

Existing garments are primarily designed for protective functions, including protection from extreme cold, physiological monitoring for emergency conditions, and wearer GPS information for emergency intervention [1,2]. Continuous miniaturization of electronic components has made it possible to create smaller and smaller electrical devices which can be worn and carried all the time [1].

Researchers have developed user friendly techniques to develop methods that will make e-textile technology available to crafters, students, and hobbyists [3]. Also possible to develop transmission lines structures screen-printed on fabrics [4] as well as photonic textile displays woven on a Jacquard loom.

Wireless communication and wearable computers coupled with clothing forms a new approach to wearable computing [5,6]. "Tool Model" and "Clothing Model" describes the different usage models of wearable systems [7]. Steve Mann has carried out extensive work in the field of making computer systems wearable [8]. Probably the best-known example of smart clothing is a textile keyboard and a synthesizer embedded into a denim jacket [9]. Intelligence in the form of electrical components has also been embedded into other pieces of clothing, e.g. gloves [10], ties [11], undergarments [12] and footwear [13]. Electro Textiles Company Limited has adopted another view in the development of smart clothes [14], active ventilation, heat transfer through garments, and reactive waterproof materials[2]. Conducting polymers are a relatively new type of material in terms of understanding, syntheses, and applications [15-18].

In this paper different type of yarn and fabric production techniques, modifications carried out in the conventional textile machines, problems faced during the production processes and the solutions were discussed. The specialty yarns and fabrics can be utilized for developing wearable electronic products.

Development of Cotton Wrapped Nichrome Yarn

There are different methods for establishing conductivity of a yarn. A combed cotton yarn of 34s was used to wrap the nichrome wire, since coarser count would enable better wrapping with high surface area and also assist in quicker dissipation and to get protection against thermal shocks. M/s. Lohia Starlinker Braiding machine is used for wrapping of nichrome wire. To produce cotton wrapped nichrome yarn of 0.17 mm diameter and the linear density of the material is 156.32denier, the nichrome wire is placed in a bobbin on the centre shaft of a circular braiding machine and 16 Bobbins/2 Delivery combinations are used (Figure 1).

Onto the cut pattern of the knitted fabric, the Nichrome wire embedded cotton fabric is integrated. The total length of the nichrome wire is 10.4 m, in sleeveless pattern pad. The sleeve knitted fabric with

***Corresponding author:** Ashok Kumar L, Professor, Department of Electrical and Electronic Engineering, PSG College of Technology, Coimbatore, Tamilnadu, India, E-mail: askipsg@gmail.com

nichrome wire embedded cotton fabric integrated into the cut pattern is shown in Figure 2.

Development of Copper Core Yarn

To develop copper core yarn, 38 SWG copper filaments of 261 Tex and cotton were used as core material and sheath material respectively, in this research work. To produce three different core-sheath ratios of conductive yarns, the Fehrer AG type DREF-3 friction spinning machine was used. With special guides, the copper core filament was fed in the first drafting unit. To increase the stability of spinning of the metal yarn, the guide device was designed and installed on the first drafting unit. To produce uniform yarn structure and which has the nominal yarn count of 328 Tex, the process parameters of DREF-3 spinning machine such as perforated drum speed and yarn delivery rate were set at 4000 rpm and 70 rpm respectively. Three different core-sheath ratio of 67/33, 80/20 and 90/10 copper/cotton yarns were produced, by varying the draft in the second drafting system. The cross sectional view of core sheath DREF-3 friction spun yarn is shown in Figure 3 and longitudinal view of core sheath DREF-3 friction spun yarn is shown in Figure 4 below. Special care was taken to wrap copper filament by adjusting the perforated drum speed, yarn delivery rate and draft in second drafting unit to produce uniform yarn structure, during the spinning process.

The conductive yarn that contained copper wire as core was fabricated by using special guide mechanism on DREF-3 spinning system. A comparison of three different core-sheath ratios of DREF-3 conductive yarns such as 67/33, 80/20 and 90/10 were produced by varying the draft in second drafting unit and it has average parent yarn linear density of 328 tex. Table 1 shows that the breaking elongation of the individual sheath component for all the samples are almost of a magnitude similar to that of the parent DREF-3 yarn, whereas in the case of the core component it breaks immediately (Table 1).

The physical characteristics of the conductive yarn were studied using different tests and found that the 67/33 core/sheath conductive yarn was having the highest tenacity of 3.27cN/tex and elongation to break 5.27% when compare to other core-sheath ratios, due to its better core-sheath interaction factor CSI_T -21.22%. It is also observed that

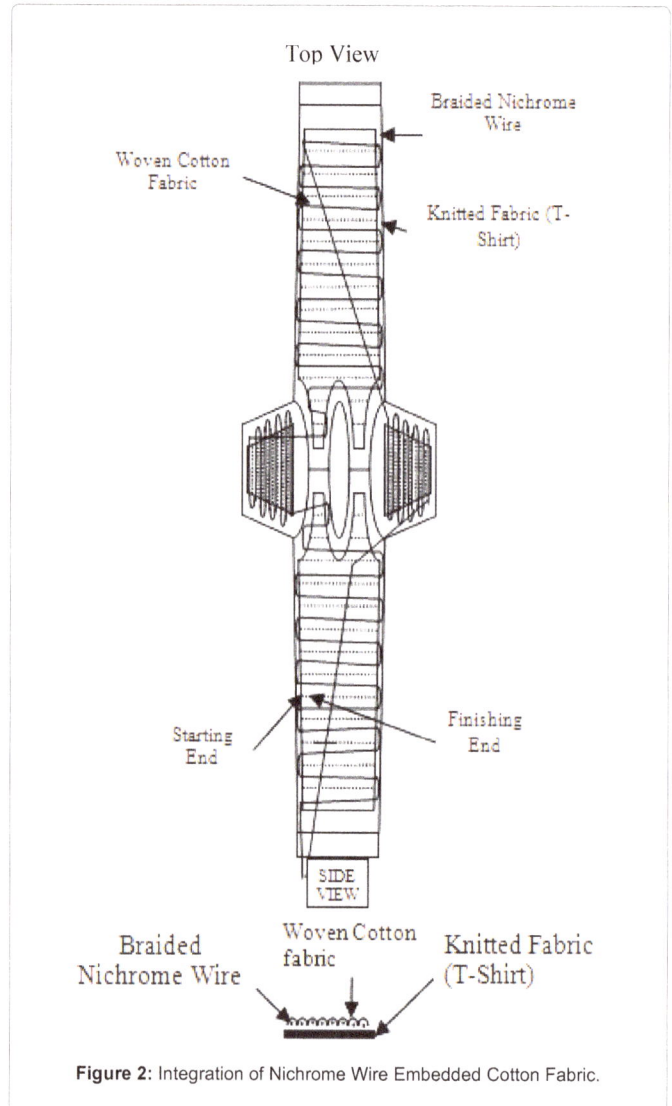

Figure 2: Integration of Nichrome Wire Embedded Cotton Fabric.

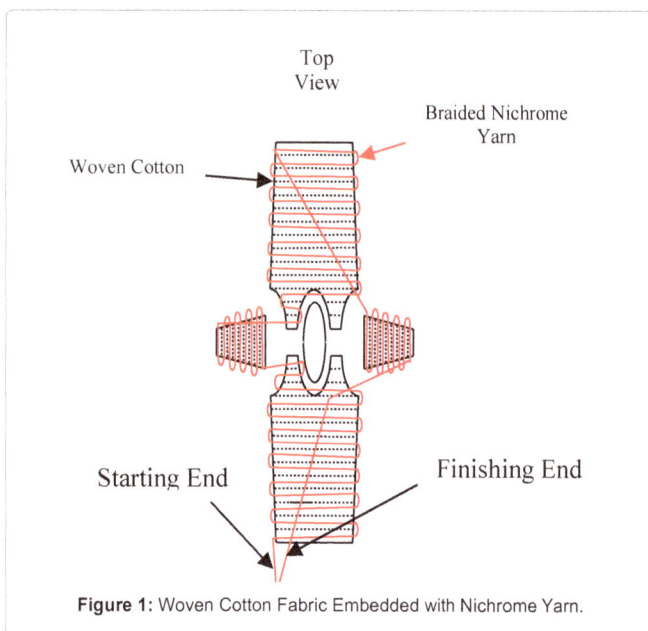

Figure 1: Woven Cotton Fabric Embedded with Nichrome Yarn.

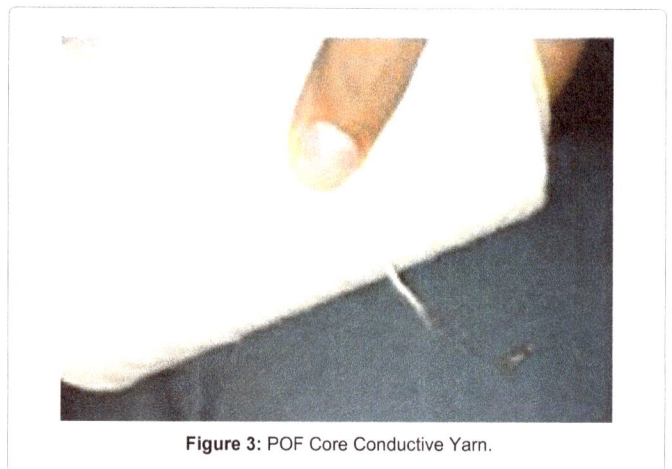

Figure 3: POF Core Conductive Yarn.

the core-sheath interaction factor of conductive yarn sample C (90/10 core-sheath ratio) was found less core-sheath interaction factor CSI_T -9.21% and it exhibits lower breaking tenacity and elongation to break, due to its behavior of copper filament.

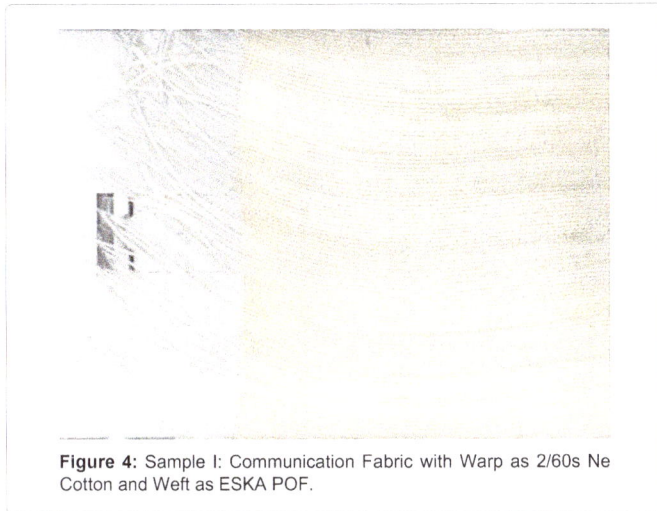

Figure 4: Sample I: Communication Fabric with Warp as 2/60s Ne Cotton and Weft as ESKA POF.

| Sample code | Tenacity (cN/tex) | | | Breaking Elongation (%) | | | Contribution Factor | | |
	Parent yarn (Z)	Core (X)	Sheath (Y)	Parent yarn	Core	Sheath	CC$_T$	SC$_T$	CSI$_T$
Sample A	3.27	1.455	1.12	5.87	1.28	2.63	44.52	34.25	21.22
Sample B	2.89	1.455	1.01	4.52	1.28	2.51	50.37	34.98	14.64
Sample C	2.54	1.455	0.85	3.85	1.28	2.27	57.31	33.46	9.21

Table 1: Tensile Properties of Parent DREF-3 Yarns and Core-Sheath Components.

The values given in Table 2 shows that, the electrical resistance of conductive yarns is varying according to the length of the yarn and current also directly proportional to the applied voltage. The electrical properties of these conductive core spun yarns were exhibited very low resistance (3-28 MΩ) at 6 V, 12 V and 24 V applied voltage. The conductive complex core spun yarn can be woven and knit into fabrics that can be used as communication garment for measuring the body temperature and for mobile charging applications (Table 2).

By using copper core conductive fabric communication garment has been developed and it is integrated with mobile phone charging circuit and temperature measurement circuit for charging the mobile phone and measuring the wearer temperature respectively.

Development of Optical Core Conductive Yarn

To develop communication fabric using optical core conductive yarn two types of POF yarn of 0.5 mm and 1 mm diameters were selected to produce two different samples. Two types of Polymeric Optical Fibres (POF) are used as core materials namely ESKA and CK-20 which is having diameter of 500 µm and 1000 µm respectively and 2/60s Ne cotton sliver as sheath materials for optical core conductive yarns. The Fehrer AG type DREF-3 friction spinning machine was used to produce the two different optical core conductive yarns. Four cotton carded slivers were fed in the first drafting unit and each sliver has 4.22 grams per meter and sliver irregularity of 3.8% (U%) respectively. The core material (optical fibre) was fed in the first drafting unit with special guides. They were designed and installed on the first drafting unit to increase the stability of the conductive yarn during spinning. To produce uniform yarn structure, the process parameters of DREF-3 spinning machine such as perforated drum speed and yarn delivery

rate were set at 4000 revolutions per min and 70 m/min respectively. The POF filament of ESKA and CK-20 type of 1024 Tex and 281 Tex were used as core material and cotton is used as sheath material in the proportion of 80:20 ratio.

During the spinning process, the optical fibre was fed from the first drafting unit with special guides. Special care was taken while wrapping optical fibre by adjusting the perforated drum speed, yarn delivery rate and draft in second drafting unit to produce uniform yarn structure. The POF core conductive yarns developed by DREF-3 spinning system are shown in Figure 3.

Development of communication fabric using optical core conductive yarn

The communications fabrics are developed using hand loom with optical core conductive yarns as weft threads and 2/60s Ne cotton threads used as warp threads. During weaving, the special care was taken while inserting the optical core conductive yarn as weft, and the speed of the loom also reduced. Figure 4 shows the sample developed for communication fabric with warp as 2/60s Ne Cotton and weft as ESKA POF. Figure 5 shows the sample developed for communication fabric with warp as 2/60s Ne cotton and weft as CK-20 POF (Figures 4 and 5).

Testing of communication fabric

To assess the physical characteristics and electrical properties of the above developed communication optical core conductive fabric samples, the various tests such as air permeability, air resistance, fabric thickness, aerial density and signal transferring capability were tested. Table 3 shows the results of the above physical tests (Table 3).

| Conductive Yarn length (m) | Electrical Resistance (mΩ) | Applied Voltage | | |
| | | 6 V | 12 V | 24 V |
		Current (µA)	Current (µA)	Current (µA)
0.5	0.06	100	200	400
1	0.1	60	120	240
2	3	2	4	8
5	20	0.3	0.6	1.2
10	28	0.21	0.42	0.84

Table 2: Electrical Properties of Core-Sheath Conductive DREF-3 Spun Yarn.

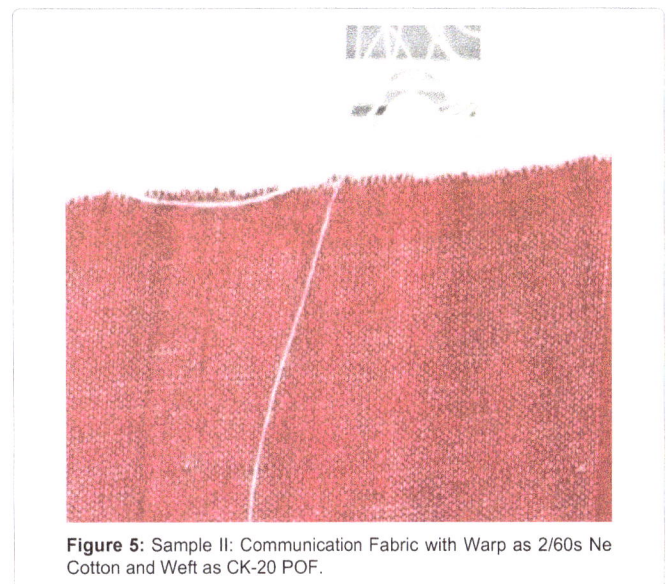

Figure 5: Sample II: Communication Fabric with Warp as 2/60s Ne Cotton and Weft as CK-20 POF.

The aerial density of the ESKA type optical core conductive fabrics are 52% heavier than the CK-20 type optical core conductive fabrics due to its higher linear density (1028 Tex) and diameter as 1000 μm. For the wearable electronic garments, CK-20 type optical core conductive fabrics are more suitable and it has higher air permeability due to the finer optical fiber (500 μm) and lesser fabric thickness (0.74 +/- 0.13 mm) when compare to the thickness of ESKA type optical core conductive fabrics (1.24 +/- 0.20 mm).

Testing of communication fabric to assess signal transferring capability

POF core conductive fabrics developed using two samples ESKA and CK-20 type were tested by different input LEDs for 15 cm length fabric. The light signal transferring of ESKA and CK-20 type optical core conductive fabrics are analyzed by passing the input LED source at one end and to receive the signal at the other end. If the fiber got damaged during weaving process, there will be interruption of these light signals. The fabric was tested using a test bench. Light source is used at the one end of the optical fiber and a photodiode is used at another end of the optical fiber. When there is a light illumination on the photodiode, the output of the photodiode produces 0 V and when there is an illumination, it shows 5 V output in the display unit. Table 4 shows test results of signal transferring capability of communication fabrics.

It was found that the signal loss in terms of light intensity was higher due to the stress to the optical fibre during spinning and fabrication processes. It is evident from the Table 4 that the signal loss is only 24% both in communication fabric developed using CK-20 and ESKA POF, when LASER light is used as input light source. But the percentage of signal loss is higher when red LED is used as input light source. This is because red LED has the minimum lux value due to the maximum wavelength of red color 9 (Table 4).

From the test results it has been concluded that the communication

Particulars	Sample I Communication Fabric (ESKAPOF – Core Cotton – Sheath)	Sample II Communication Fabric (CK-20 POF -Core Cotton Sheath)
Ends per inch Warp: 2/60s Ne Cotton	42	42
Picks per inch Weft: POF Filament Core and Cotton Sheath yarn	13	21
Aerial density (GSM)	564.2	269.4
Fabric cover factor (Kc)	20.07	18.2
Air permeability (cm³/cm².s)	311.375	259.84
Air resistance (Kpa.s/m)	0.04	0.048
Optical wire wt (g/m)	0.860	0.260
Fabric Thickness (mm)	1.24 +/- 0.20	0.74 +/- 0.13

Table 3: Physical Characteristics of Communication Fabrics.

Input Light Source	Input Voltage (V)	CK-20 POF		ESKA POF	
		Output Voltage (V)	Signal Loss (%)	Output Voltage (V)	Signal Loss (%)
Red	5.0	3.1	38	2.9	42
Infrared	5.0	3.1	38	3.2	36
White	5.0	3.6	28	3.5	30
LASER	5.0	3.8	24	3.8	24

Table 4: Test Results of Signal Transferring Capability of Communication Fabrics.

fabric can be used for the production of teleintimation garment and for applications where signal transferring is essential. The communication garment developed using optical core conductive fabric is used for signal transmission as well as to locate the number and the place of the bullet wound.

Development of Yarn for Teleintimation Fabric

In this research work four different diameters of POF such as 0.25 mm, 0.5 mm, and 1.0 mm have been chosen to develop teleintimation fabric. These optical fibres are made into fabric form by three methods namely (a) Sequential work with POF integrated in the fabric and (b) Weaving of the POF using power loom and (c) Weaving of the POF using hand loom. The teleintimation fabric is developed to produce teleintimation garment which will locate a wounded soldier quickly, perform triage, monitor vital signs and relevant physiological information and display their status locally or remotely.

Development of teleintimation fabric by sequential technique

It is a method of integrating POF with fabric by embroidery technique. For this technique, the cotton fabric of 2/40s Ne warp and 40s Ne weft woven fabric was selected which is having 140 GSM. The POF is placed over the fabric in the required matrix format and stitches are made on it to hold the POF on the fabric. The embedding/integration of POF on the fabric depend on binding of threads and number of binding points. The major problem faced in embroidery work with POF is, withdrawal of POF threads and fiber damage during integration on the fabric and also found less serviceability due to hand insertion. Figure 6 shows the fabric developed by embroidery technique with POF arranged in X-Y direction on the fabric (Figure 6).

Development of teleintimation fabric by power loom

The teleintimation fabric was fabricated using power loom, the Polymeric Optical Fiber (POF) filament is integrated with the cotton yarn in both warp and weft directions using rigid rapier loom to produce teleintimation fabric. The POF fabric developed using rapier loom is in matrix format of POF pattern, since the maximum bullet size used in military is 7.62 mm, the matrix should have the maximum pixel of 5 mm of diameter. The quality particulars of POF fabrics developed using rapier loom are 54 inches fabric width, 32 picks per inch in which every alternate picks are inserted using polymeric optical fibre filament with 24 ends per inch, which has 2 ends per inch POF filament and 2/20s Ne cotton. The weave pattern selected for the POF fabric is 1 up and 1 down plain weave and loom running speed of 170 picks per min using Picanal PGW type rapier loom.

The major problems faced during weaving of POF fabrics are warping of polymeric optical fiber on the beam due to lack of flexibility, mounting the POF beam, drawing and denting of the POF fiber, separation of the warp ends, slough-off of the POF from the spool kept for the weft insertion, slippage of the rapier while picking in higher and as well as lower speed and rubbing of the warp ends while the rapier traveling which causes damage in the outer cladding surface of the POF.

While placing the POF wound beam in the normal beam position, it was noticed that loose and slackness of POF threads in warp direction, so to avoid these problems the mounting of the beam was done with a little lift from the normal beam position. The tension in the cotton yarn and the tension in the POF are entirely different and hence the drawing of the ends became a time consuming process. There was a chance of breakages of the fiber while knotting with the cotton ends. One of the major problems with the Polymeric Optical Fiber is the slough-off of the fiber from package itself during weft insertion. Due to this slough-off of the fiber, the weft insertion is not made properly and steel rod arrangement has been made to avoid this problem. Slippage of the POF in the rapier during picking is another major problem faced during the process. While picking, the rapier could not hold the POF fiber at the middle of the race board due to the slippage. The POF is also present in the warp direction; the movement of the rapier over the warp ends will cause cladding (outer area of the POF) damage and thereby affecting the signal properties of the POF. The signal transmitting property of the POF in weft direction is also affected due to the rubbing of the rapier. The teleintimation fabric developed using rigid rapier loom with polymeric optical fibre in specific matrix format is shown in Figure 7.

Development of teleintimation fabric by hand loom

The polymeric optical fibres are inserted in warp and weft direction to produce desired matrix for detecting the signal transmission characteristics using hand loom. The quality particulars of POF fabrics developed using hand loom are 22 inches fabric width, 32 picks per inch in which every alternative picks are inserted using polymeric optical fibre filament with 24 ends per inch, which has 3 ends per inch POF filament and 2/20s Ne cotton (Figure 7).

The weave pattern selected for the POF fabric is 1 up and 1 down plain weave. To weave POF fabrics, it needs few modifications in the preparatory process namely warping, drawing and denting of the POF, separation of the warp ends, winding of the POF on the shuttle for pick insertion processes. Warping of POF on the warp beam was done by

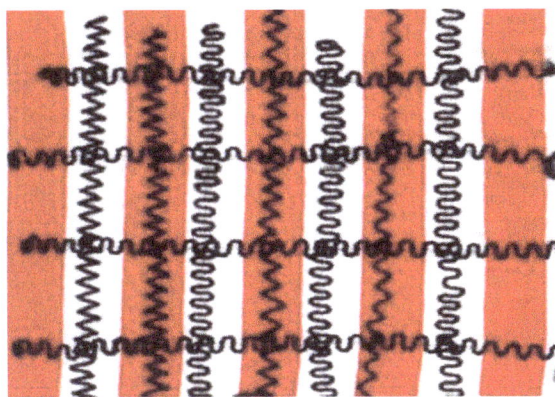

Figure 6: Integration POF in X-Y Direction on the Fabric.

Figure 7: POF Integrated Fabric Using Power Loom.

allowing the fiber manually with equal tension. The beam is marked properly so as to accommodate the POF with the cotton yarn in the fabric and this depends upon the number of warp ends and the winding of the POF on the warp beam. In order to avoid the tension variation between the POF and cotton warp ends, the drawing and denting of the warp ends are carried out manually because the width of the loom is small compared to power loom and rapier loom. The weave pattern is plain weave (1/1) and the POF is inserted for every alternate pick in the weft way. For the easy insertion of the pick the POF is wounded in the shuttle with appropriate tension. The winding is done manually and is wounded in the optimum tension to avoid the slough-off of the POF. The winding is done manually with the help of the manual winding machines for the cotton and POF. The weaving of the POF with cotton yarn in both warp and weft direction in the handloom started with the above arrangements. The pick insertion is done with alternate POF and cotton yarn of count 2/20s Ne. While the shed opening, the first color cotton yarn will be up and the next color will be down and vice versa. While shed is opened, the POF is inserted for one pick and the cotton yarn is inserted for the next pick respectively. The pick insertion is done manually in handlooms with the help of the shuttle wounded with cotton yarn and POF. The cotton yarn of count 2/20s is inserted for every alternate pick with POF and fabric width of 22 inches produced. Figure 8 represents the teleintimation fabric developed by hand loom.

Physical characteristics of teleintimation fabric

Bending rigidity: Bending rigidity measured with KES-FB2 Bending tester, is a measure of the force required to bend the fabric approximately 150°. Bending rigidity per unit fabric width value indicates greater stiffness/ resistance to bending motions. It is calculated for testing the maximum bend tolerance for the fabric with POF to transmit light energy from one end to another end. It was observed that the bending tolerance for the POF located in the warp way and weft way is 7.3 cm and 5.2 cm respectively. Beyond the bending length the attenuation of the light signals will increase resulting in loss of signals.

Tensile strength: The tensile test, done on the KES-FB1 Tensile-Shear Tester, measures the stress/strain parameters at the maximum load set for the material being tested. It measures the amount of stress applied to a material at its breaking point or the point at which it fails.

From the Table 5 it is noticed that the tensile strength is more in the warp way than in the weft way in both the POF integrated fabric. In case of the khaki fabric (Handloom)the arrangement of the POF is in such a way that for every half centimeter one POF is integrated in both warp and the weft way. Since the arrangement of the cotton yarn and the POF is compact the strength is more while in the other hand the blue fabric (Rapier loom) is integrated with POF for every one centimeter of the cotton yarn both in the warp and weft way (Table 5).

Air permeability: It was tested with KES F8 AP1 and it can be used to provide an indication of the breathability of weather-resistant and rainproof fabrics, or of coated fabrics in general, and to detect changes

Figure 8: POF Integrated Fabric Using Hand Loom.

during the manufacturing process. The air permeability of Handloom Fabric and Power Loom Fabric is 44 kPA/Sq.cm and 18 kPA/Sq.cm respectively.

Signal transferring efficiency of teleintimation fabrics: The signal transferring efficiency of the handloom, power loom and sequential integrated woven POF fabrics are analyzed using the microprocessor ATMEGA 89C51 designed with IRF14F Optical Receiver as shown in Figure 9. The digital signal is given to the microcontroller, which is being programmed to display the result in terms of voltage[1]. From the signal values measured in various places of fabrics at 15 cm, 20 cm, 30 cm optical length are made into cumulative average and their signal output voltages are 3.13 V, 3.88 V and 3.01 V for power loom, hand loom and sequential integrated POF fabrics. The signal transferring loss percentage of 0.5 mm POF at 2 feet and 10 feet fibre length with infrared light, red light and white light for various input voltage of 4.5 V and 6 V respectively were tested. From this signal transferring analysis, the signal loss percentages for power loom, hand loom and sequential integrated POF fabrics are 37.44%, 22.40% and 39.84% respectively. It is noticed that the signal transferring loss percentage of handloom fabrics is less when compared to other fabrics (Figure 9).

Conclusion

Today, the interaction of individual humans with electronic devices demands specific user skills. The cost level of important microelectronic functions is low enough and enabling key technologies are mature enough to exploit this vision to the benefit of society. Wearable Electronics are an emerging Tran's disciplinary field, bringing together concepts and expertise from a variety of disciplines, ranging from materials science, through computer engineering to textile design. This research work addresses an emerging new field of research that combines the strengths and capabilities of electronics and textiles in on, wearable electronics. In these paper different types of yarn and fabric production techniques has been developed for wearable electronic products. The Cotton wrapped nichrome yarn embedded cotton fabric integrated into knitted fabric for the development of Heating Garment. Copper core conductive yarn and Optical core conductive yarn were developed for the production of Communication Garment and POF integrated teleintimation fabric has been developed for Teleintimation Garment.

POF Fabric	POF Fabric	Tensile Strength (grams)
Handloom Fabric	Warp way	180
	Weft way	70
Power loom Fabric	Warp way	140
	Weft way	45

Table 5: Testing POF Tensile Strength.

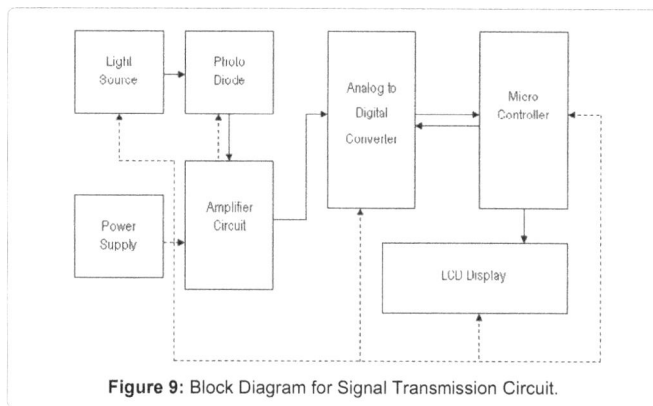

Figure 9: Block Diagram for Signal Transmission Circuit.

References:

1. Rantanen J, Ninen HN (2005) Data transfer for smart clothing: Requirements and Potential technologies. In Tao X (eds), Wearable electronics and photonics Woodhead Publishing in Textiles.

2. Sensatex, Inc. Smart Shirt physiological monitoring device.

3. Buechley L, Eisenberg L (2007) Fabric PCBs, electronic sequins, and socket buttons techniques for e-textile craf.

4. Locher I, Troster G (2007) Screen-printed Textile Transmission Lines. Textile Research Journal 77: 837-842.

5. Troster G , Kirstein T, Lukowicz P (2003) Wearable Computing: Packaging In Textiles And Clothes' Wearable Computing Lab 14th European Microelectronics and Packaging Conference and Exhibition Friedrichshafen Germany.

6. Yao J, Schmitz R, Warren SA (2005) wearable point-of-care system for home use that incorporates plug-and-play and wireless standards. IEEE Trans. Inf. Technol. Biomed 9: 363-371.

7. Mizell D (1999) Message from the Chair', Proceedings of the 3rd International Symposium on Wearable Computers. San Francisco CA, USA.

8. Mann S (1997) An historical account of the `WearComp' and `WearCam' inventions developed for applications in Personal Imaging. First International Symposium on Wearable Computers.

9. Post ER, Orth M (2000) E-broidery: Design and fabrication of textile-based computing. IBM Systems Journal 39: 840-860.

10. Perng JK, Fisher B, Hollar S, Pister KSJ (1999) Acceleration sensing glove. Proceedings of the 3rd International Symposium on Wearable Computers. San Francisco CA, USA.

11. Schmidt A, Gellersen HW, Beigl MA (1999) Wearable context-awareness component. Proceedings of the 3rd International Symposium on Wearable Computers. San Francisco CA, USA.

12. Mann S (1996) Smart clothing: The shift to wearable computing. Communications of the ACM 39:23-24.

13. Paradiso J, Feldmeier M (2005) A Compact, Wireless, Self-Powered Pushbutton Controller. Ubicomp 2001: Ubiquitous Computing, LNCS 2201, Springer-Verlag.

14. Russell DA, Elton SF, Squire J, Staples R, Wilson N (2000) First experience with shape memory material in functional clothing. Proceedings of the Avantex, International Symposium for High-Tech Apparel Textiles and Fashion Engineering with Innovation-Forum. Frankfurt-am-Main, Germany.

15. Elton SF (2000) Ten new developments for high-tech fabrics and garments invented or adapted by the research and technical group of the defence clothing and textile agency; Proceedings of the Avantex International Symposium for High-Tech Apparel Textiles and Fashion Engineering with Innovation-Forum Frankfurt-am- Main Germany.

16. Sayed I, Berzowska J, Skorobogatiy M (2010) Jacquard-Woven Photonic Bandgap Fiber Displays. Research Journal of Textile and Apparel 14: 4.

17. Ala O, Fan Q (2009) Applications of Conducting Polymers in Electronic Textiles. Research Journal of Textile and Apparel 13: 4.

18. Ashok Kumar L, Vigneshwaran C (2010) Design and Development of Signal Sensing Equipment for Measuring Light Transmission Efficiency of POF Fabrics. Indian Journal of Fibre and Textile Research 35: 317-323.

28

The Co-design Process in Mass Customization of Complete Garment Knitted Fashion Products

Joel Peterson*

The Swedish School of Textiles, University of Borås, Skaraborgsvägen 3, S-50630 Borås, Sweden

Abstract

Complete garment knitting technology is a method of producing products, generally fashion garments, ready-made directly in the knitting machine without operations such as cutting and sewing. This makes it possible to manufacture a fashion garment with fewer processes then with conventional methods. Mass customisation is a customer co-design process of products and that tries to meets the needs of an individual customer's demand. The customer can order a garment with a customised style, colour, size, and other personal preferences. Co-design is a collaborative process between the customer, the retailer, and the manufacturer by which a product is customised to fulfil the customer's requirements. This paper is based on the results of a doctoral thesis. The process of co-design and manufacture of a customised complete fashion product is examined. Research was conducted by a retail concept simulation and three case studies. A cross-case analysis was done to analyse the data. The main findings are a description of two kinds of retail concepts for knitted customized fashion products. A knitted garment can be customized, produced, and delivered to the customer in three to five hours. In the Co-design process two kinds of interactions are feasible between the company and the customer: manual or digital co-design. A manual process has advantages such as: high service level for customers, no requirement of advanced technical equipment. However, manual co-design is labour intensive, a shop assistant can only serve one client at a time. It is also only pplicable to brick-and-mortar stores and not transferable to the Internet. Digital codesign, on the other hand, encourages customers to do the customisation on their own, without the aid of sales personnel and little risk of queues. Moreover, this technique is ideal for the Internet. Disadvantages to date have included limited design options and problem of taking body measurements.

Keywords: Knitting technology; Mass customization; Co-design; Complete garment; Supply chain management

Introduction

This paper is the result of a PhD thesis at Tampere University of Technology, Finland and comprises studies in the area of knitting technology and mass customization (MC). The research objective was to investigate the possibility of combining complete garment knitting with MC. The term *mass production* was first introduced in the 1920s and is often associated with the factories of automobile manufacturer Henry Ford [1]. Since then, almost all manufacturing of textiles and garments worldwide has taken place in factories using the industrial concept of mass production. For a long time, textiles continued to be produced in Europe and elsewhere. In the 1960s labour costs increased in many countries of Western Europe, and so a great amount of domestic textile and garment production moved overseas, where manufacturing was cheaper [2]. Here we have a problem, production in low-cost countries in Asia results in long lead times from identified demand on the market to the moment the customer can buy the product. According to Hoover the supply chain needs to be time-based, customer-oriented, and agile in response to changes in demand [3].

In 1987 Stan Davis, a visionary business thinker and consultant coined the term *mass customization* for the first time. He described it as a system in which "the same large number of customers can be reached, as in mass markets of the industrial economy, but simultaneously can be treated individually, as in the era of customised markets in pre-industrial economies" [4]. This was developed further by [5], who defined it as a concept that provides such variety and individual customization that almost everyone can find what they want at prices comparable to mass-produced products. MC involves all aspects of development, manufacturing, sales, and delivery of the product [6,7]. It is a concept that comprises the whole chain from the designer's sketch to the final product received by the customer. MC allows buyers to modify products according to their taste and requirements. It exists today in a variety of areas including automobiles, furniture, food, and clothing. One advantage for the retailer is that the product can typically be sold before the manufacturing takes place. Since the customer has already purchased the product, the risk for unsold goods is lower. Customers are not always satisfied with the products they have customised and bought. For such cases, it is important to have a return policy which allows returning with a full refund. [8].

Complete garment technology (*seamless garment technology*) was introduced on V-bed flat knitting machines in 1995, having evolved from developments in the 1980s. V-bed machines have two needle beds, in a position of an inverted V and equipped with needles [9]. This is considered as a new innovative technology for the future production of knitted garments [10]. The whole garment is made in directly in the flat knitting machine without post cutting and sewing processes. This technology can make lead times shorter in the apparel industry [11].

While MC may not replace mass production of clothing, it may be a solution for certain products and niche markets. In some ways, the MC of clothing may be seen as a step back in time. We are reminded of the crafts era, when clothing was made to order as needed and produced

***Corresponding author:** Joel Peterson, The Swedish School of Textiles, University of Borås, Skaraborgsvägen 3, S-50630 Borås, Sweden
E-mail: joel.peterson@hb.se

near-by. Now this is being done again, but with modern technology - a return to clothing designed and manufactured in collaboration with the wearer. Here complete garment technology opens up new perspectives with its reduction of processes that allow a rapid response to customer demand, while the possibility of MC serves each customer individually. Fashion logistics, MC, and complete garment technology form an effective partnership. These three concepts are the focus of this study. They are relatively new and, while they have been considered separately, they have rarely or not at all been examined in combination.

The principal objective of the present article is to examine the use of complete garment flat knitting technology in the production of mass customized fashion garments.

It poses the following overall research objective: *How can complete garment knitting technology be applied in a retail concept for customised garments?* The answer is pursued in four articles containing literature reviews, simulations, and case studies.

It poses the following research questions:

Research Question One (RQ1): *How does the co-design process function in the customisation of knitted fashion garments?*

Research Question One (RQ2): *What are the advantages and disadvantages with manual co-design compared to a digital co-design process?*

Co-design is a collaborative process between the customer, the retailer, and the manufacturer by which a product is customised to fulfil the customer's requirements. According to Franke and Piller, the success of a co-design system is defined by its technological aspects (generally software-based) and how well it works in the sales environment [12]. This research is a study of how the co-design process function in combination with knitted fashion products, and the effects of which co-design system is chosen.

Methods and Materials

The Knit-on-Demand research project began as an attempt to develop a business concept utilizing complete garment knitting technology. This paper is based on the result of a doctoral dissertation with the title: Customisation of Fashion Products Using Complete Garment Technology [13]. The main methods used to compile research material for this article has been the qualitative multiple-case study defined by Yin, quantitative simulations by Banks, and action research by Näslund [14-16]. These three are the basis of the individual journal articles published in the PhD thesis.

The research gap

The research gap was identified by a literature review. Complete garment knitting technology is a niche area in textile technology. It is difficult to find research literature on the subject, especially in combination with MC. The possibilities of using complete garment machines for the production of mass customised products are discussed by Choi and Powell and Choi [10,17]. There are a number of articles and some books describing and discussing integral knitting and complete garment technique, Spencer [9], Mowbray [18] and Hunter [19], all of which presents the history, technical aspects and an overview of the subject. However, none of these sources provide a deep picture of the complete garment technique in combination with MC of fashion garments. In MC, the importance of the co-design process between the company and the customer has been a major concern for more than thirty years [20-22]. However, if we are to argue that MC provides

a complement to mass production of knitted garments, it remains a matter of concern that there are no research-based studies for this type of co-design in combination with complete garment technology.

Data analysis methods

The data analysis methods applied was cross-case synthesis. Two or more cases are performed and analysed separately but in the subsequent cross-case synthesis the results of the cases are compared and the research questions are answered [14]. A cross-case synthesis may be carried out whether the individual case studies were a predesigned part of the same study or if they were independent research studies: generalisations are sought across a number of studies.

In Articles 1, 2, 3, and 4, the case study method is applied by gathering data through simulations, interviews, and observations. Cross-case synthesis is used to analyse the data in those articles and form the basis for answering the research questions.

Article 1 [23], was the first attempt to develop a shop and production model for the Knit-on-Demand concept incorporating complete garment technology. The study was published in International Journal of Fashion Design, Technology and Education in July 2008.

Article 2 [24], was the case study of an existing MC concept for knitted products. The empirical data was collected on-site at Factory Boutique Shima in Wakayama, Japan and published in International Journal of Mass Customisation in January 2010.

Article 3 [25], was a case study comparing manual and digital co-design with most of the data supplied by Factory Boutique Shima. The study was presented at the MCP-AP Conference in Taipei, Taiwan, in December 2010 and published in Autex Research Journal in March 2011.

Article 4 [26], describes Knit-on-Demand case study from a supply chain perspective and considers design, technology, logistics, and performance and was published in Autex Research Journal in 2012.

Results

Mass customization in textiles

Before the industrial revolution, which began in the 18th century, manufacturing was largely a craft process. A product was custom made to fulfil the requirements of an individual person. It was often expensive and therefore available only to those who could afford it [27]. With the industrial revolution and the era of mass production, more goods could be obtained by more people. Today MC has emerged as a combination of craft and mass production. The textile and fashion industry was one of the first to adopt this concept. Tseng and Piller [28] refer to three aspects of textiles that must be fulfilled to apply in MC: design, function and fit. Maybe the most important of the three factors are garment fit.

Customization can be defined as a strategy that creates value by some form of interaction between the company and the client at the manufacturing stage [29]. Lampel and Mintzberg describe this as "tailored customization" [30]. This means that a company offers a prototype to the client and then change it to the demands of the customer. This is defined as a collaborative approach by Gilmore and Pine [31]. This process is cooperation between client and company in order to achieve a suitable product for the customer.

The collaborative process between the company and the customer is defined as the "co-design process" [32-34]. A client selects options in a configurator or a co-design system of some kind and then they become a co-producer or "prosumer" according to Toffler [35]. Co-design is described and defined in literature as an interaction between the company and an individual customer in order to configure a desired product [36-39].

According to Fralix [27] MC is a future direction of the fashion and textile industry but garment fit and colour selection can be a problem. It is important that the fit of the garment is accurate and this can be done by taking the body measurements in store or an on-line solution where the customer gives the measurements directly in the computer [38].

Body scanning has often been mentioned as a solution to the problem of perfect fit. Its disadvantages are three-fold: 1) an investment in specialised equipment is required, 2) not all people wish to be scanned, and 3) certain types of clothing require taking a customer's measurements manually. However, the impact can also be that some customers find body scanning exciting and like the experience of the process and that they also like to get the advantage of having accurate measurements. A manual procedure also enables a dialogue between the purchaser and the salesperson about the preferred fit of the garment, i.e., tight or roomy, an aspect often overlooked in promoting body scanning. On the negative side, taking measurements manually can be more time consuming and may raise issues of personal privacy. Catering to individual customer sizes becomes an even bigger problem in e-commerce.

There are many examples of businesses that combine manufacturing technologies with MC as shown in Table 1. The Finnish Left Shoe Company (formerly known as The Leftfoot Company), is one example of this concept. Each customer's feet are scanned by sales personnel and the collected information is then used to produce customized shoes that can be delivered to the customer within three weeks [39]. Spreadshirt, an Internet based firm sells t-shirts whose design are individually done by customers and then printed on standard t-shirt's [40].

Brooks Brothers, an upscale American apparel company founded in 1818, now offers mass customised, made-to-measure suits and shirts based on individual body sizes and preferences in partnership with Pietrafesa Corp., a private label suit manufacturing company from Liverpool based in New York [41,42]. Information technology and manufacturing processes were developed and a system called eMeasure introduced in 17 Brooks Brothers stores. The customer's dimensions are taken by a body scanner in the shop and those measurements are used to produce suits and shirts with a perfect fit. The eMeasure system also can store measurement profiles and quickly recall information for repeat customers. Many examples of MC now exist in the fashion industry, and the Internet continues to open up more possibilities for the future.

In MC the business process must be changed, from a linear to a concurrent or parallel process [43,44]. This process often starts with the co-design process and then selling the product to the customer before manufacturing starts.

Complete garment knitting

In complete garment manufacturing, the garment is ready-made in the knitting machine. The panels of the product are knitted in the right shape and knitted together with the trimmings, pockets, and other decorative elements in place as presented in Figure 1 [10]. The advantage of this technique is no waste of material (cut-loss) and no expensive post-knit operations (sewing or cutting) (Legner, 2003). Depending on the style of the garment, some minor cutting and sewing of labels or trim may still be necessary. In addition, while panels sewn together using other manufacturing techniques run the risk of having variations in colour shades between the panels because they were knitted with yarn from various dye lots, in complete garment technique, all the yarn comes from the same cones, enabling higher quality and reducing problems of colour mismatch. With seamless technology, the garment can be made to fit perfectly and be comfortable to wear. In summary, manufacturing processes are reduced and knitting is done on-demand, which can shorten production lead time considerably [11].

Shima Seiki developed the complete garment concept in 1995 and launched it under the brand name WholeGarment. Knit & Wear is the

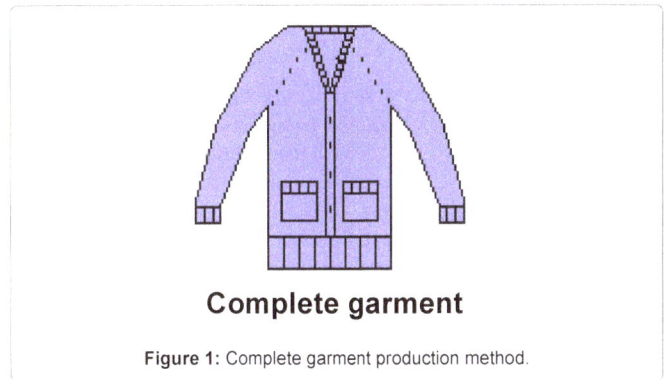

Complete garment

Figure 1: Complete garment production method.

Companies	Products	Descriptions
Tailor Store	Shirts	- Swedish on-line retailer - on-line configurator - customers take their own measurements - manufacturing in Sri Lanka - delivery to customer in 10 to 15 days
The Left Shoe Company	shoes	- Finnish on-line retailer - customers feet's are scanned in the store - delivery to customer in three weeks
Spreadshirt	t-shirts	- German on-line retailer - graphics individually designed by customers - digital printing technology is used - customers can sell their designs to other customers in Spreadshirt's on-line shopping system
Brooks Brothers	suits	- American apparel company - made-to-measure suits - eMeasure system for measurements in 17 Brooks brothers stores - body scanner in the shop - customer measurement profiles are stored and used for repeat orders

Table 1: Descriptions of MC examples.

German firm Stoll´s name of the concept of complete garment [9,10].

Complete garment technology in combination with MC in a supply chain for fashion products may result in measurable benefits such as: *manufacturing lead times reduction, close to point-of-sale production, fast deliveries and a positive shopping experience for the customer.*

There are a lot of advantages with the complete garment knitting technology but the supply chain in the production of knitted garments must be adopted to this technology to gain the benefits.

Summary of the Results

The results of four peer-reviewed published articles are used to answer the research questions stated at the beginning of this concluding article.

Article 1

Purpose and overview: Article 1 describes a design and manufacturing concept "Knit-on-Demand". This is a research project with the aim to show how the complete garment technology can be combined with MC. The manufacturing equipment is located in the retail store and the customer designs the garment in a collaborative process with store personnel. The research method for this study is the case study method combined with lead time simulations. A literature survey and information gathering from suppliers of knitting machinery are performed. The input data for the simulations were tested on both design equipment and textile machinery involved. The study involved a simulation of both the co-design process between the company and the customer and the actual manufacturing process.

Principal findings: Article 1 described the benefits achieved by combining complete garment technology with MC in a business and production system for the Knit-on-Demand concept.

We have endeavoured to show that a high fashion, customised garment may be designed, sold, and manufactured to order in two to five hours. Our findings agree with Choi and Powell [10] that complete garment technology can be effectively employed in conjunction with MC to produce knitted garments. The Knit-on-Demand concept shows an alternative way for European knit fashion producers to shift from mass production to MC, rather than outsourcing their manufacturing to low-income countries.

The present multiple-choice co-design system must be refined and expanded, and manufacturing processes have to be optimised. Products may be delivered quickly if there are no queues caused by many customers wanting to configure a self-designed product at the same time. Ideally, actual customer demand would be fulfilled on location. Where this is impossible, postal mail or express delivery may be the second best option, as is common practice in mail-order or Internet sales. Whether a delay of a few days or weeks would affect a customer's attitude towards the Knit-on-Demand concept has not been ascertained. A key success factor appears to be the quality of the shop personnel and the kind of customer service they provide. The financial aspects of the concept also need to be studied.

Article 2

Purpose and overview: Article 2 is a case study of the Japanese company Factory Boutique Shima in Wakayama. Factory Boutique Shima is a firm that combines the complete garment technology with the concept of MC. A customer can design a garment in a collaborative co-design process guided by staff in the shop. Options of styles, materials, yarns and colours are presented to the customer to support the design of the own designed product. The shop assistant collects and writes the information down to support the manufacturing of the garment when the co-design process is finished. The customer approves the design and pays for the garment before the actual production starts. By this concept the percentage of garments sold at full price (sell-through) is much higher than for a business model where garments are produced in advance of being sold [45]. The case study had an inductive approach based on company visits and interviews with shop personnel. Quantitative data were collected and a Strengths–Weaknesses–Opportunities–Threats (SWOT) analysis performed using qualitative data to identify critical success factors in fashion retailing.

Principal findings: The SWOT analyses in Table 2 indicate one strength, a positive shopping experience for the customer. The Factory Boutique Shima concept suggests how their WholeGarment technology, as the process is called by that company, can be used for MC in the future. The current system, in which one or two staff members devote their full attention to a customer during the co-design process, pleases customers. However, attending to one customer at a time is costly for the company, and so Factory Boutique Shima seeks to develop its co-design system to the point where customers do more of the customization themselves.

The SWOT analysis suggests that Internet sales may present an opportunity for the future. With an efficient co-design system on a company's web page, a vast number of customers could be served at one time. Obtaining accurate customer measurements remains a challenge. One solution is to let customers take and enter their own measurements into the co-design system, as many companies already do. The analysis showed two main areas in which Factory Boutique Shima may improve their business concept: 1) adapting MC to products that can be manufactured with complete garment technology, and 2) developing the customization concept.

Article 3

Purpose and overview: The aim of article 3 was to study and compare two different customization systems. The systems to investigate are one digital configurator and the other is a manual co-design system. Knitting technology is combined with MC in a retail store. In the fashion store the customer can look at swatches, yarn samples and different fashion styles to support the co-design process for the customer. The co-design process is supported by skilled personnel in the store. The products that can be customized are of the types cut and sew fully fashioned and complete garment. The items can be customized to correspond to each technology.

Customisation processes were studied in manual co-design and the digital ordermade WholeGarment. Both were evaluated and appropriated as models for simulation in AutoMod in order to compare their performance. Qualitative interviews with factory representatives at Shima Seiki and retail staff at Factory Boutique Shima provided additional understanding of two procedures. The data gathered from the three sources – Shima Seiki, Factory Boutique

Strengths	Weaknesses
Positive shopping experience	Risk of long queue of customer in the shop Time-consuming co-design process Little reuse of customer information
Opportunities	**Threats**
Internet sales	Limited design options available in WholeGarment Limited interest among consumers

Table 2: Strengths-Weaknesses-Opportunities-Threats (SWOT) analysis.

Shima, and SOMConcept – were used for the simulations. The results are presented in Article 4.

Principal findings: The first simulation compared customised garments made by the manual WholeGarment co-design process with products created using Ordermade WholeGarment digital co-design system. The simulation represented 200 hours and was repeated 15 times. The result for the manual WholeGarment, varied from 146 to 409 products. The variation of the result was depending on the number of shop assistants available to help the client with the co-design. For the digital alternative, Ordermade WholeGarment co-design system, the number of customised products ranged from 259 to 794, depending on the availability of one, two, or three computers for customer use.

Another simulation was done to show the difference between manual- and digital co-design. 1000 digital co-design systems were compared with the alterative with five shop assistants. The aim was to study the effect if a digital Co-design system can be accessed over the Internet. The result showed that over 8000 items can be designed in 200 hours compared to less than 1000 garments using the manual co-design alternative. This illustrates the vast possibility for the digital configurator if it can be applied over the Internet.

There are two main benefits with a configurator system: 1) customers can do a considerable amount of customisation unaided, and 2) the customization options in the co-design tool are pre-programmed to provide information to the knitting machine: when the customization process is done, knitting can begin without the need for time-consuming programming. A configurator brings the entire process a step closer to mass-production efficiency, while maintaining all the distinctive features of customization.

Article 4

Purpose and overview: The aim of Article 4 was to show the supply chain of the Knit-on-Demand concept. It examines the results of the customisation process, technology, systems, and logistics. A case study method and data from two cases were used. The purpose of the first case study was to map the Knit-on-Demand supply chain.

The method applied in this study was value-stream mapping. A technique used to identify waste in the production chain. It begins with a client entering the retail store and purchasing a garment, then follows the order back to the customer order decoupling point (CODP. The purpose of the second case study, conducted during spring 2010, was to analyse the customisation of measurements for each purchase. These measurements were then compared with the standard size tables used in Sweden. The project ended in December 2010 due to the knitting capacity constraints of the manufacturer, Ivanhoe AB.

Principal findings: The original purpose of the Knit-on-Demand project was to test and evaluate complete garment technology. However, the investment required by the participating manufacturer was rejected as being too risky. We then considered using fully-fashioned or cut & sew production methods. Those technologies involved different set-ups and placement of the CODP in the production line. Cut & sew has the fastest order-to-delivery time, since panels can be knitted and kept in stock until a customer's order is received. Value adding time in the production process was 126 minutes for a fully-fashioned product. However, when the cost of the garment is figured, the total time allotted to each operation is calculated using standard allowed minutes (SAM). This differs from actual lead time, depending on how much set-up time a process is allowed to have. In the knitting production step, the allowance is 100% due to downtime, set-up time, and problems that

might occur in manufacturing a garment. Using SAM, the total lead time equalled 179.7 minutes.

The delivery time is one or two days to the store but the product can also be delivered directly to the customer by post. Total lead time from customer order to delivery of the product varies from one to three weeks. The analysis of the co-design process viewed that most clients make minor size changes to products, supporting the market need for MC knitwear.

Discussion

Analysis of Research question 1

Research Question One (RQ1): *How does the co-design process function in the customisation of knitted fashion garments?*

In order to analyse the RQ1 a cross-case analysis method was performed. The case study method is used in Articles 1-4. Data was collected along with simulations, interviews, and observations. The result of using manual or digital co-design is illustrated in Table 3.

Crucial factors for the interaction process between a customer and the company were identified as *risk of having to wait on line (queuing), efficiency, service, co-design tools, programming of the knitting machine after point-of-sale, and internet access.* The factors are rated according to whether they affect the co-design process positively, or negatively.

The analysis shows that the positive factors in the manual co-design alternative are a high level of service provided to the customer and no need for a co-design tool. On the negative side, however, are the risk of queues, a low-efficiency level, no Internet ordering possibility, and the need for time-consuming programming of the knitting machine after the point-of-sale.

	Positive aspects	
Article	Manual co–design	Digital co-design
1	High service level	
2	High service level No need for co-design tool	
3		Low risk of queues High efficiency No knitting machine programming required after POS Internet sales possible
4	High service level No need for co-design tool	
	Negative aspects	
Article	*Manual co–design,*	*Digital co-design*
1	Risk of queues Low efficiency Need for co-design tool Knitting machine programming required after POS No internet sales	
2	Risk of queues Low efficiency Knitting machine programming required after POS No Internet sales	
3		Low service level Co-design tool needed
4	Risk of queues Low efficiency Knitting machine programming required after POS Internet access; no	

Table 3: Cross-case synthesis of data in analysis of Research Question One (RQ 1).

The positive aspects of the digital co-design process are efficiency in serving multiple customers at once, no knitting machine programming needed after point-of-sale, and the possibility of Internet ordering. Conversely, the customer is given no personal service and the retailer must invest in a sophisticated co-design tool (or several).

Analysis of Research question 2

Research Question One (RQ2): *What are the advantages and disadvantages with manual co-design compared to a digital co-design process?*

Manual co-design

Manual co-design is defined as an interaction between the client and a shop assistant in designing a garment without the aid of a digital tool. Manual co-design offers advantages, as has been shown in the study of Factory Boutique Shima, described in Article 3. Body measurements are taken to achieve a perfect fit, and colours, patterns, structures, and attachments are selected with support of staff in the store. This collaboration between customer and store personnel is a process that customers often find positive. Lampel and Mintzberg [30] define this as "tailored customization": the company shows the buyer a prototype and then modifies it to the customer's preferences. A similar dialogue with the customer is termed the "collaborative" approach by Gilmore and Pine [31].

The simulations presented in Articles 2 and 4 indicate the problem inherent in manual co-design is the limited number of customers who can be served at any one time. Perhaps this can be remedied by scheduling appointments, as people are in the habit of doing with their hairdresser or tailor. If the co-design process can be planned in advance, client frustration while waiting to be served can be minimised.

The Knit-on-Demand concept uses a manual co-design process in which the customer is allowed to change four of the garments parameters: model, fit, colour, and details (Article 4). Larsson, who has studied this approach, found that store personnel and customers both wished they had a tool that visualised the customer's choices [46].

Huffman and Kahn [47], concluded that customers prefer a limited number of *attribute-based* options that are presented by a shop assistant, rather than an *alternative-based* system in which customers are confronted with numerous possibilities to choose from on their own. Both in Factory Boutique Shima and in the Knit-on-Demand case, a manual co-design process was used with an attribute-based selection of alternatives along with the guidance of store personnel.

Digital co-design

Digital co-design incorporates a digital tool in the interaction between the client and a shop assistant in designing a garment. The interface between the company and the customer is a crucial process in MC. Customer satisfaction depends on obtaining accurate body measurements and getting the computer screen to display the true colour of the finished garment.

The MC of complete knitted garments is made more efficient through the use of a co-design configurator. Analysis of the manual and digital customization concepts and the simulations in Article 3 show the strength of such a tool, which is IT-based [48-50]. More customers can be served via computer co-design than by a manual process, reducing the number of store personnel involved and potentially lowering costs. In addition, a configuration tool enables customization over the Internet, allowing a retailer to engage in e-marketing.

The digital system examined in Article 3 offered the following advantages:

- Most customization can be done by a client working unaided with a configurator

- Limited programming of a knitting machine is required after point-of-sale

- A configuration tool makes retailing of customised garments on-line possible

The drawbacks include obtaining a client's measurements and the limited number of design options that current knitting machine technology offers. The Swedish company Tailor Store shows customers how to take their measurements on their website. Offering an increased selection of styles, colours, and materials might encourage more customers to purchase customised garments on-line.

Findings

The co-design process can take place at a shop, through an on-line configurator, or use a combination of both concepts, as described by Reichwald, Piller and Mueller [50]. Thus, a customer can be personally guided through the customization by a shop assistant, a process the authors find advantageous because of the reassuring direct interaction it allows. The same customer can later access an online tool to place a reorder. Such manual co-design is preferred by many customers for the personal service it affords, especially for clothing in the higher price range. Stores that already have an established staff may consider adopting such an MC concept if they wish to expand their business strategy.

Manual co-design does not require a great investment and do not require a digital configuration tool. Such an expense may be too much for a small company, as we found in the Knit-on-Demand project (Article 4). Larsson [26] concluded that most customers were not concerned about lead-times or price. On the other hand, offering to deliver a customised garment in 3 to 5 hours (see Article 1) might it be a considerable advantage for a retailer.

A great benefit of the digital co-design system was the pre-programmed options in the configurator. These eliminated or greatly reduced manual reprogramming of the knitting machine, thereby expediting the manufacture and delivery of the finished product. This corresponds to that all known mass customisers use systems that are to some extent IT-based as mentioned in the Frame of Reference of this thesis.

The future of mass customisation of knitted garments looks bright for co-design systems of the kind we have considered. Perhaps shops like Factory Boutique Shima or Knit-on-Demand will one day offer their clients the opportunity to design a product that is knitted in the shop and delivered to them within hours. We may also see collaborations between retailers with an insight into fashion trends and efficient knitwear manufacturers for the development and production of co-designed products. Soon the technology will be available that can facilitate the growth of stores devoted to the MC of fashion products with a minimum of help from shop staff.

Conclusions

The overall research objective of this article stated in the beginning was: *How can complete garment knitting technology be applied in a retail concept for customised garments?* A strategy for mass customised knitted garments using a manual co-design process was developed in collaboration with SOMconcept AB and Ivanhoe AB and tested in the SOMconcept store in Stockholm. Manual co-design was also examined in a case study at Factory Boutique Shima in Japan, where an analysis between this system and Shima Seiki's Ordermade system was conducted.

MC of knitted products requires specialised production facilities located in the retail store, at a near-by facility, or at a remote location linked by good shipping facilities. If production takes place in the store, it is possible to customise and deliver a garment to the customer in 3 to 5 hours. If the garment is manufactured at another location, the total lead time from customer order to delivery can range from 1 to 3 weeks. Both systems operate within a relatively short lead time compared with mass-produced products sold at ordinary fashion retailers. In the MC process itself, two kinds of interactions are feasible between the company and the customer: manual or digital co-design. A manual process, in which the customer is actively involved in the design, but guided by a shop assistant, is the basis of concepts like Knit-on-Demand and Factory Boutique Shima. Manual co-design does not require advanced technical equipment, as all the information can be entered by hand onto a customization form that is then sent to the production department. However, manual co-design is labour intensive, since a shop assistant can only serve one client at a time. It also is only applicable to brick-and-mortar stores and not transferable to the Internet.

Digital co-design, on the other hand, encourages customers to do the customization on their own, without the aid of sales personnel. If a store has an ample number of configurators, there will be little risk of queues. Moreover, this technique is ideal for the Internet. The problem of taking body measurements, however, awaits a satisfactory solution and still requires help from shop employees. If customers can be encouraged to this by themselves, as some on-line retailers like Brooks Brothers and Land's End in the US have shown, mass-customised knitted garments could be widely sold on the Internet, thereby reaching vast numbers of customers. Tailor Store is another example of an Internet MC shirt retailer whose customers take their own measurements and enter them on the company's web page. Complete garment technology is also more economical for the manufacturer: there is no material cut-loss, and a minimum of costly post-knitting processes are needed. Disadvantages to date have included limited design options and the need for custom programming of the knitting machine. (In the digital co-design system studied in this thesis, the design options were pre-programmed in the configurator and so a customised garment could be knitted without delay).

Mass customised garments are especially suited for people whose bodies do not fit standard sizes or who wish to create a garment with a unique design (Larsson, 2011). Providing a purchaser the satisfaction of a perfect fit, an original creation, rapid customization, and the opportunity to try one's hand at fashion design opens up many new retail possibilities.

References

1. Hounshell DA (1984) From the American System to Mass Production 1800-1932. John Hopkins University Press, Baltimore, MD.

2. Segerblom M (1983) Algots: en teko-koncerns uppgång och fall [Algots: a textile Group's rise and fall]. LiberFörlag Stockholm [in Swedish].

3. Hoover W, Eloranta E, Holmström J, Huttunen K (2001) Managing the Demand Supply Chain. John Wiley & Sons New York.

4. Davis S (1987) Future Perfect. Addison-Wesley Reading MA p. 243.

5. Pine BJ II (1993) Mass Customization The New Frontier in Business Competition. Harvard Business School Press Boston.

6. Kay MJ (1993) Making mass customization happen Lessons for implementation. Strategy & Leadership 21: 4-18.

7. da Silveira G, Borenstein D, Fogliatto FS (2001) Mass customization Literature review and research directions. International Journal of Production Economics 72: 1-13.

8. Lee SE, Kunz GI, Fiore AM, Campbell JR (2002) Acceptance of mass customization of apparel Merchandising issues associated with preference for product process and place. Clothing and textiles Research Journal 20: 138-146.

9. Spencer DJ (2001) Knitting Technology A Comprehensive Handbook and Practical Guide. (3rd edn) Woodhead Publishing Ltd Cambridge.

10. Choi W, Powell NB (2005) Three dimensional seamless garment knitting on V-bed flat knitting machines. Journal of Textile and Apparel Technology and Management 4: 1-33.

11. Legner M (2003) 3D-products for fashion and technical textile applications from flat knitting machines. Melliand International 9: 238-241.

12. Franke N, Piller F (2003) Key research issues in user interaction with user toolkits in a mass customisation system. International Journal of Technology Management 26: 5-6, 578-599.

13. Peterson J (2012) Customisation of Fashion Products Using Complete Garment Technology. Thesis (PhD) Tampere University of Technology Tampere Finland.

14. Yin RK (2009) Case Study Research-Design and Methods. Sage Publications London.

15. Banks J, Carson JS II, Nelson BL, Nicol DM (2004) Discrete-Event System Simulation. (4thedn) Pearson Prentice Hall Upper Saddle River NJ.

16. Näslund D (2002) Logistics needs qualitative research – especially action research. International Journal of Physical Distribution & Logistics Management 32: 321-338.

17. Choi W, N. B. Powell (2006) The Development of Specialized Knitted Structures in the Creation of Resist-Dyed Fabrics and Garments. Thesis (PhD) North Carolina State University Raleigh 253-264.

18. Mowbray JA (2002) A quest for ultimate knitwear. Knitting International 109: 22-24.

19. Hunter B (2004) Complete garments evolution or revolution (Part I) Knitting International 111: 18-21.

20. Toffler A (1980) The Third Wave. Bantam Books, New York.

21. Wikström S (1996) Value creation by company-consumer interaction. Journal of Marketing Management 12: 359-374.

22. Piller FT (2004) Mass customization: Reflections on the state of the concept. International Journal of Flexible Manufacturing Systems 16: 313-334.

23. Peterson J, Larsson J, Carlsson J, Andersson P (2008) Knit-on-Demand: Development and simulation of a production and shop model for customised knitted garments. International Journal of Fashion Design, Technology and Education 1: 89- 99.

24. Peterson J. Mattila H (2010) Mass customising of knitted fashion garments: Factory Boutique Shima – a case study. International Journal of Mass Customisation 3: 247- 258.

25. Peterson J, Larsson J, Mujanovic M, Mattila H (2011) Mass customisation of flat knitted fashion products Simulation of the co-design process. Autex Research Journal 11: 6-13.

26. Larsson J, Peterson J, Mattila H (2012) The Knit on Demand supply chain. Autex Research Journal 12: 67-75.

27. Fralix MT (2001) From mass production to mass customization. Journal of Textile and Apparel, Technology and Management 1: 1-7.

28. Tseng M, Piller FT (2003) Applying mass customisation to the fashion industry. In: Tseng M, Piller FT (eds), The Customer Centric Enterprise: Advances in Mass Customization and Personalization. Springer Verlag, Berlin.

29. Kaplan AM, Haenlein M (2006) Toward a parsimonious definition of traditional and electronic mass customization. Journal of Product Innovation Management 23: 168-182.

30. Lampel J, Mintzberg H (1996) Customizing customization. Sloan Management Review 38: 21-25.

31. Gilmore JH, Pine BJ II (1997) The four faces of mass customization. Harvard Business Review 75: 91-102.

32. von Hippel E (1998) Economics of product development by users The impact of "sticky" local information. Management Science 44: 629-644.

33. Khalid H, Helander M (2003) Web-based do-it-yourself product design. In: Tseng, M., and Piller, F. (eds.), The Customer Centric Enterprise. Springer, New York, Berlin.

34. Piller F, Schubert P, Koch M, Möslein K (2005) Overcoming mass confusion: Collaborative customer co-design in online communities. Journal of Computer-Mediated Communication: 10 article 8.

35. Toffler A (1980) The Third Wave. Bantam, New York.

36. Franke N, Schreier M (2002) Entrepreneurial opportunities with toolkits for user innovation and design. International Journal on Media Management 4: 225-234.

37. Piller FT (2004) Mass customization: Reflections on the state of the concept. International Journal of Flexible Manufacturing Systems 16: 314-315.

38. Lee SE, Chen JC (1999) Mass-customization methodology for an apparel Industry with a future. Journal of Industrial Technology 16: 1-8.

39. Sievänen M, Peltonen L (2006) Mass customising footwear: The left® foot company case. International Journal of Mass Customization 1: 480-491.

40. Reichwald R, Piller F (2006) Interaktive Wertschöpfung [Interactive Value Creation]. Gabler, Wiesbaden. [in German].

41. Rabon LC (2000) Mixing the elements of mass customization. Bobbin 41: 38-41.

42. Yeung HT, Choi TM, Chiu CH (2010) "Innovative mass customization in the fashion industry." In: Cheng TCE, Choi TM (eds.) Innovative Quick Response Programs in Logistics and Supply Chain Management. Springer Verlag, Berlin.

43. Anderson D (1997) Agile Product Development for Mass Customization: How to Develop and Deliver Products for Mass Customization, Niche Markets, JIT, Build-to-Order and Flexible Manufacturing. McGraw-Hill, New York.

44. Kincade DH, Regan C, Gibson FY (2007) Concurrent engineering for product development in mass customization for the apparel industry. International Journal of Operations & Production Management 27: 627-649.

45. Mattila H (1999) Merchandising Strategies and Retail Performance for Seasonal Fashion Products. Thesis (PhD) Lappeenranta University of Technology Lappeenranta Finland.

46. Larsson J (2011) Mass Customised Fashion. Thesis (PhD), University of Borås, Sweden.

47. Huffman C, Kahn B (1998) Variety for sale: Mass customization or mass confusion? Journal of Retailing 74: 491-513.

48. Franke N, Piller F (2003) Key research issues in user interaction with user toolkits in a mass customisation system. International Journal of Technology Management 26: 578-599.

49. Weston R (1997) Web automation. PC Week 32: 73-78.

50. Reichwald R, Piller F, Müller M (2004) A multi-channel customer interaction platform for mass customization – Concept and empirical investigation. Paper presented at the Fourth International ICSC Symposium on Engineering of Intelligent Systems.

29

Mulberry and Silk Production in Kenya

Tuigong DR*, Kipkurgat TK and Madara DS

School of Engineering, MOI University, Kenya

Abstract

Mulberry is a plant that is grown for silkworm rearing. It is the exclusive food for the silkworm, which during its larval life is reared for silk production. Mulberry forms the basic food material for silkworms. Production of mulberry leaves on scientific lines is essential for organizing sericulture on sound economic lines. It is estimated that one metric ton of mulberry leaves is necessary for the rearing of silkworms emerging out of one case of eggs which will yield about 25 kg to 30 kg of cocoons of high quality. The findings show that mulberry plant can grow and thrive very well in Kenya because of very good climatic conditions that are favorable for mulberry plant. It is worth nothing however that mulberry tree can grow in a variety of climatic conditions. As a result of successful production of mulberry, silk production training is needed for skilled labour in mulberry growing and silk worm rearing in these high production areas of the country with similar climatic condition to the experimental area of Eldoret. Sericulture has the potential of poverty eradication and economic empowerment especially for women and youth in Kenya because it is a labour intensive venture. Silk production has the potential of serving as a supplement to the textile industry in Kenya due to the the dwindling cotton production. Despite the fact that the sericulture has been going on in Kenya for more than 45 years, there has been several challenges that has crippled the success of sericulture. The major bottleneck is the lack of domestic demand for the finished products due to unclear goals in quality and minimal product awareness, lack of well established government policies and lack of capacity and insufficient technical skills on mulberry and silkworm rearing. The study recommends that proper agronomical practices should be used to increase yield, intensive research is required on the available species of mulberry in Kenya. As a result of successful production of mulberry, training is needed for skilled labour in mulberry growing and silk worm rearing for high production.

Keywords: Agronomy; Larval; Mulberry; Sericulture; Silkworm

Introduction

Historical evidence show that silk was discovered in China and from there it spread to other parts of the world. The earliest evidence is in the chronicles of Chou King (2200 B.C), where silk featured prominently in public ceremonies as a symbol of homage for the emperor. First it was kept as secret within China due to jealousy but when commercial relations was established between China and Persia and later to other countries, export of raw silk extended up to southern Europe. First to learn the secret was Korea and then Japan. War was instrumental in the spread of the silk industry especially to Japan when Semiramus, a general in the army of empress Singu_kongo invaded and conquered Korea. The other factor for the spread of the industry was migration. During the latter part of 19th century, Japan gave a serious attention to the silk industry, introducing the use of processing machinery and improved techniques and carrying out intensive research in sericulture. In Indian silkworms were first domesticated in the foothills of the Himalayas. When the British came to India they established flourishing silk trade through the British East India Company. Sericulture spread over the century from China to other parts of the country and silk became a precious commodity highly sought after [1,2].

In Kenya, introduction of sericulture was a joint venture between experts of the Overseas Technical Co-operation Agency (OTCA) from Japan and the Ministry of Agriculture of Kenya in 1974. The first task was to establish a mulberry cultivation orchard at National Horticultural Research Centre (NHRC) in Thika. In order to establish this task, preliminary research to identify indigenous mulberry cultivars was undertaken. Some high yielding cultivars were imported from India and Japan in order to assess their performance in Kenyan climatic conditions.

In 1996, the International Centre of Insect Physiology and Ecology (ICIPE), an advanced research institute situated in Nairobi, opened a sericulture unit under its commercial insects programme. The unit is developing innovative sericulture technologies geared enhancing the productivity and economic returns of small-scale land users. The unit has laid emphasis on the conservation and utilization of wild silk moth rearing of the domesticated silkworm for raw silk production. The unit has successfully developed a new domestic silkworm hybrid, which flourishes in the African environment and produces high quality silk. The unit has also developed a full package of silk technology from mulberry cultivation to weaving.

Sericulture

Sericulture, or the raising of silkworms to produce silk, involves the incubation of the tiny eggs of the silkworm moth until they hatch and become worms. Basically laying of eggs by the moths is done on special kind of cards, which specifically act as a surface for hatching of the eggs. The process of transferring the newly hatched silkworms to rearing trays is called "brushing". In order to obtain uniform hatching, eggs are kept in black boxes on the day prior to hatching. In this way, the early maturing embryos are prevented from maturing and the late maturing embryos are given time to develop and catch up with the early maturing ones. The next day they are exposed suddenly to diffused light so that the larvae hatch uniformly in response to phototropic stimulus [3].

After hatching, the worms are placed under a layer of gauze. For six weeks, the worms eat finely chopped mulberry leaves almost

---

*Corresponding author: David R Tuigong, School of Engineering, MOI University, Kenya, E-mail: davidtuigong@gmail.com continuously. At the end of this period, they are ready to spin their cocoons, and branches of trees or shrubs are placed in their rearing houses. The worms climb these branches and make their cocoons in one continuous thread, taking about 8-14 days for the process. The amount of usable silk in each cocoon is small, and about 5,500 silkworms are required to produce 1 kg of raw silk.

After the complete cocoons have been gathered, the initial step in silk manufacture is to kill the insects inside them. Thus, the cocoons are first boiled or treated in ovens, killing the insects by heat. The silk fiber is obtained from the cocoons by a delicate process known as reeling.

The cocoons are first harvested. Timing is very important in that early harvesting will produce very delicate cocoon, which can easily rapture when disturbed. On the other hand, late harvesting might lead to the emergence of the moth thus destroying the cocoon as well as the filament. After harvesting the cocoons are selected according to colour, size, fibre diameter, and even uniformity of the cocoon. All these determine cocoon quality where by the bigger and the more uniform the cocoon the better. Also the thinner the fibre, the better.

The properties of silk produced by a given cocoon (from the outer to the inner) varies as shown in Table 1.

The cocoons are then heated in boiling water to dissolve the gummy substance that holds the cocoon filament in place. After this heating, the filaments from four to eight cocoons are joined and twisted and are then combined with a number of other similarly twisted filaments to make a thread that is wound on a reel. When each cocoon is unwound, it is replaced with another cocoon. The resulting thread, called raw silk, usually consists of 48 individual silk fibres. The thread is continuous and, unlike the threads spun from other natural fibres, such as cotton and wool, is made up of extremely long fibres.

The overall efficiency in sericulture is linked to the best management of the conversion of leaf to cocoon. Leaf to cocoon ratio indicates the amount of leaf required to produce 1 kg of cocoon. The ratio 20 is recommended, although, 16 and 18 are also accepted. Sericulture is quite different from other types of farming activities in that it has a chain of interrelated and interdependent stages. It involves three distinct phases of activity namely: mulberry cultivation, silkworm rearing and silk reeling.

Silkworm Seed Production

The most important step in silkworm rearing is the production of silkworm eggs required for rearing. The silkworm eggs or seeds required for commercial rearing should be of high quality and free from diseases.

Reproductive seeds

Reproductive eggs or seeds are those intended for producing the seed cocoons, which are required in large numbers for producing commercial hybrid eggs or industrial seeds. These pure lines are difficult to rear and so special care must be taken and ideal rearing conditions provided.

Industrial seeds

Description	Near outer	Middle	Near inner
Diameter of fibres (microns)	32	28	36
Elasticity (%)	13.3	24.3	24.7
Breaking load (g)	6.25	10.2	9.0

Table 1: Silk quality according position in cocoon.

Industrial seeds are generally specific hybrids between two or more pure lines or race of silkworms and are reared by the Seri culturists for producing cocoons on a commercial scale for reeling purposes. There are four kinds of natural silk, which are commercially known and produced. Mulberry is the most important and contributes as much as 95% of world production, therefore, the term 'silk' generally refers to the silk of the mulberry silkworm. The other 3 kinds of silk apart from mulberry silk are:

- Eri silk-is produced by eri silkworms, which are reared on castor oil plant leaves to produce white or brick red silk called eri silk.

- Tasar silk-is produced by tasar silkworm, which are reared on oak leaves and allied species to produce green silk thread called tasar silk.

- Muga silk-is produced by muga silkworm, which are reared on som (machilus bombycina), to produce an unusual lustrous golden-yellow silk thread, which is very attractive and strong. These are only found in the state of Assam.

These are non-mulberry silks.

The insect producing mulberry silk belongs to a domesticated variety of silkworm referred to as the mulberry silkworms. The silkworm is reared under human care because;

a)   Its grasping power is weak, so it can hardly climb a tree, twig or a leaf of a tree; it is shaken off by even gentle breeze

b)   It has no crawling power. When there is no food in its neighbourhood, it will not be able to crawl and may starve to death. Silkworm can only move a few centimeters around itself however hungry

c)   It has a weak sense of smell such that mulberry one meter away will be too far

With all these weaknesses, it becomes a necessity that food (mulberry leaves), has to be brought close to the silkworm. They belong to the species bomyx mori. The mulberry silkworm may be further identified as:

They belong to the species bomyx mori. The mulberry silkworm may be further identified as:

- Univoltine, bivoltine and multivoltine, depending on the number of generations produced in a year under natural conditions

- Trimoulters, tetramoulters, and pentamoulters depending on the number of moults during larval growth, or

- As Japanese, Chinese, European or Indian origin, based on geographical distribution

- Univoltine, bivoltine and multivoltine, depending on the number of generations produced in a year under natural conditions

- Trimoulters, tetramoulters, and pentamoulters depending on the number of moults during larval growth, or

- As Japanese, Chinese, European or Indian origin, based on geographical distribution

Mulberry Tree

Characteristics

As said earlier, mulberry tree leaves is the exclusive food for silkworm, where the leaf of white mulberry, *morus Alba* is used. Production of mulberry leaves on scientific lines is essential for organizing sericulture on sound economic lines. It is estimated that one metric ton of mulberry leaves is necessary for the rearing of silkworms emerging out of one case of eggs which will yield about 25 kg to 30 kg of cocoons of high quality.

This plant has been cultivated in china for centuries and has been successfully adapted in other regions. It is a woody flowering deco plant, belonging to the latex bearing woody plants, among the fig and marijuana plant. The tree normally spread in the in the tropics and bears small clusters of seeds red, black, or white, termed mulberries according to the colour of the fruits. Black mulberry bears large dark purple fruit, which looks like a long, slender blackberry. The red mulberry produces red very soft fruits while the white mulberry is the one used for silk rearing. Mulberries make up the family moraceae and order *urticales*. The representative genus is *morus*. The red mulberry is classified as *morus rubra*, the Texas mulberry as *morus microphylla*, and the white mulberry as *morus alba*

Mulberry tree propegation

Mulberry trees are cheap and easy to propagate. They are very easy to transplant and the tree grows rapidly. They are perennial and their pest immunity is comparatively more dependable than a tree that has just come from the forest [4].

Uses of mulberry

Mulberry trees have a wide variety of uses. It is mainly used in silkworm rearing as their major food. The fruits are used as pigs' food. Left over mulberry leaves are used as cattle feed, not used as substitute for other forms of fodder, but even to increase milk production. The leaves can also be used as a vitamin supplement in the in the diets of poultry and can improve egg production.

Mulberry bark can be made into paper pulp and used in the paper making industry. The bark can also be used as a treatment for worms or as a purgative. The frits can be cooked into squashes, jams, jellies, and pickles. Mulberry roots are used in dyeing, tanning, and colouring processes. The berries can also be used for ornamental purposes. Mulberry wood is resilient, shock resistant and tough. Therefore, they can be crafted into hockey sticks and racquets for tennis, badminton and squash. The sticks are used on fences, or woven into baskets and silkworm rearing trays.

Mulberry agronomy in kenya

According to Kariaga [5] due to favourable climatic conditions in Kenya, mulberry is mostly cultivated under rain fed conditions. A number of mulberry cultivars (local and imported) have been under study since 1979. . In Kenya, there are four locally adapted varieties, which are:

a) Morus Alba Ex-Embu: This is characterized by short internodes, reddish bark, small size and high drought-resistance

b) Morus Alba Ex-Thika: This has large light green leaves, long internodes whitish bark and it is fairly drought-resistance

c) Morus Alba Ex-Limuru: It has small finger-shaped deeply serrated leaves. It has very thin short internodes. It is a heavy berry producer. It is not recommended for silkworm rearing and is only recommended for berry production

d) Morus Alba Ex-Ithanga: This has medium heart-shaped and smooth light green leaves. It roots easily and it is fairly drought resistant. It is suitable for both silkworm rearing and berry production

In sericulture development project, the morus Alba ex-Thika variety is the one used for silkworm rearing.

Cultivars imported from Japan are two, namely wasemindori ex- Japan and ichinose ex – Japan. The only cultivars imported from a tropical country is morus alba var kanva ex – India. A number of mulberry cultivars (local and imported) have been under study since 1979. In Kenya, there are four locally adapted varieties, which are:

- Morus Alba Ex-Embu: this is characterized by short internodes, reddish bark, small size and high drought-resistance

- Morus Alba Ex-Thika: this has large light green leaves, long internodes whitish bark and it is fairly drought-resistance

- Morus Alba Ex-Limuru: it has small finger-shaped deeply serrated leaves. It has very thin short internodes. It is a heavy berry producer. It is not recommended for silkworm rearing and is only recommended for berry production

- Morus Alba Ex-Ithanga. This has medium heart-shaped and smooth light green leaves. It roots easily and it is fairly drought resistant. It is suitable for both silkworm rearing and berry production

In sericulture development project, the morus Alba ex-Thika variety is the one used for silkworm rearing.

Conditions for mulberry growing

Land Scape: Mulberry grows well in flat fertile land, however if the slope is below 15%, it can be established in rows along the contour, where as in steeper slopes (15%-30%), the terrace system is adopted preferably on a single row. Very steep land has higher soil erosion rate, which imply that the nutrients are lacking. Furthermore, plant anchorage is also weak.

Altitude: An elevation up to 700 m above mean sea level is considered suitable for growth of mulberry. High altitude, where frost is likely to occur is unsuitable for mulberry growth.

Soils

Importance of soils to plants: Root systems usually comprise of no more than a quarter of the seed plant, but the roots are so finely divided that frequently they occupy a mass of greater than the volume of atmosphere occupied by the shoot. The result is tremendous amount of surface contact between soil and plant, which shows the plant's dependant upon the soil for anchorage, water and nutrients. Soil types affects plants in many aspects like:

- Ability of seeds to germinate

- Size and erectness of the plant

- Vigour of the vegetative organs

- Woodiness of the stem

- Depth of root systems

- Amount of pubescence

- Susceptibly to drought, pest, diseases

- Rate of growth, yield

Sufficient growth of mulberry is subject to not only full growth of roots but also to the conditions required for growth, i.e., condition and composition of the soil in connection with absorption of water and fertilizer. Quality of soil of mulberry field does not only dominate growth and yield of mulberry but also quality of leaf, rearing conditions of silkworm and thus yield of cocoon and also its quality. Mulberry requires deep, friable, well drained aerated rich soil of good water holding capacity, and should be at least 50 cm deep. Under soils, the following factors are considered:

- Soil fertility
- Depth and nature of profile
- Soil texture
- Soil pH
- Temperature of soil
- Soil moisture

Soil fertility: Soil fertility is defined as the ability of soil to provide physical conditions favourable for root growth and to supply enough water and nutrients to enable the crop to produce the most out of the environmental features of a site. Soil fertility is the most important factor in determining the yield as well as the quality of leaves produced. The major factors entailed in soil fertility are:

- Organic content
- Soil reaction and
- Availability of major and micronutrients

Depth and nature of profile: Very shallow soils are unproductive since they provide little root room for crop anchorage and extraction of nutrients and water and they are usually waterlogged or hold too little moisture. Soil depth good for mulberry growth should be at least 50 cm deep. This is because mulberry is a perennial crop and can root up to 4.5 m –6.0 m where there is profound dry season.

Soil texture: Soil texture is defined as the coarseness or fineness of the soil particles. Soil texture is classified on the basis if equivalent spherical diameter. This is a system adopted by International Society of Soil Science, and texture classification is as shown in Table 2.

Mulberry requires clayey loam sandy soils for proper growth. Mulberry requires an average of 50 mm of water once in 10 days in the case of loamy soils and same amount in 15 days in case of clay soils. To produce high yields, soils need to be well structured and accept water and rainfall without crusting, eroding or stupping. It should also hold moisture at reasonable levels.

pH of soil: Effect of pH on the plant is generally determined by the plant's root morphology. Acidity and alkalinity of the soil should be in an acceptable range. Slightly acidic soils of pH 6.2-6.8 are ideal for good growth of mulberry. Highly alkaline soils can be rectified by application of gypsum or sulphur and too acidic soils by applying lime. Table 3 below shows a ready reckoner for reclamation of acidic or alkaline soils. The quantity is given in tonnes per hectare.

Soil temperature: Growth of plants is greatly affected by soil temperature. When soils are moist, temperature is usually the dominant factor in determining the rate of germination and growth. This is because this temperature influences the root growth as well as the rate at which they take up nutrients and water, which further influences development and leaf expansion and consequently mulberry leaves

S/No.	Classification	Diameter (mm)
1	Course sand	2.0 – 0.2
2	Fine sand	0.2 – 0.02
3	Silt	0.02 – 0.002
4	Clay	Less than 0.002

Table 2: Classification of soil according to texture.

S/No.	PH range	Limestone (tonnes per hectare)	Gypsum (tonnes per hectare
1	3.5	12.5	-
2	4.5	8.75	-
3	5.5	5.0	-
4	7.4-7.8	-	2.0
5	7.9-8.4	-	5.0
6	8.5-9.0	-	9.0
7	9.1 and above	-	14.0

Table 3: Dosage for reclamation of acidity or alkalinity of soil.

yield. Recommended temperature for mulberry tree growth is 23°C to 30°C. Rate of water uptake is directly proportion to water uptake. Very high temperatures will kill the plant as well as very low temperatures. Soil temperature is normally affected by the organic matter present in the soil – they increase specific heat (due to microbial reactions) thus increase the soil temperature. It is also worth nothing that dark soils absorb more heat than light coloured soils, thus the dark soils have higher temperatures.

Soil moisture content: The moisture content is a very important factor in determining the growth of plants. The maximum amounts of capillary or hygroscopic water a particular soil can hold are determined chiefly by the kind and sizes of soil particles as well as the organic matter.

Therefore mulberry grows well on well-drained soils with an average of 50 mm of water once in 10 days in the case of loamy soils and same amount in 15 days in case of clayey soils. Generally finer soils can hold more water than course soils. Soil moisture content most suitable for full growth of mulberry is 50-60% (in weight) of the maximum water – holding capacity of the soil.

Table 4, shows water content of major soils used for mulberry cultivation. Mulberry is largely influenced by the condition of moisture distribution in the soil, partly because mulberry needs more water for its growth than the water content in its own.

Water factor: Water is a major factor for the proper growth of any plant. Though fairly resistant to drought, mulberry tree requires adequate amount of water for its growth and yield.

Importance of water to plants: In the physiology of plants, water is of paramount importance in many ways. As the closest approximation of a universal solvent, it dissolves all minerals contained in the soil. It is a medium by which solutes enter the plant and move about through the tissues. By permitting solution and ionization within the plant it greatly enhances the chemical reactivity of both simple and elaborated compounds. It is a raw material in photosynthesis. It is essential for the maintenance of turgidity without which cells cannot function actively, and indeed it is necessary for the mere passive existence of the protoplasm.

Atmospheric moisture: This is the invisible water – vapour content in the air. It is usually expressed as % relative humidity. Temperature is a more important factor in determining relative humidity. Warm air can hold more water than cold air. For proper mulberry tree growth, the relative humidity 70%-75%.

S/No.	Soil	Sandy loam	Loam	Clay
1	By weight%	48	56	61
2	By volume%	70	75	80

Table 4: Water content of soil.

Effect on intensity of solar radiation: Solar radiation is the propagation of solar energy through space. It implies the transmission of heat, light, and other rays from the sun to the earth's surface. Water in the atmosphere intercepts this energy before it reaches the surface of the plant. Strong sunlight causes excessive transpiration on the plant and might, in severe cases cause wilting and improper growth of mulberry tree. On the other hand, inadequacy of sunlight and heat under shade causes improper photosynthesis.

Rainfall: Rainfall if the main source of soil moisture for plants. Although all soil moisture is derived from precipitation, not all precipitation is equally effective in increasing soil moisture. The slower and more gentle the showers, the higher the percentage of water that soaks in the soil in relation to that lost as run off. The greater the quantity of water that falling during any one rainy period, the more it sinks below reach of surface desiccation. Efficiency of precipitation as a source of soil moisture for plants is best measured by direct studies of the degree of penetration and duration of moisture in the soil. Mulberry tree requires average rainfall of over 800 mm per annum.

Temperature Factor: There is relatively little biologic activity below $0^\circ C$ or above $50^\circ C$. Mulberry grows at temperatures between $23^\circ C$ and $30^\circ C$. Optimum temperature for germination of mulberry seed is $33^\circ C$ – $37^\circ C$ at maximum and $12^\circ C$-$13^\circ C$ at minimum.

Fertilizer for Mulberry: Components of manure required for mulberry cultivation are the same as those required for plants in general; however amount to be supplied is considerably different. Mulberry is a woody plant, and unlike annual herbaceous plants, it absorbs and utilizes fertilizer in a different way and also unlike other woody plants, such as fruit trees it is cultivate for harvest of leaves. Though for the purpose of harvesting leaves, unlike tea trees, in mulberry culture, most of the part above the earth is cut off for harvest.

Chopudhury and Giridhar [6] state that yield of mulberry leaves is strongly influenced by amount of nitrogenous fertilizer. No nitrogenous fertilizer for mulberry means no manure at all. Meanwhile the yield is reduced by about 10% if there is no phosphoric acid fertilizer. Potassium fertilizers scarcely influence the yield. This might differ a bit according to soils. It is very important to calculate the exact amount if fertilizer for mulberry field. Proper quantity of fertilizer would be different according to the target of mulberry yield, soil character, climate, training method, type of fertilizer etc. theoretically speaking, the amount of component to be supplied as fertilizer is the amount of components taken away as harvest from the field, added with the amount of components required for growth of roots minus the amount of natural supply. Ratio of components to be absorbed by mulberry is different according to each component of fertilizer, and the amount of component to be supplied as fertilizer must be calculated from the absorption rate of the manure component. The above is expressed by the following formula [7].

*Amount of fertilizer = (CH + CG _ CN) * 100 / R*

Absorption rate of manure or fertilizer components = (CH_f _ CH_{F0}) / CS

Where,

- CH – amount of components in harvest
- CG – amount of components required for growth
- CN – amount of components naturally supplied
- R – absorption rate of manure components
- CS – amount of components of manure supplied
- CH_F – amount of components in harvest with fertilization
- CH_{F0} – amount of components in harvest without fertilization

The ultimate object of mulberry cultivation is to increase the yield of mulberry leaf as the food for silkworm at the same time maintaining the good quality of leaf. It is necessary to apply fertilizer separately to mulberry for young silkworm food leaves. The quality of mulberry for young silkworm is closely related to the health of the silkworm, and the quality of mulberry for old age worms largely influences the quality of cocoon.

Silk

Silk is the solidified viscous liquid excreted from special glands or orifices by a number of insects and spiders. The only significant source of silk for textile usage is the silkworm larva, known as the silk moth from the two glands on either side of their body. Silk is available for textile use both as continuous filament yarn and staple yarn. The filament range from 300 – 1200 meters in length. It is a polymer in which the principal amino acid constituents are, glycine, serine, alanine, and tyrosine.

Chemical structure

Silk is composed of long chains of amino acids, mostly, glycine, and alanine are the main components, joined by peptide links with hydrogen bonding between parallel chains.

Composition of raw silk

Besides sericin and fibroin, raw silk contains small quantities of mineral matter, traces of fat, colouring matter and water in the compositions shown in Table 5.

Physical properties

Silk is strong with a tensile strength of 0.34 N/Tex – 0.390.34 N/Tex and an elongation at break of 20% -30%. It has a good natural crease resistance due to its good resilience and it recovers readily from deformation.

Silk is a highly hygroscopic fibre with moisture regain of 10%-15%. It is a poor conductor of heat and electricity. It characterized with lightweight, warmth and good drapability.

It is a smooth translucent fibber with a triangular cross section. Continuous filament silk has a high lustre that contributes to its aura of luxury.

Chemical properties

Silk is readily soluble in cold concentrated mineral acids. Nitric acid even in the most dilute form causes yellow discoloration. Cold concentrated solutions of caustic alkalis appear to have no effect when in contact with silk for a short time. Concentrated solutions of zinc chloride dissolve silk. Sodium chloride has no effect when exposed for a shorter time.

S/No.	Component	Amount (%)
1	Sericin	15-25
2	Fibroin	62.5-67
3	Water	10-11
4	Salts	1-1.5

Table 5: Composition of raw silk.

Fibre identification

Silk has very high lustre. It can be microscopically identified by observing the cross section. It has a triangular cross section. Silk is the only protein fibre that has no scales. Further identification is by dissolving silk in 59.5%-70% sulphuric acid. This solution dissolves it. See appendix H for the appearance of different silks.

Data Collection, Results and Analysis

Data collection

The growth of mulberry was propagated in an experimental situation at Moi University farm. Observation was carried out and the stages of sprouting and it development and elongation was measured. Table 6 shows the growth.

Figure 1 shows the growth of mulberry was measured as it shoots and length in centimeters represented against time

Results

Mulberry Tree Growth: Growth of a plant or an organ is ultimately the net result of the growth of the cells, which the plants meristematic regions are composed. There are three phases of plant growth namely; cell formation, the cell enlargement and the cell maturation. All the cells of plants are derived ultimately from the apical meristems and before maturation each cell must undergo the three phases.

Growth is confined to certain regions or growing points, which consist of meristematic cell, which are capable of cell division and giving rise to new cells. Such important cells are found at the stems and roots and are thus referred to as the apical meristems. Their activity results in primary growth i.e., the formation of the primary structure of the plant. They are responsible for all the increase in the length of the plant axis at both the stem and the roots ends for the production of stem and root appendages like root hairs, leaves and flowering part. In some cases, growth also occurs by the activity of intercalary meristems, which occur at the base of the internodes, node or at the leaf base.

The formation of new cell wall is soon followed by their enlargement. The young cell absorbs water and nutrients and as a result turgor pressure is set up and the extensible cell wall stretches. The stretching of the cell wall is made permanent by the addition of new cell wall material to the original wall, which consequently becomes thicker and still capable of further extension. In the apical meristems the maximum enlargement of the cell takes place on the side of the free apex and hence the stem and the root elongates.

The last phase of growth is the development and maturation stage. It involves the progressive transformation of the simple uniform cells with general functions into complex and diverse types, each with specialized function e.g., leaf develops from simple primary to complex mature organ. The cell wall begins to thicken on account of deposition of cellulose without stretching. It then undergoes differentiation (chemical transformation, deposition of lignin and suberin, and other complex transformations). Once differentiated the cells remain unchanged so long as they exist.

Cutting number	14	S	20	S	23	S	26	S	29	S	32	S	35
1	14	S	20	S	23	S	26	S	29	S	32	S	35
2		1	√	2	√	3	√	D	√	D	√	D	√
3		1	√	2	√	3	√	D	√	D	√	D	√
4					√	1	√	2	√	3	√	D	√
5		1	√	2	√	3	√	D	√	D	√	D	√
6	√	2	√	3	√	D	√	D	√	D	√	D	√
7					√	1	√	2	√	3	√	D	√
8							√	1	√	2	√	3	√
9					√	1	√	2	√	3	√	D	√
10					√	1	√	2	√	3	√	D	√
11							√	1	√	2	√	3	√
12			√	1	√	2	√	3	√	D	√	D	√
13							√	1	√	2	√	3	√

Where:
- S – stage of sprout
- D – development and elongation

Table 6: Growth of Mulberry propagated by cuttings.

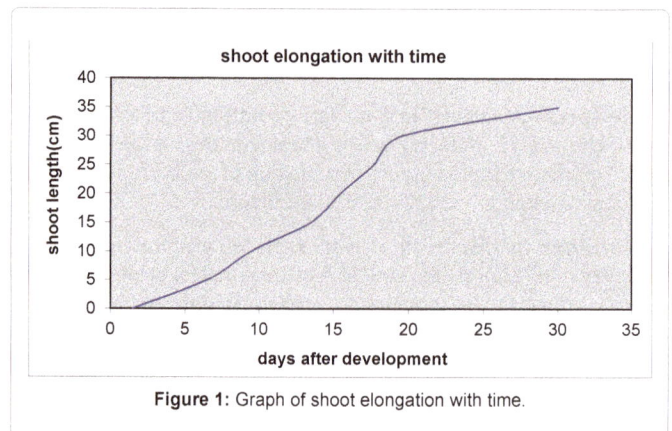

Figure 1: Graph of shoot elongation with time.

During the maturation phase, assimilation of carbohydrates is very active and the assimilation of proteins is almost insignificant.

Stages of growth of mulberry tree: For the project, mulberry was propagated by cuttings. Each cutting had at least 3 buds. On planting only one bud was left exposed. The buds had 7 – 8 scales along their circumference and contain leaves inside.

The first stage of growth was the sprouting of the buds. The first buds sprout occurred after 14 days and the rest followed. The earliest time for sprouting was found to be 14 days. A total of 13 out of 18 buds sprouted successfully. Sprouting of the buds was in three stages. The first stage was the period when the tip of the leaf appeared at the scale. The second stage was the period when a few leaves grew out of the scale and clustered together, and the third stage was the period when leaf blades appeared outside the scale and the petiole could be seen. All these occurred in a minimum of 21 days.

These sprouting stages maybe called development stage, the assimilation stage and the nutrition storing stage. The development is the initial stage and was when the branches and leaves are fostered using the stored nutrition. In the assimilation stage, the branches and leaves are fully fostered and absorbed nutrients are used to the maximum. In the last stage, the development is reduced with elongating growth checked, and nutrition assimilated in the in the leaves is transferred to the branches and roots and stored.

Conclusion and Recommendation

It is evident in this research that mulberry plant can grow very well in Kenyan because of the favourable climatic conditions and good environment. Eldoret area has the least requirements of mulberry agronomy, which are soil and rainfall. Proper fertilization or rooting chemicals were used to enable the plant to grow successfully.

It is worth noting however that mulberry tree can grow in a variety of climatic conditions. Its growth is not restricted to specific type of climate as long as there is plenty of rain. What will vary will be the yield.

It is therefore recommended that:

• The cultivation of mulberry plants is possible in Eldoret in preparation of rearing silkworms. Moi University which formed the bulk of experimental area could create suitable a research centre for sericulture

• Proper agronomical practices should be used to increase yield as well as the quality of leaves. Pruning and thinning should be practiced appropriately. These highly increase yield of leaves

• Further research is required on the available species of mulberry in Kenya

• As a result of successful production of mulberry, silk production is possible

• Training is needed for skilled labour in mulberry growing and silk worm rearing in these high production areas of the country with similar climate to the experimental area of Eldoret

References

1. Krishnaswami S, Kumararaj K, Vijayaraghavan, Kasiviswanathan K (1971) Silkworm feeding trials for evaluating the quality of mulberry leaves as influenced by variety, spacing and nitrogen fertilisation. Ind J of Sericulture: 79-89.

2. Veda K, Nagai I, Horikomi M (1997) Silkworm rearing (translated from Japanese). Science Publishers. Inc USA.

3. Rao MMM (1998) A Text Book of Sericulture. B.S. Publications. Hyderabad, India.

4. Baig M (1996) Silkworm Diseases and pests; Laboratory of silkworm pathology. Central Sericultiral Research and Training Institute Mysore.

5. Kariaga BMM (1992) Prediction of soil erosion by water and the effect of top soil depth on the productivity of Dystric Nitosols. Moi University, Eldoret.

6. Chopudhury, Giridhar K (1996) Mulberry cultivation, laboratory of mulberry agronomy. Central Sericultiral Research and Training Institute, Mysore.

7. Sheker P, Hardingham M (1995) Sericulture and silk production: A handbook. Intermediate technology publications, London.

30

The Influence of the Wet Processing Stages on Yarn Processability

El-Hadidy AM, El-Rys SM* and El-Hossiny A

Textile Engineering Department, Mansoura University, Egypt

Abstract

In this study, the influence of the wet processing stages, on cotton yarn processability was investigated. The yarn samples were collected after every treatment and tested for knitability efficiency. The total evaluation of the yarn properties was measured by using performance diagram. It was found that the highest processability index was occurred after mercerization conditions of 26°Be at temperature 19°C, vat dyeing and at temperature 19°C, reactive dyeing. While maximum processability index with reactive dyeing was occurred with mercerization condition of 28°Be` at temperature 24°C.

Keywords: Knitability; Wet processing stages; Processability index; Processability chart and performance diagram

Introduction

Single jersey knitted fabrics is generally used to make outerwear garments such as T-shirts. The knitted yarns undergo a series of different processing treatments like scouring, bleaching, dyeing, softener padding and relax drying. These processes are carried out to impart a particular property related to that process like scouring for absorbency, bleaching for whiteness, dyeing to impart color to fabric and finishing for improving softness and handle of the fabric.

Finally, there is the fact that, done right, cotton is really an excellent choice for summer weaving specially to summer knitting (ladies 100% cotton T-shirt).

A T-shirt is something to look forward to at the end of a long day. Slipping in to the soft weave of a jersey knit can take us back in time to childhood, our first rock concert or the summer we fell in love with hot pink.

The properties of knitted fabrics are influenced by various parameters like raw materials, yarn structure, fabric structure, processing stages and finishing. The amount of changes occurred in the properties of the fabric due to these parameters makes the subject complex. Further, the determination of the changes in physical and dyeing properties during different stages of wet processing is important for the control of process parameters to get the final product as the requirements of the customer.

There are limited number of studies on influence of wet processing stages and process sequences on the physical, mechanical properties of knitted yarns has been reported so far.

In this study, the influence of the wet processing stages, on the physical-mechanical-dyeing properties of cotton yarns were investigated.

On the other hand, textile finishing processes are usually used to improve the quality of materials. Among these, mercerization which improves handling and appearance of cotton fibers to simulate the superior properties of synthetic fibers.

The extent of changes that occurs depends on the processing time, caustic concentration, temperature, and degree of polymerization and source of cellulose, and the degree of tension [1].

Bleaching ensures the complete removal of impurities by destroying color matters present in the cotton fibers. The bleaching process can be applied by using hydrogen peroxide or calcium hypochlorite and sodium hypochlorite.

Dyeing is another important wet processing not only to impart colors to materials but also influences on yarn and fabric properties. Cotton and other cellulosic fibers are dyed with direct, sulphur, vat, reactive and more types than for any other fiber. Furthermore, vat dyes are one of the oldest types of dye. Vat dyes in particular give dyeing on cellulosic fibers with the best overall fastness properties [2].

In previous research works, the accessibility was investigated in terms of monolayer capacity, moisture regain, water retention values and diffusion coefficient of the congo red dye after mercerization and other chemical treatments [3].

Mehdi Akhbari [4] investigated the parameters influencing mercerization using RSM method, in order to increase the tensile strength of mercerized yarn. Nazem Samei [5] compared the effects of hot mercerization on open-end and ring spun yarns in slack and under tension conditions. Mercerized yarns were bleached and dyed with reactive dye.

Besides, the influence of different temperature of drying after mercerization and enzymatic scouring through changes in the surface properties and scouring efficiency were investigated [6]. The effect of chemical finishing treatments like scouring, bleaching, mercerization and dyeing on the properties of ring spun and compact yarns were investigated by Subramaniam [7].

Experimental Part

Before setting a machine with ladies, 100% cotton T-shirt, it is normal to check if the structure, yarn parameters and machine settings are correct. There are three separate but interconnected tasks described the processability of this structure with this yarn on this machine, in order to obtain the best productivity and quality.

*Corresponding author:** El-Rys SM, Textile Engineering Department, Mansoura University, Egypt, E-mail: samah_elrys@yahoo.com

How to check the processability of the input data is still an open field for researchers. There are three main trends: 1) Expert system, 2) physical process simulation, 3) an engineering approach. Each of these has its own power and its limitation.

The advantages of expert system are in their ability to collect and operate with a large no. of rules. knitability=ease to knit = knitting performance of the knitted structure for a given yarn [8].

In this part of the investigation the influence of wet processing stages on yarn quality will be investigated.

Experimental design

Tables 1 and 2 show the relationship between fabric weight per unit area, fabric structure and yarn count respectively.

These values may get up and down. It may be changed.

Plied combed - ring spun cotton yarn with a count of 20/2 Ne was produced. It was spun from Egyptian cotton G86 with (single twist factor αe= 3.6 and plied twist factor αe= 3.5). Besides, grey yarn without any treatment was produced for comparison.

Yarn Count (Ne) * Stitch length *GSM = "K"

Methods

Mercerization: Mercerization was done on jaeggli mercerizing yarn machine. Ten samples of yarn hanks were treated with five caustic soda concentration ranging from (26°Be` to 34°Be`) with 2°Be` intervals, and at two temperatures (19°C and 24°C), using wetting agent [(6-8) g/l of floranit - 4028 of pulcra chemical]. Both of applied tension and time of treatment were constants. To remove the excess caustic soda after the treatment, the yarn hanks were washed with hot and cold water. The hanks were then neutralized with acetic acid solution (0.5 cm/l) to remove any residual alkali. Yarns were finally rinsed with cold water, squeezed with centrifugal forces, dried at 110°C for 2 hrs.

Bleaching: The same ten samples of the mercerized yarns were bleached in an exhaustion procedure, the bleaching bath containing with (0.5%) hydrogen peroxide, (2%) of sodium hydroxide, (2%) of asbicone (detergent used as assistant factor) and (1%) of egypttool

(wetting agent). The bleaching was carried out at 100°C for 1 hr. Acetic acid was used for neutralization.

Dyeing: Half of the quantity of the mercerized and bleached samples of yarns was dyed with vat dyeing and the other half was dyed with reactive dyeing. The final no. of samples was 20 samples (10 samples of vat dyeing + 10 samples of reactive dyeing). Dyeing process was carried out by exhaustion procedure on proteks bobbin dyeing machine with liquor ratio 1:6.

Reactive dyeing: (1.5%) EcoFix.Blue.R, (5 g/l) Na_2CO_3 and (20 g/l) NaCl were adjusted at 50°C and then raised to 60°C and maintained at this temperature for 90 min to dye as reactive dyeing. Dyed samples were then washed in a soap solution, then, boiled and softened by using fatty acid, finally, squeezed and dried at 100°C.

Vat dyeing: Yarns were dyed in a solution containing (1.5%) Ind.Blue.CLF.(vat dye), (0.1%) NaOH and (5 g/l) Sodium hydro-sulphite. As known vat dyeing is based on the principle of converting water -insoluble vat dye by alkaline reduction to water-soluble leuco compound having affinity to cotton. Sodium hydro-sulphite is used to the reduction process at 60°C then raised to 80°C for 20 min, then washed. Finally, it is oxidized by using H_2O_2 at 50°C for 30 min, then washed and softened by using fatty acid.

Testing methods

Testing methods were carried on the treated yarns after each treatment and compared to grey yarns. Breaking – load was measured by using USTER @ Tensorapid 4. Color strength and degree of whiteness of yarns were observed by using Data color International SF 600 at D65. Yarn abrasion resistance was measured according to ASTM D6611. In order to measure the coefficient of friction of yarns, friction Coefficient meter F-Meter R-1183. The friction coefficient is determined on the F-meter by using the eytlwein formula as follow:

$$T2/T1 = e^{\mu\alpha}$$

Where:

T1: tension beyond the friction point.

T2: tension before the friction point.

α: friction angle in rad.

μ: the dynamic coefficient of friction.

Experimental Results and Discussion

At the end of each treatment process, the yarn samples were collected and tested for some of mechanical and color properties. To compare between the effect of mercerization and the effect of bleaching on the yarns, each of braking-load, coefficient of friction, abrasion resistance, yarn count and color properties were measured. These measured properties are considered as index for ability of yarns to work in the next processes.

Breaking-force

Figure 1 shows the effect of mercerization and bleaching on the breaking force of mercerized yarns at temperatures (19°C, 24°C) and at five different NaOH concentrations. It can be clearly seen that there was an increase in the breaking-load at temperature 24°C more than 19°C compared to unmercerized (grey) yarn. This increase may be due to the effect of mercerization which causes swelling and transforms the cross section of the fibers from bean-shaped cross section to a circular cross section.

Fabrics GSM	Plain (Single Jersey)	Rib (1*1)	Pique	Interlock
100	36.12	47.27	47.27	59.96
120	33.30	39.81	44.32	55.84
140	30.48	37.35	40.98	51.72
160	27.66	34.89	37.64	47.60
180	24.84	32.43	34.30	43.48
200	22.02	29.97	30.96	39.36
220	19.20	27.51	27.62	35.24
240	16.38	25.05	24.05	31.12

Table 1: Selection of yarn count for various GSM for different fabrics.

Fabrics	Constant "K"
Plain Jersey	12068.509
Rib 1*1	16431.497
Rib 1*2	19005.333
Interlock	24013.800

Table 2: Values of constant "K".

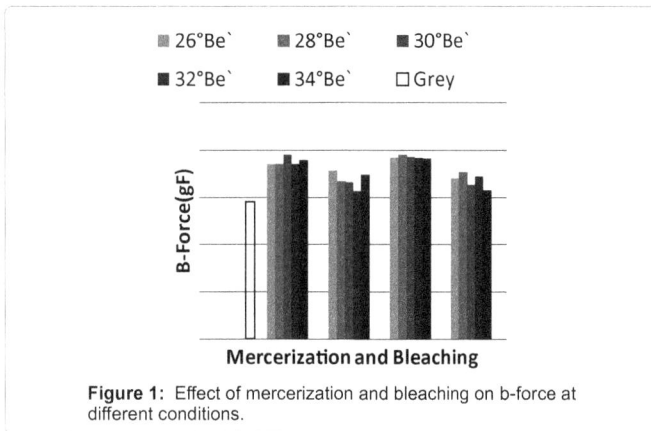

Figure 1: Effect of mercerization and bleaching on b-force at different conditions.

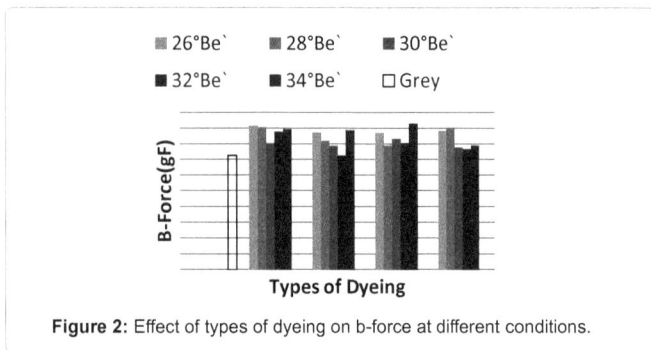

Figure 2: Effect of types of dyeing on b-force at different conditions.

Since the surface area of contact of swollen fibers becomes higher, consequently each of inter fiber friction and cohesion are improved which in turn improves the breaking-Load. In addition, raising the temperature of mercerization treatment lowers the viscosity of caustic soda solution; hence it facilitates the penetration of NaOH in to the fibers and the swelling of fibers is increased as well. As known, increasing the swelling leads to increasing the degree of mercerization.

After bleaching, there was a slightly decrease in the breaking -load compared to mercerized yarn but still more than the breaking -load of grey yarn for both cases 19°C and 24°C. Breaking-load increased from (1564 to 1786)gF with average value (1684)gF at temperature 19°C, and from (1573 to1774)gF with average value (1679)gF at temperature 24°C. This decline was occurred because the bleaching using hydrogen peroxide (H_2O_2) may leads to formation of Oxi- Cellulose which known that it decreases the tensile strength of materials.

Figure 2 shows the effect of two types of dyeing on the breaking -load of this yarn. The results showed that the increasing in the breaking-load after vat dyeing was higher than reactive dyeing in both cases 19°C and 24°C compared to grey yarn. This increase in vat dyeing is arranged during washing with alkali soap after dyeing process. In addition, the increase of these spaces between fibers with the presence of vat dyeing may be the reason for the rise of the breaking-load. After vat dyeing, the breaking -load raised from (1596 to1828)gF with average value (1753) gF at temperature 19°C, and raised from (1575 to1854)gF with average value (1686)gF at temperature 24°C. On the other hand, after reactive dyeing, the breaking-load raised from (1446 to1767)gF with average value(1631)gF at temperature 19°C, and raised from (1529 to1800)gF with average value (1642)gF at temperature 24°C.

Yarn count

Figure 3 shows the effect of mercerization and bleaching on the yarn count of mercerized yarns at temperatures (19°C, 24°C) and at five different NaOH concentrations. As shown the yarn count increased after both mercerization and bleaching compared to grey yarn, where, after mercerization it ranged between 60 tex and 64 tex in both cases 19°C and 24°C. This means that the temperature of mercerization didn't affect the yarn count here. However, this increase in yarn count was occurred generally due to the effect of mercerizing process which increases the diameter because of swelling with NaOH molecules.

Figure 4 shows the effect of two types of dyeing on the yarn count of this yarn. After vat dyeing, the yarn count raised from (57 to 59.5) tex in both cases 19°C and 24°C, whereas after reactive dyeing, it raised from (56.5 to 60) tex at temperature19°C and from (57 to 62.5) tex at temperature 24°C.

Abrasion resistance

Figure 5 shows the effect of mercerization and bleaching on the

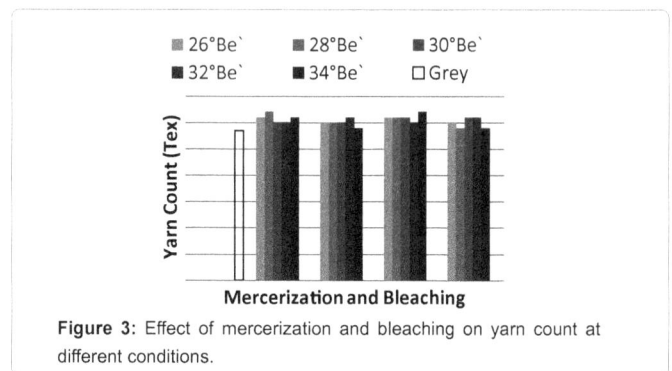

Figure 3: Effect of mercerization and bleaching on yarn count at different conditions.

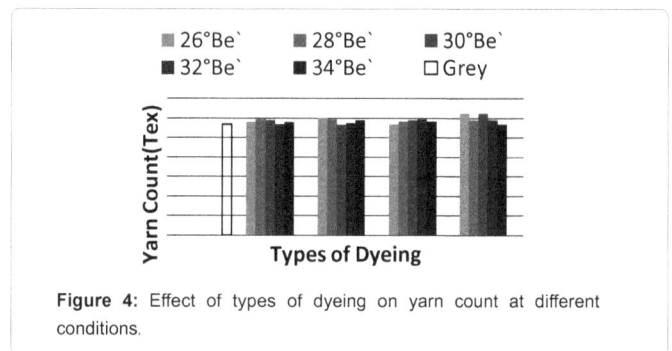

Figure 4: Effect of types of dyeing on yarn count at different conditions.

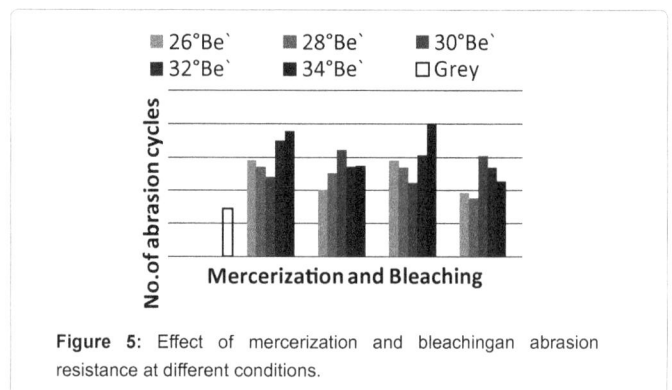

Figure 5: Effect of mercerization and bleachingan abrasion resistance at different conditions.

abrasion resistance of mercerized yarns at temperatures (19°C, 24°C) and at five different NaOH concentrations. It can be clearly seen that there was a greatly increase in the abrasion resistance of the yarns after both mercerization and bleaching compared to grey yarn. This increase in the abrasion resistance after NaOH treatment can be attributed to the degree of substitution of O-Na groups in the cotton fiber. This finding was in agreement with that of Subramaniam [7]. The abrasion resistance increased after mercerization from (48 to 76) no. of abrasion cycles with average value (61) at temperature19°C, and from (45 to80) no. of abrasion cycles with average value (60) at temperature 24°C.

After bleaching, abrasion resistance decreased compared to mercerized yarns. This is in fact normally because there is a relation between the strength and abrasion of yarn. As shown, abrasion resistance of bleached yarn increased from (41 to 64) cycles of abrasion with average value (53) cycles of abrasion at temperature 19°C, and from (35 to 61) cycles of abrasion with average value (47) at temperature 24°C.

Abrasion of the yarn is followed by the gradual removal of fibers from the yarns when they are subjected to repeated distortion. Factors affecting the cohesion of the fibers in the yarn and yarn to yarn friction will have influence on the abrasion resistance.

Figure 6 shows the effect of two types of dyeing on the abrasion resistance of this yarn. After vat dyeing the abrasion resistance raised from (89 to 155) cycles of abrasion with average value (106) at temperature 19°C, and from (31 to 97) cycles of abrasion with average value (56) at temperature 24°C.

On the other hand, after reactive dyeing, the abrasion resistance

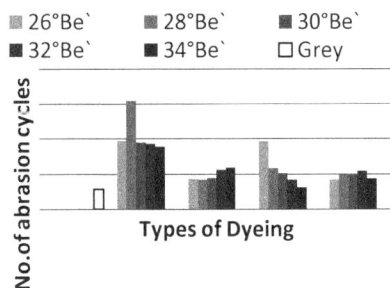

Figure 6: Effect of types of dyeing on abrasion resistance at different conditions.

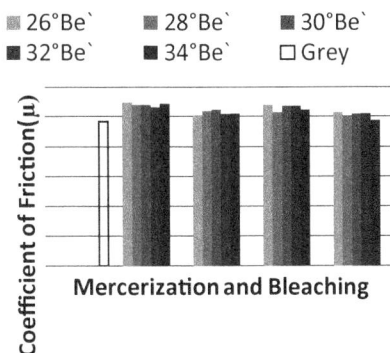

Figure 7: Effect of mercerization and bleaching on coefficient of friction at different conditions.

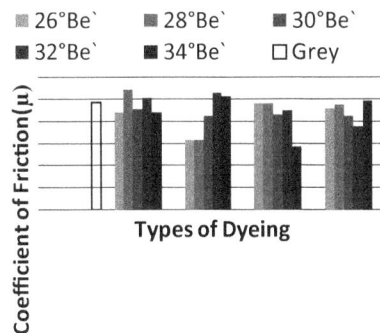

Figure 8: Effect of types of dyeing on coefficient of friction at different conditions.

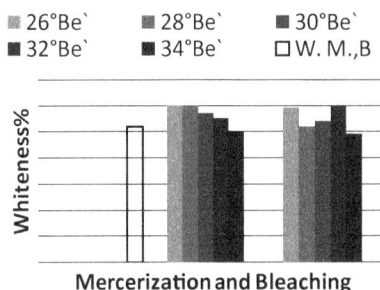

Figure 9: Effect of mercerization and bleaching on the degree of whiteness at different conditions.

increased from (42 to 59) cycles of abrasion with average value (49) at temperature 19°C, and from (42 to54) cycles of abrasion with average value (48) at temperature 24°C.

Coefficient of friction

Figure 7 shows the effect of mercerization and bleaching on the coefficient of friction of mercerized yarns at temperatures (19°C, 24°C) and at five different NaOH concentrations. The results showed that coefficient of friction increased after mercerization and bleaching compared to grey yarn. After mercerization, it ranged between 0.266 and 0.274 at temperature 19°C, and between 0.258 and 0.27 at temperature 24°C. Similarly to abrasion resistance, the coefficient of friction of yarns after bleaching decreased compared to mercerization, where, the coefficient of friction of yarns after bleached ranged between 0.252 and 0.262 at temperature 19°C, and between 0.244 and 0.258 at temperature 24°C. These findings were in agreement with those of Subramaniam [7].

Figure 8 illustrates the effect of two types of dyeing on the coefficient of friction of this yarn. After vat dyeing, it ranged between 0.22 and 0.27 at temperature 19°C, and between 0.142 and 0.24 at temperature 24°C. Besides, after reactive dyeing, coefficient of friction ranged between 0.158 and 0.264 at temperature 19°C, and between 0.188 and 0.246 at temperature 24°C.

Degree of whiteness

Figure 9 shows the effect of mercerization and bleaching on the degree of whiteness of mercerized yarns at temperatures (19°C, 24°C) and at five different NaOH concentrations. The degree of whiteness was measured compared to the standard degree of whiteness of

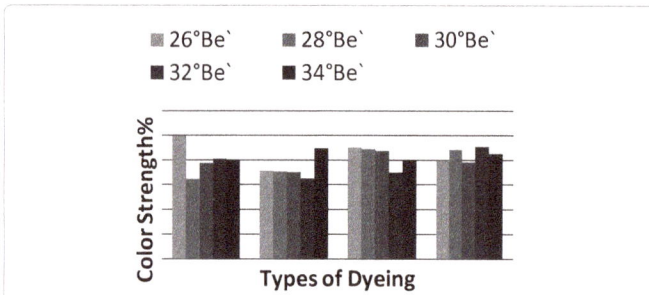

Figure 10: Effect of types of dyeing after mercerization on color strength at different conditions.

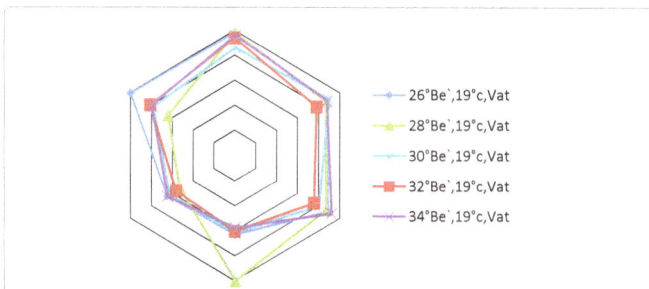

Figure 11: Overall evaluation of yarn properties at19°C, vat dyeing.

spectrophotometer tester which equals to (189). As shown the degree of whiteness improved after mercerization and bleaching more than after bleaching without mercerization. This improvement occurred due to the effect of mercerizing process which removes some color materials from cellulosic fibers.

After mercerization and bleaching, the degree of whiteness increased from (50% to 60%) at temperature19°C, and from (49% to 60%) at temperature 24°C.

Color strength

Figure 10 shows the effect of types of dyeing after mercerization on the color strength of mercerized yarns at temperatures (19°C, 24°C) and at five different NaOH concentrations. Color strength was measured compared to two dyed samples with vat and reactive dyes but without mercerization. In general, there was an increase in color strength in reactive dyeing more than vat dyeing. It can be clearly seen that after vat dyeing, color strength increased from (160% to 248%) with average value (201%) at temperature 19°C, and from (161% to 223%) with average value (183%) at temperature 24°C.

This increase may be because mercerizing process converts fibers to cylindrical shape leading to increase the spaces between the fibers, so the dye molecules can easily penetrate to the fibers and increase the color strength of this vat dye. Also, results showed an increase in color strength of reactive dyeing from (174% to 226%) with average value (208%) at temperature 19°C, and from (195% to 227%) with average value (210%) at temperature 24°C because molecules of this reactive dye are bonded with covalent bonds with the fibers.

Overall Evaluation of Yarn Properties

In order to evaluate the overall effect of chemical treatments on the final properties of the yarns, the radar chart statistical method was used. For each level of each of the three parameters of the treatment,

the processability index i.e. The ratio of the yarn polygon area before and after treatment was determined.

Figure 11 represents the processability chart for each level of NaOH concentrations at temperature 19°C and vat dyeing. It was found that the highest area of radar chart was occurred at 26°Be` as shown in Table 3 while the lowest area was occurred at 32°Be`.

Figure 12 represents the process ability chart for each level of NaOH concentrations at temperature 19°C and reactive dyeing. It was found from the determined index that shown in Table 3 the highest area of radar chart was occurred at 26°Be` while the lowest area was occurred at 32°Be`.

Figure 13 represents the processability chart for each level of NaOH concentrations at temperature 24°C and vat dyeing. It was found from the determined index that shown in Table 3 the highest area of radar chart was occurred at 34°Be` while the lowest area was occurred at 32°Be`.

Figure 14 represents the process ability chart for each level of NaOH concentrations at temperature 24°C and reactive dyeing. It was found from the determined index that shown in Table 3 the highest area of radar chart was occurred at 28°Be` while the lowest area was occurred at 30°Be. It was noticed that the vat dyeing maintain the physical properties generally more than reactive dyeing. As, the average of improvement in properties of yarns in case of vat dyeing was more

Temperature (°C)	Types of Dyeing	NaOH Conc.(Be`)				
		26	28	30	32	34
19	Vat	70.90813`	66.06878	59.0121	55.78814	64.17274
	Reactive	60.63058	55.17666	54.49425	37.75236	57.82358
24	Vat	60.00243	46.98462	53.61165	46.34582	68.10022
	Reactive	52.55869	64.52037	46.16233	52.45861	48.04597

*Actual area/ideal area.
Table 3: Total area of overall yarn properties.

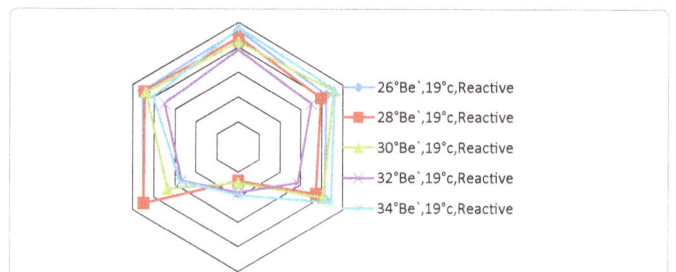

Figure 12: Overall evaluation of yarn properties at 19°C, reactive dyeing.

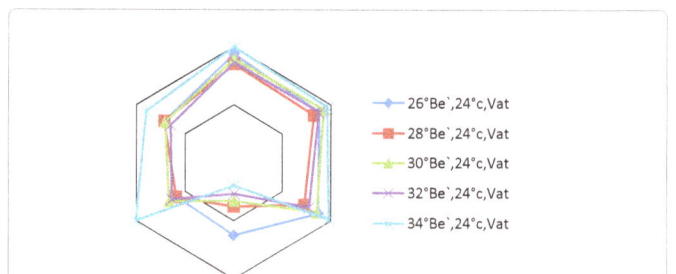

Figure 13: Overall evaluation of yarn properties at 24°C, vat dyeing.

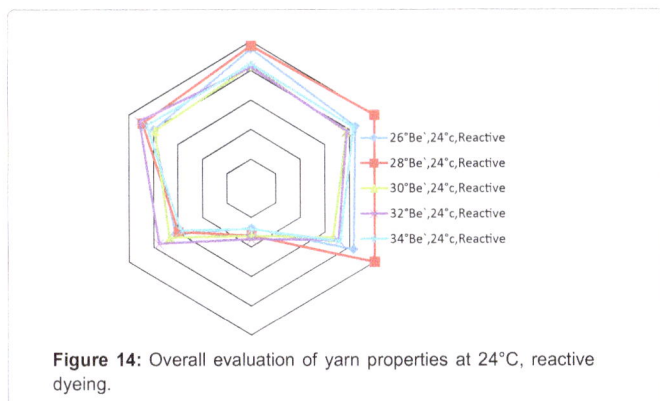

Figure 14: Overall evaluation of yarn properties at 24°C, reactive dyeing.

than that of reactive dyeing in both cases 19°C and 24°C (Table 3).

Conclusion

Mercerizing process of yarns at different NaOH concentrations, and at different temperatures, followed by bleaching and dyeing with vat and reactive dyeing have an influence on the final properties of yarns.

After mercerization, there was an increase in the breaking -load in case of 24°C more than 19°C compared to unmercerized (grey) yarn.

• The results showed that the increasing in the breaking -load after vat dyeing was higher than reactive dyeing in both cases 19°C and 24°C compared to grey yarn.

• There was an increase in the breaking force after mercerization more than after bleaching compared to grey yarn.

• The yarn count increased after both mercerization and bleaching compared to grey yarn.

• There was a greatly increase in the abrasion resistance of the yarns after both mercerization and bleaching compared to grey yarn.

• Also after dyeing the abrasion resistance increased compared to untreated yarn.

• The results showed that coefficient of friction increased after mercerization and bleaching compared to grey yarn.

• After dyeing, coefficient of friction had values close to/or less than grey yarn.

• In general, there is an increase of the color properties for both vat and reactive dyeing compared to unmercerized samples because of the mercerization effect.

• The degree of whiteness improved after mercerization and bleaching more than after bleaching without mercerization.

• From the overall evaluation of physical properties by performance diagram, It was found that the highest area of radar chart was occurred at 26°Be at temperature 19°C, vat dyeing and at temperature 19°C, reactive dyeing.

• The highest area of radar chart was occurred at 34°Be` at temperature 24°C and vat dyeing.

• The highest area of radar chart was occurred at 28°Be` at temperature 24°C and reactive dyeing.

• Vat dyeing maintain the physical properties of the fabric generally more than reactive dyeing.

References

1. Wakida T, Lee M, Park SJ, Hayashi A (2002b) Hot mercerization of cottons. Fiber 58: 304-307.

2. Arthur D, Broadbent (2001) Basic Principles of Textile Coloration. Society of Dyers and Colourists.

3. Jordanov I, Mangovska B (2011) Accessibility of mercerized, bioscoured and dried cotton yarns. Indian Journal of Fiber Textile 36: 259-265

4. Mehdi A, Abdulreza Z, Eddin BSJ (2012) Optimization of Parameters Influencing Mercerization Using RSM Method in Order to Increase the Tensile Strength of Mercerized Yarn. Fibers and Textiles in Eastern Europe 20: 30-35.

5. Samei N, Mortazavi SM, Rashidi A, Najjar SS (2008) Changes in Physical Properties of Hot Mercerized Ring and Open-end Spun Cotton Yarns. Iranian Polymer Journal 17: 937-945.

6. Jordanov I, Mangovska B (2009) Characterization on Surface of Mercerized and Enzymatic Scoured Cotton after Different Temperature of Drying. The Open Textile Journal 2: 39-47.

7. Farid F, Roshanara, Subramaniam V (2014) Effect of Chemical Treatment on the Characteristics of Regular and Compact Cotton Spun Yarns. International Journal of Research in Engineering and Technology.

8. Ibrahim S, Militký J (2015) Knitting Ability Textile Faculty. Technical University of Liberce.

31

Thermophysiological Wear Comfort of Clothing

Dinesh Bhatia[1]* and Urvashi Malhotra[2]

[1]Department of Textile Technology, Dr B R Ambedkar National Institute of Technology, Jalandhar-144011, India
[2]Department of Textile Technology, Jawaharlal Nehru Government Engineering College, Sundernagar-175018, India

Abstract

Thermophysiological wear comfort concerns with the heat and moisture transport properties of clothing and the way it helps the clothing to maintain the heat balance of the body during various level of activity. Heat and moisture flow through clothing is a complex phenomenon. So, heat and moisture transfer analysis for clothing is an important issue for researchers. This article delves into the processes which are involved in heat and moisture transmission along with mathematical models of heat, liquid and vapour transport through clothing to understand the exact phenomena of heat and moisture transmission. The reported testing methods and parameters used for determining heat and moisture are also summarized in this article. This article also describes the need of heat and moisture transmission in clothing, desired attributes for heat and moisture management and parameters affecting heat and moisture transmission in clothing.

Keywords: Absorption; Comfort; Conduction; Convection; Diffusion; Condensation; Sorption

Introduction

Comfort may be defined as a pleasant state of psychological, physiological and physical harmony between a human being and the environment. Today humans rely on clothing which protects body from cold and heat throughout full range of human activities, otherwise it leads to discomfort. Discomfort mainly results from the build-up of sweat on the skin and insufficient heat loss during overheating in hot environments and exercise [1-4].

To create a comfortable clothing a designer considers fashion and other technical factors; fiber nature and size (microfibers that have particulars properties), surface modification of fibers (hydrophobic or hydrophilic treatments), hydrophobic (Gortex' e.g.) or hydrophilic membranes fused to the textile layers, weaving or knitting patterns and abrasion of the fabric surface etc. are parameters by which we can enhance comfort of clothing [5].

Extensive research has been published in the literature on the diverse aspects of simultaneous heat and moisture transfer both theoretically and experimentally. Results shows, that the ability of clothing materials to transport moisture vapour is a critical determinant of wear comfort, especially in conditions that involve sweating. So, for satisfactory performance of clothing comfort researchers recognize that clothing comfort has two main aspects. These are thermo physiological and sensorial comfort. The first relates to the way clothing buffers and dissipates metabolic heat and moisture [6-10], whereas the latter relates to the interaction of the clothing with the senses of the wearer, particularly with the tactile response of the skin, which includes moisture sensation on the skin [11-15].

The wear comfort of clothing is affected by physical processes include heat transfer by conduction, convection and radiation, meanwhile, moisture transfer by diffusion, sorption, wicking and evaporation [16-18]. During higher activity level and/or at higher atmospheric temperatures sweat gland get activated which produce liquid as well as perspiration. When the perspiration is transferred to the atmosphere it carries heat (latent as well as sensible) thus reducing the body temperature. The fabric being worn should allow the perspiration to pass through; otherwise it will result in discomfort. If moisture transfer rate is not adequate during sweating than it may result in heat stress

due to increase in rectal and skin temperature. From last few decades the field of dynamic heat and moisture transport behaviour of clothing and their influences on clothing comfort is main interest of researchers.

Clothing by its nature has an insulating effect and resists transfer of excess heat and moisture from the body. A still layer of air confined between the skin and fabric or between two fabric layers can make the wearer extremely uncomfortable due to its barrier effect. Thus the most important purpose of clothing is to provide a stable microclimate next to skin by maximizing the rate of heat and moisture loss from the body [19].

If ratio of evaporated sweat and produced sweat is very low, moisture will be accumulated in the inner layer of the fabric system, ultimately affect the thermal insulation of clothing [20]. It means there is some correlation between heat and moisture transmission through fabrics, which play a major role in maintaining a wearer's body in comfort zone. Hence a clear understanding of heat and moisture transmission from clothing is required for designing new high performance fabrics for different application.

In this paper attempt has been made to understand the mechanism behind heat and moisture transmission along with postulated models. Details of evaluated properties and equipment used to measure heat and moisture transmission is also explained.

Processes Involved in Heat and Moisture Transmission through Clothing

Process involved in heat and moisture transport is an important factor which influences dynamic comfort of clothing. Heat can be transferred within clothing in the form of conduction, convection,

*Corresponding author: Dinesh Bhatia, Department of Textile Technology, Dr B R Ambedkar National Institute of Technology, Jalandhar-144011, India
E-mail: dineshbhatia55@rediffmail.com

radiation and latent heat transfer by moisture transport. Conduction, convection and radiation are dominated by the temperature difference between skin surface and environment and are therefore grouped as dry heat transfer. On the other hand, latent heat transfer is achieved by moisture transmission related to water vapour pressure between the skin surface and the environment.

For an unclothed body seated at rest in mild ambient conditions, the metabolic rate is about 60 W/m². If thermal comfort is assumed there is no thermoregulatory sweating and the only source of moisture loss is diffusion through the skin itself. The heat flux associated with this diffusion may be expressed in terms of the condition between skin and the ambient air as follow [21].

$$H = 4.0 + 0.12\,(P_{ssk} - P_a) \tag{2.1}$$

Where H represents evaporative heat flux in W/m², P_{ssk} is saturation vapour pressure at temperature of skin (mill bars) and P_a is ambient vapour pressure (mill bars).

For a clothed body, comfort refers to the way clothing interacts with the body, with respect to dissipation of heat and moisture generated by metabolic processes [22]. There is production of thermal energy as a by-product of physical activities. For body temperature to be stable heat losses need to balance heat production. This balance is given by following equations.

Store=(Heat production–Heat loss)=(Metabolic rate–External work)–(conduction+convection+radiation+evaporation+respiration)

If heat store is negative, more heat is lost than produced and body starts cooling. If however, heat production by metabolic rate is higher than the sum of all heat losses, heat store will be positive which means that the body heat will increase and body temperature will rise [23,24].

Fundamentals of Heat Transfer through Clothing System

Heat can transfer from textile layers by conduction, convection, radiation and wind penetration mechanisms as shown in Figure 1.

Conduction: Conduction means flow of heat through interaction or collision of adjacent molecules. Dry heat is transferred by conduction through air layers that are found on the surface of the textile layers, as well as through the air within the textile layers and through the textile fibres. Conductivity of textile fibres is much higher than air, indicating the importance of trapped air within garments to the conductive heat loss.

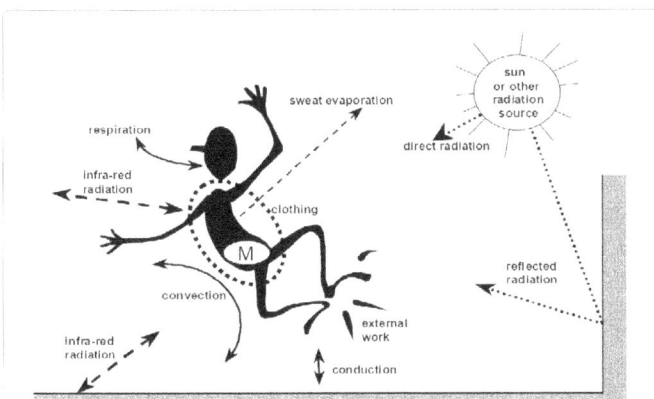

Figure 1: Schematic representation of the pathways for heat loss from the body, M: Metabolic Heat Production.

According to Fourier law, energy conducted can be expressed as [25]:

$$Q_{cond} = -KA\frac{dT}{dx} \tag{2.2}$$

Where k is proportional coefficient called thermal conductivity, A is the cross sectional area from where thermal energy passes through, T is temperature and x is thickness of material.

Convection: Convection means transfer of heat from one place to another within fluid, gas or liquid. The air is not held by textile fibres, and is able to move due to natural convection (rising of warmed air) or is forced to move by forced convection (wind, body movements creating a bellows effect); heat will be transported with the air moved defined by its enthalpy.

Appropriate heat transfer rate equation is given by [25]

$$Q_{conv} = -hA\,(T_s - T_\infty) \tag{2.3}$$

Where Q_{conv} is convective heat flow, T_s and T_∞ represent surface and fluid temperature respectively and h is convective heat transfer coefficient.

Radiation: In radiation heat flow is governed by temperature difference between the heat emitter and the heat absorber. Heat can be transported between the environment and the clothing surface by electro-magnetic radiation. This also occurs between clothing layers, and finally radiant heat transport can take place between the fibres within a textile, through the entrapped air. The more fibres, the less radiant transfer, though an optimum for overall conductivity of a textile is based on the balance between radiation and convection (denser, less radiation but more conduction through the fibre content).

Fanger derived equation for heat loss by radiation from the outer surface of the clothed body [26].

$$Q_{t,rad} = 3.97 \times 10^{-8}[(T_{cl} + 273)^4 - (T_{mrt} + 273)^4]A_{Du}\,f_{cl} \tag{2.4}$$

Where T_{mrt} is mean radiant temperature (°C), A_{Du} is the DuBois surface area and f_{cl} is the ratio of effective radiating surface to the DuBois surface area.

Wind penetration: Wind penetration is induced by air pressure between inner surface and outer surface of clothing system. Air penetration can induce the air exchange. Kerslake pointed out that the rate of air penetration through the clothing assembly V_{ap} was linearly related to the wind velocity [27].

$$V_{ap} = 3600\,A_s f_{op}v \tag{2.5}$$

Where A_s is surface area of the clothing assembly, f_{op} is the apparent portion of the surface area A_s that is open to air flow and v is wind velocity.

Fundamentals of moisture transfer through clothing system

Moisture from clothing may be transferred in vapour and liquid form. In vapour form different mechanism like diffusion, sorption, absorption, convection and condensation are involved whereas in case of liquid form wetting and wicking are two mechanisms which are generally take place as shown in Figure 2.

Diffusion: The transfer of water vapour molecules as a result of kinetic energy due to their random movement, which occurs through the air spaces between the fibers and yarns and along the fibers itself to keep microclimate dry enough to allow more sweat evaporation.

According to Fickian law, the moisture transmitted through the

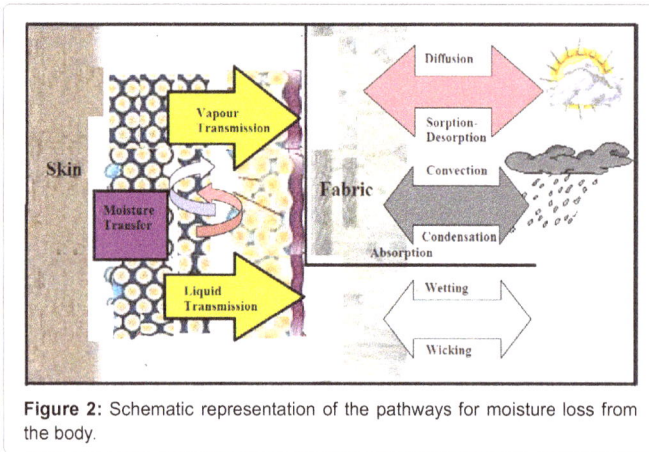

Figure 2: Schematic representation of the pathways for moisture loss from the body.

void space of the fabric through diffusion can be expressed as following [28].

$$m_{diff} = -D_{eff} A \frac{dC}{dx} \tag{2.6}$$

Where D_{eff} is effective diffusion coefficient of water vapour in fabric, A is the cross sectional area that the water vapour passes through, dC/dx is the gradient of water vapour concentration in void space.

D_{eff} is dependent on the volume fraction of water vapour (f_a) and defined as:

$$D_{eff} = \frac{D_a f_a}{\tau} \tag{2.7}$$

Where τ is effective tortuosity of the fabric for water vapour diffusion, D_a is the diffusion coefficient of water vapour in air.

Sorption-desorption: Any water bound to the textile fibres may be released as vapour again and take with it the heat of swelling plus the heat of evaporation, i.e. the reversal of absorption. This will reduce the local temperature. Sorption-desorption are an important process to maintain the microclimate during transient conditions. A hygroscopic fabric absorbs water vapour from the humid air close to the sweating skin and releases it in dry air. This enhances the flow of water vapour from the skin to the environment comparatively to a fabric which does not absorb and reduces the moisture built up in the microclimate [29,30].

Absorption (adsorption): Water vapour travelling through textiles may be absorbed by the textile fibre. All materials, when allowed to absorb vapour until an equilibrium is reached, have characteristic absorption levels (expressed as regain), which increase with relative humidity and are typically higher for natural versus man-made fibres [31]. With this absorption heat is released in the textile, composed of the heat of condensation and the heat of swelling, raising the local temperature.

Convection: Similar to dry heat loss by convection, moving air will take with it the moisture contained in the microclimate, which can then be replaced by fresh air if the convective stream actually leaves the garment (ventilation). If not, it will have an equilibrating effect on local microclimate conditions.

The rate of moisture convection (m_{conv}) can be expressed as the form of Newton's law of cooling [32]

$$m_{conv} = h_m A(C_s - C_\infty) \tag{2.8}$$

Where h_m is the moisture convective transfer coefficient, C_s and C_∞ are moisture concentration at surface and liquid.

Condensation: Condensation is a direct result of a fabric being saturated by liquid perspiration. It occurs within the fabric whenever the local vapour pressure rises to saturation vapour pressure at the local temperature [33]. Condensation normally occurs when the atmospheric temperature is very low. When the warm and moist air from the body meets the fabric, it works as a cold wall, and condensation occurs. Condensation in dry porous material takes place in three stages [34]. First of all, velocity, temperature and vapour concentration fields are developed within the material and condensation begins. In the second stage, the liquid content increases gradually, but it is still too low to move and finally, as the liquid content increases further and goes beyond a critical value, the pendulum like drops of condensate coalesce and begin to move under surface tension and gravity. When the vapour concentration at the two faces of the fabric, are at the saturation level, condensation occurs throughout the entire thickness of the fabric. If the vapour concentration at the two faces is below saturation for the local temperature, condensation occurs only over a region within the fabric.

Wetting: Wetting is the initial process involved in fluid spreading. In this process the fibre-air interface is replaced with a fibre-liquid interface. The forces in equilibrium at a solid-liquid boundary are commonly described by the Young-Dupre equation, given below [35]:

$$Y_{SV} - Y_{SL} = Y_{LV} Cos\theta \tag{2.9}$$

Where, γ represents the tension at the interface between the various combinations of solid (S), liquid (L) and vapour (V), and θ is the contact angle between the liquid drop and the surface of solid to be wetted.

Wicking: In sweating conditions, wicking is the most effective process to maintain a feel of comfort. In the case of clothing with high wicking properties, moisture coming from the skin is spread throughout the fabric offering a dry feeling and the spreading of the liquid enables moisture to evaporate easily. When the liquid wets the fibres, it reaches the spaces between the fibres and produces a capillary pressure. The liquid is forced by this pressure and is dragged along the capillary due to the curvature of the meniscus in the narrow confines of the pores. The magnitude of the capillary pressure is given by the Laplace equation [31]:

$$P = \frac{2 Y_{LV} Cos\theta}{R_C} \tag{2.10}$$

Where P is the capillary pressure developed in a capillary tube of radius Rc. A difference in the capillary pressure in the pores causes the fluid to spread in the media. Hence, a liquid that does not wet the fibres cannot wick into the fabric [36]. The ability to sustain the capillary flow is known as wickability [37]. The distance travelled by a liquid flowing under capillary pressure, in horizontal capillaries, is approximately given by the Washburn-Lukas equation [38]:

$$L = \sqrt{\frac{R_C Y Cos\theta}{2\eta}} t^{1/2} \tag{2.11}$$

Where, L is the capillary rise of the liquid in time t and η is the viscosity of the liquid. The amount of water that wicks through the channel is directly proportional to the pressure gradient.

Evaluated Properties and Equipment used to Measure Heat Transfer through Clothing

Thermal conductivity, thermal resistivity and thermal absorptivity are some of properties which are measured for heat transfer through clothing.

Thermal conductivity

Thermal conductivity is fundamental to determine the heat transfer through fabrics. For textile materials, still air in the fabric structure is the most important factor for conductivity value, as still air has the lowest thermal conductivity value when compared to all fibers (λair=0.025). Therefore, air transports a low quantity of energy by conduction and thermal conductivity decreases as well [39].

Thermal resistance

Thermal resistance expresses the thermal insulation of fabrics and is inversely proportional to thermal conductivity. In a dry fabric or containing very small amounts of water it depends essentially on fabric thickness and, to a lesser extent, on fabric construction and fiber conductivity [40].

Thermal absorptivity

Thermal absorptivity is the objective measurement of the warm-cool feeling of fabrics and is a surface related characteristic. If the thermal absorptivity is high, it gives a cooler feeling at first contact with the skin. The surface character of the fabric greatly influences this sensation [41].

Also various type of instrument has been used for measuring thermal properties of fabrics. The methods used for this purpose are discussed below.

Cooling method

In this method, a hot body is surrounded by fabric whose outer surface is exposed to air and the rate for the cooling of the body is determined. This method was used by Black and Mathew with a "kata-thermometer" [42].

Disc method

The fabric is held between a heat source and a heat sink at different temperatures and the flow of heat is measured by a thin disc. This gives the value of thermal transmissivity under particular conditions in the experiment. Since the fabric is compressed, it contains less air than under normal conditions during wear. Hence, the results that are obtained would only pertain to the particular apparatus and the pressure applied.

Measurement of propagation of waves (heat pulses)

This is a relatively new technique. It is an extension of the rate of the cooling method. In this technique, multiple waves of temperature gradients are passed through the sample and the damping of the wave is used to calculate the heat flux through the sample [43].

Constant temperature method

This method is used to get the most accurate determination of the thermal resistance. The fabric is placed on one side of an isothermal hot body that is insulated on all sides and energy required to maintain the hot body at a constant temperature is measured. The guarded hot plate is the most common form in this method which gives the most accurate determination of the thermal insulating properties of fabric. Three basic types of instruments have been tried by various workers to measure the thermal resistance of fabrics by the constant temperature method. They include

Hot cylinder type: There are many variations of the instruments that use this principle. Some useful descriptions are given by Morris [44]. Materials were wrapped around a constant temperature cylinder that was contained within another coaxial cylinder immersed in water. These instruments have the disadvantage of introducing a seam into the material, which is not recommended.

Hot semi-cylinder type (guarded): In view of the previously mentioned disadvantage, Baxter and Cassie [45] developed this instrument and provided a theory for measurements of thermal behaviour, which was based on Newton's law of cooling.

(Guarded) Hot plate type: Many workers have used this type of instrument based on the constant temperature method. The Shirley Togmeter is also based on the hot plate principle [46]. The guarded hot plate measurement technique has been employed in two modern internationally accepted standards viz. ASTM standard D1518 (2000) and BS 4745 (1974).

Evaluated Properties and Equipment Used to Measure Moisture Transfer through Clothing

As earlier, explained moisture can be transferred through clothing in two ways i.e either in vapour form or in liquid form.

Methods of measuring vapour transmission

Water vapour permeability is one major property which is measure for vapour transmission behaviour of clothing. For determination of water vapour permeability three methods are used viz. Permetest method, Cup method and MVTR cell method.

In different methods, different terms are used to express the water vapour permeability of a material [47]. Results obtained from the different available methods are not always comparable due to the different testing condition and the units used in the measurements. The most common units used for the measurement of the water vapour permeability of fabrics are [48-50] listed below:

- *The percentage water vapour permeability index:* WVP (% of turl reference fabric) is used in the evaporative disc method (BS 7209); this method uses water at 20°C and an atmospheric condition of 20°C and 65% relative humidity; this standard is based on the control dish method (CAN2-4.2-M77).

- *The moisture vapour transmission rate:* (in $g\ m^{-2}\ day^{-1}$) is used in the cup method (ASTM E 96-66); it uses air at relative humidity of 50% and a recommended water temperature of 32.2°C or a desiccant.

- *The resistance to evaporative heat transfer:* R_{et} (in m^2Pa/W) is used in the sweating guarded hot plate (ISO 11092:1993, EN 31092); it is an indirect method of measuring the vapour transmission property of a fabric. In this test method, the experiment is carried out at isothermal condition at standard atmospheric condition.

- *The Resistance:* in cm, of equivalent standard still air (in cm ESSA) is used in the holographic visualization method; in this method it is possible to measure the resistance offered by the fabric layer and the air layer separately. The resistance of the fabric (cm) can be expressed in terms of the standard still air providing the same vapour resistance.

Permetest method: This instrument works on principle of heat flux sensing [51]. The temperature of measuring head was maintained at room temperature for isothermal conditions. When water evaporates from measuring head, the heat lost from it is indirectly sensed by heat sensor. This instrument measures the heat loss from measuring head due to the evaporation of water in bare condition and with being covered by the fabric. Samples can be measured according to ISO 9920 and BS 7209 testing standard. The results of measurement are expressed by the instrument in terms of relative water vapour permeability (%) and water vapour resistance R_{et} (in m^2Pa/W). The relative water vapour permeability (p_{wv}) of the fabric sample has been calculated by the ratio of heat loss from the measuring head with fabric sample (μ_s) and without fabric (μ_o), and is determined using the following equation.

$$p_{wv}\ (\%)=100\frac{\mu_s}{\mu_0} \tag{4.1}$$

Cup method: This method directly determines the weight loss, with evaporation time (24 h) of water contained in a cup, the top of which is covered by the cover ring. In this method, test fabric is placed in an airtight manner over the top of the cup. Another cup contains the reference fabric secured in the same airtight manner and the experiment is performed in triplicate, so that three cups with sample fabric and three with reference fabric are tested. In this type of instrument samples can be tested according to ASTM E 96 (Procedure B) testing standard [52]. The results of measurement are expressed in terms of water vapour permeability index. Water vapour permeability index can be calculated by expressing the water vapour permeability (WVP) of the fabric as a percentage of the WVP of reference fabric, as shown below:

$$WVP=\frac{24\ X\ M}{A\ X\ T}g\ /\ m^2\ /\ 24h \tag{4.2}$$

Where M is the loss in mass (g); T, the time interval (h); and A, the internal area of the cup (m^2). A was calculated using the following relationship:

$$A=\frac{\Pi d^2}{4}X\,10^{-6} \tag{4.3}$$

Where d is the internal diameter of cup (mm). Water vapour permeability index (I) in % was calculated using the following equation.

$$I=\left[\frac{(WVP)_f}{(WVP)_r}\right]X\,100 \tag{4.4}$$

MVTR cell method

The Grace, Cryovac Division has developed a moisture vapour transmission cell (MVTR cell), which offers a faster and more simplified method for measuring the water vapour transmission behaviour of a fabric. In principle, the cell measure the humidity generated under controlled conditions as a function of time. The change in humidity at a time interval gives the moisture transmission rate (T) of the fabric, as shown below.

$$T=(269\ X\,10^{-7})\,X\left(\Delta RH\%\,X\,\frac{1440}{t}\right)X\,H\,g\ /\ m^2\ /\ 24h \tag{4.5}$$

Where $\Delta RH\%$ is the average difference in successive %RH values; t, the time interval in min; H, the gram water per m^3 of air at cell temperature.

Methods of measuring liquid moisture transmission

Liquid moisture transfer through clothing consists of two processes–wetting and wicking.

Methods used to determine the wettability of a textile material: Tensiometry and Goniometry are used to measure the wettability of the textile material.

Tensiometry: The Processor Tensiometer has been developed to measure the wettability of the fabric by measuring the wetting force by the Wilhelmy method. In this method the wetting force (force applied by the surface, when the liquid comes in contact with it) is measured. The contact angles are calculated indirectly from the wetting force when a solid is brought in contact with the test liquid using the Wilhelmy principle [53].

Goniometry: In this method the wettability of a material is measured by measuring the contact angle between the liquid and the fabric by a image processing method [54]. The developments of Automated Contact Angle Tester (ASTM D 5725-99), HTHP contact angle tester and drop analyser tester have been based on this principle. In the case of the drop analyser tester two processes are used, namely the static wetting angle measurement and the dynamic wetting angle measurement [55]. The dynamic contact angle is used as a boundary condition for modelling problems in capillary hydrodynamics, including certain stages of the droplet impact problem. The dynamic contact angle differs appreciably from the static advancing or receding values, even at low velocities.

Methods used to determine wicking through a textile material: After wetting the fibre, the liquid reaches the capillary, and a pressure is developed which forces the liquid to wick or move along the capillary. This capillary penetration of a liquid may occur from an infinite (unlimited) or a finite (limited) reservoir [56]. The different forms of wicking from an infinite reservoir are transplanar or transverse wicking, in-plane wicking and vertical or longitudinal wicking. A spot test is a form of wicking from a limited reservoir. In the case of a vertical capillary rise, the effect of gravity slows down the flow rate before equilibrium is reached.

There are different standards to determine the wickability (vertical wicking) of fabrics [57]:

BS 3424:1996, Method 21 - specifies a very long time period (24 hours) and is intended for coated fabrics with very slow wicking properties.

DIN 53924, 1978 specifies a much shorter time of 5 minutes maximum and is therefore more relevant to the studies of clothing comfort involving the transfer of perspiration. Testing is undertaken at the standard atmospheric condition of 20°C temperature and 65% relative humidity.

Normally terms used to measure wicking are

Amount of water wicked (AWW) gg^{-1} determines the wicking capacity of the fabric away from the absorption zone.

Surface-water transport rate (SWTR) $gg^{-1}s^{-1}$ calculates the amount of water wicked by 1 gram of fabric per second.

Wicking time (WT) s is the time in seconds for the water to wick across a specified distance (3.25 cm).

Need for Heat and Moisture Management

Energy expended by a person engaged in normal routine indoor activity is 50 watts/square meter/hour. This metabolic heat generated gets dissipated as sweat through clothing. During sporting activity e.g. tennis or cycling, the metabolic heat increases six times and perspires 14 times. There is increase in human body humidity during sweating which ultimately reduced the absorbency of the textile apparels. If humidity of fabric remains unchanged there will be no transportation to the surface for evaporation, so cooling cannot occur. So, body get warmer which results, more sweat. To maintain a uniform heat and moisture transfer the fabric worn next to skin should have two important properties. First property required is to evaporate the perspiration from the skin surface and second one is to transfer this into atmosphere and make the wearer feel comfortable. According to researcher the clothing required for such purpose should feel soft and supple and also not cause any irritation to skin such as scratching or itching. Even when the skin is wet with sweat, the clothing should not stick to the skin.

Desired Attributes of Heat and Moisture Management Fabric

A good moisture management fabric must have following positive attributes.

- Optimum heat and moisture regulation
- Absence of dampness
- Good air and water vapour permeability
- Durable

- Rapid moisture absorption and conveyance capacity
- Breathability and comfort
- Raping drying to prevent catching cold
- Easy care performance
- Dimensionally stable when wet
- Light Weight
- Soft and pleasant touch
- Smart and functional design

Influencing Factors of Heat and Moisture Management

There are various factors which affect heat and moisture management properties. Main factors which influence the comfort characteristics of heat and moisture management in fibers, yarns and fabrics are given in Table 1.

Conclusions

Clothing thermophysiological wear comfort is an important issue for general consumers, technical textiles, active athletes and clothing with varying end use application. Clothing should possess good water

Sr no.	Structure/Property	Thermal Behaviour	Moisture behaviour
1	Fibre		
1.1	Type of fibre	Little Effect	-
1.2	Moisture regain of fibre (increase)	Decreases	Lower (Swelling)
1.3	Bulk of fibre (higher)	Increase	Higher
1.4	Density of fibre (increases)	Reduces	
1.4	Specific heat of fibre (high)	Increase	-
1.5	Finer the fibre with low density material (higher surface to volume ratio)	Increase	-
1.6	Shape of fibre like Hollow fibre/flat fibre	Increase/Decrease	Increase/Decrease
1.7	Higher crimped fibres create more interstices in yarn	Increase	
1.8	Cross section of yarn (permit to form many voids)	Increase	Tortuosity Increases Ultimately Wicking
2	Yarn		
2.1	Type of yarn like textured and spun yarn hold more air	Increases	-
2.2	Yarn Twist (higher, lower will be the air volume)	Lower	Increase (Due To Capillary Action)
2.3	Fibre geometry /orientation (Parallel and Perpendicular)	Improve	-
2.4	Hairiness (surface entrapment of air)	Increases	-
2.5	Packing density (decrease)	Increase	-
3	Fabric		
3.1	Fabric structure (piled over knitted over woven)	Reduces	-
3.1	Thread Spacing (increase /woven)	Reduces	-
3.2	Fabric thickness increase (all type of fabrics)	Increases	Reduces
3.3	Density of fabric (lower)	Increases	-
3.4	GSM of fabric (increase)	Slight Increase	-
3.5	Porosity and Air permeability (Higher) and bulk density	Depend On	Increases
3.6	Surface characteristics (finishing treatments, texture)	Depend On	Not Significant
3.7	As Fibrous material compressed	First Fall Then Raises	
4	Other (Garment/multilayer)		
4.1	Gap between skin and garment increases up to 0.4 inch	Increase	-
4.2	Area of contact between fabric and hot body surface	Reduces	-
4.3	Rate of evaporation of water form skin or fabric	Depend On	-
4.4	External atmospheric condition like Temperature, R.H., Moment of surrounding air.	Depend On	-
4.5	Increase in wind velocity	Reduces	-
4.6	Pressure of surrounding air	Depend On	-
4.7	Rate of heat gain by absorption of water by fabric (moisture regain value of fibre, bulkiness)	Depend On	-

Table 1: Structure and property relation.

vapour as well as liquid moisture transmission property, along with transfer of heat generated within human body for providing the thermophysiological wear comfort. Heat and moisture transmission behaviour of clothing may be modelled mathematically for predicting their performance in actual wear condition. For such purpose heat and moisture transmission mechanism through clothing need to analyse along with material properties and other influencing parameters. Evaluation of heat and moisture transmission is very tough part. So, knowledge of instruments used to evaluate the heat and moisture transmission through clothing is of utter importance. As, there are lot of methods available to measure heat and moisture transmission through clothing but due to different testing conditions, parameters measured and unit used their results cannot be compared. As manufactures of sports and active outdoor wear, strive to improve the functionality of their collection by shifting their attention towards better heat and moisture management fabrics than existing ones. The discussion made in this article is useful for the textile researchers as a tool for further development of heat and moisture management fabrics.

References

1. Cena K, Clark JA (1981) Thermal physiology and comfort. Bioengineering Elsevier.

2. Farnworth B (1986) A numerical model of the combined diffusion of heat and water vapour through clothing. Textile Research Journal 56: 653-665.

3. Fourt L, Hollies NRS (1970) Clothing: comfort and function. Marcel Dekker, Newyork, USA.

4. Gagge AP (1981) Rational temperature indices of thermal comfort. Bioengineering. Thermal Physiology and comfort, (Eds), Elsevier, Amsterdam.

5. Fohr JP, Couton D, Treguier G (2002) Dynamic heat and moisture transfer through layered fabrics. Textile Research Journal 72: 1-12.

6. Adler MM, Walsh WK (1984) Mechanisms of Transient Moisture Transport Between Fabrics. Textile Research Journal 54: 334-343.

7. Hatch KL, Woo SS, Barker RL, Radhakrishnaiah P, Markee NL, et al. (1990) In Vivo Cutaneous and Perceived Comfort Response to Fabric Part I: Thermophysiological Comfort Determinations for Three Experimental Knit Fabrics. Textile Research Journal 60: 405-412.

8. Hollies NRS, Kaessinger MM, Bogaty H (1956) Water Transport Mechanisms in Textile Materials: Part 1. The Role of Yarn Roughness in Capillary-Type Penetration. Textile Research Journal 26: 829-835.

9. Hollies NRS, Kaessinger MM, Bogaty H (1957) Water Transport Mechanisms in Textile Materials: Part II. Capillary-Type Penetration of Yarns and Fibers. Textile Research Journal 27: 8-13.

10. Schneider A, Hoschke BN (1992) Heat Transfer through Moist Fabrics. Textile Research Journal 62: 61-66.

11. Andreen JH, Gibson JW, Wetmore OL (1953) Fabric Evaluations Based on Physiological Measurements of Comfort. Textile Research Journal 23: 11-22.

12. Gagge AP, Gonzalez RR (1974) Physiological and Physical Factors Associated with Warm Discomfort and Sedentary Man. Environmental Research 7: 230-242.

13. Goldman RF (1981) Evaluating the Effect of Clothing on the Wearer. In: Cena, K, Clark, JA (eds.) Bioengineering Thermal Physiology and Comfort, Chpt 3, Elsevier Scientific Publishing Co, Amsterdam.

14. Hollies NRS (1971) The Comfort Characteristics of Next-To-Skin Garments, Including Shirts. Third Shirley International Seminar: Textiles for Comfort. The Cotton Silk and Man-made Fibers Research Association, Manchester, England: 15-17.

15. Scheurell DM, Spivak SM, Hollies NRS (1985) Dynamic Surface Wetness of Fabrics in Relation to Clothing Comfort. Textile Research Journal 55: 394-399.

16. Li Y, Zhu QY (2003) A model of coupled liquid moisture and heat transfer in porous textiles with consideration of gravity. Numerical Heat Transfer 43: 501-23.

17. Gibson PW, Charmchi M (1997) Coupled heat and mass transfer through hygroscopic porous materials: application to clothing layers. Sen-I Gakkaishi 53: 183-194.

18. Fan J, Cheng X, Wen X, Sun W (2004) An improved model of heat and moisture transfer with phase change and mobile condensates in fibrous insulation and comparison with experimental results. International Journal of Heat and Mass Transfer 47: 2343-2352.

19. Onofrei E, Rocha AM, Catarino A (2010) Thermal comfort properties of knitted fabrics made of elastane and bioactive yarns. In: Proceedings of the Fiber Society Spring 2010 International Conference on Fibrous Materials, Bursa, Turkey: 145-146.

20. Zhang P, Watanabe Y, Kim SH, Tokura H, Gong RH (2001) Thermoregulatory responses to different moisture-transfer rates of clothing materials during exercise. Journal of the Textile Institute 92: 372-378.

21. Mcintyre DA (1980) Indoor climate. Applied Science, Publishers, London.

22. McIntyre DA, Griffiths ID (1974) Changing temperatures and comfort. Building Services Engineer 42: 120-122.

23. Haghi AK (2004) J Therm Anal Calorium 76: 1035.

24. Havenith G (1999) Heat balance when wearing protective clothing. Ann Occup Hyg 43: 289-296.

25. Bejan A (1993) Heat Transfer. John Wiley and Sons.

26. Fanger PO (1970) Thermal comfort. Danish Technical press. Copenhagen.

27. Fan J, Keighley JH (1989) A theortical and experimental study of the thermal insulation of clothing in windy conditions. IJCST 1: 21-29.

28. Sachdeva RC (2005) Fundamentals of engineering heat and mass transfer. (2ndedn), India, Publisher New Age International (P) Ltd.

29. Barnes JC, Holcombe BV (1996) Moisture sorption and transport in clothing during wear. Textile Research Journal 66: 777-786.

30. Suprun N (2003) Dynamics of moisture vapour and liquid water transfer through composite textile structures. Int J Clothing Sci and Tech 15: 218-223.

31. Chatterjee PK (1985) Absorbency. Elsevier Science Publishing Company, New Jersey.

32. Incropera FP, DeWitt DP (1996) Fundamentals of heat of mass transfer. (4thedn) John Wiley and Sons New Work, USA.

33. Ruckman JE (1997) Analysis of simultaneous heat and water vapour transfer through waterproof breathable fabrics. Journal of Coated Fabrics 26: 293-307.

34. Murata K (1995) Heat and mass transfer with condensation in a fibrous insulation slab bounded on one side by a cold surface. International Journal of Heat Mass Transfer 17: 3253-3262.

35. Kissa E (1996) Wetting and wicking. Textile Research Journal 66: 660-668.

36. Wong KK, Tao XM, Yuen CWM, Yeung KW (2001) Wicking properties of linen treated with low temperature server. Textile Research Journal 71: 49-56.

37. Harnett PR, Mehta, PN (1984) A survey and comparison of laboratory test methods for measuring wicking. Textile Research Journal 54: 471-478.

38. Kamath YK, Hornby SB, Weigman HD, Wilde MF (1994) Wicking of spin finishes and related liquids into continuous filament yarns. Textile Research Journal, 64: 133-140.

39. Oglakcioglu N, Celik P, Ute TB, Marmarali A, Kadoglu (2009) Thermal Comfort Properties of Angora Rabbit/Cotton Fiber Blended Knitted Fabrics. Textile Research Journal 79: 888-893.

40. Haghi AK (2004) Moisture permeation of clothing: a factor governing thermal equilibrium and comfort. Journal of Thermal Analysis and Calorimetry 76: 1035-1055.

41. Hes L (1987) Thermal Properties of Nonwovens. Proceedings of Congress Index 87, Geneva.

42. Black CP, Mathews JA (1934) The Physical Properties of Fabrics in Relation to Clothing, part 3-Heat Insulation by Fabrics used as Body Clothing. J Text Inst Trans 25: T197-T224.

43. Lyman F, Hollies NRS (1970) Clothing Comfort and Function, in: "Fiber Science Series. L. Rebenfeld, Series Ed.', Marcel Dekker Inc., New York.

44. Morris GJ (1953) Thermal Properties of Textile Materials. Journal of the Textile Institute 44: 449-476.

45. Baxter S, Cassie AD (1943) Thermal Insulating Properties of Clothing. Journal of the Textile Institute 34: 41-59.

46. Clulow E (1984). Comfort Indoors. Textile Horizons 4: 20.

47. Pause B (1996) Measuring the water vapour permeability of coated fabrics and laminates. Journal of Coated Fabrics 25: 311-320.

48. Congalton D (1999) Heat and moisture transport through textiles and clothing ensembles utilizing the "Hohenstein" skin model. Journal of Coated Fabrics 28: 183-196.

49. Holmes DA (2000) Performance characteristics of waterproof breathable fabrics. Journal of Coated Fabrics 29: 306-316.

50. Bartels VT (2005) Physiological comfort of sportswear. Textiles in Sport. In: Shishoo R (ed.) The Textile Institute, Woodhead Publishing Limited, Cambridge, England: 177-203.

51. Hes L, Carvalho M (1994) Diagnostic of the composition of fabrics from their thermal permeability in wet state. Indian Journal of Fiber Textile Research 19: 147-150.

52. ASTM E 96 (1995) Standard test method for water vapour transmission in annual book of ASTM standard 4.06.

53. Patnaik A, Ghosh A, Rengasamy RS, Kothari VK (2006) Wetting and wicking in fibrous materials. Textile Progress Indian Institute of Technology India.

54. Grindstaff TH (1969) Simple apparatus and technique for contact-angle measurements on smalldenier single Fibres. Textile Research Journal 39: 958-962.

55. Wei QF, Matheringham RR, Yang RD (2003) Dynamic wetting of fibres observed in an environmental scanning electron microscope. Textile Research Journal 73: 557-561.

56. Kissa E (1996) Wetting and wicking. Textile Research Journal 66: 660-668.

57. D'Silva AP, Greenhood C, Anand SC, Holmes DH, Whatmough N (2000) Concurrent determination of absorption and wickability of fabrics: A new test method. Journal of the Textile Institute 91: 383-396.

32

Investigation of the Effect of Different Plasma Treatment Condition on the Properties of Wool Fabrics

Eyupoglu S[1]*, Kilinc M[1] and Kut D[2]

[1]Engineering and Design Faculty, Department of Fashion and Textile Design, Istanbul Commerce University, Istanbul, Turkey

[2]Engineering Faculty, Department of Textile Engineering, Uludag University, Bursa, Turkey

Abstract

In this study, oxygen and nitrogen plasma treatment was carried out on wool fabrics during 5, 10 and 15 minutes at low (LF) (40 kHz) and radio (RF) (13.59 MHz) frequency. Then the effect of plasma treatment on tear strength, contact angle, whiteness and yellowness index of wool fabrics was investigated. In addition, the hydrophility of plasma treated samples was analyzed after the washing. The physical and chemical properties of wool fabrics treated with oxygen and nitrogen plasma were characterized by scanning electron microscope (SEM) and energy dispersive X-ray spectroscopy (EDX). According to the results, nitrogen plasma treatment improved the tear strength of samples. Furthermore, plasma treatment increased the yellowness of samples while decreasing the whiteness index of samples. The hydrophility of samples decreases with plasma treatment and with the increase in plasma treatment time. Moreover, SEM results showed that the plasma treatment caused to occur deformations on the surfaces. Lastly, EDX results showed that the amount of oxygen and nitrogen increased in the surfaces after oxygen and nitrogen plasma treatment.

Keywords: Oxygen plasma; Nitrogen plasma; Wool; Tear strength; Hydrophility; Whiteness; Yellowness index

Introduction

In textile industry, wool materials have been used for a long time because of their high thermal insulation, comfort and eco-friendliness, while having problems such as felting, pilling, luster and shrinkage [1-5]. In order to solve these problems, chemical and physical treatments have been recently used. Some chemical treatments used are ozone, enzyme and oxidation treatments while plasma treatment, electron beam irradiation, ion implantation and ultrasonic irradiation are among physical treatments commonly used [6-9].

Plasma technology has been applied in textile industry in order to produce a variety of surface modifications of textile materials. It also improves a wide range of textile properties such as hydrophobicity, dye exhaustion, adhesion etc. [10]. Furthermore, the use of plasma treatment provides more antibacterial, flame retardant, hydrophobic, hydrophilic, anti-pilling, electric conductivity, anti-static, scouring, anti-felting, ultraviolet protective textile materials [11-15]. Besides, plasma technology, characterized by low consumption of water, energy and chemicals, is a clean, ecologic and dry technique [6,16]. In addition to these advantages, plasma treatment does not influence textile material bulk properties [17,18].

A variety of gases have been used in plasma treatment such as oxygen, hydrogen, nitrogen and argon [16]. Each gas gives different features to textiles in relation to their chemical groups [19-21].

The aim of this study is to investigate the effects of plasma treatment and gas type on wool fabrics without any pre-treatment. In pre-treatment, wool fabrics were cleaned by Soxhlet extraction with dichloromethane rinsed with ethanol and deionized water before the plasma treatment [8]. This study differs from the others in that wool samples were not scoured. In this study, LF and RF oxygen and nitrogen plasma treatments were performed on wool fabrics for 5, 10 and 15 minutes. After the plasma treatment, the effects of plasma gas, time and frequency of treatment on the wool properties such as tear strength, hydrophility, whiteness index and yellowness were investigated. Moreover, the hydrophility of plasma treated samples

was tested to analyze the durability of their hydrophobic properties after washing. The surface of wool fabrics was analyzed by SEM. The results demonstrated that nitrogen plasma treatments improved the tear strength of wool fabrics. Moreover, after the plasma treatment yellowness of samples increased and the whiteness index of samples decreased with the increase in treatment time. The hydrophility of samples decreased with the plasma treatment. The results did not change after the washing, as plasma treated samples were hydrophobic.

Material and Methods

Fabrics

In this study, 100% wool plain weave fabric was used to investigate the different effects of plasma treatment on the properties of fabric. The fabric specimens were 1:1 plain fabrics (24 ends cm⁻¹, 40 Nm; 22 picks cm⁻¹, 48 Nm; 40 g/m²). Besides the wool fabrics were not cleaned by Soxhlet extraction with dichloromethane rinsed with ethanol and deionized water before the plasma treatment.

Plasma treatment

Plasma treatment was performed on wool fabrics with Diener vacuum plasma with oxygen and nitrogen gas. In this study, the effects of different gases, frequency intensity and plasma treatment time on wool fabrics were investigated. Wool fabrics were treated with oxygen and nitrogen plasma for 5, 10 and 15 minutes. After a lot of

*Corresponding author: Seyda Eyupoglu, Faculty of Engineering and Design, Department of Fashion and Textile Design, Istanbul Commerce University, Kucukyali E5 Crossroad, Istanbul 34840, Turkey, E-mail: scanbolat@ticaret.edu.tr

pre-treatment were carried out by vacuum plasma device, optimum conditions were determined as 40 kHz frequency to LF and 13.59 MHz frequency to RF at the power of 100 W and at the pressure of 0.4 mbar. The properties of fabrics exposed to plasma treatment were given in Table 1.

Tear strength measurements

Tear strength of samples was analyzed by using SDL ATLAS M008E Digital Elmandorf with ballistic pendulum method according to ISO 13937-1:2000 Standard [22] and 64000 mN load was attached to pendulum. The measurements of tear strength were iterated three times.

Contact angle measurements

Contact angle of samples was measured with an optical contact angle measurement instrument such as Attension by Ksv Instrument. The velocity of digital camera was adjusted so as to take 80 images per second and the volume of drop was 4-6 cm³. Distilled water was used to analyze the contact angle. Contact angle of samples was measured by the images. The measurements of contact angle were iterated four times.

Washing

The washing of plasma treated samples was carried out according to ISO105:C06 at 40°C for 30 minutes [23].

Whiteness index and yellowness measurements

Whiteness index of samples was determined by using Konica Minolta Spectrophotometer CM-3600d according to Stendsby Method and whiteness values were calculated using illuminant D65 and 10° standard observer values. Yellowness of samples was measured with Konica Minolta Spectrophotometer CM-3600d according to ASTM D 1925 [24] and yellowness values were calculated using C-10° standard observer values.

Physical and chemical properties of plasma treated wool fabrics

Physical and chemical structures of samples were analyzed with SEM and EDX by using ZEISS/EVO 40 Electron Microscope.

Results and Discussion

Results of tear strength measurements

The results of the measurements of tear strength of samples treated

Sample code	Properties of fabric
O₂ 5' LF	Oxygen plasma treatment at low frequency for 5 minutes
O₂ 10' LF	Oxygen plasma treatment at low frequency for 10 minutes
O₂ 15' LF	Oxygen plasma treatment at low frequency for 15 minutes
O₂ 5' RF	Oxygen plasma treatment at radio frequency for 5 minutes
O₂ 10' RF	Oxygen plasma treatment at radio frequency for 10 minutes
O₂ 15' RF	Oxygen plasma treatment at radio frequency for 15 minutes
N 5' LF	Nitrogen plasma treatment at low frequency for 5 minutes
N 10' LF	Nitrogen plasma treatment at low frequency for 10 minutes
N 15' LF	Nitrogen plasma treatment at low frequency for 15 minutes
N 5' RF	Nitrogen plasma treatment at radio frequency for 5 minutes
N 10' RF	Nitrogen plasma treatment at radio frequency for 10 minutes
N 15' RF	Nitrogen plasma treatment at radio frequency for 15 minutes

Table 1: The properties of fabrics exposed to plasma treatment.

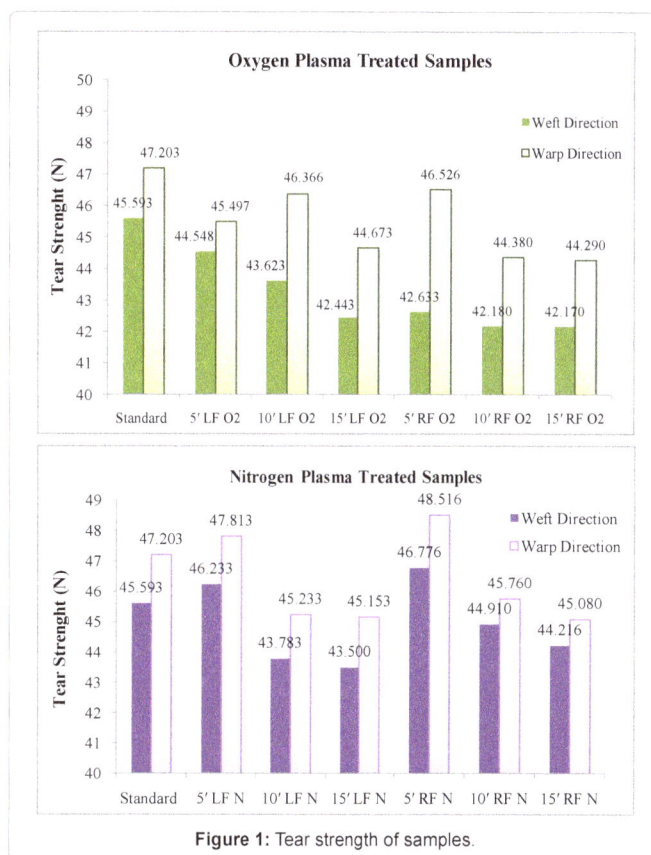

Figure 1: Tear strength of samples.

with oxygen and nitrogen plasma were given in Figure 1.

According to Figure 1, the results showed that the tear strength of warp direction of all samples was higher than that of weft direction of all samples as the warp density of all samples was higher than their weft density. The increase in the number of yarns led to the increase in the strength density of warp direction of all samples. Furthermore, the results demonstrated that oxygen plasma treatment resulted in the decrease in the tear strength of samples. After oxygen plasma treatment, deformation on surface increased with the increase in oxygen treatment time.

The comparison of tear strengths of samples in terms of plasma gas demonstrated that tear strength of samples treated with nitrogen plasma was higher than those treated with oxygen plasma. According to these results, compared to nitrogen plasma, oxygen plasma treatment was considered to cause greater damage on the wool fibers since oxygen has higher electronegativity than nitrogen, therefore, causing more damage than nitrogen.

According to results, the tear strength of samples that were applied LF plasma was lower than those with RF plasma treatment. This is because LF plasma treatment was considered to remove the wax layer slower than RF plasma treatment. Furthermore, it was considered that RF plasma treatment removed less wax layer than LF plasma treatment. Such removal arose from the collusion and scattering of the ionized electrons in different ways due to their high frequency. As the RF collusion occurred more rapid than LF collusion, less ionized electrons were deemed to interact with the samples.

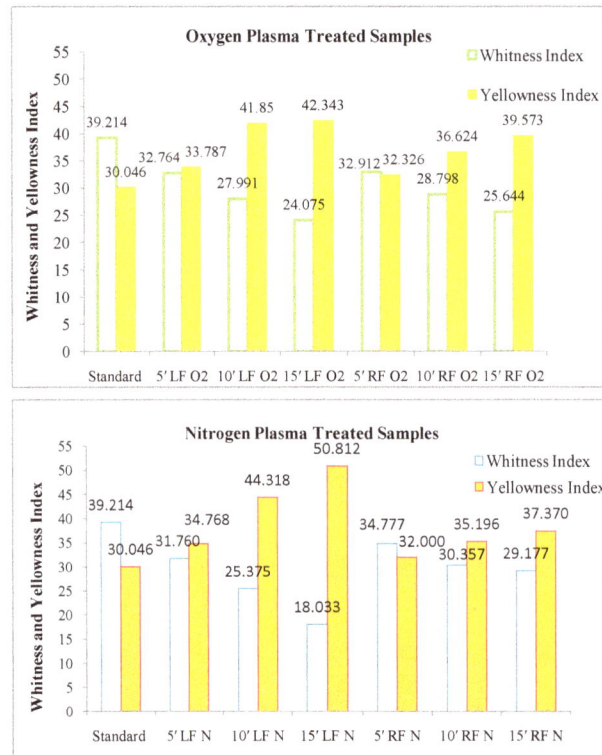

Figure 2: Whiteness and yellowness index of samples.

Results of whiteness and yellowness measurement

Figure 2 shows the measurements of whiteness and yellowness results of the samples treated with oxygen and nitrogen plasma.

According to the measurements, oxygen and nitrogen plasma treatment decreased whiteness index of samples while increasing their yellowness. Whiteness measurements led to the consideration that the plasma treatment of oxygen and nitrogen gas caused deformation on surface. After the plasma treatment, the roughness of surface increased. As is known, reflections of light from rough surfaces are less than smooth ones [25]. Furthermore, increasing treatment time of plasma application resulted in the decrease in whiteness index of samples and the increase in their yellowness values. Such result arises from the consideration that the etching effect of plasma treatment on wax layer of wool fabrics increased due to the increase in plasma treatment time. For this reason, the yellowness of samples increased with the plasma treatment [26]. Besides, comparison of the results in terms of types of plasma gas demonstrated that, at low frequency, nitrogen plasma treatment reduced whiteness index of wool samples while increasing yellowness index. However, at radio frequency, oxygen plasma treatment yielded the same results at a better rate. Such results led to the consideration that each type of gas affected the structure of wool differently [25].

Results of contact angle measurements

The results of contact angle measurements of samples treated with oxygen and nitrogen plasma were given in Table 2.

The results of Table 2 departed from other results in that oxygen and nitrogen plasma treatment led to the increase in hydrophobicity

of wool samples [27]. Such result was based on the consideration that samples used in this study were not treated with pre-treatment. These results were considered that the wax layer of surfaces of wool fabric could not be removed, and therefore wax layer was penetrated the wool fabric surfaces with plasma treatment. Furthermore, when the results

Sample	Contact Angle
Standard	111.69
5' LF O_2	131.90
10' LF O_2	130.60
15' LF O_2	129.33
5' RF O_2	131.14
10' RF O_2	128.83
15' RF O_2	125.00
5' LF N	136.840
10' LF N	126.570
15' LF N	116.24
5' RF N	122.773
10' RF N	118.110
15' RF N	117.320

Table 2: Contact angle measurements of samples.

Sample	Contact Angle
Standard	111.69
15' LF O_2	111.783
15' RF O_2	125.703
15' LF N	116.19
15' RF N	112.297

Table 3: Contact angle measurements of samples.

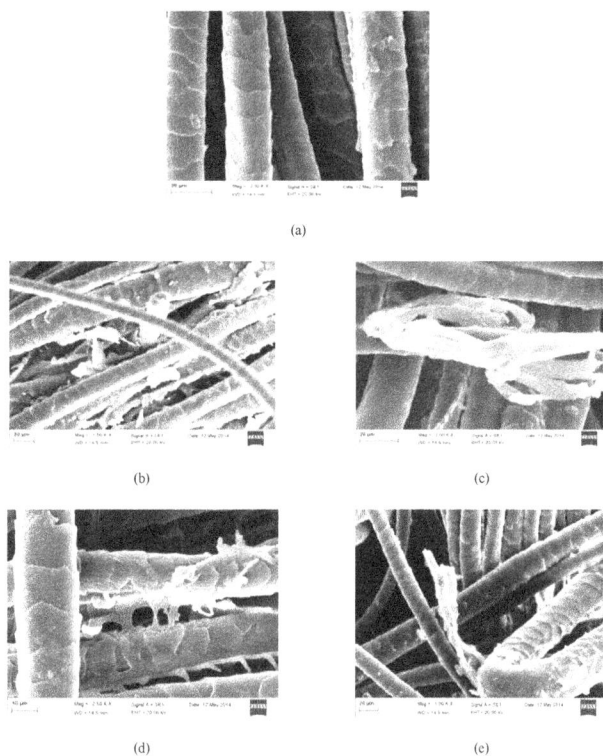

Figure 3: The SEM images of samples ((a) The SEM images of raw wool, (b) The SEM images of sample treated with LF oxygen plasma for 15 min, (c) The SEM images of sample treated with RF oxygen plasma for 15 min, (d) The SEM images of sample treated with LF nitrogen plasma for 15 min, (e) The SEM images of sample treated with RF nitrogen plasma for 15 min).

were compared in terms of the plasma gas, the results demonstrated that the contact angle of samples treated with oxygen plasma was higher than that treated with nitrogen plasma. The reason of this result considered that oxygen gas caused to penetrate more oil on the surface than nitrogen gas because of being more electronegative gas.

According to the results, contact angle of the samples treated with RF plasma was higher than that of the samples treated with LF plasma. The ground for this result was considered that the collusion of RF plasma is higher than the collusion of LF plasma on the surface of wool fabric due to higher frequency of RF plasma. For this reason, it was deemed that the penetrated wax on the samples increased.

Contact angle measurements after the plasma treatment

The hydrophility of plasma treated samples was achieved after the washing to analyze durability of hydrophobic properties of plasma treated samples. The contact angle measurements of plasma treated samples were carried out after washing. The results of contact angle measurements of samples were given in Table 3.

According to the results, the contact angle of plasma treated samples was higher than standard sample after the washing. The hydrophobic properties of plasma treated samples did not significantly decrease after washing when compared to plasma treated samples. Owing to the results, the penetrated wax layer on the wool surface after the plasma treatment was not removed with washing.

Scanning Electron Microscope images

Surface characterization of samples was analyzed with SEM. The SEM images of samples treated oxygen and nitrogen plasma for 15 minutes were shown in the Figure 3.

SEM results led to the conclusion that oxygen and nitrogen plasma treatment gave rise to micro cracks in the surface of samples. In addition, the increase in the duration of treatment of plasma application increased micro cracks in the surface.

When the results compared with each other in terms of plasma gas, oxygen plasma treatment caused more damage than nitrogen plasma treatment, which was attributed to higher electronegativity of oxygen compared to nitrogen.

Results of energy dispersive X-Ray spectroscopy

The EDX results of samples treated with oxygen and nitrogen plasma for 15 minutes were given in Figure 4.

According to the results, the amount of carbon decreased with plasma application. The amount of carbon in the sample treated with oxygen plasma is lower than that with nitrogen plasma. Furthermore, the oxygen and nitrogen plasma treatment led to the increase in the amount of oxygen, nitrogen and sulfide in the samples as each type of gas and treatment frequency is deemed to affect the structure of wool differently [27].

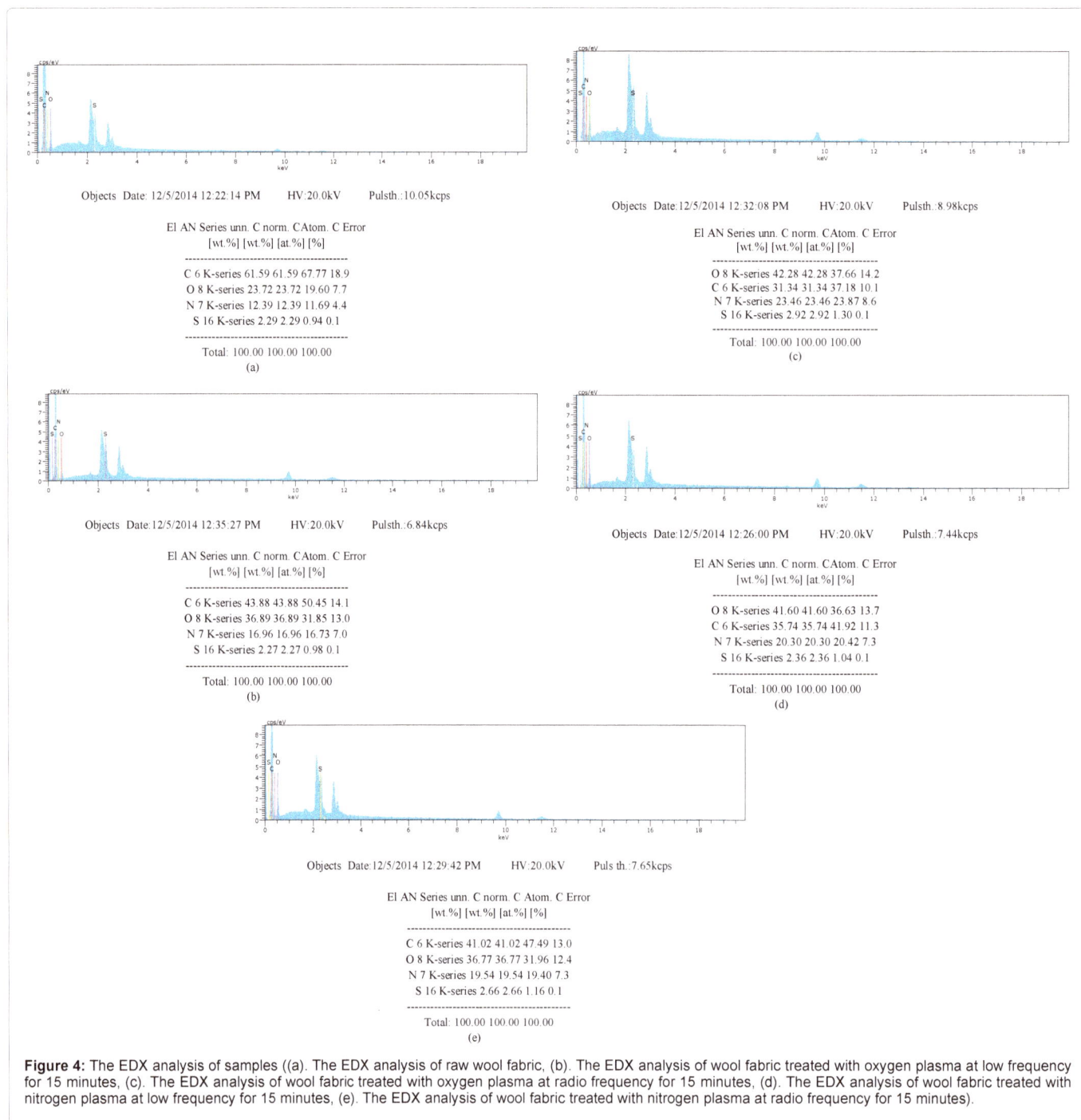

Objects Date: 12/5/2014 12:22:14 PM   HV:20.0kV   Pulsth.:10.05kcps

El AN Series unn. C norm. CAtom. C Error
[wt.%] [wt.%] [at.%] [%]

C 6 K-series 61.59 61.59 67.77 18.9
O 8 K-series 23.72 23.72 19.60 7.7
N 7 K-series 12.39 12.39 11.69 4.4
S 16 K-series 2.29 2.29 0.94 0.1

Total: 100.00 100.00 100.00
(a)

Objects Date:12/5/2014 12:32:08 PM   HV:20.0kV   Pulsth.:8.98kcps

El AN Series unn. C norm. CAtom. C Error
[wt.%] [wt.%] [at.%] [%]

O 8 K-series 42.28 42.28 37.66 14.2
C 6 K-series 31.34 31.34 37.18 10.1
N 7 K-series 23.46 23.46 23.87 8.6
S 16 K-series 2.92 2.92 1.30 0.1

Total: 100.00 100.00 100.00
(c)

Objects Date:12/5/2014 12:35:27 PM   HV:20.0kV   Pulsth.:6.84kcps

El AN Series unn. C norm. CAtom. C Error
[wt.%] [wt.%] [at.%] [%]

C 6 K-series 43.88 43.88 50.45 14.1
O 8 K-series 36.89 36.89 31.85 13.0
N 7 K-series 16.96 16.96 16.73 7.0
S 16 K-series 2.27 2.27 0.98 0.1

Total: 100.00 100.00 100.00
(b)

Objects Date:12/5/2014 12:26:00 PM   HV:20.0kV   Pulsth.:7.44kcps

El AN Series unn. C norm. CAtom. C Error
[wt.%] [wt.%] [at.%] [%]

O 8 K-series 41.60 41.60 36.63 13.7
C 6 K-series 35.74 35.74 41.92 11.3
N 7 K-series 20.30 20.30 20.42 7.3
S 16 K-series 2.36 2.36 1.04 0.1

Total: 100.00 100.00 100.00
(d)

Objects Date:12/5/2014 12:29:42 PM   HV:20.0kV   Puls th.:7.65kcps

El AN Series unn. C norm. C Atom. C Error
[wt.%] [wt.%] [at.%] [%]

C 6 K-series 41.02 41.02 47.49 13.0
O 8 K-series 36.77 36.77 31.96 12.4
N 7 K-series 19.54 19.54 19.40 7.3
S 16 K-series 2.66 2.66 1.16 0.1

Total: 100.00 100.00 100.00
(e)

Figure 4: The EDX analysis of samples ((a). The EDX analysis of raw wool fabric, (b). The EDX analysis of wool fabric treated with oxygen plasma at low frequency for 15 minutes, (c). The EDX analysis of wool fabric treated with oxygen plasma at radio frequency for 15 minutes, (d). The EDX analysis of wool fabric treated with nitrogen plasma at low frequency for 15 minutes, (e). The EDX analysis of wool fabric treated with nitrogen plasma at radio frequency for 15 minutes).

Conclusion

The use of plasma treatment in textile industry has long been discussed. Plasma treatment provides many advantages in terms of water, energy, chemical consumption and ecological production for textile industry. In this study, wool fabrics were treated with oxygen and nitrogen plasma for different duration of treatment. After plasma treatment, tear strength, contact angle, hydrophility, yellowness, whiteness index and surface properties of wool fabrics were investigated so as to determine the effect of plasma treatment. The results demonstrated that nitrogen plasma treatment led to the increase in the tear strength of wool fabrics whereas oxygen plasma treatment resulted in the decrease in the tear strength of wool fabrics. In addition, oxygen and nitrogen plasma treatment increased the hydrophobicity of wool fabrics, which, in turn, decreased with the increment in the duration of application of plasma treatment. Moreover, nitrogen and oxygen plasma treatment increased the yellowness of samples. The increase in the duration of plasma treatment gave rise to the decrease in the whiteness index of samples. Besides, SEM images demonstrated that nitrogen and oxygen plasma treatment created micro cracks in the surface of wool fabrics. As a result of this study, plasma treatment can be used in textile industry for the modification of textile materials.

References

1. Wan A, Dai XJ, Magniez K, Plessis JD, Yu W, et al. (2013) Reducing the pilling propensity of wool knits with a three-step plasma treatment. Textile Research Journal 83: 2051-2059.

2. Erra P, Molina R, Jocic D, Julia MR (1999) Shrinkage Properties of Wool Treated with Low Temperature Plasma and Chitosan Biopolymer. Textile Research Journal 69: 811-815.

3. Kan CW,Yuen CWM, Hung ON (2013) Improving the pilling property of knitted wool fabric with atmospheric pressure plasma treatment. Surface and Coatings Technology 228: 588-592.

4. Ceriaa A, Rombaldonib F, Roveroa G, Mazzuchettib G, Sicardi S (2010) The effect of an innovative atmospheric plasma jet treatment on physical and mechanical properties of wool fabrics. Journal of Materials Processing Technology 210: 720-726.

5. Kan CW, Chan K, Yuen CWM, Miao MH (1998) The effect of low-temperature plasma on the chrome dyeing of wool fibre. Journal of Materials Processing Technology 82: 122-126.

6. Shishoo R (2007) Plasma Technologies for Textiles. Woodhead Publishing, Cambridge, UK.

7. Bozzi A, Yuranova T, Kiwi J (2005) Self-cleaning of wool-polyamide and polyester textiles by TiO2-rutile modification under daylight irradiation at ambient temperature. Journal of Photochemistry and Photobiology A: Chemistry 172: 27-34.

8. Kan CW,Yuen CM (2007) Plasma technology in wool. Textile Progress 39: 121-187.

9. Simpson WS, Crawshaw GH (2002) Wool: Science and technology. Woodhead Publishing, Cambridge, UK.

10. Mehmood T, Kaynak A, Dai XJ, Kouzani A, Magniez K, et al. (2014) Study of oxygen plasma pre-treatment of polyester fabric for improved polypyrrole adhesion. Materials Chemistry and Physics 143:668-675.

11. Caschera D, Mezzi A, Cerri L, Caro T, Riccucci C, et al. (2014) Effects of plasma treatments for improving extreme wettability behavior of cotton fabrics. Cellulose 21: 741-756.

12. Shahidi S (2014) Novel method for ultraviolet protection and flame retardancy of cotton fabrics by low-temperature plasma. Cellulose 21:757-768.

13. Mirjalili M, Nasirian SS, Karimi L (2011) Effects of corona discharge treatment on some properties of wool fabrics. African Journal of Biotechnology 10: 19436-19443.

14. Canal C, Erra P, Molina R, Bertrán E (2007) Regulation of Surface Hydrophilicity of Plasma Treated Wool Fabric. Textile Research Journal 77: 559-564.

15. Shahidi S, Ghoranneviss M (2014) Effect of Plasma Pretreatment Followed by Nanoclay Loading on Flame Retardant Properties of Cotton Fabric. Journal of Fusion Energy 33: 88-95.

16. Janca J, Czernichowski A (1998) Wool treatment in the gas flow from gliding discharge plasma at atmospheric pressure. Surface and Coatings Technology 98:1112-1115.

17. Garg S, Hurren C, Kaynak A (2007) Improvement of adhesion of conductive polypyrrole coating on wool and polyester fabrics using atmospheric plasma treatment. Synthetic Metals 157: 41-47.

18. Kan CW, Yuen CWM (2006) Low Temperature Plasma Treatment for Wool Fabric. Textile Research Journal 76: 309-314.

19. Osenberg F, Theirich D, Decker A, Engemann J (1999) Process control of a plasma treatment of wool by plasma diagnostics. Surface and Coatings Technology 116-119: 808-811.

20. Sun D, Stylios GK (2006) Fabric surface properties affected by low temperature plasma treatment. Journal of Materials Processing Technology 173: 172-177.

21. Chi-Wai K, Kwong C, Chun-Wah MY (2004) The Possibility Of Low-Temperature Plasma Treated Wool Fabric For Industrial Use. AUTEX Research Journal 4: 37-44.

22. http://techeduhry.nic.in/syllabus/TEXTILE-TECHNOLOGY/6.1.pdf

23. ISO 105-C06 (2010) Textiles-Tests for Color Fastness - Part C06: Color Fastness to Domestic and Commercial Laundering.

24. (1999) Test Method for Yellowness Index of Plastics, ASTM D-1925.

25. Gregorski KS, Pavlath AE Fabric (1980) Modification Using the Plasmod1 the Effect of Extensive Treatment in Nitrogen and Oxygen Plasmas at Low Pressure. Textile Research Journal 50: 42-46.

26. http://global.britannica.com/EBchecked/topic/637995/wax.

27. Karahan HA, Ozdogan E, Demir A, Kocum IC, Oktem T, et al. (2009) Effects of atmospheric pressure plasma treatments on some physical properties of wool fibers. Textile Research Journal.

33

Effects of Dyeing Temperature and Molecular Structure on the Dye Affinity of Polyurethane Films containing Polyethylene Glycol Segments

Enomoto M[1]*, Tokino S[2] and Ishihara H[3]

[1]*Department of Applied Life Studies, College of Nagoya Women's University, Japan*
[2]*Industrial Technology Center of Wakayama Prefecture, Japan*
[3]*Ryukoku University, Japan*

Abstract

The dyeing behavior of segmented polyurethane containing copolymerized polyethylene glycol soft segments was investigated. Four types of segmented polyurethane were prepared by controlling the aggregation of the hard segments and the mobility of polyethylene glycol in the soft segments. Dyeing rate curves and sorption isotherms were obtained for direct dyes at various temperatures along with the swelling ratios of the different segmented polyurethane samples. The sorption of CI Direct Red 28 by these samples is shown to decrease with increasing dyeing temperature, with the swelling ratio of the segmented polyurethane film also decreasing. Changes in the tensile strength of the films before and after dyeing with CI Direct Red 28 are shown to be related to the degree of aggregation of the hard segments, as well as to the amount of dye used.

Keywords: Direct dye; Aggregate structure; Polyurethane; Dye affinity; Polyethylene glycol

Introduction

The high flexibility of polyurethane resin, even at low temperatures, and the fact that its molecular structure can be modified to produce fibers or films makes it an ideal material for cloth and interior linings. Segmented polyurethane (SPU) is commonly used as a raw material in the production of polyurethane elastic fibers as well as artificial and synthetic leather, but its hydrophobic nature means that disperse dyes are generally used despite their color fastness and colorability being unsatisfactory. The ability to dye SPU with water-soluble dyes is particularly important for artificial and synthetic leather, particularly in the single-bath dyeing of artificial leather (nonwoven nylon fibers impregnated with polyurethane resin) or dyeing of synthetic leather fashion garments. In industry, SPU has been rendered responsive to acid dyes by synthesizing soft segments through copolymerization with polyethylene glycol (PEG) [1,2] or by introducing dyeing sites [3,4] through copolymerization with tertiary amino groups or tertiary nitrogen chain extenders.

In a previous study [5] SPU films with soft segments prepared using different relative contents of PEG and polytetramethylene glycol (PTMG) were investigated in relation to their affinity to acid dyes. The tensile strength of these films was found to decrease with increasing PEG concentration, whereas the elongation to breaking tended to increase. The affinity of the SPU films for acid leveling-type dyes was low, but the films with higher PEG contents were readily stained with milling-type dyes. Elsewhere [6] dyeing rates and sorption isotherms for direct and milling-type acid dyes were obtained for SPU films with various PEG to PTMG composition ratios. From this, a dyeing mechanism was proposed involving interactions between the dye and hydrophilic PEG groups in SPU soft segments [6] in which dyeing is presumed to occur through via van der Waals forces that exist between the dye and hydrophobic urethane moieties.

Based on this knowledge, the present study was conducted by first investigating the effect of dyeing temperature on the interaction between a dye and the hydrophilic PEG groups in the soft segments. For this, a SPU sample was prepared using only PEG for the soft segments in order to maximize its affinity for direct dyes. Since the aggregation of hard segments in SPU restricts the movement of PEG chains, the

impact of this on dyeing was also investigated. Finally, SPU samples were prepared using different combinations of PEG and PTMG for the soft segments, and their tensile strengths and breaking elongations were compared. The measured dye uptake for each SPU and swelling ratios of the SPU films in water makes it possible to optimize the dyeing conditions for acid and direct dyes used in industry, meaning that this study should greatly contribute to improving the industrial dyeing of elastic fibers and artificial/synthetic leather.

Experimental Methods

Chemicals and reagents

The PTMG and ethylene glycol used in this study were procured from Mitsubishi Chemical Corp.; dimethylformamide (DMF) was obtained from Mitsubishi Gas Chemical Company, Inc.; PEG was purchased from NOF Corp. and MDI was procured from Tosho Corp. The 1,4-butanediol (1,4BD), 1,6-hexanediol (1,6HG) and CI Direct Red 28 were purchased from BASF Corp., UBE Industries, Ltd. and Wako Pure Chemical Industries, Ltd., respectively.

Preparation of the SPU solution

A total of five SPU solutions were prepared as follows: Solution 1 was prepared using only PEG for the soft segments; Solutions 2, 3 and 5 were prepared with a PEG to PTMG weight ratio of exactly 80:20, while for Solution 4 a weight ratio of 78:22 was used. As shown in Table 1, SPU was synthesized in each instance by extending the chain after pre-polymerization, with all reactions being performed in DMF under a nitrogen flow. The solid matter content and viscosity of all the SPU

*Corresponding author: Masao Enomoto, Department of Applied Life Studies, College of Nagoya Women's University, Japan
E-mail: enomoto@nagoya-wu.ac.jp

Solution	1		2		3		4		5	
Prepolymer reaction[a] (soft segments)	PEG2000 MDI	0.50 1.00	PEG2000 PTMG2000 MDI	0.40 0.10 1.00	PEG2000 MDI	0.40 0.30	PEG600 MDI	1.20 0.80	PEG2000 PTMG2000 MDI	0.40 0.10 1.00
					PTMG2000 MDI	0.10 0.40	PTMG2000 MDI	0.10 1.00		
Chain extension[a] (hard segments)	1,4BD MDI	1.50 1.00	1,4BD MDI	1.50 1.00	1,4BD MDI	1.50 1.30	1,4BD MDI	1.50 1.00	ethylene glycol 1,6HD MDI	0.75 0.75 1.00
PEG/PTMG (wt%)	100/0		80/20		80/20		78/22		80/20	
NCO (%)	10.28		10.28		10.28		13.40		10.28	

PEG: polyethylene glycol; PTMG: polytetramethylene glycol; MDI: 4,4'-diphenylmethanediisocyanate; 1,4BD: 1,4-butanediol; 1,6HG: 1,6-hexanediol

[a]all values in mol.

Table 1: Soft- and hard-segment contents and reaction methods used to prepare the five segmented polyurethanes investigated here.

solutions were adjusted to 30 wt% and 50 Pa·s (at 30°C), respectively. Figure 1 shows the chemical structure of the PTMG- and PEG-based soft segments, as well as that of the 1,4BD-based hard segments.

Preparation of cast films

The five pre-prepared SPU solutions were individually flow cast onto release paper using a Baker-type applicator (Tester Sangyo Co., Ltd.) to form solution layers of a specific thickness. These layers were then placed in a dryer for 20 minutes at 120°C to obtain cast films. Using this method, films of two different thicknesses (50 and 60 µm) were obtained.

Measurement of dyeing rate curves and sorption isotherms

As all of the cast films were found to swell in water, they were immersed in boiling water for 10 minutes before dyeing with the direct dye CI Direct Red 28, the chemical structure of which is shown in Figure 2. The dyeing rates for the films cast from Solution 1 were measured at dyeing temperatures of 30°C, 50°C, 70°C and 90°C by immersing for 120 minutes in a dye bath prepared with a material-to-liquor ratio of 1:200, and dye and sodium chloride concentrations of 0.1 and 1 g/L, respectively. Films 2-5 were dyed at 90°C for 80 minutes. All dyed films were washed twice for 1 minute in room-temperature water at a material-to-liquor ratio of 1:2000, after which the samples were washed with water and dried in preparation for dye affinity tests. The dye content of each cast film was obtained by dissolving each film in room temperature DMF solution and, using a calibration curve prepared in advance from DMF solutions of the dyes, deducing the dye concentration from the absorbance of the solution. The dye concentration in each film was then calculated based on its absolute dry mass, as measured with an infrared electronic moisture meter (AD-4715, A&D Co. Ltd.) [6].

Sorption isotherms for the films cast from Solution 1 were measured at 30°C, 50°C, 70°C and 90°C using a dyeing time of 120 minutes; i.e., the sorption equilibration time determined from the dyeing rate curves. For the films cast using Solution 2-5, the dye temperature was set to 90°C and the samples were immersed for 80 minutes in the same dye liquor described above. The samples were left to swell and were washed using the same procedure as for the measurement of the dyeing rates. The sorption isotherms were derived from the sorbed dye concentration, $[D]_f$, and the dye concentration in the bath after dyeing, $[D]_s$; the former was determined by colorimetry after dissolving the films in DMF, while the latter was deduced from the initial concentration in the bath and $[D]_f$. The thermodynamic parameters were calculated by applying the standard method for cellulose fiber systems interacting with direct dyes [7]. The effective volumes of the samples were calculated from the measured area swelling ratios of the

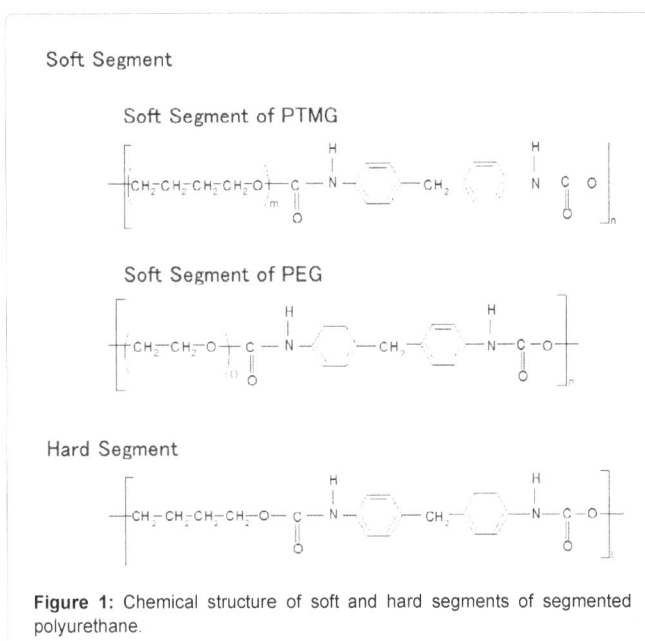

Soft Segment

Soft Segment of PTMG

Soft Segment of PEG

Hard Segment

Figure 1: Chemical structure of soft and hard segments of segmented polyurethane.

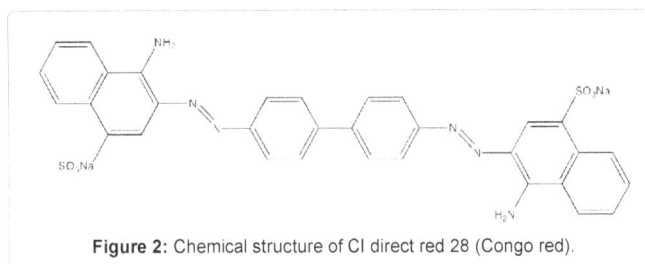

Figure 2: Chemical structure of CI direct red 28 (Congo red).

cast films, as described in the following section. The films had an initial thickness of 60µm and water-swelling was assumed to occur uniformly in all directions.

Measurement of water-swelling ratio

Water-swelling measurements performed by first cutting each 60-µm-thick cast film into square 50 × 50 mm specimens. For Solution 1, the specimens were immersed in tanks of deionized water held at different temperatures and photographed using a digital camera (Optio 750Z, Pentax). For Solution 2-5, the specimens were similarly photographed after immersion for 10 minutes in glass petri dishes filled with deionized water at 20°C. Image analysis software (Image-Pro Plus v. 6.3, Media Cybernetics) was used to measure the film area in each image, and this was then compared with the area of the dry films to determine the water-swelling ratio at different temperatures.

Measurement of tensile strength, breaking elongation and thermal softening

The tensile strength and breaking elongation of the 60-μm thick cast films were measured before and after dyeing with CI Direct Red 28. Reference films were also prepared under the same temperature and wetting conditions, but were immersed in dye-free water. Test specimens were prepared from the cast films using a type 2 cutting die, as stipulated in the JISK7113 standard. In light of the sensitivity of the specimens to temperature and humidity, they were held at 20°C and 50 % relative humidity for more than 24 hours prior to carrying out the tensile tests. These tests were performed using an Instron Model 5569 testing system (Instron Japan Co., Ltd.) operated at 400 mm/minute with a chuck spacing of 60 mm. Measurements were not performed for samples prepared from Solution 1 and 2, as the strength and ductility of these were already known from a previous study [5].

Thermal softening experiments were performed with a TMA-50 thermo-mechanical analyzer (Shimadzu Corp.) on 50-μm-thick cast films prepared from Solutions 2-5 using the needle insertion method. Measurements were taken under a nitrogen atmosphere and 0.05 N load, with the temperature being increased from 25°C to 250°C at 10°C/minute.

Results and Discussion

Affinity of the cast films for direct dye

Dyeing rate curves: Figure 3 shows the sorption curves for films produced from Solution 1 when dyed with CI Direct Red 28 at different temperatures. Note that the dyeing rate increases with temperature, with an apparent equilibrium being reached at 40, 20, and 10 minutes at 30°C, 50°C, 70°C and 90°C, respectively. The equilibrium concentrations of adsorbed dye are, however, higher at lower temperatures.

Sorption isotherms: Figure 4 shows the sorption isotherms obtained for cast films of Solution 1 at different temperatures. As

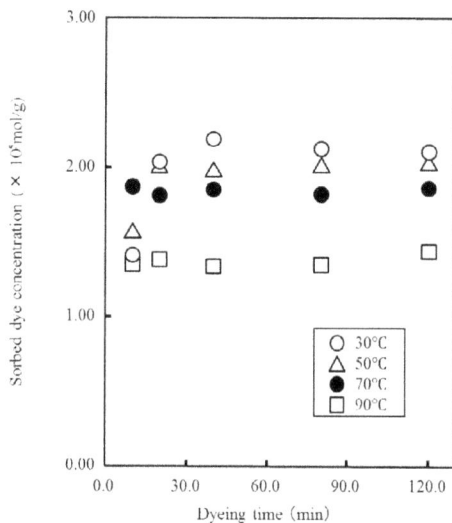

Figure 4: Adsorbed dye concentration, $[D]_f$, as a function of dye concentration in the bath, $[D]_s$, for cast films of Solution 1 (composition listed in Table 1) when dyed for 120 minutes with CI Direct Red 28 at different temperatures.

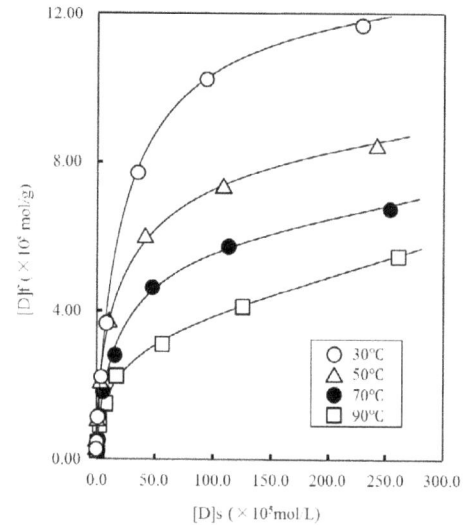

Figure 5: Concentration of adsorbed dye (CI direct red 28) as a function of dyeing time for cast films of Solution 2-5 (compositions listed in Table 1) dyed at 90°C.

reported previously [6], the saturation concentration is not reached at any of these dyeing temperatures (from 30°C to 90°C). The dye concentrations in the films are higher at lower temperatures, with the separation between the isotherms increasing with $[D]_s$.

Effect of cast film structure on dye affinity

Dyeing rate curves: Figure 5 shows the dyeing rate curves derived from the sorbed dye concentration (CI Direct Red 28) measured for cast films of Solution 2-5. The apparent equilibrium sorbed dye concentration is reached after about 40 minutes for each film, and is higher for those samples produced from Solution 5 than from Solution 3, 2 and 4 (in that order). This suggests that the presence of higher-order structures in the polymer reduces the dyeing rate.

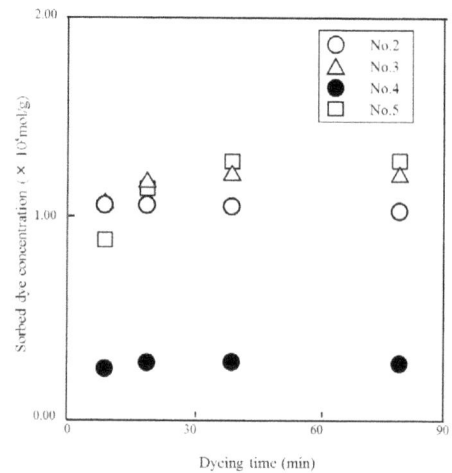

Figure 3: Concentration of adsorbed dye (CI direct red 28) as a function of dyeing time for cast films of Solution 1 (composition listed in Table 1) dyed at different temperatures.

Sorption isotherms: Figure 6 shows the sorption isotherms for cast films of Solution 2-5 when dyed with CI Direct Red 28. No saturation is observed in the case of Solution 1 and all sorbed dye concentrations are low, particularly with Solution 4. The dye concentrations for Solution 5 are slightly higher than those for Solution 3 and 2, in that order.

Thermodynamic parameters: The thermodynamic parameters calculated for the five cast films dyed with CI Direct Red 28 are listed in Tables 2 and 3. We see from this that films prepared from Solution 1, 2, 3 and 5 share a similar affinity for CI Direct Red 28, but that of Solution 4 is much lower. In terms of the structure of SPU in each film, the degree of hard-segment aggregation is the lowest with Solution 5, the mobility of hydrophilic soft segments containing PEG is increased with Solution 3, while Solution 2 produces a film with a standard composition.

Water-swelling ratio of the cast films

Figure 7 shows that the swelling rate of 60-μm-thick films cast from Solution 1 decreases as the temperature of the water increases. This behavior is similar to that of non-ionic surface-active agents with PEG hydrophilic groups exhibiting a clouding point, which suggests that more coordinated water molecules discharge from the soft segments of the SPU (composed of PEG and MDI) when the temperature is increased.

Figure 8 shows the measured water swelling rates for cast films of Solution 2-5, in which it can be seen that the degree of hard-segment aggregation and PEG mobility in the soft segments differ. The films are

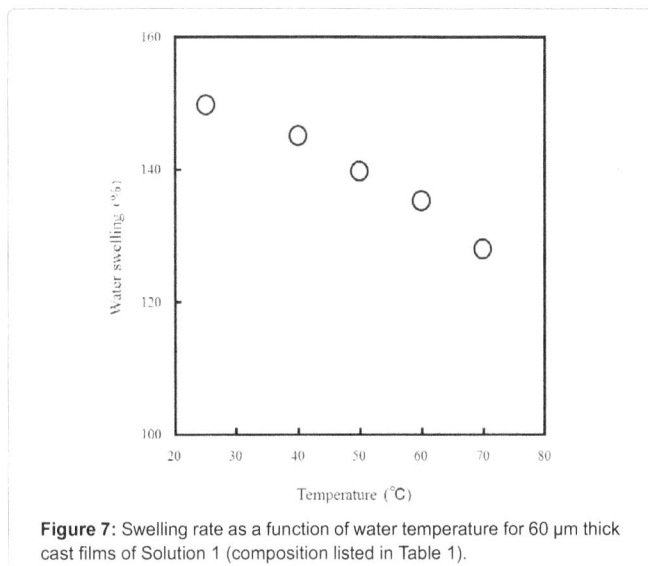

Figure 7: Swelling rate as a function of water temperature for 60 μm thick cast films of Solution 1 (composition listed in Table 1).

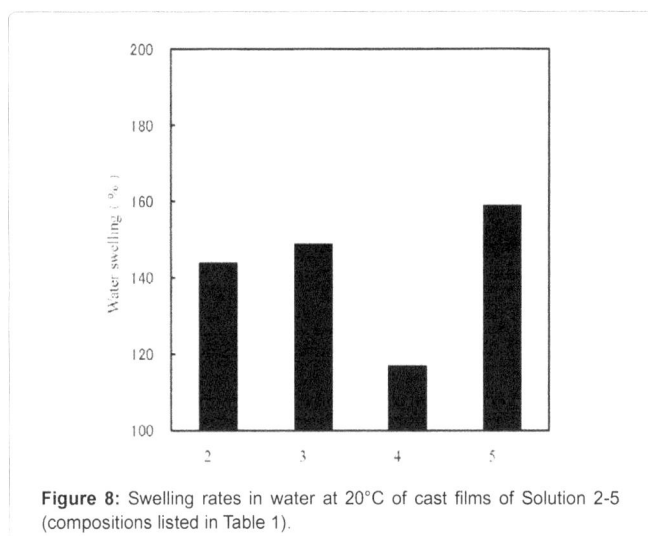

Figure 8: Swelling rates in water at 20°C of cast films of Solution 2-5 (compositions listed in Table 1).

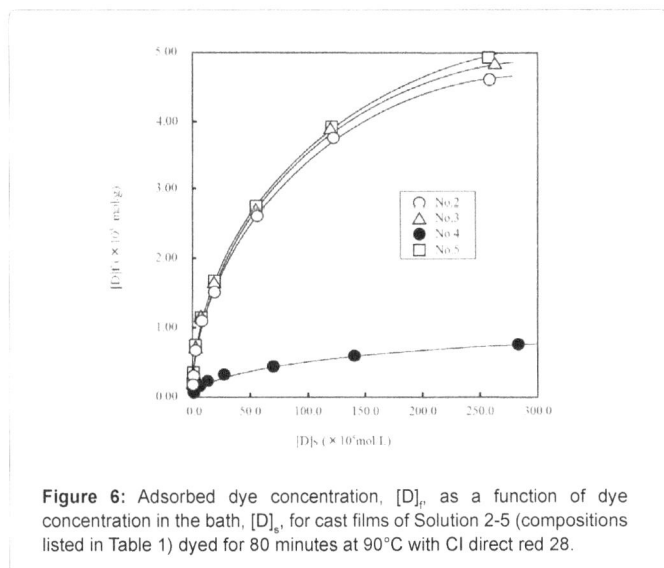

Figure 6: Adsorbed dye concentration, $[D]_f$, as a function of dye concentration in the bath, $[D]_s$, for cast films of Solution 2-5 (compositions listed in Table 1) dyed for 80 minutes at 90°C with CI direct red 28.

Film	−Δμ° (kJ/mol)				−ΔH° (kJ/mol)	−ΔS° (J/K)
	30°C	50°C	70°C	90°C		
1	11	10	9	8	24	43

Table 2: Thermodynamic parameters calculated for dyeing films of Solution 1 (composition listed in Table 1) with CI Direct Red 28.

Solution	−Δμ° at 90°C (kJ/mol)
2	8
3	8
4	2
5	8

Table 3: Standard affinities of CI Direct Red 28 for segmented polyurethane films.

ordered 5 > 3 > 2 > 4 in terms of their swelling rates in water, which suggests that the concentration of higher-order structures (which varies with the design of the polymer) influences its swelling properties in water.

Tensile strength, breaking elongation and thermal softening

The tensile strength and breaking elongation for cast films of Solution 2-5 before dyeing are shown in Figures 9 and 10, respectively. Similar stress-strain curves were obtained for films produced from solution 2 and 3, and these have the same composition but differ in terms of the reaction method used for their soft segments, this may be due to their identical MDI contents. In the film produced from Solution 4, the soft segments (PEG oligomers with a molecular weight of 2300) contain polyurethane bonds arising from the reaction between PEG (molecular weight = 600) and MDI. Since this film has a higher content of MDI, it therefore has a higher tensile strength than the other films but a lower breaking elongation. With Solution 5, the lower tensile strength and higher breaking elongation of the film suggests a low level of aggregation, which may stem from the random combination of single-chain glycol and methylene groups of varying lengths in the hard segments.

The tensile strengths of the films decreases after dyeing, more so for Solution 2 and 3 than for 4 and 5 (Figure 9), but the change in their breaking elongation is not significant (Figure 10). This presumably reflects the effect of CI Direct Red 28 on the hard-segment aggregates in the films, in that the dye can generally penetrate both the interior of the soft segment and the surface of the hard segment. In other words, the films produced from Solution 2 and 3 were affected by dye molecules on their hard segment surface, resulting in a decrease in tensile strength. The composition of the Solution 5 film moderates aggregation, resulting in a low tensile strength that is only slightly reduced by dyeing. The similarly small decrease observed with the Solution 4 film is, on the other hand, the result of a low dye uptake, which is hindered by the restricted mobility of the PEG-containing hydrophilic soft segments in this sample. It is also believed that the difference in dye concentration between each film has a significant effect on their tensile strength. For

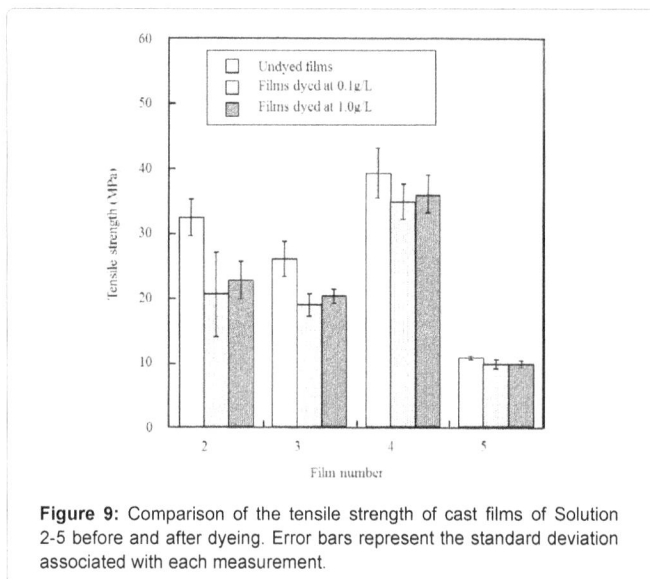

Figure 11: Thermal softening curves for cast films of Solution 2-5.

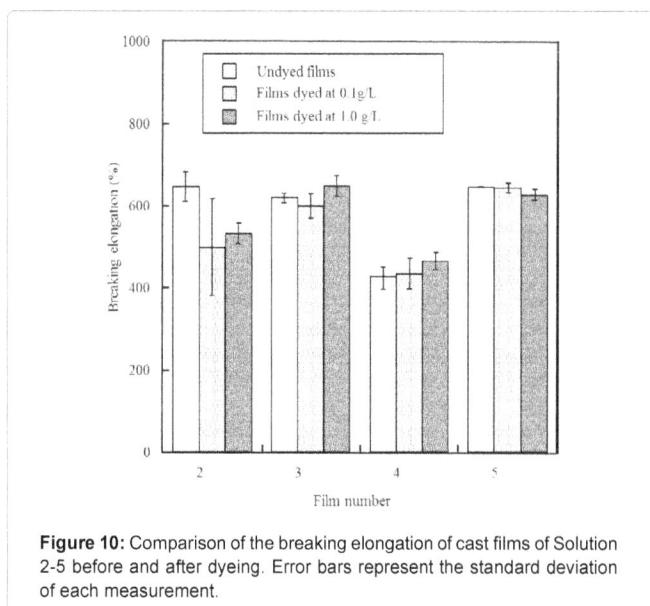

Figure 9: Comparison of the tensile strength of cast films of Solution 2-5 before and after dyeing. Error bars represent the standard deviation associated with each measurement.

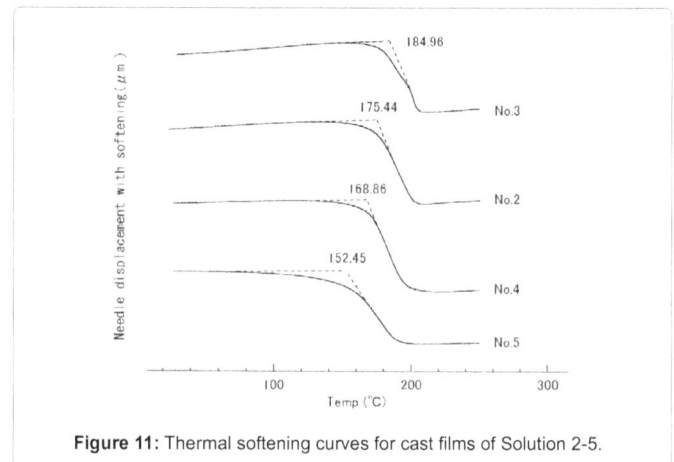

Figure 10: Comparison of the breaking elongation of cast films of Solution 2-5 before and after dyeing. Error bars represent the standard deviation of each measurement.

example, the Solution 2 film has lower tensile strength after dyeing than that of a film that is not dyed, which is likely due to the influence of the dye molecules on the hard segment surface. The higher tensile strength when a concentration of 1.0 g/L is used rather than 0.1 g/L could be due to dye molecules penetrating the soft segment. Since a greater quantity of dye has a more pronounced influence on the film due to an increased interaction with dye molecules, variation in the tensile strength is to be expected. Thus, undyed films had the highest tensile strength, followed by cast films dyed with 1 g/L and 0.1g/L of dye.

The thermal softening curves obtained for the different SPU films in Figure 11 show a clear difference in the softening behaviors of films produced from Solution 2 and 3, which contrasts with the similarity in their stress-strain properties (Figures 9 and 10). The higher softening temperature recorded for the Solution 3 film suggests that block copolymerization may have occurred and progressed to a higher order in this SssPU, resulting in greater hard segment aggregation than with Solution 2. Similarly, the lower softening temperature of the Solution 4 and 5 films reflects the lower aggregation propensity of their hard segments. In the case of Solution 4, this can be attributed to the urethane bonds in the soft segments, whereas with Solution 5 it is due to the random combination of single-chain glycol and methylene groups of varying lengths in the hard segments.

Previous studies of SPUs with the same compositions as those investigated here found that the number of hydrogen bonds in the hard segments decreases as the proportion of PEG in the soft segments increases, which is due to a reduction in crystallinity and a restriction of the molecular chains [8,9]. This difference in hydrogen bonding was confirmed by infrared spectroscopy based on the relative intensities of the absorption bands arising from urethane and hydrogen-binding urethane groups [8,9]. Figure 12 shows schematic diagrams based on this model of SPU aggregates with low and high PEG contents in the soft segments.

The different tensile strengths, breaking elongations, and thermal softening behaviors of the SPU samples may be interpreted as follows. Based on the results of the aforementioned study [8], the similar PEG contents in the soft segments of films produced from Solution 2-5 means that the moderation of hard-segment aggregation should also be similar. Nevertheless, the different physical properties of these samples presumably reflect different extents of hard-segment aggregation. The structural models proposed for these samples are shown in Figure 13.

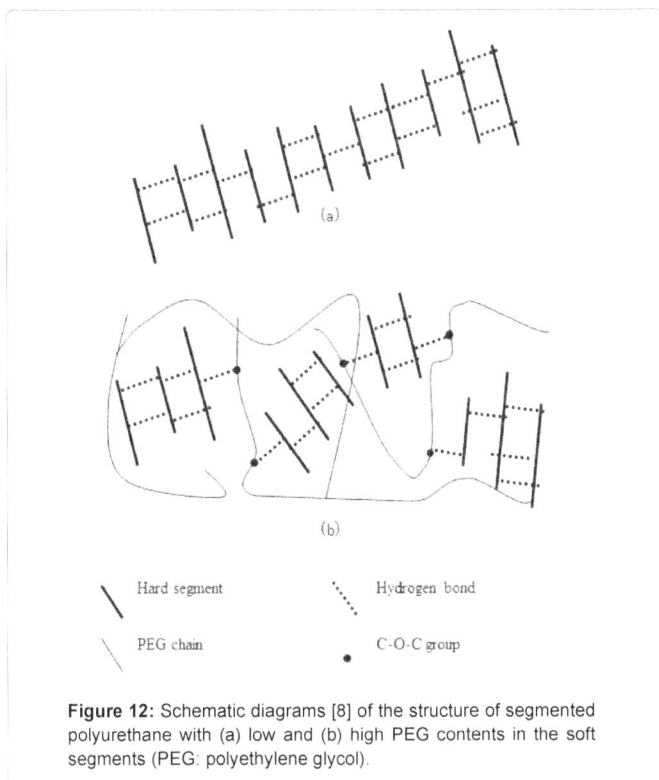

Figure 12: Schematic diagrams [8] of the structure of segmented polyurethane with (a) low and (b) high PEG contents in the soft segments (PEG: polyethylene glycol).

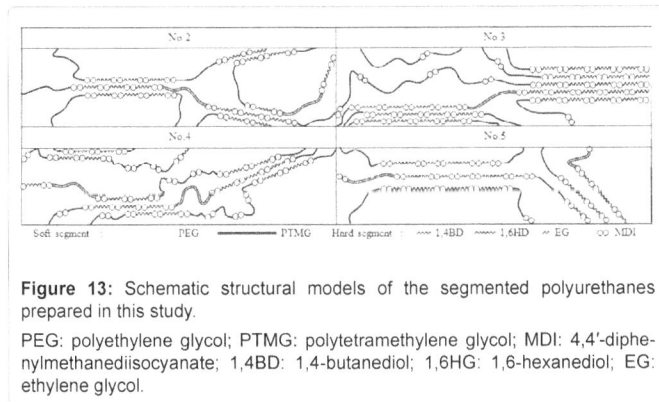

Figure 13: Schematic structural models of the segmented polyurethanes prepared in this study.

PEG: polyethylene glycol; PTMG: polytetramethylene glycol; MDI: 4,4'-diphenylmethanediisocyanate; 1,4BD: 1,4-butanediol; 1,6HG: 1,6-hexanediol; EG: ethylene glycol.

Conclusion

The dyeing rates and sorption isotherms of SPU films prepared by copolymerization of PEG in the soft segments and dyed with CI Direct Red 28 indicate on one hand a correlation between the extent of hard segment aggregation, and on the other a connection between the amount of adsorbed dye and the swelling rate in water. Coordinated water molecules tend to discharge from the SPU soft segments composed of PEG and MDI, with this increasing at higher dyeing temperatures. The different dyeing behaviors of the four types of SPU prepared with different higher-order structures confirm that dye uptake is also governed by how restricted the movement of the PEG chains is, which has been controlled here through the composition of the samples and the reaction methods used.

The tensile strength and breaking elongation of the films with high dye uptakes and low-level hard-segment aggregation, as well as those with low dye uptakes and restricted soft-segment mobility, vary little

after dyeing. In contrast, the tensile strength of the film with a high dye uptake and highly aggregated hard segments decreases significantly as a result of dyeing. This indicates that CI Direct Red 28 affects the degree of aggregation of SPU hard segments. The thermodynamic parameters of these SPUs also reveal that their affinity for CI Direct Red 28 depends on the level of aggregation of the hard segments, as well as on the mobility of the PEG-containing hydrophilic soft segments.

In conclusion, this study demonstrates a correlation between the dye (CI Direct Red 28) uptake, water-swelling ratio, and degree of hard-segment aggregation of different SPU compositions. These results should prove valuable from an industrial perspective, as the strength of dyed products is a clear marker of their quality. As color fastness is also crucial, this will be investigated in a future study into these SPUs.

Funding

This research was sponsored in part by a special research subsidy provided by Nagoya Women's University.

Conflicts of Interest

The authors have no conflicts of interest to declare.

References

1. Kuraray Co. Ltd. (1989) Leather-mode sheetlike material having high durability. Patent 1532434, Japan.

2. Kuraray Co. Ltd. (1989) Leather-mode sheetlike material having durability. Patent 1532435, Japan.

3. Kuraray Co. Ltd. (2000) Polyurethane and leather-like. Patent 3128373, Japan.

4. Yoshihara N, Ishihara H, Yamada T (2002) Structure and properties of segmented poly (urethane-urea) elastic fibers part 4: Improvement of dye-affinity using tertiary nitrogen compounds. J Text Mach Soc Japan 55: 27-32.

5. Enomoto M, Tokino S, Masuda H, Yoshihara N, Ishihara H, et al. (2009) Change of dyeing behaviors depending on soft segment component in segmented polyurethane. J Text Eng 55: 119-124.

6. Enomoto M, Tokino S, Ishihara H (2011) Dyeing properties of segmented polyurethane containing poluethylene glycol as hydrophilic group. J Text Eng 57: 45-49.

7. Wakida T, Lee M, Niu S, Yanai Y, Yoshioka H, et al. (1995) Dyeing properties of cotton fibres treated with liquid ammonia. J Soc Dyers Colourists 111: 154-158.

8. Shibaya M, Dobashi H, Suzuki Y, Ishihara H, Enomoto M, et al. (2005) Structures and moisture permeability of polyurethane films. J Text Mach Soc Japan 58: 115-122.

9. Shibaya M, Suzuki Y, Doro M, Ishihara H, Yoshihara N, et al. (2006) Effect of soft segment component on moisture-permeable polyurethane films. J Polym Sci 44: 573-583.

34

T-Shirt with Propping Effect for Natural Ventilation: Design Development and Evaluation of its Functionality by Thermal Manikin in Standing and Walking Motions

Chupo Ho[1]*, Jintu Fan[2], Edward Newton[3] and Raymond Au[4]

[1]*Institute of Textiles and Clothing, the Hong Kong Polytechnic University, Hung Hom, Kowloon, Hong Kong SAR, China*
[2]*College of Human Ecology, Cornell University, New York, USA*
[3]*Institute of Textiles and Clothing, The Hong Kong Polytechnic University, Hung Hom, Kowloon, Hong Kong SAR, China*
[4]*Hong Kong Design Institute Vocational Training Centre 3, King Ling Road, Tseung Kwan O, N.T., Hong Kong SAR, China*

Abstract

To improve the natural ventilation of garments is often a crucial task in designing functional garments. Normally, the ventilation can be improved by using appropriate types of fabric, or applying functional design details such vented panels on certain parts of a garment. Apart from these two means, are there any other alternative ways that clothing designers can utilize? Designing is a means to solve problems. Through a patent search, it was found that some inventors have adopted different means of designing to improve natural ventilation. Putting mesh panels on the garments was the most common way to achieve such aim. This design method had been proven by scholarly researches on its contribution on heat and moisture transfer. In addition, some designers noted that the air gap between the garment and body could be a key to affect such natural ventilation. For example, Moretti proposed putting additional spacer objects under the shoulder areas to create a gap so as to improve the natural ventilation of the wearer. Although this new design was claimed to have contribution to enhance natural ventilation for the wearer, the testing results were not provided. In this study, T-shirt designed with larger air gaps on the shoulders was developed. In order to test the effectiveness of this design method on heat and moisture transfer, a series of testing were conducted by using a movable thermal manikin in a chamber with the control of temperature and relative humidity. In order to test the functionality of the chimney/pumping/ventilation effect, the testing was conducted under no wind and windy conditions with a thermal manikin in a standing position, and under a simulated "walking" motion (walking speed of 1.24 km/h). The test results showed that the T-shirt with a larger air gap on the shoulders can significantly improve moisture vapor resistance during windy conditions.

Keywords: Functional clothing; T-shirt; Design; Natural ventilation; Pumping effect

Introduction

Clothing is not only a commodity that meets our basic needs or an item of aesthetic appreciation, but also a portable environment that helps us to face different external conditions every day. In order to do so, a true functional garment is one that can protect our body from acute changes in the environment. However, the body will generate heat, which will eventually result in sweating. If heat and sweat cannot be efficiently transmitted out through our clothing and released into the external environment, they not only cause heat stress and affect our performance, but in extreme cases, would even lead to death. Since clothing has been found to be an obstacle that defers a certain amount heat and moisture transfer [1-5], the maintaining of heat and moisture transfer to a consistent level is one of the key elements that clothing designers have to bear in mind. Therefore, fabrics that have moisture management can be a simple way to solve the problem. Moisture management materials are fabrics that can absorb sweat and vaporize moisture away from the skin surface from one side to the other side for faster evaporation. Many types of sports clothing that use these types of fabrics have the ability to maintain dryness for the wearers. However, moisture management fabrics are not able to enhance natural ventilation for the wearer. The attaching of devices in the clothing system such as cold pads or fans can actively increase ventilation; however, this is not feasible for daily wear due to the extra weight and bulkiness. Thus, natural ventilation can be increased through clothing design only, which will be more simple and direct. In this regard, the placement of vented panels, such as openings or through the use of mesh fabrics to allow for better natural ventilation to the wearer, has become very common in sports clothing designs. The

advantages and the effect of similar garment types have been reported by many researchers [6-8]. However, simply placing vented panels onto clothing has been carried out by many designers/brands for many years, and it seems that there are a lack of breakthroughs on improving natural ventilation by using other design methods. Apart from placing vented panels onto garments, are there any other options for designers in developing functional garments to improve natural ventilation? The objective of this paper is to explore an alternative design method which may have potential in improving natural ventilation for wearers. According to this design concept, a set of garment prototypes will be created and tested by objective measurements. The data will be analyzed to determine the effectiveness of the new design on improving natural ventilation. Finally, a recommendation will be provided for further design development.

Pumping Effect and Clothing Design

For a clothed person, there will be more or less air entrapped

---

*Corresponding author: Chupo HO, Institute of Textiles and Clothing, the Hong Kong Polytechnic University, Hung Hom, Kowloon, Hong Kong SAR, China
E-mail: tccpho@polyu.edu.hk

between the clothing layers, or between the skin and fabric layer if he is wearing a garment with only a single layer fabric. As the thermal conductivity of still air is low, solutions should be sought for improving exchange between the inner warmer microclimate and the outside environment, and this is known as air ventilation. One of the most common ways to do this is through open apertures in clothes in order to increase the exchange between the air inside the clothing and the ambient air. Moreover this ventilation will be further increased by the pumping effect of body movement. Clothing with openings involves "indirect" and "direct" ventilation. "Indirect ventilation" refers to the exchange of the garment microclimate through the materials itself; while "direct ventilation" means that the microclimate between garment and ambient air is formed by openings in the garment [9]. Direct ventilation, known as forced ventilation, can carry away warm air from the skin and fabric layer and also assist in evaporating sweat and dissipating warm vapor. When people move, the air of the garment microclimate can be influenced by the ambient environment through the openings of the garment or pores in the fabric [2,10-13]. The air space between the skin and the inner fabric layer changes over time, depending on the level of activity and movement. During body movements, air must go in and out through openings (such as the collar and cuffs) as the fabric moves outward and inward towards the skin surface, thus leading to ventilation. This ventilation of the microclimate of the clothing is called the "pumping effect" [14].

When a garment is properly worn, it tends to hang on the shoulders and thus the fabric touches the skin surface due to natural gravity. As warm air rises, similar to a chimney effect, one may feel warm on the upper chest and upper back due to the trapping of warm air in these areas. To address this issue, two questions can be raised. If there are vented panels on both shoulder areas, will the natural ventilation improve? Second, if there are vented panels on top of both of the shoulder areas, when the wearer swings his or her arms (viz. walking), will heat and moisture be released through the vented panels more significantly through the pumping effect? These two questions will be the key considerations that are kept in mind when exploring the different methods for clothing design in this research work.

Designing is a means to solve problems. Through a patent search, it was found that some inventors have adopted different means of designing to improve natural ventilation. According to Peteu and Helvenston Gray [15], patents are one of the best ways to explore how various products are designed to solve particular problems. Many studies have used patents as an important source of information on clothing development [15-17]. This research uses a patent search to review some of the design examples with special design features that improve natural ventilation. Although sometimes the patents showed descriptions and sketches by the applicants, the test results of the invention are not provided and the data are not shown. Thus, it is difficult to determine the functionality of the final product from the patents themselves [15]. However, a patent search is still useful because we can see how the potential proposal/design may solve specific problems. In terms of designs that will improve natural ventilation, the following are some examples taken from the patent search.

Putting mesh on the garments is not a new technique for improving ventilation. In 1944, Zelano [18] used lace fabrics and put them into the garment for ventilation purpose. In his design, he put lace as stripe-like openings on various positions of the garment such as the front, back panel, sleeves and trousers. For the suit, the lace was not only put on the shell fabric but also on the lining layer. Lace, in this case, could release the body heat and moisture vapor away because of the porous structure of this fabrication.

Bengtsson et al. [19] invented a garment for vigorous physical activity in warm environments. The inventors placed cords underneath a garment to create many piles of vertical air channels. Warm body air and moisture vapor could be released through the gaps between each cord channel.

Gioello [20] developed a ribbed ventilating undergarment which is worn beneath a non-porous outer garment. In this design, she used a series of parallel raised ribs or cords to form air channels. The channels would come into contact with the base of the outer garment so that a wider air gap between the channels can be formed. By creating a wider gap between the undergarment and the outer garment, the inventor claimed that body heat and moisture from the undergarment could be transferred and released through this air gap.

Lemoine [21] invented a garment for beekeepers. By placing a layer of polyurethane open cell foam that was approximately 0.4-0.6 cm in thickness underneath net fabric, bees cannot penetrate the material and sting the skin surface. At the same time, Lemoine claimed that this invention could allow ventilation between the skin of the wearer and the outside air.

Moretti [1] made use of a similar concept and invented a garment in which spacer objects are placed under the shoulder areas to create a gap. With the aid of vented panels on the shoulders, the accumulated warm air from the shoulders could be released to the outside, by assuming that warm air tends to rise in the space between the garment layer and body skin surface. He also indicated that solely creating an air gap between the garment and skin surface might be inadequate. It could be possible that the inside air would not be released from the shoulder areas, so that the air gap area increases in heat and moisture vapor resistance, and the wearer would experience an increase in discomfort. Thus he created openings on the shoulder parts of the garment. Warm air could dissipate, much like a chimney effect. The allowance of space between a garment and the skin could create air circulation around the body. However, it should be mentioned that if such air is damp, warm air may accumulate and thus create even more thermal insulation for the body. Thus, Moretti [1] pointed out the importance of air movement between a garment and the skin. This design approach seems to be a potential design that use vented panels to improve the natural ventilation for the wearer. More important, this design is easy to make and more light weight, as only two extra objects are needed on shoulders.

Design Development of T-Shirts

Design of objects that prop garments

According to the designs searched from the patent, the designers got the similar idea of the garment. They pointed out that keeping distance between fabric layer and skin surface could improve the air circulation, as the propping up effect could reduce the area that the fabric layer stuck to the skin directly. With reference to the design per Moretti, spacer shoulder pads were placed underneath the shoulder areas to refrain the garment from contact with the body. To enhance the pumping effect, piles of spacer material were cut to construct "shoulder pads" in the first trial. The shoulder pads could be used for "propping up" the garment, but if excessive piles of spacer material were stacked together (Figure 1), it was considered to be possibly too heavy for the wearer. Furthermore, the thickness of the material could increase thermal insulation by reducing air flow, although the spacer material is fabricated with porous material. Therefore, this method had to be modified.

To cope with the problem of the lack of air flow within spacer objects, "U-shaped" members (Figures 2A and 2B) were chosen to address this issue. In the construction of the U-shaped members, wider space was created between the layers of the spacer materials so that theoretically, this would discourage still air from accumulating, thus increase the flow of air penetration, which would contribute to the transfer of heat and moisture vapor away from the skin. However, there were two main problems that were found once this design was created. The first problem was the collapse of the U-shaped members (Figure 2C). Although the spacer material used for the experiment was ideal in its stiffness, it could not hold the shape of the U-waves once the T-shirt was placed onto the body without distortion. This suggests that when the U-shape wave design was placed onto the body, gravity and the weight of the T-shirt pressed downwards onto the U-shape waves. This resulted in reduced distance between the garment and the skin surface, i.e., reduced space and air circulation. Thus, the construction had to be further developed to minimize the amount of collapsing. The second problem was the bending of the spacer material, i.e. its tendency to be flattened (Figure 2D), thus making it difficult to control the U-shaped waves of the spacer material.

In order to hold the U-shape of the waves without distortion or bending, loops were continuously formed so as to reduce the space between each individual wave (Figure 3). To solve the bending problem, a piece of flat spacer material was placed between the loops and T-shirt fabric to provide a stronger base. In addition, horizontal stitches were used to gather the loops together, while other vertical

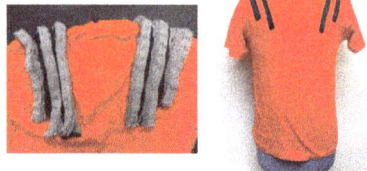

Figure 1: Spacer shoulder pad. (Left: inner construction; Right: T-shirt).

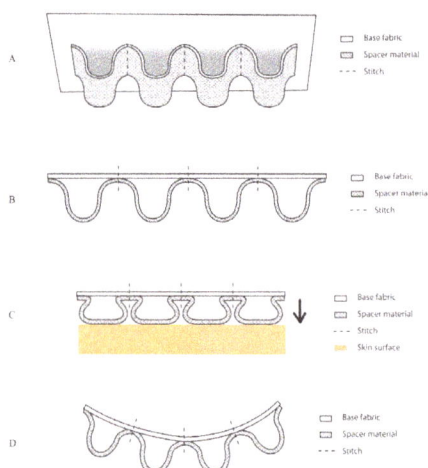

Figure 2: The design of "wave" construction of the spacer piles.

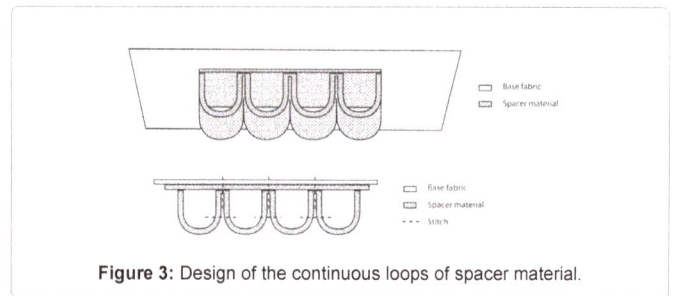

Figure 3: Design of the continuous loops of spacer material.

stitches were used to attach the loops and T-shirt fabric together. These two types of stitches eliminated the recovery ability of the spacer loops, so that the problem of bending was solved. Additionally, the sides of each U-shaped wave were stitch gathered together to reinforce them so that they were strong enough to support the weight of the garment, and a desirable distance could be maintained between the shell fabric and the surface of the body skin with minimum distortion.

T-shirt designs

The placement of mesh panels has been proven to have different effects on natural ventilation of garments. In this study, placement of mesh panels was carried out to determine its relationship with propping up effect on the improvement of pumping effect. The T-shirt designs were categorized into two groups: those without spacer loops (Group 1) and with spacer loops (Group 2) underneath the shoulders. Each group had five T-shirt styles, with different placements of mesh panels as previous studies have indicated that the positioning of vented panels could affect the natural ventilation of garments. The aim was to test the performance of the T-shirts with the placement of vented panels, whether this can work with a prop-up effect. The prototypes were made into the following five styles: NM (no meshes on the T-shirt), MS (meshes on the shoulders), MSC (meshes on the shoulders and chest area), MSS (meshes on the shoulders and two sides in the front) and MO (meshes on shoulders, chest area, and two sides in the front). Figure 4 shows all of the T-shirt styles, and the vented panels are presented as shaded areas. Each T-shirt was constructed with a zipper at the back for ease of placing the T-shirt onto the thermal manikin. Figure 5a shows one of the prototypes (Group 2–MO) and its inner construction (Figure 5b). NM of Group 1 was a basic T-shirt, and used as the control piece in the testing for comparison of the performance of heat and moisture transfer with the other T-shirts. For the main body of the T-shirts, a basic fabric typically available in the market was used, which was a 100% cotton fine gauge jersey with a weight of 200 g/m². The mesh fabric was also common in the market, 100% polyester with a weight of 66.3 g/m².

Measurement method

Wearer trial by subjects is not considered in this stage of product development. Since the fabric quality of the spacer loops are still not the best up to this stage, the results of the subjective measurement may not be reliable because more or less the subjects may have bias due to the quality of this spacer material. In order to obtain a more objective measurement to evaluate the design, a walkable thermal manikin may be the most ideal method. These thermal manikins are widely used to evaluate the thermal comfort of specific types of functional clothing, such as flight suits [22] and body protection during horse-riding [23]. In the same environmental setting, thermal manikins measure heat loss in a relevant, reliable and accurate way. In order to test the functionality of the chimney/pumping/ventilation effect, the testing was conducted

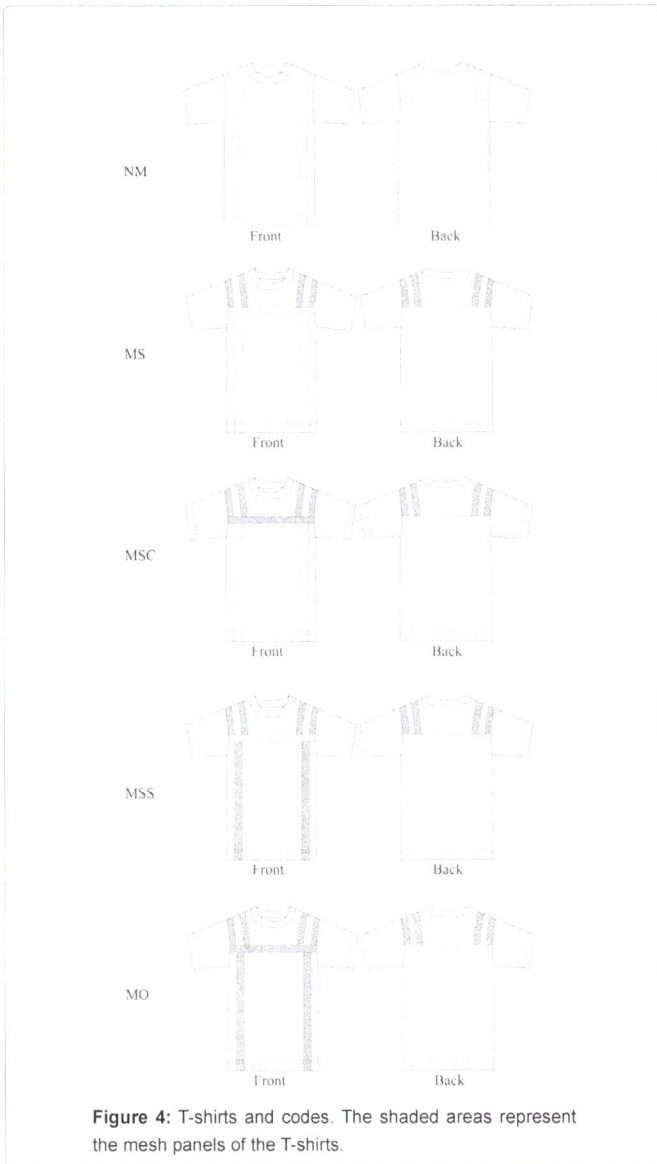

Figure 4: T-shirts and codes. The shaded areas represent the mesh panels of the T-shirts.

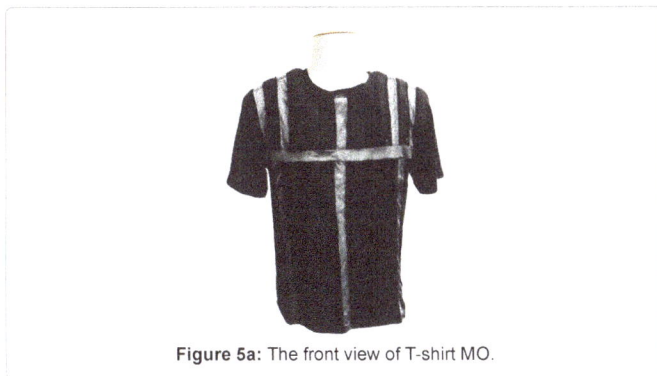

Figure 5a: The front view of T-shirt MO.

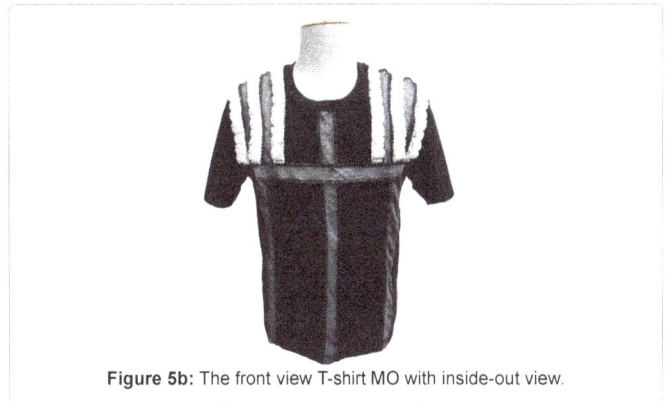

Figure 5b: The front view T-shirt MO with inside-out view.

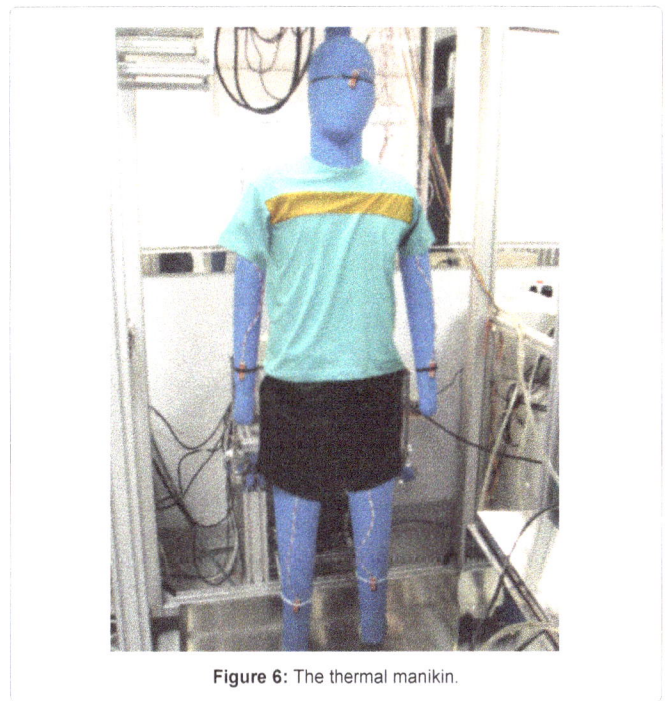

Figure 6: The thermal manikin.

under no wind and windy conditions with a thermal manikin in a standing position, and under a simulated "walking" motion (walking speed of 1.24 km/h). In the walking mode, the arms and legs of the manikin were swinging like the walking posture of a real human. The experiment was conducted in a climatic chamber of 20.0 ± 0.5°C and 65.0 ± 2% relative humidity (RH) with an air velocity of 0.5 ± 0.3 m/s (no wind condition) and 2.0 ± 0.3 m/s (windy condition). The core temperature of the manikin was set at 37°C. For all of the tests with the various T-shirt prototypes, the same pair of short trousers (made of 100% cotton) was placed on the manikin. Each T-shirt was tested three times, from which the mean value was taken. In order to avoid the effects of dirt and grime, which would possibly affect the testing results, the T-shirts were washed once by the laundry machine before undergoing testing. Clothing thermal insulation (R_t) and moisture vapor resistance (R_{et}) are the two most important parameters for thermal comfort. Therefore, a sweating fabric manikin (Figure 6) was used, which allowed these two parameters to be effectively determined [24]. T-test will be used to evaluate the performance of the reduction of thermal insulation and moisture vapor resistance.

According to the specifications of the sweating thermal manikin, the total thermal insulation, including the insulation of the clothing and surface air layer, is calculated as:

$$R_t = \frac{A_s(T_s - T_a)}{H_s + H_p - H_e},$$ (1)

where A_s is the total surface area of the manikin (A_s=1.79 m²), T_s is the mean skin temperature, T_a is the mean temperature of the environment, H_s is the heat supplied to the manikin or the heat generated by the heaters, H_p is the heat generated by the pump, and H_e is the evaporative heat loss from water evaporation. The evaporative heat loss can be calculated by using

$$H_e = \lambda Q \qquad (2)$$

where λ is the heat of the evaporation of water at the skin temperature, and Q is the perspiration rate or water loss per unit time, which can be measured by measuring the water supply.

The total moisture vapor resistance, including that of the clothing and the surface air layer, can be calculated by using

$$R_{et} = \frac{A_s (P^*_s - RH_a P^*_a)}{H_e} - R_{es} \qquad (3)$$

where P^*_s is the saturated water vapor pressure at the skin temperature, which is the water vapor pressure of the water film on the skin. RH_a is the relative humidity of the surrounding environment as a fraction, P^*_a is the saturated water vapor pressure in the surrounding environment, and R_{es} is the moisture vapor resistance of the skin.

Results and Discussion

The results from all the T-shirts are listed in Table 1 for both the no wind and windy conditions, respectively. The coefficients of variation for the repeated tests are less than 5%. T-shirts with lower R_t and R_{et} values are preferred because they indicate that the design has a greater ability to release body heat and moisture vapor out into the ambient environment.

Thermal insulation under no wind condition

Table 2 shows the percentage change for each garment under the no wind condition. As mentioned, the basic T-shirt (NM) of Group 1 was the control piece for comparison purposes. From the testing results, it can be seen that the new designs cannot significantly reduce the thermal insulation of the garment under a no wind condition. In a standing posture, MSC (t (4)=1.201, p=0.296), MSS (t (4)=1.331, p=0.254) and MO (t (4)=1.549, p=0.196) of Group 2 (with spacer loops) reduce only 4% of the thermal insulation compared to the control piece. In a walking posture, MSS and MO of Group 2 achieve the best performance in reducing thermal insulation, although the percentage of change is only 2.59% (t (4)=1.441, p=0.223) and 3.45% (t (4)=1.922, p=0.127), respectively. The findings indicate that placing spacer loops in a continuous manner underneath the shoulders cannot significantly reduce the thermal insulation, irrespective of standing or walking when there is no wind.

Thermal insulation under windy condition

Table 3 shows the percentage change for each garment compared with the control piece (NM of Group 1) during standing and walking under a windy condition. From the results, it appears that generally T-shirts in Group 2 perform better with lower thermal insulation compared to those in Group 1 under a windy condition. In a standing position, MSC (t (4)=80573, p=0.001), MSS and MO of Group 2 show more than a 10% reduction in thermal insulation compared to the control piece, although when they are compared to the same styles in Group 1, the difference between the two is minimal (approximately 1.5% to 3%).

In the walking mode, the T-shirts in Group 2 are not able to demonstrate a significant reduction in thermal insulation in comparison with those in Group 1. MSC, MSS and MO of Group 2 reduce thermal insulation by 6.35% (t (4)=4.899, p=0.008), 7.94% (t (4)=6.124, p=0.004) and 7.94% (t (4)=6.124, p=0.004), respectively, relative to the control piece. However the same designs in Group 1 also produce a 6.35% reduction in thermal insulation compared with the control piece. Thus the difference is very minimal.

The results for thermal insulation in the two positions during no wind and windy conditions show that although the new designs of Group 2 could reduce a certain amount of thermal insulation compared

			Standing (no wind)		Walking (no wind)		Standing (windy)		walking (windy)	
			Mean	S.D	Mean	S.D	Mean	S.D	Mean	S.D
Group 1	NM	R_t	0.126	0.006	0.116	0.003	0.067	0.001	0.063	0.001
		R_{et}	20.85	0.45	18.51	0.3	7.52	0.23	7.57	0.23
	MS	R_t	0.122	0.005	0.113	0.004	0.064	0.001	0.061	0.003
		R_{et}	18.82	0.65	17.643	0.37	7.35	0.13	6.861	0.33
	MSC	R_t	0.121	0.005	0.115	0.003	0.062	0.01	0.059	0.002
		R_{et}	18.72	0.55	17.301	0.45	6.52	0.2	6.902	0.28
	MSS	R_t	0.119	0.005	0.113	0.004	0.062	0.001	0.059	0.001
		R_{et}	18.85	0.47	17.5	0.47	6.48	0.13	6.646	0.23
	MO	R_t	0.119	0.005	0.114	0.002	0.061	0.001	0.059	0.001
		R_{et}	18.77	0.44	17.439	0.13	6.4	0.24	6.309	0.14
Group 2	NM	R_t	0.126	0.005	0.116	0.003	0.063	0.002	0.063	0.002
		R_{et}	20.42	0.5	18.354	0.37	7.52	0.1	7.115	0.05
	MS	R_t	0.123	0.004	0.116	0.003	0.062	0.001	0.062	0.001
		R_{et}	18.02	0.5	17.632	0.55	7.36	0.15	6.8	0.07
	MSC	R_t	0.121	0.004	0.114	0.004	0.06	0.001	0.059	0.001
		R_{et}	18.21	0.29	17.341	0.21	6.42	0.22	6.199	0.25
	MSS	R_t	0.12	0.005	0.113	0.002	0.06	0.001	0.058	0.001
		R_{et}	18.51	0.46	17.284	0.53	6.32	0.2	6.152	0.22
	MO	R_t	0.12	0.003	0.112	0.002	0.06	0.001	0.058	0.001
		R_{et}	18.17	0.54	17.282	0.42	5.9	0.14	5.845	0.17

Table 1: R_t and R_{et} values of standing and walking modes in both no wind and windy conditions.

	Group 1					Group 2				
	NM	MS	MSC	MSS	MO	NM	MS	MSC	MSS	MO
Standing R_t (%)	0%	-3.17%	-3.97%	-5.56%	-4.76%	0%	-2.38%	-3.97%	-4.76%	-4.76%
Walking R_t (%)	0%	-2.59%	-0.86%	-2.59%	-1.72%	0%	0%	-1.72%	-2.59%	-3.45%

Table 2: Percentage change in total thermal insulation for all T-shirts for standing and walking postures during no wind condition.

	Group 1					Group 2				
	NM	MS	MSC	MSS	MO	NM	MS	MSC	MSS	MO
Standing R_t (%)	0%	-4.48%	-7.46%	-7.46%	-8.96%	-5.97%	-7.46%	-10.45%	-10.45%	-10.45%
Walking R_t (%)	0%	-3.17%	-6.35%	-6.35%	-6.35%	0%	-1.59%	-6.35%	-7.94%	-7.94%

Table 3: Percentage change in total thermal insulation for all T-shirts for standing and walking postures under windy condition.

	Group 1					Group 2				
	NM	MS	MSC	MSS	MO	NM	MS	MSC	MSS	MO
Standing R_t (%)	0%	-9.74%	-10.22%	-9.59%	-9.98%	-2.06%	-13.57%	-12.66%	-11.22%	-12.85%
Walking R_t (%)	0%	-4.68%	-6.53%	-5.46%	-5.79%	-0.84%	-4.74%	-6.32%	-6.62%	-6.63%

Table 4: Percentage change of moisture vapor resistance for all T-shirts for standing and walking postures under no wind condition.

	Group 1					Group 2				
	NM	MS	MSC	MSS	MO	NM	MS	MSC	MSS	MO
Standing R_t (%)	0%	-2.26%	-13.30%	-13.83%	-14.89%	0%	-2.13%	-14.63%	-15.96%	-21.54%
Walking R_t (%)	0%	-9.37%	-8.82%	-12.21%	-16.66%	-6.01%	-10.17%	-18.11%	-18.73%	-22.79%

Table 5: Percentage change of moisture vapor resistance for all T-shirts for standing and walking postures under windy condition.

to the control, the performance is not that remarkable if compared with the same designs in Group 1 (viz. a propping effect). Hence, the new design in Group 2 does not attain a large amount of reduction in thermal insulation in comparison with that of Group 1. For the tests, each T-shirt utilized was made of only a single layer of fabric, so its original thermal insulation value was already quite low. Perhaps this is the primary reason that the designs that use propping may not be able to obtain an even lower value. Hence, performance in moisture vapor resistance could instead be the key factor for determining the degree of natural ventilation.

Moisture vapor resistance under no wind condition

Table 4 shows the percentage change of each garment compared with the control piece (NM of Group 1) under a no wind condition. Under a no wind condition and standing posture, the designs of Group 2 including MS (t (4)=7.287, p=0.002), MSC (t (4)=8.541, p=0.001), MSS (t (4)=6.298, p=0.003) and MO (t (4)=6.604, p=0.003) show significantly reduced moisture vapor resistance relative to the control by about 11% to 13%. In comparison, the designs of Group 1 reduce moisture vapor resistance by nearly 10% compared to the control piece. This means that the open mesh panels alone on the T-shirt already reduce a certain amount of moisture vapor resistance, and the placing of additional spacer loops underneath the shoulders in Group 2 T-shirts is only a minor contribution (the difference is approximately 1% to 3%) when compared with the designs of Group 1. In the walking posture, all T-shirts recorded lower R_{et} than the control piece, and thus the designs of Group 2 do not have a significant advantage over the designs of Group 1.

Moisture vapor resistance under windy condition

The percentage change of the R_{et} values is listed in Table 5. For standing postures, MSC, MSS and MO of Groups 1 and 2 show over 10% lower moisture vapor resistance than the control piece. Among

these, MO of Group 2 has the best performance with a reduction in the R_{et} by 21.54% (t (4)=5.836, p=0.004) relative to the control piece. The data also show that for a standing posture under windy conditions, open meshes on the shoulders are not as effective in releasing moisture vapor, irrespective of the presence of spacer loops. This suggests that the convection or chimney effect is not as effective for this design, unlike other designs with mesh panels placed across the chest, or two front vertical side panels near the side seams, even when both panels were kept open. It can be seen that when the meshes were placed only on the shoulders, approximately 2% lower R_{et} was recorded as opposed to the control piece (t (4)=1.009, p=0.37). Once the mesh panels of other body parts were opened, the percentage of reduction increased from approximately 2% to 13-14%, and the differences were significant (MSC: t (4)=5.986, p=0.004; MSS: t (4)=6.818, p=0.002; MO: t (4)=5.836, p=0.004). This indicates that vented designs placed solely on the shoulders cannot achieve a chimney effect, although moisture vapor would still be released from the porous fabric. However, if mesh panels were placed on other body parts at the same time, the ventilation effect would be more relevant. This result showed that dry air could enter the T-shirt through these mesh panels, and carry away the moisture vapor by natural convection.

In a walking posture, again, MO of Group 2 shows the lowest R_{et} value compared to the others, which is 22.79% (t (4)=10.447, p=0.000) lower than the control piece, and 6.13% (t (4)=3.649, p=0.022) lower than MO in Group 1 (-16.66%). MSC and MSS of Group 2 show lower values for R_{et} that is, 18.11% (t (4)=6.99, p=0.002) and 18.73% (t (4)=7.717, p=0.002), respectively. Body motion thus contributed to a further reduction in moisture vapor resistance. The test result showed that the new design could have better performance to reduce moisture vapor resistance, when the mesh panels on other body parts were placed, and the wearer was moving arms in a windy condition.

Conclusions

The experiment in this research work shows that T-shirts with newly designed spacer objects could reduce moisture vapor resistance relative to a normal T-shirt in certain testing conditions. T-shirt is a simple garment not intended for thermal protection, its thermal insulation is less important than its moisture vapor resistance with respect to thermal comfort. Therefore, the sole application of meshes or other vented panels on the shoulders is not the best solution for reducing the total thermal insulation and moisture vapor resistance, especially when the manikin was in a no wind condition. Perhaps the air gap between the shoulders and fabric layer is too limited in the prototypes, or the choice of spacer material defer the heat and moisture transfer from the body, the "pumping effect" of the thermal manikin could not actively improve the natural ventilation in no wind condition. However, the test results provided foundation for product development in the future. The placement of mesh panels on the body is vital. Together with the propping up and create opening on the shoulder areas, the garment may have different effects on the improvement on moisture vapor transfer. If this design concept can be applied on other functional wear, thus the user may be able to adjust the level of thermal comfort according to different environmental conditions, especially the change of relative humidity. In the future, the T-shirts with propping up effect should be evaluated by wearer trial. Apart from the analysis of the new design concept under real situation of doing exercise, it is also vital to investigate to what extent the presence of the spacer material may cause discomfort to the wearer.

References

1. Moretti MP (2002) Ventilaed item of clothing, US Patent 6442760B2.

2. Haghi A (2004) Moisture permeation of clothing: A factor governing thermal equilibrium and comfort. J Thermal Analysis and Calorimetry 76: 1035-1055.

3. Levine L, Sawka MN, Gonzalez RR (1998) Evaluation of clothing systems to determine heat strain. American Industrial Hygiene Association J 59: 557-562.

4. Nielsen R, Gavhed DC, Nilsson H (1989) Thermal function of a clothing ensemble during work: Dependence on inner clothing layer fit. Ergonomics 32:1581-1594.

5. Parsons KC (1993) Human Thermal Environments, the effects of hot, moderate and cold environments on human health, comfort and performance: the principles and the practice. Taylor and Francis, London.

6. Ho C, Fan J, Newton E, Au R (2008) Effects of athletic T-shirt designs on thermal comfort. Fibers and Polymers 9: 503-508.

7. Ruckman J, Hayes S, Cho J (2001) Development of a perfusion suit incorporating auxiliary heating and cooling system. Int J Clothing Science and Technology 14: 11-24.

8. Zhang X, Li J, Wang Y (2012) Effects of clothing ventilation openings on thermoregulatory responses during exercise. Indian J Fibre and Textile Research 37: 162-171.

9. Qian X, Fan J (2006) Prediction of clothing thermal insulation and moisture vapor resistance of the clothed body walking in wind. Annals of Occupational Hygiene 50: 833-842.

10. Bouskill LM, Havenith G, Kuklane K, Parsons KC, Withey WR (2002) Relationship between clothing ventilation and thermal insulation. AIHA J 63: 262-268.

11. Havenith G, Heus R, Lotens WA (1990) Resultant clothing insulation: a function of body movement, posture, wind, clothing fit and ensemble thickness. Ergonomics 33: 67-84.

12. Nielsen R, Olesen BW, Fanger PO (1985) Effect of physical activity and air velocity on the thermal insulation of clothing. Ergonomics 28: 1617-1631.

13. Vokac Z, Kopke V, Keul P (1973) Assessment and Analysis of the Bellows Ventilation of Clothing. Textile Research J 43: 475-482.

14. Olesen BW, Sliwinska E, Madsen TL, Fanger PO (1982) Effect of body posture and activity on the insulation of clothing: measurement by a movable thermal manikin. ASHARE Transactions 88: 791-805.

15. Peteu MC, Helvenston Gray S (2009) Clothing invention: Improving the functionality of women's skirts, 1846-1920. Clothing and Textiles Research J 27: 46-61.

16. Farrell-Beck J, Poresky L, Paff J, Moon C (1998) Brassieres and women's health from 1863-1940. Clothing and Textiles Research J 16: 105-115.

17. Helvenston, Gray S, Peteu MC (2005) Invention, the angel of the nineteenth century: Patents for women's cycling attire in the 1890s. Dress 32: 27-42.

18. Zelano J (1944) Ventilated clothing, US Patent US2391535A.

19. Bengtsson AG, Mullsjo KE (1980) Garment for use in vigorous physical activities. US Patent 4195364.

20. Gioello DA (1984) Ribbed ventilating undergarment for protective garments. US Patent 4451934.

21. Lemonine PG (1991) Ventilated beekeeper suit. US Patent US4985933

22. Crown E, Ackerman MY, Dale JD, Tan Y (1998) Design and evaluation of thermal protective flight suits. Part II: Instrumented mannequin evaluation. Clothing and Textile Research Journal 16: 79-87.

23. Dlugosch S, Hu H, Chan A (2013) Thermal comfort evaluation of equestrian body protectors using a sweating manikin. Clothing and Textile Research J 31: 231-243.

24. Fan J, Chen YS (2002) Measurement of clothing thermal insulation and moisture vapor resistance using a novel perspiring fabric thermal manikin. Measurement Science and Technology 13: 1115-1123.

Permissions

List of Contributors

Sisay Awoke and Yirga Adugna
Department of Chemistry, College of Natural Science, Wollo University, Ethiopia

Redwan Jihad
Department of Textile Engineering, Wollo University, Ethiopia

Habtam Getaneh
Department of Textile Biology, Wollo University, Ethiopia

Kadi N and Karnoub A
Department of Textile and Spinning, Faculty of Mechanical Engineering, University of Aleppo, Syria

Gloy YS
Institut für Textiltechnik der RWTH Aachen University, Aachen, Germany

Renkens W
Renkens Consulting, Aachen, Germany

Herty M
Mathematics (IGPM), RWTH Aachen University, Aachen, Germany

Gries T Hasan NB
Senior Lecturer, Green University of Bangladesh, Private University in Dhaka, Bangladesh

Begum AR
Professor and Head, Bangladesh University of Textiles, University in Dhaka, Bangladesh

Islam A
Executive, Quality Assurance Department, Delta Spinning Ltd, Bangladesh

Parvez M
Senior Lecturer, Atish Dipankar University of Science and Technology, Bangladesh

Amir Kiumarsi
Department of Chemistry and Biology, Ryerson University, Toronto, Canada

Mazeyar Parvinzadeh Gashti
Department of Textile, College of Engineering, Yadegar-e-Imam Khomeini (RAH) Branch, Islamic Azad University, Tehran, Iran

Kozar T, Rudolf A, Cupar A, Jevšnik S and Stjepanović Z
Faculty of Mechanical Engineering Department of Textile materials and Design, University of Maribor, Slovenija

Muhammad Qamar Tusief
School of Textile and Design, University of Management and Technology, C-II Johar Town Lahore54770, Pakistan
Department of Fiber and Textile Technology, University of Agriculture, Faisalabad, Pakistan

Nabeel Amin and Mudassar Abbas
School of Textile and Design, University of Management and Technology, C-II Johar Town Lahore54770, Pakistan

Nasir Mahmood and Israr Ahmad
Department of Fiber and Textile Technology, University of Agriculture, Faisalabad, Pakistan

Helmy HM and Kamel MM
National research centre, Textile research division, Dokki, Giza, Egypt

Shakour AA
National research centre, Air pollution department, Dokki, Giza, Egypt

Rashed SS
The Islamic museum, Cairo, Egypt

Samuel C Ugbolue and Yong K Kim
Department of Bioengineering, University of Massachusetts, Dartmouth, USA

Olena Kyzymchuk
Kyiv National University of Technology and Design, Ukraine

Mohy uddin H.G, Jin Z and Qufu W
Key Laboratory of Eco-textiles, Jiangnan University, Wuxi, Jiangsu, China

Teli MD, Javed Sheikh and Pragati Shastrakar
Department of Fibres and Textile Processing Technology, Institute of Chemical Technology, Mumbai-400019, India

Azari Z
Laboratory of Biomechanics, Polymers and Structures, ENIM, 57000 Metz, France

Karnoub A, Makhlouf S and Kadi N
Faculty of Mechanical Engineering, University of Aleppo, Syria

Motavalli MR
Department of Communication Technique, Faculty of Engineering, University of Qom, Qom, Iran

Solbach K
Department of Microwave and RF-Technology, Faculty of Engineering, University of Duisburg-Essen, Duisburg, Germany

Abeer S Arafa
Cotton Research Institute, Agricultural Research Center, Egypt

Chattopadhyay DP
Department of Textile Chemistry, Faculty of Technology and Engineering, The M S University of Baroda, Vadodara, India

Patel BH
Ichetaonye SI, Ichetaonye DN, Adeakin OAS, Yibowei ME and Dawodu OH
Department of Polymer and Textile Technology, Yaba College of Technology, Yaba, Lagos, Nigeria

Tenebe OG
Department of Home Economics, Federal College of Education, Kontagora, Nigeria

Das S and Bhowmick M
Central Institute for Research on Cotton Technology (ICAR), Adenwala Road, Matunga, Mumbai

Prasad RK, Jannat F and Ali A
Department of AMT, BGMEA University of Fashion and Technology, Dhaka, Bangladesh

Hemdan Abo-Taleb, Heba El-Fowaty and Aly Sakr
Textile Engineering Department, Mansoura University, Mansoura, Egypt

Htike HH, Kang J, Yokura H and Sukigara S
Kyoto Institute of Technology, Kyoto, Japan

Gawande NG, Deosarkar DV and Kalyanker SV
Department of Agricultural Botany, Vasantrao Naik Marathwada Krishi Vidyapeeth, Parbhani, India

Bagwan AS and Rajput A
Center for Textile Functions, Mukesh Patel School of Technology, Management, Engineering, Shirpur, India

Dalal S and Aakade A
Maral Overseas Khalghat, Nimrani, India

Ibrahim W
Department of Postgraduate Studies, National Textile University, Faisalabad, Pakistan

Sarwar Z, Abid S, Munir U and Azeem A
Department of Textile Processing, National Textile University, Faisalabad, Pakistan

Fangueiro R, Carvalho R, Silveira D and Sampaio S
Center for Textile Science and Technology, University of Minho, Guimarães, Portugal

Ferreira N, Ferreira C and Monteiro F
A Ferreira e Filhos S.A., Vizela, Portugal

Diane E
Faculty of Sports Science, Leisure and Nutrition, Liverpool John Moores University, UK

Shunsuke Otoyama
Division of Environmental Engineering Science, Gunma University, Tenjin-cho, Kiryu 376-8515, Japan

Yutaka Kawahara
Division of Environmental Engineering Science, Gunma University, Tenjin-cho, Kiryu 376-8515, Japan
The Center for Fiber and Textile Science, Kyoto Institute of Technology, Matsugasaki, Sakyo-ku, Kyoto 606-8585, Japan

Kazuyoshi Yamamoto and Noboru Ishibashi
Research Lab., Carbo-tec. Co. Ltd., 305, Creation Core Kyoto Mikuruma, Kajii, Kamigyo-ku, Kyoto 602-0841, Japan

Hiroyuki Wakizaka
North Eastern Industrial Research Center of Shiga Prefecture, 27-39, Mitsuyamotomachi, Nagahama, Shiga 526-0024, Japan

Yutaka Shinahara
Nippon Felt Co. Ltd., Saitama Mill, 88, Haramamuro, Kounosu, Saitama 365-0043, Japan

Hideki Hoshiro
Kuraray Co. Ltd., North Umeda Hankyu Building Office Tower, 8-1, Kakudacho, Kita-ku, Osaka 530-8611, Japan

Norio Iwashita
National Institute of Advanced Industrial Science and Technology, 16-1, Onokawa, Tsukuba, Ibaragi 305-8569, Japan

Ashok Kumar L
Department of Electrical and Electronic Engineering, PSG College of Technology, Coimbatore, Tamilnadu, India

Joel Peterson
The Swedish School of Textiles, University of Borås, Skaraborgsvägen 3, S-50630 Borås, Sweden

Tuigong DR, Kipkurgat TK and Madara DS
School of Engineering, MOI University, Kenya

El-Hadidy AM, El-Rys SM and El-Hossiny A
Textile Engineering Department, Mansoura University, Egypt

Dinesh Bhatia
Department of Textile Technology, Dr B R Ambedkar National Institute of Technology, Jalandhar-144011, India

Urvashi Malhotra
Department of Textile Technology, Jawaharlal Nehru Government Engineering College, Sund-ern-agar-175018, India

Eyupoglu S and Kilinc M
Engineering and Design Faculty, Department of Fashion and Textile Design, Istanbul Commerce University, Istanbul, Turkey

Kut D
Engineering Faculty, Department of Textile Engineering, Uludag University, Bursa, Turkey

Enomoto M
Department of Applied Life Studies, College of Nagoya Women's University, Japan

Tokino S
Industrial Technology Center of Wakayama Prefecture, Japan

Ishihara H
Ryukoku University, Japan

Chupo Ho
Institute of Textiles and Clothing, the Hong Kong Polytechnic University, Hung Hom, Kowloon, Hong Kong SAR, China

Jintu Fan
College of Human Ecology, Cornell University, New York, USA

Edward Newton
Institute of Textiles and Clothing, The Hong Kong Polytechnic University, Hung Hom, Kowloon, Hong Kong SAR, China

Raymond Au
Hong Kong Design Institute Vocational Training Centre 3, King Ling Road, Tseung Kwan O, N.T., Hong Kong SAR, China

Index